FIELDBUS SYSTEMS AND THEIR APPLICATIONS 2003
(FET 2003)

*A Proceedings volume from the 5th IFAC International Conference,
Aveiro, Portugal, 7 – 9 July 2003*

Edited by

D. DIETRICH
*Institute of Computer Technology,
Vienna University of Technology,
Vienna, Austria*

J.P. THOMESSE
LORIA CNRS, Nancy, France

and

P. NEUMANN
*Institute for Automation and Communication,
Magdeburg, Germany*

Published for the

INTERNATIONAL FEDERATION OF AUTOMATIC CONTROL

by

ELSEVIER LTD

ELSEVIER Ltd
The Boulevard, Langford Lane
Kidlington, Oxford OX5 1GB, UK

Elsevier Internet Homepage
http://www.elsevier.com

Consult the Elsevier Homepage for full catalogue information on all books, journals and electronic products and services.

IFAC Publications Internet Homepage
http://www.elsevier.com/locate/ifac

Consult the IFAC Publications Homepage for full details on the preparation of IFAC meeting papers, published/forthcoming IFAC books, and information about the IFAC Journals and affiliated journals.

First edition 2003

Library of Congress Cataloging in Publication Data

A catalogue record for this book is available from the Library of Congress

British Library Cataloguing in Publication Data

A catalogue record for this book is available from the British Library

ISBN: 9780080442471

To Contact the Publisher

Elsevier welcomes enquiries concerning publishing proposals: books, journal special issues, conference proceedings, etc. All formats and media can be considered. Should you have a publishing proposal you wish to discuss, please contact, without obligation, the publisher responsible for Elsevier's industrial and control engineering publishing programme:

Christopher Greenwell
Publishing Editor
Elsevier Ltd
The Boulevard, Langford Lane
Kidlington, Oxford
OX5 1GB, UK

Phone:	+44 1865 843230
Fax:	+44 1865 843920
E.mail:	c.greenwell@elsevier.com

General enquiries, including placing orders, should be directed to Elsevier's Regional Sales Offices – please access the Elsevier homepage for full contact details (homepage details at the top of this page).

Printed and bound in the United Kingdom

Transferred to Digital Print 2011

5th IFAC INTERNATIONAL CONFERENCE ON FIELDBUS SYSTEMS AND THEIR APPLICATIONS 2003

Sponsored by
International Federation of Automatic Control (IFAC)
Technical Committee on Components and Instruments

Co-sponsored by
IFAC Technical Committees on
- Computers for Control
- Manufacturing Plant Control

Organized by
Instituto de Engenharia Electrónica e Telemática de Averio (IEETA)

Hosted by
Universidade de Aveiro, Portugal

Supported by
FCT Fundação para a Ciência e a Tecnologia, Ministério da Ciência e do Ensino Superior, Portugal

Preface

The IFAC International Conference on Fieldbus Systems and their Applications, FeT 2003, was organised by the University of Aveiro, in Aveiro, Portugal, from the 7th to the 9th of July 2003. It was sponsored by the IFAC Technical Committee on Components and Instruments, and co-sponsored by the Committees on Computers for Control and Manufacturing Plant Control as well as by APCA – *Associação Portuguesa de Controlo Automatico*, the national IFAC organization. It represented the 5th edition of the FeT conference, following the previous editions of Vienna (95 and 97), Magdeburg (99) and Nancy (01). The initial issues of the FeT conference were held in the axe Austria-Germany and were carried on in German language. The recognition of the importance achieved by the research, development and use of fieldbuses led to the decision to make the conference more international and to organize it in the scope of IFAC.

Therefore, the 5th issue, FeT 2003, aimed at bringing together researchers and industrialists to discuss communication systems, fieldbuses and applications supported on them. Distributed embedded systems were one of the current hot topics. The list of topics in the call for papers were services and protocols, profiles, wireless networking for field applications, system integration, management, dependability, real time, fieldbus based systems, security, education and emerging trends.

Besides FeT, another IFAC event, the International Symposium on Intelligent Components and Instruments for Control Applications (SICICA) has traditionally addressed the fieldbus systems area. As the organisers of FeT were also successful candidates to organize SICICA, a decision was made to organize both events (curiously, both in their 5th edition) in conjunction. Therefore, it was possible to organize a joint session on communication issues during the 9th of July, the start day of SICICA (9th to 11th of July). This session was dedicated to Web technologies and fieldbus applications. Besides that session, FeT participants could attend that full SICICA day, even if they were not registered in both events. Besides the referred session they could participate in a SICICA communications session and a SICICA invited lecture entitled *Intelligent instrumentation: the data source for technical management of industrial plants - end users' targets*, presented by Dr. Dario Fantoni.

Consequently, the two joint events lasted for one whole week during which 90 papers were presented, 46 in FET 2003 (4 of which in the special joint session) and 44 in SICICA 2003. The authors came from 20 countries mostly from Europe and America (north and south). The specific program of FeT included two invited lectures. The first one was entitled *In Car Embedded Electronic Architectures: how to Ensure Their Safety* and it was presented by Prof. Françoise Simonot – Lion. She is the head of the TRIO (*Temps Réel et InterOpérabilité*) team of the LORIA (*Laboratoire lorrain de recherche en informatique et ses applications*) in Nancy, France, and is the responsible for many research projects with the French automotive industry. The second lecture was entitled *The Necessity of an Upgrade in Industrial Communications* and it was presented by Mr. Roland Heidel from SIEMENS, Germany. Mr. Heidel is the leader of a research and development department for common technologies with the focus on methods and products for distributed applications in the field and has been responsible for several EU projects and fieldbus standardisation activities. Beyond the invited lectures, which generated a high interest and discussion between the speakers and the audience, the program also included one special session dedicated to the R-FIELDBUS European project. In this session an overview of the project which consists in introducing wireless R-F based communications in field applications based on the PROFIBUS fieldbus, was given. Three additional papers on project details were also presented in that session.

The remaining regular sessions of FeT 2003 covered a diversity of topics including *building automation and artificial intelligence in building automation, automotive systems, wireless sensor networks, CAN / CANOPEN, new protocol mechanisms or extensions, system modelling, fault tolerance and QoS, TDMA and predictability, scheduling and performances*. The papers therein presented are included in this volume. From those, five got recommendations for possible publication in the IFAC journal Control Engineering Practice. During the technical and the social part of the conference there were quite enthusiastic technical discussions and contacts leading to new projects and networks preparation. It was clear that the research community behind the conference is growing and reinforcing links steadily. FeT 2005 will certainly be a great success.

The FeT 2003 IPC Chair

Jean-Pierre Thomesse
LORIA, Nancy, France

The FeT 2003 NOC Chair

José Alberto Fonseca
University of Aveiro, Aveiro, Portugal

CONTENTS

RFIELDBUS PROJECT
(Special Session)

FAULT TOLERANCE AND QoS

PLENARY PAPER II

TDMA AND PREDICTABILITY

SCHEDULING

AUTOMOTIVE SYSTEMS

CAN / CANOPEN

PERFORMANCES I

NEW PROTOCOL MECHANISMS OR EXTENSIONS

PERFORMANCES II

WEB TECHNOLOGIES AND FIELDBUSES

IN CAR EMBEDDED ELECTRONIC ARCHITECTURES: HOW TO ENSURE THEIR SAFETY

Françoise Simonot-Lion

LORIA – INPL (UMR CNRS 7503)
2, avenue de la Forêt de Haye – 54516 Vandoeuvre- lès-Nancy Cedex France
Tel. +33 3 83 59 55 79
Fax. +33 3 83 59 56 62
simonot@loria.fr

Abstract: The part of software based systems in a car is growing. Moreover, in the next years will emerge the X-by-Wire technology that intends to replace mechanical or hydraulic systems by electronic ones even for critical function as braking or steering. This requires a stringent proof that these new vehicles will ensure the safety of driver, occupants, vehicle and environment. In this paper, we intend to list certain activities and key points for ensuring the development of a safe and optimized embedded system. More precisely, we propose two main axis that contribute to establish a design methodology of such systems. The first one identifies the generic components of an embedded system while the second one details how to model and validate the embedded system throughout the different steps of the development process. *Copyright © 2003 IFAC*

Keywords: embedded systems, safety, real time, validation, architecture description language

1. INTRODUCTION

While automobile production is likely to increase slowly in the coming years (42 millions cars produced in 1999 and only 60 millions planned in 2010), the part of embedded electronics and more precisely embedded software is growing. The cost of electronic systems was $37bn in 1995 and $60bn in 2000, with an annual growth rate of 10%.

The reasons for this evolution are technological as well as economical. The cost of hardware components is decreasing while their performances and reliability are increasing. The emergence of embedded fieldbuses leads to a significant reduction of the wiring cost. Finally, software technology facilitates the introduction of new functions whose development would be costly or even not feasible if using only mechanical or hydraulic technology and allows therefore satisfying the end user requirements in terms of safety and comfort.

Who is concerned by this evolution? First the *vehicle customer*, for which the requirements are on the one hand, the increase of performance, comfort, assistance for mobility efficiency (navigation), ... and on the other hand, the reduction of vehicle consumption and cost. Furthermore he requires a reliable embedded electronic system that ensures safety properties. Secondly, the stakeholders, *car makers* and *suppliers*, who are interested in the reduction of time to market, development cost, production and maintenance cost. Finally this evolution has a strong *impact on the society*: legal restrictions on exhaust emission, protection of the natural resources and of the environment, ...

In this presentation, we intend to list certain activities and key points for ensuring the development of a safe and "optimal" embedded system that is suited to the above mentioned requirements. For this purpose, we develop the context and the problematic in section 2. Then in sections 3 and 4, we propose two main axis that contribute to establish a design methodology of such systems. The first one identifies the generic components that compose an embedded system while the second one details how the embedded system is modelled and validated along the different steps of the development process.

2. CONTEXT

2.1. Several domains and specific problems

Traditionally, an in-car embedded system is divided in four domains that correspond to different functionalities, constraints and models. Two of them are concerned specifically with safety : "power train" and "chassis" domain. The third one, "body", is emerging and presently integrated in major of cars. And finally, "telematic" and "Human Machine Interface" domains take benefit of continuous progress in the field of multimedia and internet.

Power train

This domain represents the system that controls the motor according to explicit solicitations of the driver (speeding up, slowing down, ...), implicit solicitations of the driver (driving facilities, fuel consumption, ...) and environmental constraints (exhaust pollution, noise, ...). Moreover, this control has to take into account requirements from other parts of the embedded system as climate control or ESP (Electronic Stability Program). Traditional tools dedicated for general control command applications are used for the development of power train systems (Matlab / Simulink, for example and simulation approach).

In this domain, the main characteristics are:
- at a functional point of view: different control laws, complex control laws (multi-variables), different sampling periods, ...
- at a hardware point of view: specific sensors (minimisation of the criteria "cost / resolution"), high computation power, high storage capacities, dedicated coprocessors (floating point computations),
- at an implementation point of view: several tasks with different activation rules (different periods), stringent time constraints imposed to task scheduling, mastering safe communications with other systems and with local sensors / actuators.

Chassis

It gathers systems as ABS (Antilock Braking System), ESP (Electronic Stability Program), ASC (Automatic Stability Control), 4WD (4 Wheel Drive), ... that control the chassis components (wheel, suspension, ...) according to requirements as steering, braking solicitations and several forces (ground, wind, ...).

The characteristics of the chassis domain and the underlying models are similar to those presented for power train domain.

Body

Wipers, lights, doors, windows, seats, mirrors are more and more controlled by software based systems. This kind of functions make up the body domain. In this case, there are numerous functions, some of them being critical. They imply globally many communications between them and consequently a complex distributed architecture. There is an emergence of the notion of sub-system or sub-cluster based on low cost fieldbuses as, for example, LIN that connect modules realized as integrated mechatronic systems. On another side, the body domain integrates a central subsystem, termed the "central body electronic" whose main functionality is to ensure message transfers between different systems or domains. This system is recognized to be a central critical entity.

Body domain implies mainly discrete event applications. Their design and validation rely on state transition models (as SDL, Statecharts, UML state transition diagrams). These models allow, mainly by simulation, the validation of a functional specification. Their implementation implies a distribution over a complex hierarchical hardware architecture. High computation power for the central body electronic entity, fault tolerance and reliability properties are imposed to the body domain systems. A challenge in this context is first to be able to develop exhaustive analyze of state transition diagrams and second, to ensure that the implementation respects the fault tolerance and safety constraints. The problem here is to achieve a good balance between time triggered approach and flexibility.

Telematic and HMI (Human Machine Interface)

Next generation of telematic devices provides new sophisticated Human Machine Interfaces to the driver and the other occupants of a vehicle. They enable not only to communicate with other systems inside the vehicle but also to exchange information with the external world. Such devices will be in the future upgradeable and for this domain, a "plug and play" approach has to be favoured.

These applications have to be portable and the services furnished by the platform (operating system and / or middleware) has to offer generic interfaces and downloading facilities. The main challenge here is to preserve the security of the information from, to or inside the vehicle. Sizing and validation do not relies on the same methods than for the other domains. Here we shift from considering messages, tasks and deadline constraints to fluid data streams, bandwidth sharing and multimedia quality of service.

2.2. A cooperative development process

Strong co-operation between suppliers and carmakers in the design process implies the development of a specific concurrent engineering approach. In order to specify this process, synchronisation points (rendezvous) across the co-operative development model have to be identified and the information exchanged at these points must be characterised. Furthermore, an unique syntax of the exchanged information has to be defined and the development of reusable components is a main way for cost reduction.

2.3. The emergence of X-by-Wire technology

At present some critical functions are realised by software-based systems, as braking assistance, active suspension, steering functionalities etc.. They are subject to stringent timing constraints and more generally to dependability constraints. In the close future, these constraints will be more critical with the generalisation of X-by-Wire technology. X-by-Wire is a generic term used when mechanical and / or hydraulic systems are replaced by "electronic" ones (intelligent devices, networks, computers supporting software components that implement filtering, control, diagnosis, ... functionalities). For example, we can cite brake-by-wire, steer-by-wire, that will be shortly integrated in cars for the implementation of critical and safety relevant functions.

Therefore the development of such systems must define an feasible system, i.e. satisfying these constraints. Conventional mechanical and hydraulic systems have stood the test of time and have proved to be reliable; it is not the same for critical software based systems. In aerospace / avionic industries, X-by-Wire technology is currently employed; but, for ensuring safety properties, are used specific hardware and software components, specific fault tolerant solutions (heavy and costly redundancies of networks, sensors and computers) and certified design and validation methods. Now there is a challenge to adapt these solutions to automotive industries that impose stringent constraints on component cost, electronic architecture cost (minimisation of redundancies) and development length. Consequently, there is a real need for mastering the cooperative development

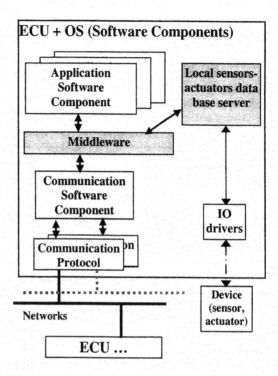

Fig. 1. Generic component – generic architecture

process of critical low cost electronic embedded architectures.

3. FROM DEDICATED COMPONENTS TO SYSTEM ARCHITECTURE

Nowadays, most embedded sub-systems (hardware and software) are separately defined and developed. Each one is dedicated to a single functionality and is designed and tested as closed systems by a supplier according to the carmaker requirements.

On the one hand, this is a bar to the reusability of solutions in other projects and on the other hand, this leads to oversize the resources (eg. number of ECUs, memory size, ...). In order to solve this problem, several proposals have been made for the structuring of an embedded system into generic components and generic architecture and, also, for defining the "perimeter" of the what is re-usable and / or portable (a component, a set of component, ...)

For example, the French AEE project (AEE, 1999) and the European ITEA project EAST-EEA (EAST, 2001) characterised formally the basic embedded components and defined the perimeter of the reusable ones. Furthermore they provided a generic architecture for an Electronic Control Unit (ECU), i.e.

a station connected to one or several network(s) and supporting the embedded application.

3.1. Component classes

The generic model presented in Fig. 1 enables the development of components that are independent of a specific ECU. That is the case for *Devices* (sensors and actuators) and *Application Software Components*. This independence is provided by two kinds of components:

- *middleware* and *communication software component* that hide the distribution and specifically the ECU supporting an information producer; in fact, the middleware actually implements a global real time database,

- *sensors – actuators data base server* whose function is to manage local data produced or consumed locally by devices directly connected to the ECU

These two components furnish a common interface to Application Software Components and on the other hand they implement fault tolerant services.

The communication between ECU and external actors are done thanks to *Communication Protocol Components* (data link layer for each connected network) and *I/O Drivers* that manage Input and Output for each device.

All the components, except devices, are implemented in software; they are supported by *ECU hardware* and managed by an Operating System. So the last class of component is *OS Software Components*.

3.2. Critical components

The safety properties of an embedded application depends on two points:

- reliability of the hardware architecture,

- properties on application behaviour.

For the first point, several metrics are available in order to evaluate this reliability. For example, starting with values given by furnishers of hardware components, we can compute the probability that a hardware architecture fails in one hour or the mean time between failures. Consequently, it allows to verify that these values are less than an imposed one; for example, a probability of failure occurrence in one hour that is less than 10^{-9} ensures that the given architecture is relevant to SIL (System integrity Level) class 4 (IEC, 1997).

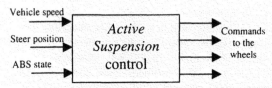

Fig. 2. Functional "Active Suspension Control" component

Note that this kind of evaluation is a necessary activity but not a sufficient one and the second point must be studied. It requires to have a quantified and precise model of the software behaviour: tasks, messages, scheduling policies, protocols, ... For example, if we consider the power train or the chassis domain, the control laws impose stringent constraints that must be respected by an embedded architecture. For illustrate this, we analyse an "active suspension" that realises a balance between comfort and control. This function takes three input data: vehicle speed (VS), steer position (SP) and state of the ABS system (ABS) and delivers four consistent commands to the four wheels (FL, FR, RL, RR) as shown in Fig. 2. The temporal properties that an implementation of this function must respect, assumed here to be realized by only one task, are of the following types (Fig. 3):

- deadline on the response time of the algorithm implementing the control law,

- freshness of the input data when consumed by this algorithm,

- deadline on the response time of the message transmitting the output data to the wheel,

- temporal consistency of the three input data,

- temporal consistency of the four output data that have to be deliver in the same temporal window.

How to specify and how to ensure these properties ? We list below some open issues.

- *Time Triggered Approach*. Of course, the underlying products and techniques used in this approach allows a deterministic proof of the distributed application (TTP, 1999; Kopetz, 2002). Furthermore, by this way, fault detection and fault tolerant mechanisms are easier to introduce. Nevertheless, some problems must be addressed. In fact, if the specification of task activation is naturally deterministic and periodic for a control law, it is not the case for the sampling of driver solicitations (lights up, ...). The overload due to systematic periodic task activation or message transmission might be compatible with the strong cost pressure imposed by the automotive industry.

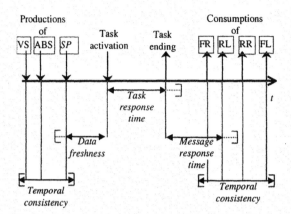

Fig. 3. Temporal constraints

There is a lack of flexibility and ability to extend and modify the embedded architectures. Finally, a significant part of an embedded application is not subject to strict hard real time constraints. So a compromise between ensuring "hard real time" constraints and "soft real time" constraints has to be found. A first response is given by protocols that allow a certain flexibility, as Flexray (Flexray, 2001), TTCAN (Führer *et al.*, 2000; Hartwich *et al.*, 2000), FTTCAN (Almeida and Pedreiras, 2000) or by Operating Systems as OSEK Time (OSEK-VDX, 2001) that keeps slots for non deterministic tasks. A second one relies on specific scheduling techniques as the *(m,k)* firm approach that guaranties always *m* satisfied deadline among *k* (Hamadoui and Ramanathan, 1995). Here the problem is the determination of *m* and *k*. Finally, deterministic approach and stochastic approach must be simultaneously used.

- *Robustness*. A vehicle, in particular the embedded electronic, is subject to non permanent failures due to electromagnetic perturbations; the safety properties must be kept in these situations. The problem is here to identify and model realistic perturbation scenarios (Navet *et al.*, 2000) and to propose a robust architecture (Gaujal and Navet, 2003) that is an architecture respecting the safety constraints even under these perturbations.

- *Portability vs Safety*. As we saw previously, the middleware is a central component for the portability of any embedded applications. Its main purpose is to act as a server of a global real time data base. Furthermore, for economical reasons, it has to offer a support for portable components, extensible applications, ... Therefore, an in-car embedded middleware has to be sized and adapted to the application requirements. The solution of this problem is relevant to complex discrete optimisation techniques under real time constraints. These problems are known to be NP-complete and, consequently, efficient heuristics have to be developed (Santos-Marques *et al.*, 2003).

4. ARCHITECTURE REFERENCE MODELS AND DEVELOPMENT PROCESS

4.1. Development of a safe electronic embedded system

The quality of an embedded system depends obviously of the quality of its development. It is well known that only testing the final product for ensuring that it respects its functional requirement under the required properties is not the best way to develop a complex system because it implies costly back loops on former design steps. For an efficient process, validation and verification activities must take place at each step, from end user requirements to final implementation.

This leads to dispose of models of this system according to specific validation / verification techniques and consequently to the suited formalisms. Two remarks have to be done:

- *Several models*. Validation / verification techniques are not the same at each level of the process; they don't use the same modelling language or formalism and are not applied to same entities. For example, at end user requirement specification, a V&V activity can consist to prove the consistency of these requirements. The used formalism can be UML Use Cases, Message Sequence Charts, UML Collaboration Diagrams, ... No information about software or hardware components or about distribution are used for this activity. On the other hand, to verify that a proposed implementation will respect performance and temporal properties needs to develop, just before coding, models that can represent event occurrences in a quantified way. For this purpose, well suited formalisms are Timed Automata, Temporal Petri Nets, Queuing Systems, ... The modelled objects are tasks, frames, protocols, schedulers, memories, ...

- *Modelling complexity*. Designers that have a good knowledge of electronic embedded systems and specific constraints in automotive domain are usually not specialist of the above mentioned languages or formalisms. This is a bar to the massive use of formal techniques.

A way to take account of these two points is to provide a domain oriented language that helps the designer to describe its system with its proper vocabulary and add a strict semantic to this language in order to furnish an automatic generation of formal models. At the present a lot of works are engaged in this way and main of them propose specific UML Profiles for achieving this purpose (OMG, 1999a; OMG, 1999b; Apvrille *et al.*, 2001; Cavaliere *et al.*, 2001). Fig. 4 shows schematically how these principles can be used.

For the description of component and architecture of component, computer science community engaged studies for the development of Architecture Description Languages (Taylor and Medvidovic, 1997). Unfortunately, these languages are restricted to software domains. Few of them are concerned with discrete event applications (Luckham, 1996; Allen and Garlan, 1997) or real time properties of an implementation (Vestal, 1993; Vestal,1995).

4.2. Reference Architecture Models

As explained in the previous section, a unique domain oriented language allows the representation of a system at each level of its development. This language is a declarative language for the modelling of components and architecture of components. At each design step, must be defined which component must be represented (classes and attributes) and how these components can be "composed" to form an architecture (relation, interaction, composition, …).

Furthermore, for traceability and model consistency purposes, the language has to provide a way for a formal description of the relation between components and architectures at different steps.

In order to illustrate these concepts we propose below an abstract of such a language, AIL_Transport (Elloy and Simonot-Lion, 2002) that was developed in the AEE Project (AEE, 1999). AIL_Transport has been defined as a modelling language dedicated to the specification of architectures described as an assembly of standard components. Every component is an instance of class belonging to a generic model, and it includes all pertinent characteristics necessary to the subsequent analysis of the whole architecture: interaction consistency, logical behavior, real-time performances, fault tolerant properties, … AIL_Transport is formalized using UML syntax.

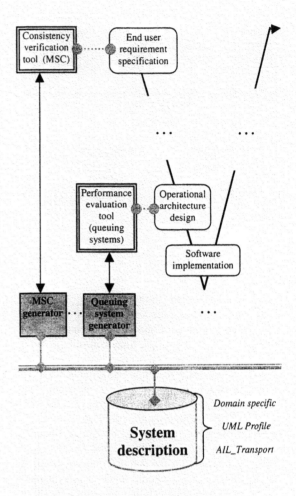

Fig. 4. A unique language - several validation activities

Using AIL_Transport, any designer describes the representation of an embedded architecture according to five different levels of abstraction. Each level models a particular point of view of the architecture. Entities specified at high levels ("vehicle project" level and "functional" level) are abstract components. Entities specified at low levels are, on the one hand, ECUs and communication protocols at the "hardware" level, on the other hand, the software and the devices (sensors and actuators) in the "software" level and the "operational" level. At each level, the designer describes the architecture as an assembly of objects instanced from predefined classes; then he specifies interactions between these objects using predefined connection types. The Fig. 5 illustrates these levels and their relationships. The development process associated to this representation links components introduced at different levels.

Vehicle Project Level. The upper level describes an embedded application from a vehicle point of view. At this level, objects shall represent functionalities offered by a vehicle (ABS, cruise control, air conditioning, etc), the different variant of these services (manual or automatic climate control system, for example) and the set of "on the shelf" vehicle versions. Five main classes are used at this architecture level: Vehicle project, Vehicle type, Vehicle, Service, Variant. Mainly, these classes document the architecture and support the project validation in terms of model consistency.

Functional Level. After building a validated vehicle model, the functional level describes one (or several) graph of elementary functional components realizing the services specified at the vehicle project level. Every graph of these components is specified disregarding the distribution and implementation aspects. The model supports a hierarchical specification of functions and flows. Function, Functional Flow and Functional Architecture are the main classes used to build graphs of elementary functions. Documentation, formal validation and formal test generation are the main process activities fulfilled at this step.

Hardware Level. This level models the electronic components of architecture as a set of processors, micro-controllers, electronic devices connected by networks. The main classes are
- *Operating Hardware.* Objects whose main subclasses are ECU (Electronic Control Unit for the computation nodes) and Network
- *Dependent Software Component.* These components are closely linked to hardware devices: Network Protocol, OS Software Components, I/O Drivers
- *Hardware Architecture* that specifies how each node is connected on one or several networks.

Software Level. At this level, two sets of classes are used. The first one is derived by class refinement of the functional architecture: the Application Software Components, the Software Flow, the Instrumentation Hardware Objects (Sensor and Actuators) and the Software Architecture. The second set of classes models the distribution of all software entities. For this, Software Component are decomposed in Logical Task communicating using Software Input and Software Output which are linked to Software Flow of the functional level. Moreover, the activation policies of Logical Tasks are specified (timed or event triggered).

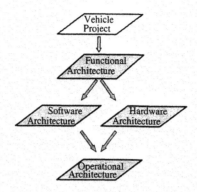

Fig. 5. Reference architectures during the design process

In order to validate the approach, a prototype implementing the access to AIL_Transport compliant data base has been realised. Several model generators were developed: synchronised timed automata based model for performance evaluation of an operational architecture and graph-based model for automatic allocation and scheduling of periodic distributed activities.

5. CONCLUSIONS

The introduction of software based systems are growing in automotive domain. This new technology has to be mastered in an efficient way. First it must preserve the know-how of each actor (car makers and suppliers), ensure the ability of system extensions or modifications and allow the portability of components. Furthermore, several constraints are applied to their design process: on the one hand, the development length must be shortened and, on the other hand, the safety of the system has to be formally proved as soon as possible before the real implementation. In this paper, we presented two axis for research and development activities in this context: the first one is concerned by on line architectures and mechanisms while the second one is relevant to off line a priori modelling, validation and verification.

The author would like to thank the actors of AEE Projecs, in particular, Jörn Migge, Franck Gasnier, Jean-Pierre Elloy, Yvon Trinquet, Xavier Hanin, Evelyne Silva, Bernard Bavoux, Philippe Germanicus for their essential contributions to the definition of the AIL_Transport language and Nicolas Navet for the fruitful debates that we had on this topic.

6. BIBLIOGRAPHY

TTP, (1999) http://www.tttech.com

Kopetz H., (2002), Time-Triggered Real-Time Computing, Proceedings of 15th IFAC World Congress, IFAC B'02, Barcelone, 21-26 juillet 2002.

OSEK-VDX, www.osek-vdx.org/

Navet N., Song Y-Q., Simonot F., (2000), Worst-Case Deadline Failure Probability in Real-Time Applications Distributed over CAN (Controller Area Network)", Journal of Systems Architecture, Elsevier Science, vol.46, n°7, 2000.

Gaujal B., Navet N., (2003), Optimal Replica Allocation for TTP/C Based Systems, to appear in Proc. 5th FeT IFAC Conference, FeT'2003, Fieldbus Technology, Aveiro (Portugal), 7-8 Juillet 2003.

Santos-Marquez R., Navet N., Simonot-Lion F., (2003), Frame Packing under Real-Time Constraints", to appear in Proc. 5th FeT IFAC Conference, FeT'2003, Fieldbus Technology, Aveiro (Portugal), 7-8 Juillet 2003

Flexray, (2001), http://www.flexray-group.com/

Führer T., B. Müller, W. Dieterle, F. Hartwich, R. Hugel, (2000) M. Walter, Time Trigger Communication on CAN (Time Trigger CAN-TTCAN) 7th International CAN Conference (ICC), Amsterdam, Netherlands, Oct. 24-25, 2000

Hartwich F., B. Müller, T. Führer, R. Hugel, (2000) CAN Network with Time Triggered Communication, http://www.can.bosch.com

Pedreiras P., L Almeida (2000), Combining Event-Triggered and Time-Triggered Traffic In FTT-CAN : Analysis of the Asynchronous Messaging System 2000 IEEE International Workshop on Factory Communication Systems (WFCS), Porto, Portugal, Sept. 6-8, 2000

Hamadoui M., Ramanathan P., (1995), A Dynamic Priority Assignment Technique for Streams with (m,k) firm deadlines, IEEE Transactions on Computers, 44(4), 1443-1451, December 1995.

IEC (1997), International Electrotechnical Commssion, 61508-1, Functional Safety of Electrical : Electronic / Programmable Electronic Safety-Related Systems.

AEE (1999), Architecture Electronique Embarquée, http://aee.inria.fr

EAST EEA (2001), European ITEA Project - Embedded Electronic Architectures, www.east-eea.net/docs

OMG, (1999a), White Paper on the Profile mechanism v.1.0», Analysis and Design Platform Task Force, Report ad/99-04-07, Avril 1999.

OMG, (1999b) UML Profile for Scheduling, Performance, and Time , RFP ad/99-03-13, 1999.

Apvrille L., P. de Saqui-Sannes, C. Lohr, P. Sénac, J.-P. Courtiat, "A New UML Profile for Real-time System Formal Design and Validation", Proceedings of the Fourth International Conference on the Unified Modeling Language (UML'2001), Toronto, Canada, October 2001.

Cavaliere D., Simonot-Lion F., Song Y.-Q., Hembert O., A Component Model Approach for Modelling and Validation of an Automated Manufacturing System, in Actes de 8th IEEE International Conference on Emerging Technologies and Factory Automation, Antibes - Juan les Pins, 15-18 octobre 2001.

Taylor R. N. and N. Medvidovic, (1997). A Framework for Classifying and Comparing Architecture Description Languages, *Technical Report*, Department of Information and Computer Science, University of California, Irvine.

Luckham D. C., (1996). Rapide: A Language and Toolset for Simulation of Distributed Systems by Partial Orderings of Events, *in Proceedings DIMACS Partial Order Methods Workshop IV*, Princeton University.

Allen R. and D. Garlan, (1997). A Formal Approach for Architectural Connection, *PhD thesis*, school of Computer Science, Carnegie-Mellon University, Pittsburgh.

Vestal S., (1993). Scheduling and Communicating in MetaH, *in Proceedings of Real -Time System Symposium*, Rleigh-Durham (NC), p.194-200.

Vestal S., (1995). MetaH Reference Manual, *Technical Report*, Honeywell Technology Center.

Elloy J.-P., Simonot-Lion F., *An Architecture Description Language for In-Vehicle Embedded System development*, Proceedings of 15th IFAC World Congress, IFAC B'02, Barcelone, 21-26 juillet 2002.

ELSEVIER

IFAC

PUBLICATIONS
www.elsevier.com/locate/ifac

DYNAMIC INTERCONNECTION OF CONSUMER ELECTRONICS AND HOME AUTOMATION NETWORKS

Michael Ziehensack * **Manfred Weihs** *

* *Institute of Computer Technology,*
Vienna University of Technology,
Gusshausstr. 27-29/E384,
A-1040 Vienna, Austria
{zie,weihs}@ict.tuwien.ac.at

Abstract: In this paper the dynamic interconnection between consumer electronics and home automation networks is discussed. As there are different networks in the home the user has to deal with various user interfaces and inter-network applications are not possible. The idea of network interconnection is to overcome the borders between the networks and to provide functionality, which is available in one network, to devices of the other network. A three layer service architecture is introduced to compare consumer electronics and home automation networks. Based on this architecture some characteristics of network interconnections will be described. As an example an implementation of a one-way control gateway which allows HAVi based consumer electronic devices to use home automation services provided by the field bus LON will be presented. *Copyright © 2003 IFAC*

Keywords: Home Networks, Interconnection, HAVi, LON

1. INTRODUCTION

There are different kinds of networks in a home to connect devices of consumer electronics, home automation, and computer equipment, like HAVi (HAVi Inc, 2001) based on IEEE 1394 (IEEE Computer Society, 1995), IP/x, EIB (Dietrich *et al.*, 2000) or LonWorks (Dietrich *et al.*, 1997). Each of these home networks is optimized for some kind of devices. Therefore they differ significantly with respect to some features such as bandwidth, safety, security, maximum distance, possible number of devices, configuration effort, transport medium etc.

Network interconnections enables applications on devices of one network to use services provided by devices of the other network. Another way to overcome the heterogeneity in the home would be to establish a new super network. But as this new network would have to deal with trade-offs due to different requirements of home applications this network would pale beside specialized networks. Thus specialized networks with interconnections seem to be the solution for the smart home. This enables to use the UI capabilities of consumer electronics to control home appliances. But also novel applications which use devices from both networks are possible.

This paper discuss the interconnection between consumer electronics and home automation networks. In the next section a three layer service architecture will be established. According to this service architecture an overview of consumer electronics and home automation networks will be given. The following section deals with the characteristics of network interconnection. Considering the theoretical part an implementation of an inter-connection between the multimedia home network HAVi and the field bus LON will be presented.

2. SERVICE ARCHITECTURE

Although there are different networking specifi-cations to connect devices in the home there is a common set of characteristics. Connecting devices requires solutions for addressing, discovery, de-scription, control, notification, and presentation. While these requirements are also used by the UPnP specification (Microsoft, 2000) to describe the role of utilized protocols their meaning will slightly differ from that used in this paper. The service architecture consists of services required for networking. As shown in Fig. 1 the services are grouped in three layers whereas a higher layer requires the services of the lower layer.

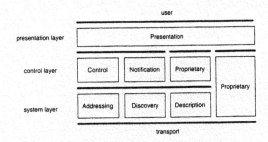

Fig. 1. Service Architecture

Addressing is responsible for the unique identifi-cation of a communication object. This includes assignment of a unique and possibly structured address and in some cases also address translation facilities. For communication the knowledge of communication objects and their addresses are required. This is the job of *Discovery*. While discovery makes communication objects visible, *Description* is responsible for specifying proper-ties and capabilities of communication objects and defines a mechanism for obtaining this in-formation. Addressing, discovery and description provide system services and build therefore the system layer. *Control* defines how a communi-cation object ("service user") is able to use a service of another communication object ("service provider"). Thereby the communication is initi-ated by the service user. Beside the definition of the communication process also an API definition of each service is required. *Notification* describes how a service provider is able to report a state change to a service user which previously regis-tered for such notifications. In this case the com-munication is initiated by the service provider. Control and notification provide services for de-vice control and form the control layer. *Presenta-tion* is concerned with the user interface of service providers and makes the presentation layer.

If control layer services are not defined by an API which is publicly available the control layer is considered as proprietary. The consequence is a loss of interoperability as the services cannot be used by applications of other vendors. In certain cases the API definition of simple services can be moved to the presentation layer by displaying the user well known keywords like "play", "stop" or "on", "off". In case the definition of a system layer service is missing or not publicly available both, the system and control layer are considered as proprietary. If so there is no interoperability as applications require proprietary information to use the services.

Mainly there are three areas of home networking: home automation, consumer electronics and per-sonal computer environment including telecom-munication. The following two subsections de-scribe the characteristics of home automation and consumer electronics networks according to the service architecture defined above.

2.1 Home Automation

European Installation Bus (EIB) (Dietrich *et al.*, 2000) and Local Operating Network (LON) (Dietrich *et al.*, 1997) are two widely used buses in the field of home automation and therefore two good representatives of home automation sys-tem standards. For both buses the assignment of unique addresses and the introduction of new devices have to be done manually. Thus address-ing and discovery requires human interaction. The description of services is left to so called profiles (Echelon, 2002; LonMark, 1996) and interoper-ability guidelines (LonMark, 2002; LNO, 2000). Control and notification are generally based on a link between two or more communication ob-jects which have to be configured during setup. The link consists of an address (unicast or mul-ticast), a defined message format and an action which shall be performed after the message was received (control) or had been performed before the message was sent (notification). Presentation is mainly performed via additional and easy to use hardware like switches, dimmers or simple LCD displays with buttons. The home automation systems works therefore in the background after the bus has been configured manually.

2.2 Consumer Electronics

Audio Video Control (AV/C) (1394 Trade Asso-ciation, 1998) and Home Audio Video interop-erability (HAVi) (HAVi Inc, 2001) are two rep-resentatives of network protocols for consumer electronic devices. Both are based on the Se-rial Bus IEEE 1394 which supports data rates of currently up to 1600 Mbit/s (IEEE Computer Society, 2002). Some parts of addressing, discovery and description are performed directly by IEEE 1394. While AV/C is based on the client/server principle, HAVi provides real peer-to-peer networking.

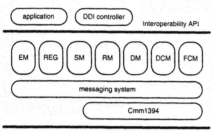

Fig. 2. HAVi Software Architecture

Fig. 2 shows the HAVi software architecture. Within HAVi so called software elements (SE) provide services by standardized interfaces (Interoperability API). The messaging system is responsible for message exchange between communication objects (software elements) and assigns each SE a unique address based on the GUID of the IEEE 1394 bus interface (addressing)[1]. The event manager (EM) provides an event notification services, whereas an event is a status change which a SE reports to the event manager. The registry (REG) stores attributes of local SE and provides a query interface to search for SE in the entire network. A new SE registers its attributes at the registry, which stores the data (description) and posts an event to inform other SE of the new SE (discovery). Control is performed via software elements abstracting device functionalities: device control module (DCM) and functional component module (FCM). Device services are based on a well defined API for different consumer electronic device types, like tuner, audio video disc, amplifier or camera. Notification is performed by the event manager or directly by the related SE. The stream manager (SM), resource manager (RM) and dcm manager (DM) are optional modules which provide services for managing audio/video transmissions, reservation of FCMs for exclusive use and the installation of DCM/FCMs for IEEE 1394 devices which are not executing a HAVi stack. Beside a HAVi application which uses device specific GUI capabilities, HAVi defines two special GUI concepts for presentation, namely Data Driven Interaction (DDI) and Havlets. DDI consists of a GUI renderer (DDI Controller) and a device control module extension which provides the renderer UI descriptions of the controlled device. Havlets are HAVi applications written in Java which assume a device with a Java virtual machine and well defined TV-friendly UI capabilities. HAVi provides users an easy to use access to functionalities of devices on the bus without manual device configuration.

3. INTERCONNECTION

There are several characteristics of network interconnections. In this section some of them are discussed based on the three layer service architecture which has been introduced in section 2 (Fig. 1).

Network interconnection enables applications on devices of one network to use services provided by devices of the other network. But this approach must not be mixed up with a device which is part of two or more networks and which provides services from the directly connected networks via the local presentation layer. These kind of device will be named *multihomed non transforming device*, see Fig. 3. A digital TV which contains a HAVi stack as a link to other consumer electronic devices, an EIB module which provides access to home appliances and a common user interface will be an example of such a device. Here the usage of remote services from different networks is limited to the multihomed device. In contrast the main point of interconnection is that remote services can be used like services provided by the own network. The interconnection device (hereinafter called gateway) has therefore to perform some transformation. A multihomed non transforming device does not perform transformation and is therefore not a gateway.

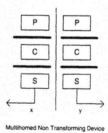

Multihomed Non Transforming Device

Fig. 3. Multihomed Non Transforming Device

Some characteristics of network interconnection:

(1) transformed service layer
(2) direction
(3) transformation steps and involved devices
(4) manual setup effort

The first interconnection characteristic depends on the transformed service layer and will be shown in Fig. 4. A device which performs a transformation of the presentation layer will be called *presentation gateway*. An example of this gateway type is the HAVi DDI to WAP gateway described

---

[1] Beside addressing the messaging system performs also parts of the other system and control layer services as all communication is done via the messaging system.

in (Nikolova *et al.*, 2002). If the transformation is performed on the control layer, it will be called *control gateway*. This kind of gateway is the most important one, as a control layer transformation allows to control a device via an API which is different to the original one. Therefore many gateway implementations are of this type, e.g. Jini-LON gateway (Kastner and Leupold, 2001), EIB-UPnP gateway (Wischy and Gitsels, 2000), or the HAVi-LON gateway which is described in section 4. A *system gateway* is a device which performs a transformation of the system layer. For example two system gateways can be used to interconnect two remote HAVi networks via a TCP/IP network.

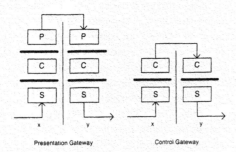

Presentation Gateway   Control Gateway

Fig. 4. Gateway Types

A further interconnection characteristic is the direction of network interconnection. If services provided by devices of one network can be used from devices of the other network and the other way around, it is called a two-way interconnection. Otherwise it will be called a one-way interconnection.

The third interconnection characteristic is related to the transformation steps, involved devices and the required number of interconnections. A direct interconnection between two networks requires one transformation step and one gateway device. But if there are n networks and each network requires a connection to each other, the required number of two-way interconnections is $n * (n - 1)/2$. This seems unlikely for a certain home, but may be more probable considering all available networks and the different environments. By using two-way interconnections to a central network instead of direct interconnections between each network, the number of interconnections can be reduced to $n - 1$. But for non direct interconnections the number of transformation steps is two. Cho (2002) proposed a home networking framework which is based on this approach. There the central network is implemented on a single device. Another implementation is described by Nakajima *et al.* (2002) whereas the central network is called Virtual Overlay Network and consists of different devices which perform the control layer transformation.

The fourth interconnection characteristic deals with the manual set up effort of the gateway. If no (or only very little) additional effort is required beside the effort to setup the related networks the gateway will be called a dynamic gateway, otherwise a static gateway.

4. HAVI-LON GATEWAY

The HAVi-LON gateway is a dynamic one-way control gateway which enables HAVi software elements to use home automation services provided by the field bus LON (see Fig. 5).

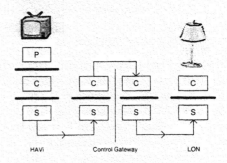

HAVi   Control Gateway   LON

Fig. 5. HAVi-LON Gateway

4.1 Gateway HW/SW Architecture

The HW of the gateway consists of a Windows PC with an Echelon PCI-LON-Interface card and Free Topology Transceiver and a Unibrain PCILynx IEEE 1394 card. The SW of the gateway is split into two processes:

(1) LNS process: application which uses Echelon LNS (LonWorks Network Services) Server (Echelon, 2000) to access the LON,
(2) HAVi process: application based on dmn HAVi stack (dmn, 2002) including an embedded HAVi Device Control Module (DCM) and Functional Component Module (FCM).

The communication and synchronization between the two processes is performed via Shared Memory and Semaphores.

4.2 Control Layer Transformation

On the LON side control and notification is mainly based on so called network variables (Echelon, 2002) and a binding between these variables which has to be configured during setup. In the HAVi network devices are represented as DCMs and its functional components as FCMs. Therefore a mapping between network variables and DCM/FCMs is required.

At the EIB-UPnP-Gateway implementation of Wischy and Gitsels (2000) each EIB device is represented by a UPnP device which provides UPnP services related to abstract device functions (e.g. relay, analog input, analog output). While this mapping is reasonable from an engineering point of view, it is not very convenient for the user. The user expects services and does not want to deal with devices, networks and their internal structure (Rose, 2001). A home automation device which provides 4 relay outputs to control lights shall be represented by 4 independent light services and not by one device which provides 4 abstract relay services. Therefore it is very important to consider that a home automation service should offer the user home automation functions instead of device functions which are used to implement home automation functions.

Table 1 shows how LON network variables can be mapped to DCM/FCMs according to a three level fragmentation of the home automation network. The lowest level consists of home automation functions which are accessible via LON network variables. Home automation services are based on one or more of these functions and establish the second level. The complete home automation network is represented by the third level.

Table 1. Mapping of LON network variables to HAVi DCM (D) and FCMs (F).

level/option	1	2	3	4	5	6
HA network	D/F	D	D	-	-	-
HA service	-	F	-	D/F	D	-
LON NV	-	-	F	-	F	D/F

The result of the evaluation of the different options can be itemized as follows:

- At option 1, 4 and 6 each DCM is assigned a single FCM which represents the functionality of the home automation network up to its level. Therefore inter-level functionalities and informations are not supported or available. In contrast to this option 2 enables a DCM to list all its FCMs each representing a HA service.
- As at option 1 the complete home automation network is represented by a single DCM and FCM, the HAVi registry cannot be used to query for location or type of a HA service. A separate query interface would be required.
- At option 1, 3 and 6 the relation of LON network variables to home automation services is not possible as there is no mapping on the HA service level.
- Option 3, 5 and 6 will lead to a high number of software elements which may confuse the user. The UI of a HAVi TV will typically have an option to show all DCMs of the

network which becomes very complex if there are many DCMs. It does not make sense that each light in the home will be represented by an own DCM with one single FCM.
- As option 4, 5 and 6 have no mapping at the HA network level a common DDI target (HAVi UI level 1) for the complete home automation network cannot be provided.

As there are less complain about option 2 it seems to be the best solution. The complete LON network is assigned to a single DCM (HA-DCM) and all network variables which belong to one HA service are assigned to an FCM (HA-FCM). The name, location and type of a home automation service are some of the FCM attributes stored at the HAVi registry. Therefore the query capabilities of HAVi can be used to find services by means of these attributes.

For the mapping the gateway needs some information for every home automation service provided by the field bus:

(1) name, location, type, identifier and
(2) all LON network variables related to the service and for each a suitable function name and function type.

As this data can be provided during the required manual LON setup and will be supported by a GUI, the additional effort for the HAVi interconnection is kept at a minimum. Therefore the gateway can be considered as a dynamic gateway. All relevant data are stored in a gateway definition file which will be loaded during startup of the LNS application (LON part of the gateway).

The LNS application is responsible:

- to notify the HA-DCM of new or gone home appliances,
- to return the current value of a specified network variable on request of the related HA-FCM,
- to change the value of a specified network variable on request of the related HA-FCM.

The HA-DCM will be installed during startup of the HAVi part of the gateway. It is responsible to install or remove HA-FCMs according to the services currently provided by home appliances on the field bus. Each HA-FCM will be assigned all data necessary for requesting the LNS application to access the LON.

4.3 Home Automation Extension for HAVi

As HAVi does not specify home automation services a Home Automation Extension for HAVi has been specified (Ziehensack, 2002c). This extension defines a Home Automation FCM (HA-FCM) which provides home automation services and is based on a field bus to DCM/FCM mapping according to option 2 of Tab. 1. To distinguish different services, 5 service types and 2 location types each with a number of subtypes have been defined. The HA-FCM API consists of services to retrieve HA-FCM properties and two generic services to access home automation functions, namely *SetValue* and *GetValue*. As a HA service may consist of more than one HA function the first service parameter specifies the desired HA function. Additionally the HA-FCM API defines one HAVi FCM attribute indicator (*currentValue*) which can be used by a SE to subscribe for notifications of attribute changes.

While a simple light service offers only one HA function, some services consist of many or complex HA functions; see (Kim *et al.*, 2001) for white goods. For interoperability all HA functions offered by a HA service must be defined. As there are many different and very special HA services, they are not defined within the HAVi HA Extension document, but in separate HA service specifications (Ziehensack, 2002*b*; Ziehensack, 2002*a*). A HA service specification must specify all its HA functions and classify the provided service according to the service type defined in the HAVi HA Extension document. Thereafter it is assigned a unique service identifier from a central authority (e.g. HAVi Inc). This identifier is part of the information which must be provided during installation of a new home appliance and becomes an attribute of the HA-FCM. Therefore all HA functions of a HA-FCM are defined and the related specification can be uniquely identified by the service identifier.

Each software element on the HAVi network is able to access home appliances from different manufacturers on the home automation network via HA-FCMs running on the gateway. For example a HAVi application is able to switch on all lights in the living room by querying the HAVi registry for the attributes service location ("living room") and service type ("light")[2] and by invoking the service SetValue of all HA-FCMs returned by the registry.

In certain cases a user may also be able to use unknown HA services if the HA functions are given well known descriptions (e.g. "blinds up", "blinds down") as function names and if they are shown to the user via the presentation layer.

The HAVi application developer has to deal with three types of specifications to use home automation services: HAVi spec., HAVi HA Extension spec. and HA service spec.

---

[2] Registry attribute names and values have been simplified: ATT_LOCATION_ROOM = LIVINGROOM and ATT_SERVICE_TYPE = LIGHTING/LIGHT.

5. CONCLUSION AND FURTHER WORK

This paper described a dynamic interconnection between consumer electronics and home automation networks. As there are different networks in the home the user has to deal with various user interfaces and inter-network applications are not possible. The idea of network interconnection is to overcome the borders between the networks and to provide functionality, which is available in one network, to devices of the other network.

To compare different networks in the home a three layer service architecture has been introduced. Based on this architecture it has been shown that consumer electronic and home automation networks differ significantly in the way they implement services. After describing some interconnection characteristics an implementation of a one-way control gateway between HAVi and LON has been shown.

The HAVi-LON gateway performs a translation on the control layer and therefore a mapping between LON network variables and HAVi DCM/FCMs. As a user expects services and does not want to deal with devices, networks and their internal structure, HAVi DCM/FCMs representing LON devices shall provide home automation functions (light on, blinds down, ...) instead of device functions (relay, analog input, ...). HAVi does not specify home automation services and therefore a Home Automation Extension for HAVi has been specified. To deal with a huge number of different home appliances, the service specifications are separated from the HAVi extension and assigned a unique identifier. The result is that each software element on the HAVi network is able to access home appliances from different manufacturers on the LON via HA-FCMs running on the gateway.

Currently the HAVi home automation extension has to be proofed by implementations with more and different LON devices and especially by implementations for other field buses (e.g. EIB). Aside from this also two-way interconnections and interconnections between other home networks will be investigated.

REFERENCES

1394 Trade Association (1998). *AV/C Digital Interface Command Set, General Specification, Version 3.0*. 1394 Trade Association. Austin, Texas, USA.

Cho, Song Yean (2002). Framework for the composition and interoperation of the home appliances based on heterogeneous middleware in residential networks. *IEEE Transactions on Consumer Elecetronics* **48**, 484– 489.

Dietrich, Dietmar, Dietmar Loy and Hans-Jörg Schweinzer (1997). *LON-Technologie. Verteilte Systeme in der Anwendung.* Hüthig.

Dietrich, Kastner and Sauter (2000). *EIB Gebäudebussystem.* Hüthig.

dmn (2002). *HAVi SDK Professional, Programmers Guide Vers. 1.2.* dmn Software-Entwicklung GmbH.

Echelon (2000). *LNS for Windows Programmers Guide, Vers. 3.0.* Echelon Corporation. http://www.echelon.com.

Echelon (2002). *LonMark SNVT Master List.* Echelon Corp. http://www.echelon.com.

HAVi Inc (2001). *Specification of the Home Audio/Video Interoperability Architecture.* HAVi, Inc.

IEEE Computer Society (1995). IEEE standard for a high performance serial bus. Standard IEEE 1394-1995. IEEE Computer Society. New York, USA.

IEEE Computer Society (2002). IEEE standard for a high performance serial bus–amendment 2. Standard IEEE 1394b-2002. IEEE Computer Society. New York, USA.

Kastner, Wolfgang and Markus Leupold (2001). Discovering internet services: Integrating intelligent home automation systems to plug and play networks. In: *Innovative Internet Computing Systems, International Workshop, IICS 2001, Ilmenau, Germany.*

Kim, D. S., G. Y. Cho, W. H. Kwon, Y. I. Kwan and Y. H. Kim (2001). *Home network message specification for white goods and its applications.*

LNO (2000). *Standard for the Implementation of Interoperable Applications, LNO AK II.* LON Nutzer Organisation e.V (LNO). http://www.lno.de.

LonMark (1996). *Functional Profile: Switch.* LonMark Interoperability Association. http://www.lonmark.org.

LonMark (2002). *LonMark Interoperability Guidelines.* LonMark Interoperability Association. http://www.lonmark.org.

Microsoft (2000). *Universal Plug and Play Device Architecture, Version 1.0.* Microsoft Corporation. http://www.upnp.org/.

Nakajima, T., I. Satoh and H. Aizu (2002). A virtual overlay network for integrating home appliances. In: *Symposium on Applications and the Internet (SAINT 2002).* pp. 246–253.

Nikolova, M., F. Meijs and P. Voorwinden (2002). Remote mobile control of home appliances. In: *International Conference on Consumer Electronics, 2002. ICCE. Digest of Technical Papers..* pp. 100–101.

Rose, Bill (2001). *Consumer Requirements for Home Networks.* CEA R7 Home Network Committee.

Wischy, Markus Alexander and Martin Gitsels (2000). *EIB-Universal Plug & Play Gateway.* Siemens, ZT SE2, Munich.

Ziehensack, Michael (2002a). *HAVi HA Service Specification - Blind, Draft 0.3.* Institute of Computer Technology, Technical University of Vienna.

Ziehensack, Michael (2002b). *HAVi HA Service Specification - Light, Draft 0.8.* Institute of Computer Technology, Technical University of Vienna.

Ziehensack, Michael (2002c). *Home Automation Extension for HAVi, Draft 0.3.* Institute of Computer Technology, Technical University of Vienna.

EVALUATION CRITERIA FOR COMMUNICATION SYSTEMS IN BUILDING AUTOMATION AND CONTROLS SYSTEMS

Peter Fischer

University of Applied Sciences Dortmund
Fachbereich Nachrichtentechnik
P.O. Box 50 10 18
D-44047 Dortmund
Germany

Abstract: Since mid of the eighties digital communication systems in building automation and controls systems (BACS) are used. In the beginning there were proprietary systems, but later on more and more communication standards have been used. The special requirements for the building automation and control systems regarding the application, i. e. the technical processes, had some impact to the communication systems. In this paper the possibilities for classifying communication systems are shown. With these classifications it is possible to specify requirements for the communication systems based on certain situations within technical (building) processes and to draw someone own's conclusion from the description of a communication system regarding the use and benefits of the system in a certain application. *Copyright © 2003 IFAC*

Keywords: Communication protocols, communication systems, control systems

1 COMMUNICATIONS IN BUILDING AUTOMATION AND CONTROLS

In the last 10 years the amount of communication systems in building automation and controls environments are decreasing, especially the proprietary ones. Nevertheless the choice which one should be used in an given project is still open and there are no rules nor guidelines for making the right decisions.

In the following sections an overview is given regarding the evaluation criteria of communication systems in building automation and controls systems (BACS). This overview is based on the results of a thesis (Fischer, 2002).

2 MODELS OF BAC

For describing technical systems the model approach is often used. In this case there are four different models:

- System model
- Process model
- Communication model
- Reference point model.

Each of these models is used in practice for describing the BAC application and the related communication and in most cases more than one of it in parallel.

2.1 System Model

The system model describes the functional levels of BAC systems, the management, automation / control and the field level.
Management functions are performed by the software of a BACS. The plant / application specific management functions are for the activity of a user taking decisions for supervision of plant and evaluating energy use and operational costs. The required functionality at this level is:

- Communications with devices of the control functions network
- Communications for data exchange with dedicated special system to provide operator and management functions
- Recording, archiving and statistical analysis
- Energy management

Automation / control functions are part of the plant / project specific application software and parameters which provide monitoring, interlocks, closed loop control and optimization of building services in real-time within self-contained controllers [automation stations]. At this level the required functions are:

- Physical input and output functions
- Communication input and output functions
- Monitoring
- Interlocks
- Closed loop control

- Calculation / optimization
- Room control functions (e. g. individual zone control, lighting control, shades / blinds control).

Field devices are generally sensors and actuators, coupling units and local override / indication devices that are connected to input / output interfaces of controllers. Field devices can be connected to controllers via field bus network or direct wiring. The field devices perform connection to the physical items of plant providing the necessary information about the conditions, states and values of the processes and effect the programmed operations. Functions supported by field devices are:

- Switching
- Positioning
- State monitoring
- Input for counting
- Measuring .

In the past the functional levels have been related to specific communication systems (Fischer, 1995). Nowadays one standardized communication systems is able to handle all the information exchange needed depending on the amount of datapoints and the complexity of the applications in in a building. Therefore there is no close relation anymore between the communication level (related to protocols) and the functional level (related to BACS functions).

2.2 Process Model

The technical processes in building automation and controls systems are well-defined and will not be a matter of discussion in this paper. The link between the technical process and the above mentioned functions are the information lists defined in the ISO draft "Building Automation and Control Systems (BACS), Part 3: Functions" (ISO FDIS 16484-3, 2001).

2.3 Communication Model

The different functional levels and applications in BACS require different kinds of communications. This does not mean that different protocols are needed because most of the BACS communication systems have the possibilities to use these different kinds depending on the functionality needed. The communication models

- Client-server
- Publisher-subscriber and
- Producer-consumer

will be shown in the following sections.

The client-server model is well-known and used in BAC communication systems based on a LAN infrastructure. The disadvantage of the client-server communication is the overhead because of the request-re-

sponse mechanisms which can become a problem, especially in the field level applications.

In the publisher-subscriber model the client can subscribe for a service at a server under certain conditions. If this condition is given the server will send the information to all clients subscribed for this service. The subscription is valid until the client cancels it or after the time for it is gone by. The server informs all subscribers with a single communication which reduces the communication on the network. But there is an administration overhead needed for controlling the subscribers.

In field devices this can be a disadvantage because of the restricted CPU power and memory size. In that case the producer-consumer model could fit. The "producer" of information sends it over the network (as multi-cast or broadcast) and all "consumers" who are interested in this information receive it. The receiving of information is dependent on the type of data the consumer wants to get. The relation between the type of information, the producer and the consumers is build during the commissioning phase of the project. During operation there is no administration regarding addresses, registration of devices etc.

2.4 Reference Point Model

The idea of this model is based on a German preliminary standard (DINV 32734, 1992). This model was enhanced and describes the interfaces between devices and communication systems within the same and between functional levels (Fig. 1). The boxes represent devices on the functional levels management (M), automation/controls (A) and field (F). The circles are the reference-points with an index showing the relation: "R_{FF}" means reference-point between devices with field level functionality; the devices can be from the same or from different manufacturers and the communication systems can be identical or different. "R_{MA}" means reference-point between devices with management and automation level functionality; the devices can be from the same or from different manufacturers and the communication system can be identical or different.

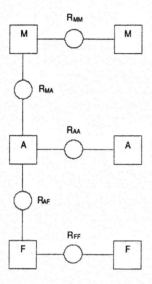

Fig. 1. Enhanced reference-point model

This reference-point model is independent from the functional levels used in a project. It doesn't matter whether there are all three functional levels or two. The reference points describe in any case the communication over the interface.

Table 1 Summary communication models

	Client-Server	Publisher-Subscriber	Producer-Consumer
Form	dialog	multi-cast	broadcast
Kind	connection-oriented	connection-less	connection-less
Physical Structure	point-to-point	multipoint	multipoint
Logical Structure	master-slave, peer-to-peer	peer-to-peer	peer-to-peer
Application	request, set value, para-meterize	change of value/state	alarm/event message, change of value/state

Table 1 gives a summary about the three communication models regarding the form, the kind of the connection, the physical and the logical structure and the application mostly used.

3 TYPES OF COMMUNICATIONS

In BACS there are three types of communication (Fig. 2):

- Engineering communication (E) for the interaction between field level processes and engineering-related processes in a commissioning or service tool

- Management communication (M) for the interaction between field level processes and a building management application or supervisory application

- Process communication (P) for the exchange of data between distributed field level processes.

The breadth of the arrowhead shows the amount of data received by the area where it is pointed to.

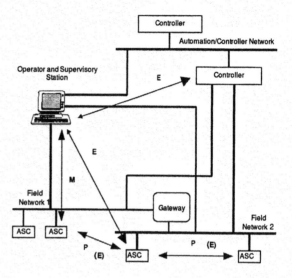

Fig. 2 Types of communication

3.1 Engineering Communication

In the working draft standard (ISO (WD) 16484-1, 2001) the term "engineering" is defined as follows:

The process of configuring and commissioning the various parts of a Building Automation and Control System. The tasks to perform are project- and system-specific.

The engineering communication is used during the engineering phase of a project ending with the "commissioning" which is also defined in the above mentioned standard as:

The process of calibrating field devices, testing data points, functions and system software for the various parts of a Building Automation and Control System application. Commissioning reports are proof for the completeness of work. The tasks to perform are project- and system-specific.

3.2 Management Communication

This type of communication describes the access by a monitoring or supervisory station to process information. In most cases this is an interaction between different functional levels, e. g. read operations for showing values on the display or write operations for setting new temperature setpoints. Event and alarm messages sent from the field functional level are also part of the management communication.

3.3 Process Communication

The process communication is the data-exchange between processes and applications on the same or on different functional levels. Especially in distributed systems on field bus level there is a lot of interaction between the field bus devices.

4 ENHANCED REFERENCE POINT MODEL

The enhanced reference point model of Fig. 1 will be described in more detail using the the example of communication types in BACS (see Fig. 2); Fig. 3 shows the model. The white circles are reference points within the same functional level, the shaded ones are communication between different levels. If you look to this figure you can see several communication paths via the reference points between devices in functional level networks using protocols. The vendor of the devices, the functional level and the communication protocol can be identical or different. Therefore several different scenarios are possible.

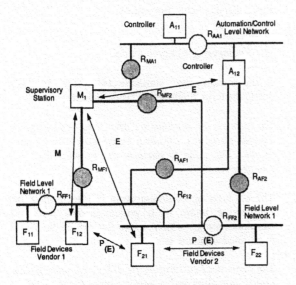

Fig. 3. Communication types shown in an enhanced reference point model

The more different vendors, protocols and functional levels are involved the more increases the probability that problems in the application may occur because of problems in the communication. This sounds like: "Use proprietary systems!", which in a few cases may be the best solution. But in most projects the functions needed are so common to most of the BAC devices that the field engineers and system integrators are aware of the possible problems and know how to handle them.

The communication over the reference points can be read operations, write operations, alarms and event messages. Even if the communication protocol on both sides of the reference point is identical it is required that the same services and objects with the same properties are used by the communicating devices. Otherwise some of the information gets lost. Table 2 the types of communication and the corresponding services.

Table 2 Types and services of communication

Type / Service	Engineering	Management	Process
Read Operation	Engineering data	Value or state of an object	Value or state of an object
Write Operation	Download of engineering data	Setting values, states or parameters	Writing values or states
Event Message	-	Event message	Triggering off processes
Alarming	-	Alarming	Triggering off processes

As an example for the reference point model two communications on the field level are shown in the following sections.

4.1 EIB in the Reference Point R_{FF}

The European Installation Bus EIB (Dietrich, 2000) is one of the most installed communication systems in homes and buildings. It is part of the European pre-standard ENV 13154-2:1998. The given scenario is a communication between two field devices via the reference point R_{FF}. The EIB application layer offers the following application services for the 'interoperation" of application processes in the different devices. The following communication relations are possible:

- Point-to-multipoint (multi-cast), connectionless
- Point-to-multipoint (broadcast), connectionless
- Point-to-point, connectionless
- Point-to-point, connection-oriented.

To reach the most efficient degree in interoperability both devices should use the same services and the same objects with the same properties. Fig. 4 shows a schematic drawing of the exchange of an APDU between two EIB devices via the R_{FF}.

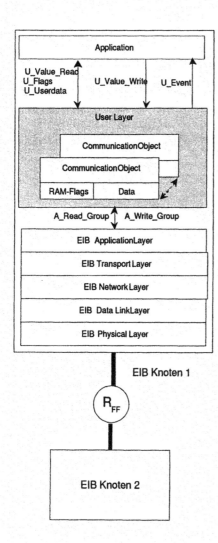

Fig. 4. Reference point on field level with EIB

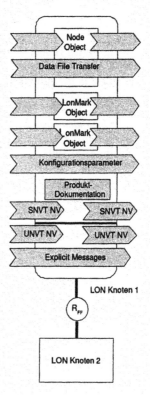

Fig. 5. Reference point on field level with LON

4.2 LON in the Reference Point R$_{FF}$

The LonWorks technology (Dietrich, 1999) is one of the most installed communication systems, too – not only for homes and buildings but also for other applications in automation systems. It is as well part of the European pre-standard ENV 13154-2:1998. The LonWorks communication offers five services on the application layer:

- Network variable propagation
- Generic message passing
- Network management messages
- Network diagnostic messages
- Foreign frame transmission.

The service 'Network Variable Propagation" is used for the process communication. The corresponding interface has an interoperable part and non-interoperable one. The interoperable interface contains the node object, standardized LonMark objects, standardized network variables, data file transfer, configuration parameters and product documentation. The other part can be used for proprietary information

5 CONCLUSION

The evaluation of a communication system in a building automation environment is dependent on many factors as

- Dimensions of the building or buildings
- Installed industries (e. g. heating, air conditioning, lighting, security)
- Complexity of the processes to be controlled by the BACS
- Number of devices installed and connected via the communication link
- Applications
- Customer requirements
- and others more.

The different communication requirements needed in a realization of a BAC project should be presented to the field engineers and system integrators in a kind of recommendation, technical report or part of a standard series (like the ISO 16484 series). With the knowledge about the possibilities a communication system offers the integration of different systems and the "inter-industries communication" should become more efficient and should lead to a satisfied customer.

REFERENCES

CEN ENV 13154-2:1998 E, *Data Communication for HVAC Application - Field Net - Part 2: Protocols* - Brussels: CEN, June 1998

Dietrich, D. (Hrsg.) u. a. (2000). *EIB - Gebäudebussystem.* Hüthig Verlag Heidelberg 2000

Dietrich, D. (Hrsg.) u. a. (1999). *LON-Technologie - Verteilte Systeme in der Anwendung.* 2. Auflage, Hüthig Verlag Heidelberg 1999

DINV 32734 (1992). *Digitale Automation für die Technische Gebäudeausrüstung - Allgemeine Anforderungen für die Planung und Ausführung* - Berlin: Beuth, April 1992.

Fischer, P. (1995). *Ein universelles Profil für Feldbus-Protokolle in der Gebäudeautomation.* FeT'95, ÖVE 1995.

Fischer, P. (2002). *Analyse und Bewertung von Kommunikationssystemen in der Gebäudeautomation.* Dissertation, Technische Universität Wien.

ISO (WD) 16484-1, 2001. *Building Automation and Control Systems (BACS), Part 1: Overview and Vocabulary.* ISO/CEN 2001, Brussels

ISO FDIS 16484-3, 2001. Building Automation and Control Systems (BACS), Part 3: Functions. ISO/CEN 2001, Brussels.

ELSEVIER

IFAC
PUBLICATIONS
www.elsevier.com/locate/ifac

INTERFACING WITH THE EIB/KNX
A RTLINUX DEVICE DRIVER FOR THE TPUART

Wolfgang Kastner and Christian Troger

Technische Universität Wien
Institut für Rechnergestützte Automation
E-mail: k@auto.tuwien.ac.at

Abstract: This article presents the basis for a variety of advanced Linux software
applications for configuring, monitoring and controlling the European Installation
Bus: interfaces and their corresponding device drivers. Thus, we shortly outline
connections realized between a PC and the well-known bus coupling units BCU1
and BCU2 and then go into detail for a realization based on the rather new
TPUART chip. Using this chip requires the system to meet hard deadlines, which
cannot always be kept by standard Linux. To address the timing constraints, up-to-
date real-time Linux alternatives are briefly conferred with a special scope whether
they are freely available and have an embedded version. Once done, we point out
the design of a RTLinux-TPUART device driver implemented as kernel module
and finally compare it to an existing solution for standard Linux by analyzing
benchmark tests. *Copyright © 2003 IFAC*

Keywords: fieldbus, real-time operating systems, home and building automation

1. INTRODUCTION

The importance of applications for (remote) services, maintenance, monitoring and visualization
in the area of building automation is rapidly
growing. For the realization of such applications,
an interconnection between fieldbus systems for
building automation and computers is necessary,
whereby the latter ones are responsible for the
execution of the applications. For this reason,
the computers must provide a minimal and cost-
efficient hardware interface to the fieldbus systems
of choice. Furthermore, dedicated drivers have to
accomplish the data transfer between the applications and the fieldbus systems.

The fieldbus system of our choice is the *European Installation Bus* (EIB), nowadays forming
the EIB/KNX alliance together with Batibus and
European Home Systems. EIB is a decentralized,
structured bus system, built by bus lines that

connect EIB compliant devices that allow a user
to perform simple tasks like lighting and heating
but also to control air-conditioning or power management processes joining them into an integrated
system. The communication does not rely on a
single master node but is managed by all devices
that run the EIB-protocol offering decentralized
control. The EIB-protocol is based upon the well-
known ISO/OSI reference model. Unlike most
fieldbus systems that only implement the layers
1, 2 and 7, the EIB-protocol additionally includes
the layers 3 and 4 (for further details, see for
instance (Sauter *et al.* 2001)).

An EIB device can be addressed in two ways: on
the one hand by using its *physical address*, on
the other hand by specifying one or more *group
addresses* allocated to that device. The physical
address is related to the location of the device
and composed of three parts, called area, line
and device number. Uniquely assigned during the

installation of the device, the physical address is mainly used to download new applications, to set new parameters and to attach group addresses to the corresponding node.

In contrast to the physical addressing, group addressing combines several devices that fulfill a common task and may be located anywhere on the bus by assigning them the same group address. Members of a group communicate by means of a *group telegram*. When such a telegram is sent by a certain member, it is simultaneously received and processed by all other members (thus implementing a multicast communication). In addition, EIB supports a special group addressing at layer 2, the *polling group*, where one node (the so-called *polling master*) is able to request data from a series of up to 15 devices (*polling slaves*). The data returned by the slaves can only consist of a single byte and must be transmitted within a specific time slot shortly after the receipt of the initial poll request.

As stated above, besides a minimal and inexpensive hardware interface to the fieldbus system, one needs device drivers that allow data transfer to and from the applications. The operating system of our choice to design and implement such device drivers is Linux. Once disregarded as a gimmick, Linux nowadays is more and more utilized for applications in automation, too. This is on the one hand due to the fact that it is free available (and thus independent from company policies). On the other hand – thanks to the availability of its source code – it can be changed and adjusted easily (and thus quickens the interest for the market of limited resources and embedded systems). In case of home automation it might be the number one possibility to build up future low-cost residential gateways.

For these reasons we have been busy developing EIB device drivers for the Linux operating system during the last years. All device drivers are implemented as kernel modules and offer (at the lowest level) a programmer application interface relying on the standard file systems calls:

- `open()` to connect to the driver and initialize the underlying hardware and `close()` to disconnect from the driver.
- `write()` to write messages to the bus and `read()` to read messages from the bus.
- `poll()` to query the availability of new messages.
- `ioctl()` to change the behavior of the device driver (e.g. switch between normal mode and busmonitor mode).

Our device drivers have been the basis for advanced applications, like the integration of EIB systems into dynamic networks (Kastner and Wabel 2001) or software architectures relying on a so-

called Virtual Shared Group Object Space (Kastner and Tumfart 2002). Now, the forum of FET'03 seems the ideal platform to present the whole suite of device drivers – however, with a special scope on the newly developed one for the TPUART IC.

Thus, the paper is structured as follows: Section 2 scrutinizes the work done so far, when connecting computers running the Linux operating system to the EIB. Apart from a brief presentation of the device drivers for the BCU1 and the BCU2, we introduce one that rests upon the before mentioned TPUART IC. Using this chip requires the system to meet hard deadlines, which cannot be met by standard Linux. Hence, real-time solutions have to be taken into account. Thus, Section 3 summarizes up-to-date Linux real-time systems and their features. Next, Section 4 provides an overview of our TPUART real-time Linux device driver. Section 5 compares it to one running on standard Linux operating system. Finally, Section 6 summarizes and provides some outlook on future work.

2. EIB INTERFACES AND THEIR LINUX DEVICE DRIVERS

For reasons of component configuration, device testing, monitoring and control of EIB systems, it is necessary to provide an access to the fieldbus system by software running on external computers. One possibility is to use the well known *bus coupling units* (BCU) type 1 and 2. Via their *physical external interface* (PEI) and a plugged on serial data interface an asynchronous serial communication can be setup to a standard PC. Although easy to setup, these variants have a big disadvantage: they are bound to the command set of the used BCU type. In contrast, interfacing with the EIB via the TPUART IC grants more freedom, since the device driver has to care for most parts of the EIB-protocol stack. Thus, for instance, up to 2^{16} physical addresses and group addresses can be served via the same TPUART interface.

The next subsections introduce the distinct interfaces to the EIB and briefly present the design of their corresponding Linux device drivers.

2.1 BCU1 device driver

According to the PEI Type 16 for message exchange between a PC and a BCU1, 4 protocol steps have to be followed:

(1) *Communication request.* By means of the RTS and CTS signals a hardware handshake takes place on each byte transfer.

(2) *Software handshake.* At the beginning, a bi-directional transmission of the first byte of the message (i.e., the length byte) takes place. If one of the partners has nothing to send, it transmits the bit sequence 0xFF, signaling that it will act as receiver. In case of a simultaneous desire of the PC and the BCU1 to start a transmission, the BCU1 has higher priority and is granted access to the medium. Thus, the "loosing" PC may only request a new data transfer, after the complete receipt of the message the BCU1 sent.

(3) *Data transfer.* Once the communication relationship has been setup, the initiator sends its user data according to the so-called *external message interface* (EMI)[1]. In parallel, the receiver responds with bytes of value 0x00.

(4) *Pause.* When a complete message has been transferred, a new access to the medium may take place after the medium has been idle for approximately 3ms.

The design of the BCU1 device driver can be roughly divided into three parts (Fig. 1):

(1) A *state machine*, which is mainly interrupt controlled, handles the protocol with the BCU1. Note, that in state 3 and 6 the driver has to wait until the transmit buffer is empty. Since there is no interrupt to signalize this event, the driver has to poll the transmitter empty bit. However, polling in kernel space may lead to lost interrupt signals. To solve this dilemma, a special kind of Linux interrupt handler, a so-called bottom half, was chosen.

(2) A *buffer system* separates the hardware part (state machine) and the application layer part.

(3) A *character driver interface* is responsible for the communication with the application.

A detailed description of the driver can be found in (Kastner and Thallner 2002).

2.2 BCU2 Device Driver

Regarding the message exchange between a PC and a BCU2, a pure software handshake has to be followed. The protocol is defined by the PEI Type 10 and allows simultaneous communication via three transmission services:

(1) The SEND_UDATA/ACK-service serves for the transmission of up to 23 byte user data from one station (PC or BCU2) to the other (BCU2 or PC) and uses a frame of variable length; the acknowledgment is formed by the

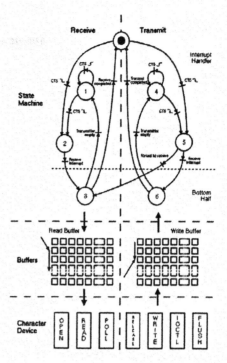

Fig. 1. BCU1 device driver

single character frame. In case of a transmission failure, the message is discarded and no acknowledgment is returned.

(2) The SEND_RESET/ACK-service causes the reset of the protocol automata in the PC and BCU2, respectively. The reset request is transmitted by a frame of fixed length, the acknowledgment uses the single character frame. In case of a transmission failure, no acknowledgment is returned.

(3) Using the REQ_STATUS/RESP_STATUS-service the protocol status of the communication partner is requested. Both, the request and the response are coded by a frame of fixed length. In case of a transmission failure, no response is sent.

A transmission service is completed, when it is finished successfully by receiving an answer message (ACK- or RESP_STATUS-message) or after the occurrence of an error. As long as transmission failures cause messages already sent to be repeated, new transmission services are delayed.

For the BCU2 device driver, we selected a cascaded structure (Fig. 2). The correct handling of the protocol is realized by so-called *server processes* (daemons) that care for some buffers in between the user space and the Linux standard serial interface device driver:

• The *writing server process* de-queues a message from the output buffer, then initiates a SEND_UDATA/ACK-service by directing the message to the driver of the serial interface, repeats it in case of a missing answer and completes the service with success or error.

---

[1] i.e., the above mentioned command set of the BCU1.

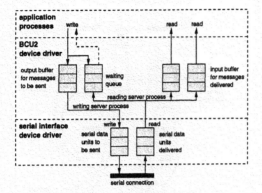

Fig. 2. BCU2 device driver

- The *reading server process* copies delivered messages into the income buffers of all registered application processes. In addition, it sends ACK- or RESP_STATUS-messages for BCU2-initiated services or informs the writing server process or writing application processes about the completion of their services upon the receipt of an answer by de-taching them from a waiting queue, respectively.

A detailed description of the driver can be found in (Kastner and Tumfart 2002).

2.3 TPUART Device Driver

The *Twisted Pair Universal Asynchronous Receive Transmit* (TPUART) IC is a transceiver that primarily supports the connection of host controllers (i.e. micro-controllers) to the EIB. Its basic components are:

- An *analog* part with a transmitter, a receiver and a power supply. The power supply is a linear regulator with a controlled minimum current consumption of 3mA. Additionally, it permits monitoring of the IC's temperature and voltage drop. In case of failures, a reset signal is generated. The receiver is preconnected to a bandpass filter. The transmitter regulates the transmission amplitude to a typical value of 7.5V.
- A *digital* part responsible for medium access and tasks of the layer-2 protocol of the EIB. Beyond others, these tasks include parity control, immediate acknowledge (ACK, NACK, BUSY), repetition of packets in case of failures and observing of timing constraints (e.g. bus idle times). The host controller has to care at least for the rest of the layer-2 protocol of the EIB (e.g. checksum calculation, management of tables for physical addressing, group addressing and polling).

The TPUART (software) interface provides the following services:

- The RESET-service forces the TPUART into an initial state.
- The STATE-service determines the internal communication state of the TPUART.
- The ACTIVATEBUSMON-service activates the busmonitor mode, where all data transmitted over the EIB are forwarded to the host controller, transparently.
- The ACKINFORMATION-service permits the host controller to indicate whether it has been addressed (ACK) or it is currently busy (BUSY), or – in case of a faulty packet – cannot handle a frame (NACK).
- The services LDATASTART, LDATACONT and LDATAEND allow for a byte-wise transmission of user data from the host controller to the TPUART that finally is responsible for subjecting it to the EIB.
- If a polling control byte has been received, the POLLING-service informs the TPUART about the slot-number and hands over the requested data. Once done, the TPUART has to care for inserting the packet at the assigned time (corresponding to the specified slot number).

The services are completed by the TPUART with:

- Indications for the RESET- and STATE-services.
- Confirmations for the LDATA-services. Thus, a simple software handshake protocol is used for exchanging data between the host controller and the TPUART.
- In busmonitor mode all packets including illegal ones and any kinds of acknowledgments (ACK, NACK, BUSY) are passed to the host controller.
- In the normal operating mode layer-2 frames without acknowledgments are delivered to the host controller.

In (Stocker and Grzemba 2001) a TPUART device driver for the standard Linux operating system has been presented consisting of three parts (Fig. 3):

(1) A *hardware layer* with an interrupt service routine and corresponding bottom halves, that carry out the slow parts of message processing (e.g. checksum calculation).
(2) A *data layer* comprising of two queues and two ring-buffers (one per each direction) that are responsible for queuing messages and (dis-)assembling them after receipt or before transmission.
(3) A *character driver interface* between the driver and the application.

The device driver works as follows: When a user application wants to transmit a message it uses the write() system call that simply stores the message into the TX-queue, activates the driver

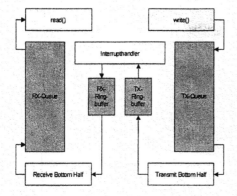

Fig. 3. TPUART device driver

and returns control to the user application. The transmit bottom half takes the message out of the queue, translates it according to the TPUART-protocol into TPUART services and stores it into the TX-ring-buffer. Finally, it causes the interrupt handler to transfer the bytes to the TPUART.

Upon the receipt of a message the driver has to match the destination address of the incoming message against the set of physical addresses or – in case of a multicast communication – against the set of group addresses. If it has been addressed, it reacts with a corresponding acknowledgment information, otherwise it drops the rest of the message. The remaining data processing is done by the receive bottom half that assembles the incoming bytes out of the RX-ring-buffer into the RX-queue where it can be taken out via a common `read()` file system call.

By default, the timeout between two EIB messages is 5.42ms long. During this bus idle time, the TPUART inserts diagnostic and error information useful for the host controller, but reducing the original timeout significantly (Fig. 4). To be sure, that a complete EIB message has been received, the device driver must detect a timeout of approximately 2ms after the last packet of the message. While this detection is normally not a problem for up-to-date used host controllers, a PC or micro-controller with parallel execution of applications cannot give any warranties. Due to system overload interrupts can be delayed (or even lost). Thus, a measured timeout of more than 2ms must cause the device driver to drop the whole message, if the packet has not been the last one of an expected EIB message. To cope with this problem, guaranteed detection is only possible with a real-time variant of the device driver.

3. REAL-TIME LINUX

Linux is not a real-time operating system by default. It is optimized for maximal throughput. Therefore, it is not possible to give any predictions

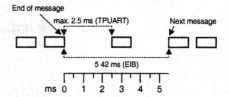

Fig. 4. EIB and TPUART time constraints

about runtime behavior. Two different approaches have been established to solve this drawback:

- *Preemption improvement*. Additional synchronization points are inserted to achieve a shorter time delay in non-interruptible code parts. Although this improves the situation, increased maintenance effort and a higher bug risk are unavoidable due to necessary kernel code modifications.
- *Interrupt abstraction*. An additional layer between the hardware and the standard Linux kernel is introduced (Fig. 5). This layer has complete control over the hardware interrupts and simulates the interrupts up to the standard Linux kernel. This allows the kernel to run nearly unmodified as a low-priority real-time task. As in the first method it is still necessary to modify the Linux kernel although the changes are very small by comparison. Another disadvantage is that real-time programming in user space is not fully implemented, currently (i.e. real-time tasks have to be implemented as kernel modules).

Today's real-time Linux variants can roughly be divided into commercial and open source distributions (Dankwardt 2003). Another important characteristic is their ability to support hard real-time in contrast to firm or soft real-time. Hard real-time capability is mainly determined by fast interrupt response. Other criteria for a quick classification of real-time Linux distributions may include (Tab. 1):

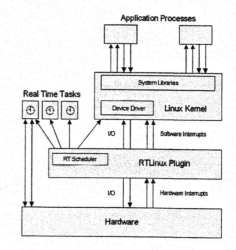

Fig. 5. Real-time Linux based on interrupt abstr.

Table 1. Real-time Linux distributions

	Commercial				Open Source						
	MontaVista Hard Hat Linux	TimeSys Linux/RT	Lineo Embedix RealTime	REDSonic RED-ICE Linux	RTLinux	RTAI	RTAI + LXRT	RED-Linux	KURT	Linux-SRT	Molnar Patch
Separate Real-time Kernel	with RTLinux	with RTAI	with RTAI	with RTAI	•	•	•				
User space Programming	•	•	•	•	•		•	•	•	•	•
Improved Linux Kernel Preemption	•	inside resource kernel (RK)		•					•		•
Interruptible Linux Kernel	announced	announced									
Precise Scheduling	•	with RK, without RTAI		•				•	•	•	
Solutions for Priority Inversion		with RK, without RTAI									
Interrupt Responses <1ms	with RTLinux	with RTAI	with RTAI	with RTAI	•	•	•				
Interrupt Responses <5ms	•	inside RK		•					•		•
Higher Precision for Timer Functions		with RK, without RTAI		•				•	•		

- *User space programming* is possible. This feature enables applications to execute real-time tasks in user space.
- The Linux kernel is fully *interruptible*.
- Only the *priority* is used for a scheduling decision (cf. standard Linux takes more factors into account).
- Solutions for the *priority inversion* problem are included.
- *Higher precision* for timer functionality is available, whereby the standard resolution of 10ms must be improved significantly.

Only RTAI (Mantegazza *et al.* 2003) and RTLinux (Yodaiken 2003) support hard real-time in the open source area at the moment. Both distributions use the interrupt abstraction approach and are basis for according embedded version projects, namely AtomicRTAI (RTAI) and Mini-RTL (RTLinux). Unfortunately, AtomicRTAI is no longer an open source project and sooner or later this will also be true for RTAI. Therefore, RTLinux respectively MiniRTL was used for our real-time device driver development.

4. TPUART RTLINUX DEVICE DRIVER

The original driver can be run in RTLinux without any changes. In this case, the driver shows of course no real-time capabilities. Therefore, it is necessary to apply changes to essential code parts. RTLinux recommends to keep the amount of real-time parts small and to modify only those parts where real-time functionality is really needed.

Adapting the interrupt handler is the most important task to provide real-time features. Essential is the usage of the real-time interrupt service routine for real-time interrupts. The normal interrupt handlers of the original driver can be used only for soft interrupts that are forwarded by RTLinux if no real-time interrupt handlers are assigned. For hardware interrupts concerning real time, the corresponding real time interrupt service routine (RT-ISR) of the RT kernel will be called.

The usage of a handler for hardware interrupts

```
static void int_handler (int irq,
  void *filp, struct pt_regs *regs)
{
  ...
}
```

is initialized and released in conventional Linux systems as follows:

```
request_irq (IRQ, int_handler, Flags,
  DeviceName, DeviceID);
free_irq (IRQ, DeviceID);
```

Hardware interrupts must be handled different in RTLinux. First of all a real-time task has to be created:

```
return pthread_create (&thread,
  Attributes, rt_thread, Argument);
```

and removed when the driver is unloaded:

```
pthread_cancel (thread);
pthread_join (thread, Return value);
```

Since the parameters of the original interrupt handler are not used in our real-time task, they can be ignored (otherwise they had to be buffered in the real-time handler):

```
static void int_handler_new ()
{
    original interrupt handler code...
}
```

The created real-time task just waits for a signal and then executes the code of the original interrupt handler:

```
void *rt_thread(void *arg)
{
    while (1) {
        pthread_suspend_np(pthread_self());
        int_handler_new ();
    }
}
```

Initializing and realizing the interrupt handler is necessary as in a normal system:

```
rtl_request_irq (IRQ, rt_int_handler);
rtl_free_irq(IRQ);
```

The real-time hardware interrupt handler has two important tasks:

(1) It has to send a signal to the waiting real-time task. Interrupts are deactivated in the real-time interrupt handler. Since only a notification for the real-time task is necessary, the time with disabled interrupts is very short.
(2) It must explicitly (re-)enable the corresponding hardware interrupt.

```
static unsigned int rt_int_handler
    (unsigned int irq,
    struct pt_regs *regs)
{
    pthread_wakeup_np (thread);
    rtl_hard_enable_irq (IRQ);
    return 0;
}
```

Fig. 6 shows the difference between interrupt handling in standard Linux and RTLinux.

Of particular importance is the priority of the created real-time task that runs the original interrupt code. Only a high priority can assure the timeli-

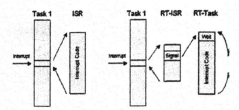

Fig. 6. Interrupt-handling in Linux and RTLinux

ness of code execution. This is appropriate when other real-time modules are used. It is obvious, that if the driver is the only real-time part of the system, the priority is insignificant.

Code that is part of real-time tasks or the real-time interrupt handler must be adopted to preserve real-time characteristics. Some relevant modifications have been:

- *Task Queues.* Task queues (i.e., bottom halves in older Linux versions) should not be processed immediately. The task queue of the scheduler should be used instead. Thus, RTLinux provides the macro `rtl_schedule_task(Task)`.
- *Spinlocks.* The functions for this type of code synchronization must be replaced by corresponding functions in RTLinux.
- *Timers.* RTLinux provides high-resolution timers. The functions in the original driver have to be replaced accordingly. Keep in mind, that this affects not only the real-time driver. Applications that use the new driver must deal with the more exact timer values, too.

In addition, we used the RTLinux subset MiniRTL to create an embedded solution. The difficulties arise from the demand of a minimal configuration in an embedded solution. Since no development environment is included in the target system, the driver and the applications must be developed on standard RTLinux and then copied to a MiniRTL image. Thus, the kernel versions and library versions must be equal on RTLinux and MiniRTL. This can be a challenging task if the embedded version is only available for an older kernel version or if older library version are used.

5. EVALUATION

For evaluation purposes, the timeout detection of 2ms is predestined to compare the original driver with the real-time alternative. Generating an increasing system load should provoke the original driver to produce wrong measurements. This should not affect the real-time version. To compare both drivers, we implemented a simple blocking module as a real-time module for RTLinux that can simulate an increasing system load. It can be used to test both drivers because the original driver (without real-time characteristics) is working in RTLinux, too. The system load is generated with a periodic scheduled task that performs more or less calculations according to the desired utilization.

The amount of calculation for the maximal producible system load must be determined empirical. Another challenge is to explore the limits of RTLinux. Overloading the system results in system crashes, since everything takes place in kernel code. For reasons of reproducibility we created a simple test-script responsible for loading the blocking module, permanently sending messages to the EIB system and logging of the timeout values measured by the drivers.

Fig. 7 shows a characteristic test result for the original driver. The y-axis holds the measured timeout values in ms. 2ms is the boundary where a driver detects a timeout. The individual measurements are plotted on the x-axis. One EIB message sent to the bus should cause 8 timeout detections performed by the driver. Accordingly each x-value should consist of 8 data points. Furthermore, the system load is increased on the x-axis. It can easily be seen that a larger load results in an increasing deviation from the ideal values. Thus, the standard driver will detect timeouts that have not occurred (but are due to the system load). In case of a "faulty" timeout detection, the device driver stops further measurements, discards the message and has to re-synchronize to a new message begin. This leads to less than 8 data points per x-value with increasing system load.

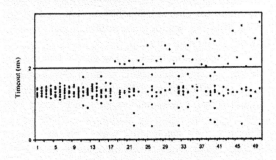

Fig. 7. Benchmarks for the Linux driver

In contrast, Fig. 8 shows a test result for the real-time driver. Fluctuations that appear already with the standard driver when no system load is present (cf. the origin of the x-axis at Fig. 7) are minimal in this diagram. This means, the real-time system produces basically more exact values. Furthermore, the deviations with increasing system load are completely eliminated. The real-time driver operates independently from the produced workload. Comparing the two results, it is obvious that using the real-time variant of the driver is reasonable.

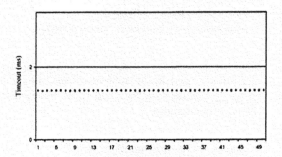

Fig. 8. Benchmarks for the RTLinux driver

6. CONCLUSION

This article presented a whole suite of Linux device drivers for the EIB. So far all our device drivers have been licensed under the GPL. The interested reader may check our homepage http://www.auto.tuwien.ac.at/eib4linux and download their source code. For the embedded solution we provide an image of a boot-able floppy with MiniRTL, the real-time driver and some demo applications. In the future, we can again concentrate on high-level applications for the EIB. Thus, we are currently busy developing an OSGi gateway for it.

REFERENCES

Dankwardt, K. (2003). Comparing real-time Linux alternatives. *http://www.linux-devices.com/articles/AT4503827066.html.*

Kastner, W. and B. Thallner (2002). Connecting EIB to Linux and Java. *Proceedings IEEE AFRICON* pp. 273–277.

Kastner, W. and M. Wabel (2001). JAPE. *Kongreßband zur Fachtagung der GI/ITG Fachgruppe APS+PC zum Thema Pervasive Ubiquitous Computing* pp. 159–167.

Kastner, W. and W. Tumfart (2002). Remote control of EIB systems based on virtual shared group objects. *Proceedings 4th IEEE International Workshop on Factory Communication Systems* pp. 63–71.

Mantegazza, P., E. Bianchi, L. Dozio, S. Papacharalambous, S. Hughes and D. Beal (2003). RTAI: Real-time application interface. *http://www.linuxdevices.com/articles/-AT6605918741.html.*

Sauter, T., D. Dietrich and W. Kastner (Eds.) (2001). *EIB Installation Bus System.* Publicis-Verlag.

Stocker, R. and A. Grzemba (2001). A Linux device driver for the TPUART interface. *Proceedings EIBScientific* pp. 110–119.

Yodaiken, V. (2003). An introduction to RTLinux. *http://www.linuxdevices.com/articles/-AT3694406595.html.*

MODELISATION OF A DISTRIBUTED HARDWARE SYSTEM FOR ACCURATE SIMULATION OF REAL-TIME APPLICATIONS

Mikaël Briday * Jean-Luc Béchennec *
Yvon Trinquet *

* IRCCyN, UMR CNRS 6597,
1 rue de la Noë - BP92101
44321 Nantes Cedex 3 - France
{Mikael.Briday,
Jean-Luc.Bechennec,Yvon.Trinquet}@irccyn.ec-nantes.fr

Abstract: This paper presents a validation approach for real-time applications using simulation techniques. Simulation concerns the operational architecture and takes into account the executable code of real-time tasks and the execution support model: set of processors, network and basic software (real-time executive and communication system). We show in this paper the performances that can be reached for the analysis of the temporal behaviour of real-time programs in a distributed system. *Copyright ©2003 IFAC*

Keywords: real-time systems, distributed systems, timing analysis, CAN network, cycle accurate simulation, instruction set simulation.

1. INTRODUCTION

The context of the work presented in this paper is that of real-time embedded systems. These systems, used for process control have to react on stimulus emitted by the process, and reactions have to respect timing constraints of the process. During the development of a real-time system one is led to define two main architectures: the software architecture representing the implementation of application functions and the execution support architecture. Software architecture is made up of a set of cooperative and concurrent activities (Event Triggered execution model). Execution support architecture consists, in the most general case, of a set of processors, a network and the basic software (real-time executive and communication system on each node). Application timing constraints can be of various kinds: absolute or relative constraints, throughput

constraints on inputs-outputs. They can concern simple treatments (i.e. maximum alarm response time) or a complete chain of treatments (end to end timing constraints). The missions of a real-time system can be critical. So if timing constraints are not met that can possibly have serious consequences on financial or human aspects. As a result of which the development of such systems requires so the use of specific methods and techniques, particularly to prove the respect of timing constraints.

Verification of a real-time application is a huge and difficult problem. It must be led throughout the development cycle on functional and temporal aspects. The operational architecture is needed for the verification of timing constraints. This one can be seen as the result of the mapping of the software architecture on the execution support. As long as the complexity of the operational architecture model - which depends on the abstraction level - remains low, formal techniques can be used to verify functional and sometimes temporal aspects (Alur and Dill, 1994; Berthomieu and Diaz, 1991). But as soon as one wants to take into account in a more detailed way the execution support formal techniques are not appropriate any more. Nevertheless, with semi formal techniques one can conduct a validation taking into account the temporal behaviour of the real-time executive and the system of communication (Faucou et al., 2002). Lastly, when the executable code is available one can carry out a validation using a very accurate model of the execution support architecture. It is a step before the final test on the real target. All these various validation levels have different objectives. They are not conflicting: they are complementary.

Our contribution is related to the last phase of the development process, before the final test on target. It consists of the analysis of the execution of the code (application code and basic software) on a very detailed model of the hardware architecture. The simulation is used because the other analysis techniques would be completely ineffective. Our objectives are the analysis of the executable code temporal behaviour in specific conditions in order to verify timing constraints and to get time characteristics of the software (i.e. Worst Case Execution Time or WCET). Interests of simulation techniques (Bedichek, 1990; Rosenblum et al., 1995; Albertsson and Magnusson, 2000) at this step of the development - with regard to the final test on the target - are numerous:

- no need to have the target available;
- no limitations of the final target: executable code in Rom memory, difficult control of task execution, difficult control of the environment behaviour;
- analysis is not intrusive: no disturbance is brought
- delicate situations can be replayed - indefinitely and without difficulties - to study some problems on timing behaviour;
- it is quite easy to set up observers for verifying that timing constraints are met, as well as to set up various time measurements.

This paper presents our work on the modelling of typical hardware architecture for embedded systems in the field of automotive. One presents the modelling of the Infineon C167CR processor (Infineon Technologies, 2000) and the associated CAN communication controller (Bosch GmbH, 1991). The technique used for the modelling is an essential element because it conditions the performances of the simulation. So it constitutes the main part of this paper.

2. MODELISATION

2.1 Working environment

The modelisation and simulation platform environment is built upon *SystemC* (Synopsys Inc, n.d.). *SystemC* has been launched in september 1999 and is backed by a large community of over 45 company from semiconductors, systems, IP, embedded software and EDA industries. *SystemC* is an Open Source object-oriented framework written with C++ that enables from Register Transfer Level (RTL) to system level design and simulation. A set of cooperating classes and templates allows to design boxes, ports and signals to link the boxes. Basic features of *SystemC* can be compared to those found in Hardware Description Language (HDL) like VHDL or Verilog but since version 2, *SystemC* offers high level features (channels, interfaces and events) which allows a higher level of design. *SystemC* has better performances than standard HDL too.

Choosing *SystemC* as a working environment has many advantages:

- the models could be easily connected to models provided by hardware vendors.
- any C or C++ library can be included in a *SystemC* model.
- the availability of the source code allows to customize the simulation engine when needed.

2.2 Modelisation of the C167 core

The main core of the CPU consists of a 4-stages instruction pipeline, a 16-bit arithmetic and logic unit (ALU) and dedicated Specific Function Registers (SFR). Additional hardware is provided for a separate multiply and divide unit, a bit-mask generator and a barrel shifter. In the real time context, it is necessary to model the CPU to reproduce not only the same behaviour, but also the time constraint.

Two model schemes are studied:

Instruction Set Simulator (ISS): The simulator reproduces the same fonctionnal behaviour as the real processor: it reads the assembly code written for the architecture and simulates it on the host machine. In this case, an Instruction Set Simulator (ISS) (Bedichek, 1990) is used. However, ISS does not consider time and is not cycle accurate especially for pipelined or superscalar architectures;

Cycle accurate simulation: The cycle accurate simulation have to model the internal architecture of the CPU to respect the timings of the real system. The *C167* uses a 4 stage pipeline and a Jump cache that have to be modeled. The data bus have to be modeled too because programs are essentially located in external memory, so a code fetch and a memory access uses the same 16 bits external bus.

Both simulation schemes are interesting. A cycle accurate simulator is slow but essential for real-time systems simulation. An ISS has better performance. To take advantage of both schemes, A modular simulator where the ISS is separated from the timing simulator has been designed. The ISS can run alone or connected to the timing simulator. The mode can be switched dynamically during the simulation. This way one can focus on some parts of the program with the timing simulator and reach these parts quickly with the ISS.

A processor pipeline splits instructions in elementary steps. In the C167, the pipeline stages are:

Code Fetch: Instruction code selected by IP (instruction Pointer) and CSP (Code Segment Pointer) is fetched from memory (RAM or ROM).

Decode: The instruction is decoded. The addresses of the operands are calculated and the operands are fetched. The Stack Pointer (SP) is updated if necessary. For branch instructions, IP and CSP are updated when a branch is taken;

Execute: operations are performed using the ALU (Arithmetic and Logic Unit). SFR (Specific Function Registers) are accessed at this stage;

Write Back: All external operands and remaining operands within the internal RAM space are written back.

Executing an instruction imply going through each pipeline stage. It can be very effective to make a distinction between instructions and time consideration.

Each of the 245 instructions (Infineon Technologies, 2001) is implemented with a C++ class. Each class uses a constructor which decodes the instruction and one method to execute the instruction (memory accesses, execution and write back). All the memory accesses are then stored in buffers and deferred at the appropriate pipeline stage when the timing simulator is used (see figure 1) or immediatly otherwise.

The Write Buffer contains all the data that will be written in the *Write Back Stage*. A write buffer entry is associated with each instruction in the pipeline. A *Read After Write* (RAW) hazard is easily detected by checking the Write Buffers before a read access.

The Read Buffer is necessary with external accesses in order to wait for the external bus. As this buffer is used only in the *decode stage*, only one buffer is needed in the pipeline (no conflit can occur).

Each memory access (read, write or both) can be associated with an *action* (a piece of code that is triggered when the access is performed). It is very useful for interfacing the CPU with other components and for debugging purpose. A special class is used and inherits the action to interface with external devices (such as CAN controller) to centralize external environment actions.

2.3 Wrapping the C167 model in a SystemC module

In order to communicate with other devices such as CAN controllers, it is necessary to wrap the processor model in a *SystemC* module. The module has an input clock that is connected to a method that simulate one cycle (i.e. *exec_one()*). Additionnal hardware can still be easily connected through external devices. Memory access are not modelised at the RTL (through pin WR, RD, CS, ...) because the CAN controller is on-chip. Anyway, the transactions are kept cycle accurate.

2.4 Simulating the CAN controller at a higher level

The CAN protocol, created by Bosch in 1986 is widely used in automotive industry. It is based on the producer-consumer model. Its main caracteristics are:

- asynchronous multi-master bus. There is no shared memory, no traffic controller, no global clock. A node can send a message as soon as the bus is idle;
- a priority between messages. Collisions are resolved without destroying the message which has the higher priority (in fact the lower value);
- an acknowledgment from every node that received a message without error. There is an error detection through a CRC;
- a distinction between temporary errors and permanently errors;
- a message identifier that defines data (vehicle speed in km/h, motor rotation in rpm,..). 11 or 29 bits identifier (standard or extended frame) are available;
- 0 to 8 bytes of data for the messages.

The bus state may be in two states, the *recessive state* (logical 1) and *dominant state* (logical 0). If at least one node sent a *dominant bit*, all the others nodes connected will receive a *dominant*

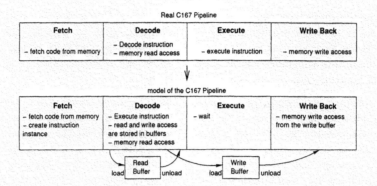

Real C167 Pipeline			
Fetch	**Decode**	**Execute**	**Write Back**
– fetch code from memory	– Decode instruction – memory read access	– execute instruction	– memory write access

model of the C167 Pipeline			
Fetch	**Decode**	**Execute**	**Write Back**
– fetch code from memory – create instruction instance	– Execute instruction – read and write access are stored in buffers – memory read access	– wait	– memory write access from the write buffer

load | Read Buffer | unload   load | Write Buffer | unload

Fig. 1. The model of the C167 Pipeline differs from the original one to allow a separation between code emulation and cycle accurate simulation.

bit. It results that if a node receives a *recessive* bit, no *dominant* bit are sent by other nodes.

A message that is sent by a node is received by all the other ones (in case of no error), bit after bit. Then each node can apply a filter (on identifier) to select which message is important, store the message and inform CPU that a new data is available.

As there is no global clock, each node resynchronizes itself on *dominant* to *recessive* edges. In case of monotonic sequence, the insertion of an opposite bit is necessary to resynchronize clocks (*bit stuffing*). As a consequence, bit stuffing can increase frame length, and so transmission time.

In the real-time context, calculating the transmission time which is proportionnal to the frame length 'F' (in bits) is required: $F = 32 + 8 \times D + 20 \times E + B$ where 'D' is the number of data bytes, 'E' is equal to 1 if the frame is extended (identifier is 29 bits long) and 0 if the frame is standard (identifier is 11 bits long), and 'B' is the number of bits added by bit stuffing. And the intervalle of B is: $0 \leq B \leq \frac{(32 + 8 \times D + 20 \times E)}{5}$ As the number of bits added is data dependent, it is necessary to produce the whole message to know its exact length. It is not necessary to model the simulator as a more accurate level (each bit is divided in *time quantum*), as electrical aspects are not considered.

The goal of the CAN simulator is to model a CAN components such as Philips SJA1000, Intel i527 or Infineon C167CR embedded controller. Each CAN component can be splitted in 2 parts, a *specific part* that manages message filtering, message storage and CPU interface, and a *generic part* that implements the protocol part.

Simulation of the CAN generic component using SystemC. The CAN generic component is a *SystemC* module that takes place between the CAN specific and the CAN bus components. This separation between CAN specific and CAN generic does not exists in the real component, the in-

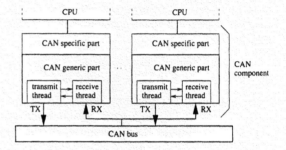

Fig. 2. model of the CAN bus using *SystemC*

teraction is then realised with a standard C++ API (Application Program Interface) that offers higher level functions than direct bit access such as *sendMessage()* or *nodeError()*.

The interaction with the CAN bus is done using the *SystemC* library. Each node connected to the bus is using 2 ports carrying boolean signals (see figure 2). The first one which carries a common signal to each node is called *RX*. It represents the state of the bus. The second one (*TX*) represents the signal that is sent by the node. In this case, the CAN bus is a *SystemC* module which has $(n+1)$ ports for n nodes. The first n ports receives the state of each node. It performs a AND logic operation (as a dominant bit is a logic 0 and a recessive bit is 1) at each bus cycle and applies the resulting signal to the *RX* port of each node:

$$RX = \bigwedge_{i=0}^{nbnode} TX_i$$

The CAN generic module uses 2 *SystemC* threads. The first one is listening to the bus and the second one is used to send bits (messages, error frames and Acknowledgment). The receive and transmit parts communicate because each bit transmitted is also received for error handling. When receiving, an Acknowledgment bit is sent if the message that was just received contains no error.

Simulation of the CAN specific component. The CAN specific component is only an interface between the CPU and the CAN generic part. The

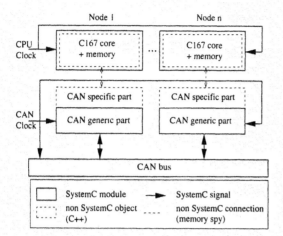

Fig. 3. model of the whole architecture

connection with CAN generic is the API previously introduced.

In the Infineon C167 processor, the CAN controller is embedded on the chip. To access the CAN controller, the CPU is accessing the memory in the range 0xEF00 to 0xEFFF. The connection to the CPU is done using memory actions (see section 2.2) for control registers. This part implements message filtering and storage in 16 messages objects, as these two caracteristics are not standardized. It will support in the future material interruption.

2.5 Simulating the whole architecture

The whole architecture consists of n C167 processors communicating through the CAN network (figure 3). Only one clock is used for all processors as the accuracy of the model is set on the CPU clock cycle.

About CAN controllers, using only one global clock prevent us to model the resynchronisation (on the recessive to dominant edge) of each clock. The behaviour at a bit accuracy is the same.

3. IMPROVEMENT

3.1 Instruction cache

In the implementation of the first simulator version, each instruction is modelled using an object that is created (in the *Fetch* stage), executed (through *exec* method) and deleted (in *Write Back* stage) to prevent memory leak. The problem is that creation and deletion of an object requires access to memory allocation functions which slows down simulation. An instruction cache (Hennessy and Patterson, 1996) can be used to prevent multiple memory allocations.

Instead of having a memory allocation for each new instruction that enter in the pipeline, the algorithm first checks for the availability of the instruction in the cache. The overload is quite unsignificant compared to a memory allocation. It is particularly interesting for program loops when the instruction cache is large enough.

The instruction cache have been implemented as in complex processors. The cache is a 256 entries size, 2-ways set associative, with an 'LRU' (last recently used) replacement policy. The number of entries can be easily increased, but a pretty good hit ratio (numbers of hits / numbers of access) with these values - over 99% with the benchmark - has been obtained.

3.2 pulling C167 model out of the SystemC framework

In the section 2.3, The C167 simulator has been wrapped into a *SystemC* module. The new simulator has been compared with the one that doesn't use *SystemC* . 10 million cycles (half a second with a 20 MHz clock) were simulated with a standalone application, without the instruction cache. The simulation without the *SystemC* layer was 822% faster than with the *SystemC* layer.

With such a result, it is interesting to take the simulator out of the *SystemC* framework, but with this approach, processors and the *SystemC* framework don't have the same time reference. When an external access is required, it is imperative to resynchronize *SystemC* framework and the corresponding processor.

Detection of external accesses. The first thing to do is to detect external accesses. Each external device (CAN controller for instance) is connected to the CPU through a centralised *External controller* that uses memory actions (section 2.2). As interruptions are not yet simulated, the external accesses are always initiated by the CPU.

Each time the cpu wants to access an external device, a resynchronisation is needed. The previous **exec** method is replaced with a new one:

New exec function with external devices
input: max the maximum of cycles to simulate
return: nb the real number of simulated cycles
$nb = 0$
while external access not required and nb < max **do**
  exec_one()
  $nb = nb + 1$
end while
returnnb

In that case the parameter is not the number of cycle to simulate, but the maximum number of cycle until an external access is required.

Resynchronisation of processors. The example considers a network of 3 CPUs, with *Tmax* cycles to simulate. Since the simulation of a CPU runs

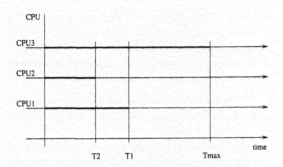

Fig. 4. Example of synchronisation between processors and *SystemC* framework. The processors are only synchronised with *SystemC* in case of external access (such as CAN access requirement) or at the *Tmax* date.

until a synchronisation is needed, all the CPU can run independantly (fig 4).

In this example, CPU1 wants to access the CAN controller at date *T1*, the CPU2 at date *T2* and the CPU3 doesn't need external access and stops at *Tmax*. *SystemC* is then started until *T2* (the minimum date). At that time, the second processor and the *SystemC* network are synchronised: the external access can occur. The second processor is started again and the algorithm restarts. With this algorithm, the only date when all processors and *SystemC* framework are synchronised is at *Tmax*.

The whole algorithm is:

input: max the maximum of cycles to simulate
initialisation
for all Processor **do**
    $nb\_cycle\_done_{processor} = 0$
    $wait_{processor} = false$
end for
main loop
while mini $\neq$ nb_cycle **do**
    for all Processor **do**
        run processor until end of required simulation or external access needed (i.e. synchronisation)
        $nb\_cycle\_done_{processor} = exec(nb\_cycle - nb\_cycle\_done_{processor})$
    end for
    detect the minimum of cycle done
    $mini = \min_{i=1}^{n} nb\_cycle\_done_i$
    for all Processor **do**
        check for processors that need an external access
        if $mini = nb\_cycle\_done_{processor}$ **then**
            $wait_{processor} = true$
        end if
    end for
    $Start\ SystemC for\ (mini - simu)\ cycles.$
    $simu = mini$
    for all Processor **do**
        if *wait* **then**
            Synchronisation is Ok
            Processor can execute its external action.
            do_action
        end if
    end for
end while

4. RESULTS

Different configurations of the architecture have been simulated to show the benefits on computation time.

The host is a 1GHz Pentium III computer with 128 MB of RAM running GNU/Linux with a 2.2.20 kernel.

A small program named *CanEcho* is the workload for the simulations. One node (the transmitter) sends a message with an identifier that ends with 0xFF. All the other nodes with the *CanEcho* program receive this message, modify the identifier with their unique node number *NodeId* and then resend the new message. All the nodes have the same processor, CAN node and program (except the last one which is the transmitter). In this way they react nearly at the same time and generate conflicts on the bus. The node with the lowest *NodeId* has the most priority, the other will get the bus after (see frame example 5). When the transmitter has received all the frames, it resends another message to keep a high traffic on the bus.

The interest of the *CanEcho* program is that it can be run with an arbitrary number of nodes and it generates a lot of traffic on the bus.

The architecture is simulated until the first processor (which run *CanEcho*) reach 10 millions instructions. It represents approximatively 40 million cycles, *i.e.* 2 seconds with a 20 MHz clock, and about 1500 CAN frames. One instruction needs about four cycles to be executed because of C167 memory accesses that require 3 cycles with the Phytec KitCon C167 development kit. In order to minimize the variations on the computation time due to simulator execution (Linux Operating System Influence), 10 computations are performed. The 'User time' is stored.

When results are linear in function of the number of nodes, The approximative line in the meaning of the least squares criterion is drawn.

4.1 Pipeline versus emulation

The simulator computation time at a cycle accurate level (pipeline) has been compared with the one at Instruction Set Simulator level (emulator).

Figure 6 shows that emulation is really faster than detailled simulation. And so, as previously mentionned, it can be useful to simulate the pipeline only in the time critical part of an application. The emulator is more than 3 times faster (independent of the number of nodes), except for the first node where the influence of the *SystemC* kernel is the most important.

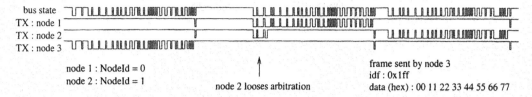

bus state
TX : node 1
TX : node 2
TX : node 3

node 1 : NodeId = 0
node 2 : NodeId = 1

node 2 looses arbitration

frame sent by node 3
idf : 0x1ff
data (hex) : 00 11 22 33 44 55 66 77

Fig. 5. Bus trace. There are 2 nodes with *CanEcho* program, and a third sends a message. Then, the 2 nodes tries to resend the echo. The first node has the most priority. The second node sends its echo just after.

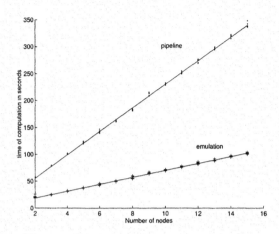

Fig. 6. Simulation of the architecture in pipeline and emulation mode. Time of computation in emulation mode is much faster than in pipeline mode. The processor is out of the *SystemC* framework and instruction cache is enabled.

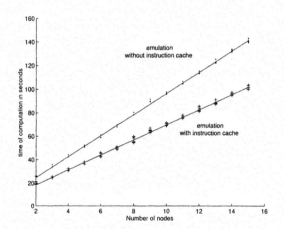

Fig. 7. Influence on the computation time provided by the instruction cache in emulation mode. The processor is out of *SystemC* framework.

4.2 Interest of the Instruction cache

In a second part, The performance gain provided by the instruction cache can be measured. The influence of the cache have been simulated with both pipeline (fig. 8) and emulation (fig. 7). In emulation mode, the gain is between 32% and 40% except the architecture with 2 nodes where the gain is only 24%.

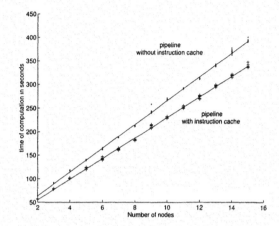

Fig. 8. Influence on the computation time provided by the instruction cache in pipeline mode. The processor is out of *SystemC* framework.

With this particular program, processors are mostly in an active wait state, *i.e.* The program runs in a loop to read the status register of the CAN controller and checks for changes. The loop uses not many instructions and can be entirely stored in instruction cache. That's why the cache *hit ratio* is 99,96%.

When using the pipeline, the instruction cache gain is included between 12% and 17%. That's a little less than in emulation mode because of pipeline computation overload.

4.3 Interest of pulling processors out of SystemC framework

Finally, The improvement obtained by running processors out of the *SystemC* framework is tested. As shown in figure 9, the algorithm considerably improve performances of the simulator.

The gain in computation time is not depending on the number of processors. It can be justified by the fact that the simulator uses only one clock for all the processors and thus pulling processors out of the *SystemC* framework remove only one clock, which is not depending of the number of nodes. Then, gain in computation time (figure 10) decreases according to the number of nodes, but is still more than 1.6 with 15 nodes (compared to the 3.9 of gain with 2 nodes).

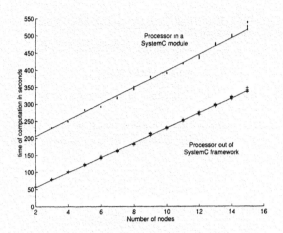

Fig. 9. Influence on the computation time provided by the processor model place: inside/outside *SystemC* framework. Instruction cache is enabled.

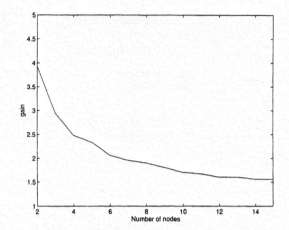

Fig. 10. Gain in computation time: processors are out of *SystemC* framework.

5. CONCLUSION AND FUTURE WORK

This paper have presented a simulation platform based on *SystemC* , an Open Source industry standard HDL C++ framework. An accurate model of the C167CR processor which includes a CAN controller has been designed. An ISS and a cycle accurate simulator have been studied and 2 optimisations to improve the simulators have been proposed. The impact of accuracy and optimisations on the performances of the simulator for a network of 2 to 15 processing nodes has been shown. When the accurate simulation is used, the simulator - running on a standard PC - is about 10 times slower than a real 20MHz C167CR (per node). When the emulation mode is used, it is about 3 times slower. Another interesting feature is the possibiliy to switch dynamically between the 2 modes. This result shows that such a platform is suitable to simulate a distributed real time architecture (hardware and software) and could allow the developer to test and debug the software without being intrusive. Many areas remain to be investigated:

(1) validation of the platform against a real network of processors (the real hardware platform is under construction);
(2) more performances improvements;
(3) debugging features to track variables and OS events in a distributed system, to measure time between events, to observe and track timing constraint violations...;
(4) controlled process modelisation and integration with the simulation

REFERENCES

Albertsson, Lars and Peter S. Magnusson (2000). Using complete system simulation for temporal debugging of general purpose operating systems and workloads. In: *MASCOTS 2000*.

Alur, R. and D. Dill (1994). A theory of timed automata. *Theoretical Computer Science B* **126**, 183–235.

Bedichek, Robert (1990). Some efficient architecture simulation techniques. In: *Winter 1990 USENIX Conference*. pp. 53–63.

Berthomieu, Bernard and Michel Diaz (1991). Modeling and verification of time dependent systems using time petri nets. *IEEE transactions on software engineering* **17**(3), 259–273.

Bosch GmbH (1991). *CAN Specification version 2.0*.

Faucou, S., A.M. Déplanche and Y. Trinquet (2002). Timing fault removal for safety-critical real-time embedded systems. *10th ACM SIGOPS European Workshop "Can we Really Depend on an OS?", St Emilion, France* pp. 247–251.

Hennessy, John L. and David A. Patterson (1996). *Computer Architecture A Quantitative Approach*. Morgan Kaufmann Publishers, Inc.

Infineon Technologies (2000). C167cr derivatives 16-bi t single-chip microcontroller, user's manual v3.1.

Infineon Technologies (2001). Instruction set manual for the c167 family of infineon 16-bit single-chip microcontrollers , user's manual v2.0.

Rosenblum, Mendel, Stephen A. Herrod, Emmett Witchel and Anoop Gupta (1995). Complete computer system simulation: The simos approach. In: *IEEE Parallel and Distributed Technology: Systems and Applications*. IEEE. pp. 34–43.

Synopsys Inc (n.d.). SystemC 1.1 user's guide.

ELSEVIER

**IFAC
PUBLICATIONS**
www.elsevier.com/locate/ifac

HARDWARE DESIGN OF A HIGH-PRECISION AND FAULT-TOLERANT CLOCK SUBSYSTEM FOR CAN NETWORKS *

**Guillermo Rodríguez-Navas, José-Juan Bosch
and Julián Proenza**

*Departament de Matemàtiques i Informàtica
Universitat de les Illes Balears, Palma de Mallorca, Spain
email: vdmigrg0@uib.es (G. R-N.), dmijpa0@uib.es (J. P.)*

Abstract:

One reported weakness of the Controller Area Network protocol is its lack of a clock synchronization service. In this paper, we present the architecture of a hardware clock subsystem which provides CAN networks with such a clock synchronization service. Our architecture presents significant advantages in front of previously suggested solutions. First, it is orthogonal to any CAN network. Second, it is compatible with timer-driven as well as clock-driven real-time distributed systems. Third, it achieves high-precision clock synchronization. And fourth, it presents a fault-tolerant behaviour. *Copyright ©2003 IFAC.*

Keywords: Distributed control, Embedded systems, Real-time systems, Fieldbus, Clock synchronization, Fault tolerance, Controller Area Network.

1. INTRODUCTION

The Controller Area Network protocol (ISO, 1993) is one of the most popular solutions for implementing the communication subsystem in distributed embedded systems. Figure 1 shows the architecture of a CAN network. CAN nodes are made up of three basic components: a processor, typically a microcontroller, which executes the application software; a CAN controller, which implements most of the CAN protocol; and a CAN transceiver, which adapts the transmission and reception signal to the transmission medium.

One reported weakness of the Controller Area Network protocol is its lack of a clock synchronization service (Törngren, 1995). Due to this, and because of the importance of having a synchronized time base, CAN networks usually include some kind of mecha-

---

* This work has been partially supported by the Spanish MCYT grant DPI2001-2311-C03-02, which is partially funded by the European Union FEDER programme

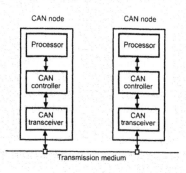

Fig. 1. Components of a CAN node

nism for clock synchronization. These clock synchronization mechanisms are implemented in two different ways: in the software executed by the processors or by means of a specific circuit.

Implementation by means of a specific circuit has significant advantages in front of software solutions. First, a hardware circuit can be integrated at a low level in the communication subsystem. This significantly reduces the impact of communication latencies and therefore allows a clock synchronization of higher precision. Second, hardware solutions reduce the computation overhead as processors are released from executing the clock synchronization algorithm. And third, hardware solutions allow implementation of fault tolerance techniques at a low level of the system architecture, which also causes a reduction of both the computation and the communication overheads.

In spite of the aforementioned benefits, hardware solutions also present an important drawback. They usually require services that are not implemented in standard CAN network components. This implies that standard components must be replaced by other components which do provide such services. This lack of orthogonality to the available CAN networks makes hardware solutions be often considered as too expensive and impractical, and thus software solutions are preferred.

Nevertheless, an orthogonal hardware solution for clock synchronization in CAN networks can be designed. Our solution achieves this orthogonality by implementing the clock synchronization mechanism in a set of additional hardware modules, which are attached to the existent CAN nodes and, therefore, do not require the replacement of any network component. Moreover, the architecture we have designed presents other important properties. It provides a clock service which is compatible with the two paradigms currently used in real-time distributed systems: timer-driven and clock-driven (Mullender, 1993; Veríssimo, 1995). It achieves a clock synchronization of very high precision. And it presents a fault tolerant behaviour. This paper describes how our architecture achieves these four important properties.

2. CLOCK SYNCHRONIZATION IN CAN NETWORKS

This Section briefly discusses the current solutions for clock synchronization in CAN networks; further discussion on this issue can be found in (Rodríguez-Navas and Proenza, 2003). First, the specific properties of the CAN protocol which facilitate clock synchronization are discussed. After that, the software solutions which have been proposed for CAN networks are introduced. Finally, the hardware mechanism that the Time-Triggered CAN protocol uses for clock synchronization is described.

2.1 Some useful properties of the CAN protocol

Arbitrated non-destructive access to the medium.
CAN protocol implements a prioritized access to the medium. Each message has a unique identifier, which determines its priority with respect to the other messages. When the bus is idle, any CAN controller is allowed to transmit. If more than one controller starts

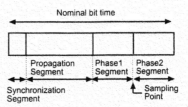

Fig. 2. Bit synchronization in CAN networks

to transmit simultaneously then a *bitwise* comparison of the message identifiers is performed. Finally, only the highest priority message is transmitted, whereas the rest of nodes give up and do not try to transmit until the bus is idle again.

Simultaneous view of the bits. In order to allow the bitwise comparison of message identifiers, CAN controllers must have a simultaneous view of each bit on the channel. This implies, first, having a specific relation between bit rate and bus length (Etschberger, 2001), and second, having a bit synchronization mechanism which ensures that all CAN controllers sample each bit at a quasi-simultaneous instant (ISO, 1993). The location of the sampling point within a bit is depicted in Figure 2. The *Synchronization Segment* and the *Propagation Segment* have constant length, whereas *Phase 1 Segment* and *Phase 2 Segment* can be enlarged or shortened, respectively, following the next rules. A receiving CAN controller must enlarge *Phase 1 Segment* if its local oscillator is faster than the local oscillator of the transmitting CAN controller. In contrast, a receiving CAN controller must shorten *Phase 2 Segment* if its local clock is slower than the local oscillator of the transmitting CAN controller. By acting in this manner, deviation among local oscillators is compensated and hence the points at which the various CAN controllers sample the same bit can only differ by a little amount. As remarked in (Rodrigues *et al.*, 1998), this difference is in the order of the propagation delay (10%-30% of a bit time).

Atomic broadcast. Thanks to the error-detection and error-signaling capabilities of the CAN protocol, any frame that is sent through a CAN network is, in principle, consistently received by all the controllers of the network or by none of them. Moreover, because of the automatic frame retransmission that CAN implements, any message issued to a CAN network will eventually arrive to its destination, as long as the transmitter controller remains non-faulty. Assuming that CAN controllers are fail-silent, and therefore cannot violate the CAN protocol, this behaviour would correspond to *atomic broadcast*, as remarked in (Rufino *et al.*, 1998). Nevertheless, it is a proven fact that in the presence of errors in the channel, scenarios exist in which CAN protocol does not provide atomic broadcast (Rufino *et al.*, 1998; Proenza and Miro-Julia, 2000). Despite of this fact, CAN controllers can be assumed to provide this property as techniques that solve this problem at the controller level have been already proposed, e.g. in (Proenza and Miro-Julia, 2000).

Tightness. The combination of the last two properties allows CAN networks to guarantee that as long as the sender remains non-faulty, the last retransmission of the same message is delivered to any two correct processors at real time values that differ, at most, by a known interval Δ (Rodrigues *et al.*, 1998). The value of Δ depends on the delay caused by the latency of the message processing and is a significant source of imprecision in most of the software-implemented algorithms that have been designed for clock synchronization in CAN networks.

2.2 Software clock synchronization in CAN networks

Some software algorithms which rely on the properties discussed above can be found in the literature. The algorithms suggested in (Gergeleit and Streich, 1994) and (Rodrigues *et al.*, 1998) take advantage of the tightness of the CAN protocol in order to achieve clock synchronization of high precision. However, the former does not provide fault tolerance whereas the latter provides fault tolerance by means of a protocol that causes high computation and communication overhead. In (Turski, 1994), a very simple algorithm is described which does not address fault tolerance and relies on the use of a specific CAN controller. Nevertheless, a significant contribution of this algorithm is the way in which the clocks of the processors are sampled. Instead of sampling its clock after indication of a message reception —like in (Gergeleit and Streich, 1994) and (Rodrigues *et al.*, 1998)— each processor samples its clock at the sampling point of the *Start Of Frame* bit (ISO, 1993) of a periodic synchronization message. By doing this, and because of the quasi-simultaneous view of the bits that CAN provides, the imprecision due to reception latencies is significantly reduced and higher precision can be achieved.

2.3 Clock synchronization in TTCAN networks

Time-Triggered CAN (TTCAN) is a higher-layer extension of the CAN protocol which defines the mechanisms required to operate a CAN network in a time-triggered mode (Führer *et al.*, 2000; Hartwich *et al.*, 2000). Two possible implementations of TTCAN exist, each one of them providing a different level of synchronism. TTCAN Level 1 provides a cyclic time base which is relative to a periodic message, whereas TTCAN Level 2 provides the former relative time base together with a synchronized global clock which measures absolute time and is implemented in hardware.

The clock synchronization algorithm which TTCAN Level 2 defines is based on a centralized scheme of synchronization. A node named *time master* imposes

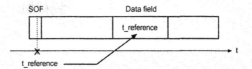

Fig. 3. Function of the reference message in TTCAN

its time view to the rest of nodes (or *slaves*). The time master spreads its time view by periodically sending a specific message, which is named *reference message*. Each slave samples its clock at the sampling point of the SOF bit of the reference message, as defined in (Turski, 1994). Then, since the data field of the reference message contains the value of the time master's clock at said instant (see Figure 3), the slaves can synchronize their clocks after receiving the reference message. This algorithm causes low communication overhead but it implies that the data field of the reference message has to be overwritten once the transmission of the message has already started. Since overwriting a message that is being transmitted is not allowed by standard CAN controllers, TTCAN Level 2 has to be implemented by means of a specific circuit, which is called TTCAN controller. This lack of compatibility with standard CAN controllers is one of the main disadvantages of TTCAN Level 2. Moreover, TTCAN controllers require time-triggered communication, and therefore they cannot be used in most of the CAN networks, which are event-triggered.

In order to measure absolute time, TTCAN Level 2 defines a *Network Time Unit* (NTU) which is known throughout the network and corresponds to a certain physical duration (i.e. measured in seconds). Every TTCAN controller translates the duration of the NTU to a certain number of local oscillator ticks, which is named Time Unit Ratio (TUR). Thus, every TUR ticks of local oscillator, each TTCAN controller increases its counter of NTUs by one unit. Nevertheless, due to the drift of the local oscillators, the NTU counter of every TTCAN controller will accumulate a difference from the one maintained by the time master. This difference is called *offset* and can be corrected as soon as the reference message is received, by simply adding or substracting the offset to/from the NTU counter. The time between the reception of two consecutive reference messages is named the *synchronization interval*. In order to avoid accumulating the same offset during the next synchronization interval, each TTCAN controller carries out the following drift correction:

$$TUR = TUR_{prev} \frac{L\_ref - L\_ref_{prev}}{M\_ref - M\_ref_{prev}} \quad (1)$$

Where $L\_ref - L\_ref_{prev}$ represents the duration of the synchronization interval as measured by the local oscillator; and $M\_ref - M\_ref_{prev}$ represents the duration of the synchronization interval, as measured by the time master. Once TUR is recalculated, it remains unchanged for the entire synchronization interval. Note that having a constant TUR implies accumulating some error in every NTU, as TUR cannot perfectly adapt to the fractional part of Equation 1.

In order to tolerate the failure of the time master, TTCAN declares a number of nodes as being *spare time masters*. As soon a spare time master detects that the reference message is not received on time, it tries to transmit its own reference message. Even though more than one spare time master may simultaneously try to send the reference message, this conflict is solved by the arbitration mechanism of CAN as each spare time master sends a reference message with a different identifier. In this way, the node which sends the reference message with highest priority is the one that succeeds in transmitting and becomes the new main time master.

An in-depth analysis of the mechanisms for fault tolerance which TTCAN defines shows that they are incomplete. In particular, two problems concerning the detection of failures of the time masters exist. The first problem is related to the fact that time masters are assumed to fail only by not sending the reference message. However, this assumption is not substantiated by the architecture of the TTCAN controllers, since they are not provided with any mechanism that prevents the time masters from suffering other kind of failures, like Byzantine failures (e.g. a faulty time master that takes the identity of the main time master and sends an absurd value of time). The second problem is that the actual availability of the spare time masters is unknown to the rest of the network as only the failures of the main time master can be detected.

3. REQUIREMENTS OF A CLOCK SERVICE FOR CAN NETWORKS

In this section, the requirements of our clock service are introduced. The way to fulfill each one of these requirements is further discussed in the following sections.

3.1 Orthogonality to the CAN network

An important requirement of our clock service it that it must be orthogonal to the existent CAN network in two different ways. First, it must not require the replacement of any network component. And second, it must be compatible with both event-triggered and time-triggered communication. An architecture that fulfills this conditions is presented in Section 4.

3.2 Compatibility with real-time distributed systems

Real-time distributed systems can manage time by means of two different techniques: *timers* or *global clocks* (Mullender, 1993; Veríssimo, 1995). Timers are adequate to measure local relative time whereas global clocks are adequate to measure distributed durations and absolute positions in a time line. Distributed systems that rely on local timers are known as *timer-driven*, and distributed systems that rely on a global clock are named *clock-driven*.

A desirable property of any clock service is to be compatible with timer-driven as well as clock-driven systems (Kailas and Agrawala, 2000). In order to fulfill this requirement, our clock service provides the microcontroller with both a global clock and a local timer. Moreover, in order to improve the applicability of our solution, the clock service also provides a synchronized *global tick*, of programmable frequency, which can be used like the tick of a regular clock (for instance to clock external timers). Section 4.1 discusses how this functionality is implemented.

3.3 High-precision clock synchronization

High precision is one of the most important requirements of any clock synchronization algorithm. In fact, the algorithm implemented in this work is inspired by the algorithm used in TTCAN Level 2 because of the high precision it achieves. The main difference between the TTCAN algorithm and our algorithm is that the latter allows slight variations of the TUR during a synchronization interval. By allowing these variations, the clock synchronization can achieve higher precision as the length of the NTUs can better adapt to the value which exactly corrects the drift (see Equation 1). The implementation of the clock synchronization algorithm is further discussed in Section 4.2.

3.4 Fault tolerance

The two problems related to fault tolerance that we said TTCAN exhibits have been solved in the present work. First, the assumption that time masters can only fail by not sending the reference message has been substantiated by means of a *duplication-with-comparison* scheme. This scheme provides *crash failure semantics* (Cristian, 1989). Second, a *membership protocol* which allows knowing the availability of every time master has been designed. In addition, and to avoid the fast *redundancy attrition* which the duplication-with-comparison scheme may cause, a mechanism to recover from transient failures has been also designed. These fault tolerance mechanisms are further discussed in Section 5.

4. ARCHITECTURE OF THE CLOCK SUBSYSTEM

In order to achieve orthogonality, the clock service is implemented by means of a set of additional hardware modules, which are named *clock units*. As depicted in Figure 4, each one of these clock units is attached to

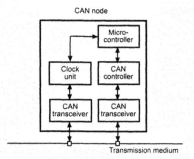

Fig. 4. Location of the clock unit

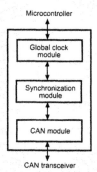

Fig. 5. Structure of the clock unit

a node of the existent CAN network. The function of the clock unit is to provide its microcontroller with a clock which is transparently synchronized to those of all the clocks units throughout the network. Note that a CAN transceiver must be also incorporated to the node so that the clock unit is able to send and receive messages independently. The subnetwork constituted by all the clock units together with their corresponding transceivers will be called *clock subsystem* hereafter.

As depicted in Figure 5, each clock unit is made up of three modules, which are named global clock module, synchronization module and CAN module. The global clock module contains the synchronized clock and supplies the microcontroller with the services that were introduced in Section 3.2 (i.e. a global clock, a local timer, and the global tick).

The synchronization module is the core of the clock unit as it implements most of the clock synchronization algorithm. Moreover, this module is intended to manage the communication between the clock unit itself and the rest of the clock subsystem.

The CAN module includes the circuitry to send and receive the reference messages which the synchronization modules generate. A standard CAN controller cannot be used instead of this CAN module because the algorithm implemented in the synchronization module requires two additional services which a standard CAN controller does not provide. First, the CAN module must indicate to the synchronization module the sampling point of the SOF bit of any message. And second, the CAN module must allow the synchronization module to write the value of time in the data field of the reference message while it is being transmitted.

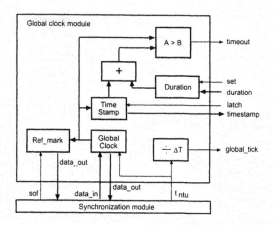

Fig. 6. Global clock module

4.1 Global clock module

Figure 6 shows the structure of the global clock module as well as its interfaces to the synchronization module and the microcontroller. The global clock module counts absolute time by means of a counter, named Global Clock, which is clocked by the signal t_{ntu}. A pulse of the signal t_{ntu} is generated by the synchronization module every time that an NTU elapses. The synchronization module is allowed to read the content of Global Clock because, as explained in Section 5, the absolute time is required to trigger the communication among the clock units. Furthermore, the synchronization module is allowed to modify the value of Global Clock when required. Finally, the synchronization module indicates by means of the signal *sof* the sampling point of each SOF bit. When this signal is activated, a sample of the Global Clock is latched in the Ref_mark register. The content of the Ref_mark can be read by the synchronization module as this value is used in the synchronization algorithm.

The value of Global Clock can be latched and read by the microcontroller. This allows the microcontroller to accurately timestamp events and gives support to clock-driven real-time systems. Furthermore, a local timer is provided for timer-driven real-time systems. The microcontroller sets this timer by writing in the register named Duration. If the time that Duration indicates elapses, the signal *timeout* is activated. To increase even more the applicability of the clock subsystem, the global clock module issues the output *global_tick*, which allows the clock unit to be used as a regular clock (e.g. to clock external timers). The global tick is obtained from the signal t_{ntu} and has a period ΔT, which can be configured by the microcontroller at the initialization of the clock unit.

4.2 Synchronization module

The function of the synchronization module is to implement the clock synchronization algorithm. This clock synchronization algorithm is based on the one defined by TTCAN Level 2. Thus, some of the clock

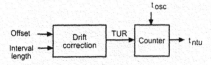

Fig. 7. Drift correction scheme

units act as time masters whereas the rest of clock units act as slaves. The function of the time masters is to periodically send a reference message which is similar to the one described in Section 2.3. This means that the data field of each reference message contains the clock value sampled by the transmitter time master at the sampling point of the SOF bit.

Despite this common features, some relevant differences exist between the algorithm defined by TTCAN and the one implemented in our clock subsystem. The most important difference is that our algorithm allows slight variations of the TUR during the same synchronization interval. This is done by recalculating the length of every NTU taking into account the offset accumulated during the last synchronization interval (see Figure 7). For instance, if the local clock is faster than the time master clock then some NTUs are enlarged (by increasing their TUR). The greater the difference, the greater the number of NTUs which are enlarged. If the local clock is slower then NTUs are shortened. This technique achieves better drift correction that having a constant TUR (like in TTCAN) because a constant TUR causes some error in every NTU, whereas a variable TUR is able to reduce this error.

5. FAULT TOLERANCE

In order to improve the fault tolerance of our clock subsystem, some specific mechanisms have been designed. They are described next.

5.1 Crash failure semantics

The correctness of the error detection mechanism described in TTCAN relies on the assumption that the time master can only fail by not sending the reference message. This assumption is actually substantiated in the present work by implementing the *duplication with comparison* architecture of Figure 8 for all the clock units (both time masters and slaves). This architecture is made up of two duplicated clock units and one specific circuit which manages the introduced redundancy; the Redundancy Manager. Note that the CAN transceiver do not have to be duplicated as their faults are tolerated by the CAN protocol.

This design assumes that faults in the duplicated clock units are independent. Then, whenever the duplicated clock units behave differently, the Redundancy Manager interprets that one of the replicas has a fault

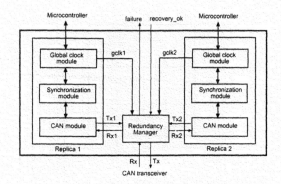

Fig. 8. Duplication with comparison architecture

and isolates the duplicated clock units from the clock subsystem. This technique guarantees that, from the point of view of the clock subsystem, the only way in which a clock unit can fail is "by doing nothing". This property corresponds to *crash failure semantics* (Cristian, 1989).

The main drawback of implementing this kind of mechanism is that it causes a fast redundancy attrition as a single transient fault in a replica is enough to disable the duplicated clock unit forever. This problem can be overcome if an error recovery mechanism is designed. The function of this mechanism is to re-establish, if possible, a correct state in each replica of the clock unit. As soon as this state is reached, the duplicated clock unit can be connected again to the clock subsystem. In the present work, the provision of this error recovery mechanism is left to the microcontroller.

The Redundancy Manager carries out two different comparisons to detect if the duplicated clock units behave differently. The first one is a comparison of the bits which each replica issues to the network. The second one is a bit-by-bit comparison of the global clock which each replica maintains. These comparisons are possible because both replicas use the same local oscillator. The function of the Redundancy Manager corresponds to the following logical equations:

$$discr = (tx_1 \neq tx_2) \lor (gclk_1 \neq gclk_2) \quad (2)$$

$$failure = discr \lor \neg recovery\_ok \quad (3)$$

$$tx = tx_1 \lor failure \quad (4)$$

$$rx_i = rx \lor failure \quad (5)$$

In Equation 2 the internal signal *discr* is activated if the values issued to the network by the replicas (signals tx_1 and tx_2) or the values of the global clocks maintained by the replicas (signals $gclk_1$ and $gclk_2$) differ. Such discrepancies are considered as failures, as remarked in Equation 3; signal *failure* is activated by signal *discr*, and continues activated until the microcontroller indicates that a correct internal state has been re-established (signal *recovery_ok*). As long as signal *failure* remains activated, the values transmitted to and received from the network are logical '1'

(Equations 4 and 5), which in CAN corresponds to *recessive* values (ISO, 1993) and is equivalent to disconnect the clock unit from the clock subsystem .

5.2 Design of a membership service

The second problem of the error detection mechanism defined in TTCAN is that the actual availability of the spare time masters is unknown to the rest of the system. This problem is solved in our architecture by forcing all the time masters to play the role of main time master one after the other, following a *round robin* scheme. Therefore, whenever a time master has a failure, this will be detected by the rest of clock units as soon as the time master does not send the reference message in its assigned turn.

In order to tolerate faults of the main time master, all the spare time masters also try to transmit its own reference message in every turn, though they send a reference message that has lower priority than the one sent by the main time master. Therefore a spare reference message can be received before the main reference message only if the main time master has actually failed. Table 1 shows how the identifiers of four time masters are assigned so as to guarantee this condition. Note that each time master uses two different identifiers for the reference message; one for its own turn and another one for the rest of turns.

The detection of failures of the time masters is guaranteed to be *consistent* because the atomic broadcast property, which we assume for the CAN protocol, ensures that if a clock unit does not receive a frame because of a time master failure then no clock unit will actually receive this frame. Moreover, the duplication with comparison architecture of the clock units reinforces this assumption, as it implies that no clock unit is allowed to violate the CAN protocol (Rufino *et al.*, 1998).

Table 1. Assignment of the reference message identifiers (4 time masters).

	Main id.	Spare id.
Time Master 1	00000000000	0000000001000
Time Master 2	00000000001	0000000001001
Time Master 3	00000000010	0000000001010
Time Master 4	00000000011	0000000001011

Once all the clock units can consistently detect whether a main time master has sent its reference message, the rest of the membership algorithm is simple. Each clock unit keeps a count of the omissions that each time master has shown, which is named *penalty count*. Whenever an omission of a reference message is detected, the penalty count of the corresponding time master is increased by a given amount. And, whenever a time master does not fail and sends the reference message in its turn, then its penalty count is decreased by a lower amount. If the penalty count of a time master reaches a certain value, then all the clock

units consistently consider that it has a permanent failure and eliminate its turn from the round robin. This mechanism does not eliminate a time master immediately after its first omission. Thus, a time master that does not send its reference message because of a transient fault has the opportunity to rejoin the round robin after recovering from the fault. How this is done is the subject of the next section.

5.3 Rejoining the round robin after a transient fault

Our membership algorithm can only work as long as all the clock units have a consistent knowledge of the membership information. However, note that consistency is only guaranteed for non-faulty clock units. If a clock unit suffers a transient fault then it may not receive a reference message and, therefore, the correctness of its membership information is no longer guaranteed. Due to this, an additional mechanism to obtain the membership information from the other time masters is required.

The mechanism we have implemented consists in sending the membership information (as well as the time value) within the data field of any reference message. Thus, once a reference message is received, the consistency of the membership information is guaranteed again. And, thereby, a just-recovered time master can know, on the one hand, the state of the other time masters and, on the other hand, in which state the other time masters consider it is. If the just-recovered clock unit is not considered in a permanent failure then it joins the round robin. In contrast, if the rest of time masters consider it as having a permanent failure, then it does not join the round robin and does not send a reference message anymore.

Since a just-recovered clock unit does not have consistent membership information until a reference message is received, then it is not able to know which is the main time master of the current turn. Therefore, if the main time master failed and the reference message of a spare time master was received, then the just-recovered clock unit would not be able to increase the corresponding penalty count. To avoid this situation, the spare time masters send the membership information with the penalty count of the main time master already increased. Thus, as the reference message sent by a spare time master can only be received if the main time master has actually failed, the received membership information will be correct in any case.

Sending the membership information within the reference message limits the number of time masters of the clock subsystem (though not the number of slaves). In this work, eight time masters are used, and the penalty count of each one of them is codified with three bits. In this way, the membership information occupies three bytes in the data field of the reference message.

6. FUTURE WORK

A complete evaluation of the implemented synchronization algorithm is currently being performed. In particular, the imprecision that clock amortization may cause should be analyzed. First calculations show that at 1Mbps this design allows the provision of a clock of granularity 10 μsecond and a precision of 2 μseconds when it is synchronized every second.

As explained in section 5.1 the duplicated clock units share a single clock. This means that faults of this clock cannot be detected by comparison. Therefore, clock units can actually show *performance failures*. The design of a mechanism to tolerate these failures is being currently addressed. Furthermore, and in order the clock subsystem to achieve more independence of the microcontroller, a hardware recovery function must be implemented in hardware within the Redundancy Manager.

7. CONCLUSIONS

A hardware solution for clock synchronization in CAN networks has been presented. In opposition to software solutions, our solution can provide a fault-tolerant clock service of high precision with reduced computation and communication overheads.

In order to improve its applicability our architecture achieves four important properties. First, it has been designed so as to be orthogonal to the existent CAN network in two differente ways: it does not require the replacement of any component of the network and it is compatible with both even-triggered and time-triggered communication. This orthogonality has been achieved by implementing the clock service in a set of additional hardware modules, which are named clock units. Second, the service provided by our architecture is compatible with the two paradigms currently used in distributed real-time systems; it provides a local timer which can be used by timer-driven systems as well as a global clock which can be used by clock-driven systems. Third, this architecture implements a clock synchronization algorithm which takes advantage of the good timing properties of the CAN protocol in order to achieve a very high precision. This algorithm is based on the one defined by TTCAN Level 2, though some changes concerning the drift correction have been made. Fourth, the clock units have been provided with a fault tolerant behaviour. This property has been achieved by means of a duplication with comparison scheme for the clock units and a distributed algorithm which allows knowing the actual availability of all the relevant clock units.

REFERENCES

Cristian, F. (1989). Questions to ask when designing or attempting to understand a fault-tolerant distributed system. *Proceedings of the 3rd Brazilian Conference on Fault-tolerant Computing, Rio de Janeiro, Brazil.*

Etschberger, K. (2001). *Controller Area Network.* IXXAT Press. Weingarten.

Führer, T., B. Müller, W. Dieterle, F. Hartwich, R. Hugel, M. Walther and Robert Bosch GmbH (2000). Time Triggered Communication on CAN. *Proceedings of the 7th International CAN Conference, Amsterdam, The Netherlands.*

Gergeleit, M. and H. Streich (1994). Implementing a Distributed High-resolution Real-Time Clock using the CAN-bus. *Proceedings of the 1st International CAN Conference, Mainz, Germany.*

Hartwich, F., B. Müller, T. Führer, R. Hugel and Robert Bosch GmbH (2000). CAN network with Time Triggered Communication. *Proceedings of the 7th International CAN Conference, Amsterdam, The Netherlands.*

ISO (1993). ISO11898. Road vehicles - Interchange of digital information - Controller area network (CAN) for high-speed communication.

Kailas, K. and A. Agrawala (2000). An Accurate Time-Management Unit for Real-Time Processors. *Proceedings of 8th International Conference on Advanced Computing and Communications (ADCOM 2000), Cochin, India* pp. 79–86.

Mullender, S.J., Ed.) (1993). *Distributed Systems, 2nd edition.* ACM Press, Addison Wesley.

Proenza, J. and J. Miro-Julia (2000). MajorCAN: A modification to the Controller Area Network to achieve Atomic Broadcast. *IEEE Int. Workshop on Group Communication and Computations. Taipei, Taiwan.*

Rodrigues, L., M. Guimaraes and J. Rufino (1998). Fault-tolerant Clock Synchronization in CAN. *Proceedings of the 19th IEEE Real-Time Systems Symposium, Madrid, Spain.*

Rodríguez-Navas, G. and J. Proenza (2003). On the Design of a Clock Service for CAN Networks. Technical Report 1-2003. Dept. Matemàtiques i Informàtica, Universitat Illes Balears.

Rufino, J., P. Veríssimo, G. Arroz, C. Almeida and L. Rodrigues (1998). Fault-tolerant broadcasts in CAN. *Digest of papers, The 28th IEEE International Symposium on Fault-Tolerant Computing, Munich, Germany.*

Törngren, M. (1995). A perspective to the Design of Distributed Real-time Control Applications based on CAN. *Proceedings of 2nd International CAN Conference, London-Heathrow, United Kingdom.*

Turski, K. (1994). A global time system for CAN. *Proceedings of the 1st International CAN Conference, Mainz, Germany.*

Veríssimo, P. (1995). Ordering and timeliness requirements of dependable real-time programs. *Journal of Real-Time Systems, Kluwer Eds.* 7, 105–128.

ELSEVIER

IFAC

PUBLICATIONS

www.elsevier.com/locate/ifac

A RADIO PROTOCOL FOR LOW POWER WIRELESS SENSOR NETWORKS

Gunnar Stein, Klaus Kabitzsch

Dresden University of Technology
Institute for Applied Computer Science
Dresden, Germany

Abstract: Radio transmission may be a good alternative to the wire based communication for a lot of applications in building automation (Rauchhaupt and Hähniche, 1999). The advantage is an easy, fast and low cost installation scenario. Nevertheless, the radio specifics require a very specialized protocol stack that affects the whole system architecture. Starting from the requirements of building automation, the paper presents a new radio protocol for low power sensor networks. A special medium access control protocol meets the requirements in terms of energy saving and application layer functionality. *Copyright © 2003 IFAC*

Keywords: protocols, field bus, wireless sensors, building automation

1. INTRODUCTION

Field buses are high reliable automation systems to meet real time requirements and for transmitting a high amount of short messages. Compared to the widespread LAN technologies, the response time periods are much more important than throughput. Most of the LAN systems are optimized for high data rates. In wireless automation, the bandwidth demands are moderate. Other features like long service life and high reliability become important. Thus, most of the LAN technologies are not applicable to wireless automation without pro-found changes (Hähniche, 2000).

A lot of ways have been suggested to integrate wireless devices at the field bus level, e.g. in (Cavalieri and Panno, 1997), (Mahlknecht, 2002). Less attention was paid to the service lifetime of the devices in building automation. A low energy adaptive clustering hierarchy is presented in (Heinzelman *et al.*, 2000). But this hierarchy is focused only on energy saving without any regard to the potential application layer functions.

Building automation systems are often designed for a service life of 10 to 30 years. Hence, battery powered wireless field bus devices should feature a minimal operation time of 5 years. In (Stein and Kabitzsch, 2002) an architecture concept for a balanced combination of wireless and wire based devices was proposed as a solution. In the present paper, the radio protocol is explained in detail.

2. CORE CONCEPTS

The only way to achieve a long term operation for a battery powered device is to use it as seldom as possible. It has to spend most of its life time in a so called sleep mode. During sleep mode the power consumption is normally reduced to a few microamperes. But since the radio decoder is not active either, the device is unable to receive data frames. In comparison to sending, the receiving process is much more difficult. Sending could be performed event oriented or time triggered. But for receiving, additional information is needed about the time point at which the receiver has to be switched on.

It would be advantageous, if a device could observe the medium during the sleep mode and wake up if a transmission occurs. However, the sensing of the medium requires decoding normally. Otherwise a mechanism for selective wake up is needed that can be performed at a minimal energy expense.

There are certain technical problems associated with low power selective wake up. For this reason the protocol presented in this paper only uses time or event driven self wake up. A standard protocol like Ethernet (Tanenbaum, 2003), Profibus or ASI is not useable. They require devices that are able to receive and process frames at any time. Furthermore they are based on a simple broadcast concept. Indeed in a radio cell, it is not sure, that each device is able to receive all other devices. To illustrate this, Figure 1 shows a simple example of three devices A, B and BS. The device B is not able to receive device A and vice versa. However the device BS can communicate with both devices. Natural, the specifics of the radio transmission medium are completely different from such of wires. A wire performs a closed physical medium. In addition, a device is strictly separated from other devices that are connected to different wires.

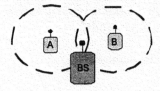

Fig. 1. Not all nodes are able to hear each other

Under radio conditions the bus nature is broken. It is not sure that all sensors are able to hear a message. Hence common medium access schemas like CSMA/CD will not work very well. Moreover, neighbouring radio cells could overlap and mutual interferences can occur.

In (Stein and Kabitzsch, 2002) a wireless building architecture was presented that is based on the classification of devices into high power devices (Hpd's) and low power devices (Lpd's). Lpd's have a battery whereas Hpd's have a wire based power connector for auxiliary energy. The basic idea was to find an asymmetric protocol that mostly uses the energy of the Hpd's and saves the energy of the Lpd's. The protocol presented in this paper is based on this classification. There are sensors and base stations. Sensors are typical low power devices. They spontaneously send if events occur. Base stations are high power devices and therefore they are able to receive at any time. Both complement each other to safe energy on the sensor side.

A number of sensors and one base station are organized within a radio cell. Figure 2 illustrates

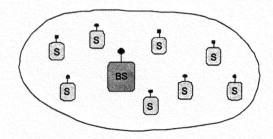

Fig. 2. Basis structure of a radio cell

the cell structure. Each radio cell has got a base station (abbreviated BS) and a number of sensors (abbreviated S). Batteries power the sensors whereas auxiliary energy powers the base station.

3. PROJECT SPECIFICATIONS

Based on the items discussed above a radio protocol at 429.25 - 429.75 MHz ISM frequency band was developed (Fettweis, 2002). The goal was to design a stable protocol that is able to deal with common interferences and interferences caused by neighbouring radio cells. All radio cells have to be separately workable. Interlinking of the base stations over an additional wire based network for mutual cooperation is an optional feature and should not be necessary for proper operation. Further demands for the protocol stack are:

- up to 255 sensors per radio cell
- sensor live cycle of 5 years
- daily 2-3 Alive Check messages per sensor
- short event recognition time

For high security reasons all sensors should have to perform Alive Checks in short intervals (may be every minute). Due to the energy saving requirements this is impossible. Therefore, the system focuses on transmitting event messages as fast as possible. The technical specifications listed below were the preconditions for protocol design.

- 46 channels with 1200 bps data rate
- a frame consists of a sync word and data (Figure 3)
- constant frame length for simple decoding
- sensors and base station are equipped with a narrow band send/ receive unit
- sending and receiving at the same time is not possible
- the channel number is adjustable for each sending/receiving action
- base station has an additional wide band receiver for the sync word detection

Two special features are to point out. First, 46 channels are available within a radio cell. Second, the base station is equipped with a wide band receiver. The wide band receiver enables to monitor all channels at the same time. However, the wide band receiver is only able to detect sync

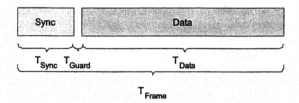

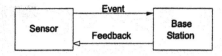

Fig. 3. MAC frame format

words, not to decode a whole frame. If a sync word is detected, the narrow band receiver will be adjusted to the regarding channel for decoding. Only one channel can be decoded, even if there are receivable frames on other channels.

4. COMMUNICATION SCHEME

Each transaction is caused by a sensor. It sends a Request and the base station answers with a Response. Figure 4 illustrates a single Request/Response cycle. A sensor initiates a transaction, if an event occurs or to send an Alive Check. Under no circumstances the base station is able to request a transaction.

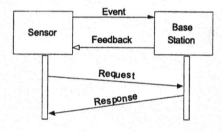

Fig. 4. Request/Response cycle

The basic assumption for the protocol is the ability of the base station to sense the medium permanently. It is necessary for the detection of the sporadic transmissions originating form the sensors.

The base station is able to send and receive at any time. Sensors can also send at any time, but can receive only after a sending activity. Responses are used to transmit data to the sensors. But messages from the base station to the sensors can not be considered as normal event messages. From an application point of view they have to be messages that are valid during a long time or such one that are immediately calculated after a Request. For instance the Response can be used to send a feedback to the sensor based on the received Request. Therefore such messages are called Feedback Messages. The messages from a sensor to the base station are called Events Messages correspondingly. Figure 5 illustrates these two types.

Fig. 5. Event and Feedback Messages between a sensor and a base station

5. MEDIUM ACCESS CONTROL

Periodic MAC cycles are used for the medium access control. The MAC cycles time period $T_{MACcycle}$ is constant. Within one cycle the transmission of a complete frame is performed. All sensors in a radio cell normally work with the same time reference for the beginning of a new MAC cycle. Therefore, all sensors are synchronized to the base station. Figure 6 shows the MAC cycle scheme. All cycles are constant.

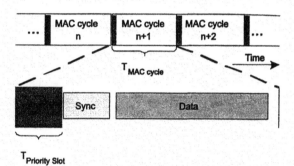

Fig. 6. Periodic cycles are used for medium access control

Like discussed above it is not sure, that each sensor is able to receive all other sensors within a radio cell. Under such circumstances CSMA/CD will not work very well. However, the available power at the transmitter of the base station is higher. Hence it can be assumed for the sensor network that all active sensors can receive all frames transmitted by the base station. The idea is to assign priority to the base station. A sensor can only send if the base station does not use the channel. Thus, destroying the base station frames by own sensors will be avoided. Therefore, each MAC cycle starts with a priority slot. If the base station wants to send, it will use this priority slot. All other active sensors on this channel will detect the sync word and defer its Request. This means:

- the base stations sends immediately at the beginning of the MAC cycle
- because of the power and central positions of the base station all active sensors will detect the sync word and defer their transmission activity

So it is guaranteed that base station frames are not destroyed by own sensors. Figure 7 illustrates the send activities of the base station dependent on the MAC cycle.

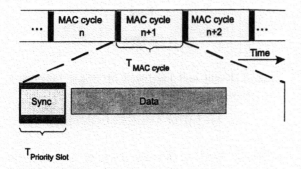

Fig. 7. Base station transmit activity

The sensors on the other hand have to sense the priority slot and will only send, if they do not detect a sync word:

- a sensor senses the priority slot to detect a sync word from the base station or jam
- if the medium is free, the sensor sends immediately after the priority slot
- otherwise the transmission is deferred

Since more than one sensor can decide to send on a channel within a slot, frames may collide. But this behaviour must be accepted. Below the channel assignment is discussed to remedy this problem. Figure 8 illustrates the send activities of a Sensor dependent on the MAC cycle.

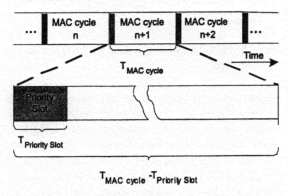

Fig. 8. Sensor transmit activity

Carrier sensing is reduced to the priority slot. Therewith only the base station responses are protected. Sensor requests may collide. Additionally a sensor can use the sensing results to determine the channel state. If there is a lot of noise/interference, it can defer the transmission. This avoids useless sending over a jammed channel.

6. CHANNEL ASSIGNMENT

If a sensor wants to send, it uses a randomly chosen channel number. In conjunction with the randomly chosen MAC cycle for retransmissions, the collision probability is very low. Furthermore a side stepping of jams is achieved. If a request is destroyed by jamming, the sensor will use an other channel for the retransmission. The base station has to sense all channels using the wide band receiver. If a sync word is detected by the wide band receiver, the narrow band receiver will be adjusted to the appropriate channel.

Collisions of sensor frames are very seldom because of the randomly chosen channel number. If two sensors want to use the same MAC cycle for sending, it is unlikely that both use the same channel. Admittedly, the base station is only able to receive a limited number of frames at the same time, depending on the number of narrow band receivers. If the number of received frames exceeds the number of narrow band receivers, frames will be lost, but not destroyed. This is a positive effect of the multiple channel usage. On one channel, destroying of frames significantly reduces the data throughput.

7. RETRANSMISSIONS

Nevertheless, if two or more sensors send on the same channel at the same time, the frames will be destroyed. The result is the same as if the transmission is jammed by interferences. Both transmission problems are be handled in the same manner. After sending a request, a sensor initializes a timeout value. If the timeout expires without receiving the corresponding response, the sensor tries to retransmit the request. For the retransmission a sensor randomly selects the next MAC cycle. Thus, Alive Checks are deferred for a longer time than Event Messages.

8. MAC CYCLE SYNCHRONIZATION

The synchronization of the starting time point of the MAC cycles to all sensors is important for effective response behaviour of the base station. The base station has to respond to each sensor request. If all sensors send uncoordinatedly, this task is difficult to perform. Thus, synchronization is needed for effectiveness but not necessarily required. The base station senses all channels at every time (unless during a transmit process).

The synchronization mechanism works as follows. Each sensor holds an internal MAC cycle counter. If a sensor sends a request, the base station detects the beginning of the sync word and calculates a correction value. This correction value is included in the response frame. The sensor uses the correction value to adjust its MAC cycle counter. It does not matter if the counter is not adjusted for the first request. After the second request, the sensor will be synchronized.

9. SIMULATION MODEL

All investigations are based on a simulation model that was established to examine the protocol. A prototype validation of systems with up to 255 nodes is possible by the model. For the implementation of the simulation model the Java programming language was used. All simulation components were implemented as classes based on Java standard class hierarchy. No specialized framework for protocol development and simulation were used. The physical layer model is encapsulated within a class named CMedium. It models the transmission delay and the collision behaviour of the radio medium. The CModem class that models the radio modems of the sensors and the base station uses CMedium. Figure 9 gives an overview of the use case relations between the core elements of the simulation model.

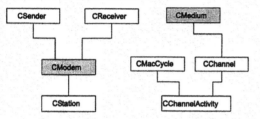

Fig. 9. Use case relations between the core elements of the simulation model

For controlling the simulation process a simple user interface was implemented. There are start/stop buttons as well as a display showing the current simulation time. There is also a monitoring window for the protocol actions. Figure 10 shows the control window and the statistical evaluation window of the simulator

Fig. 10. Control window and statistical evaluation window of the simulator

10. PROTOCOL PERFORMANCE

The simulation bases on some working assumptions for the time values. T_{Frame} is the frame transmission duration. It depends on the length of the sync word, the guard duration between the sync word and the data frame and finally the length of the data frame.

$$T_{Frame} = T_{Sync} + T_{Guard} + T_{Data} \qquad (1)$$

Based on a data frame with 16 byte = 128 bit (doubled by coding), a sync word length of 40 bit, an estimated guard time T_{Guard} of 16 bit times and a transfer rate of 1200 bps the frame transmission duration is as follows:

$$T_{Frame} = \frac{312\ bit\ s}{1200\ bit} = 260\ ms \qquad (2)$$

$T_{Priority}$ is the priority slot time. It should be long enough to detect a sync word. A sync word lasts about 33 ms. Thus, a priority slot time of 38 ms is a good choice. $T_{MACcycle}$ is the MAC cycle period time. It has to be longer than the sum $T_{Priority}$ and T_{Frame} :

$$T_{MACcycle} > T_{Priority} + T_{Frame} \qquad (3)$$

A MAC cycle period of 300 ms is appropriate. Two MAC cycles are necessary for sending and receiving. It is assumed that the base station needs two cycles after receiving a Request, to check the frame and create a Response. Thus, at minimum four MAC cycles are necessary for a complete request/response cycle. The time duration results in:

$$4 * T_{MACcycle} = 4 * 300ms = 1.2\ s \qquad (4)$$

All sensors randomly generate frames holding an underlying exponential distribution for the inter event time. The average event rate is equal for all sensors. For certain numbers of sensors and various inter event times the average number of necessary transmissions was determined. In this regard the transaction times per event are also important. The results are given in Figure 11 and 12. It is considered to use two narrow band receivers in the base station to improve its response time for sensor requests. More over, two receivers could reduce the frame loss if more than one frame arrive at different channels, but at the same time.

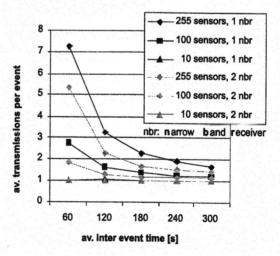

Fig. 11. Average transmissions per event

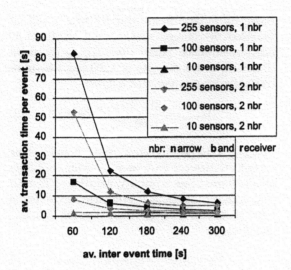

Fig. 12. Average transaction time per event

Up to a number of 10 sensors the transmissions per event are nearly 1. This results in a short transaction time. For 100 and more sensors, the number of retransmissions significantly increases if the inter event time will be less than 120 s. This is the result of the limited number of narrow band receivers in the base station and an increasing collision rate. The more sensors send within a MAC cycle, the more frames may be collide or getting lost. However, as the diagrams illustrate, the protocol has got a good performance even if 255 sensors are present as long as the inter event time is greater than 240 s. This means, with the maximum number of sensors, the system is able to transmit one frame per sensor every 240 s. Reworded, the whole system can cope with an average frame rate of nearly one frame per second. This is roughly equivalent to the minimum duration for a complete request/response cycle discussed above!

11. CONCLUSIONS

In this paper a protocol stack was designed to meet the specific requirements of a low power sensor system. Many problems were bypassed by the usage of simple principles. The result is a consistent, but very specific solution. It is only usable for simple sensor systems but not for real radio field applications. For investigations a simulation model was established. Apart from the simulation, some physical prototypes for real testing purposes are also available.

REFERENCES

Cavalieri, S. and D. Panno (1997). On the integration of Fieldbus traffic within IEEE 802.11 wireless LAN. *Birches. J.* **35**, 131–138.

Fettweis, G.P. (2002). Smart Wireless Sensor Network. Interim report. Asahi Kasei.

Hähniche, J. (2000). Funkgestütze Kommunikation - Eine Herausforderung für die Automatisierungstechnik. In: *VDI Berichte 1580.* Chap. Einführung und Motivation, pp. 1–12. VDI Verlag GmbH. Düsseldorf.

Heinzelman, W., A. Chandrakasan and H. Balakrishnan (2000). Energy-Efficient Communication Protocols for Wireless Microsensor Networks. In: *Proc. Hawaiian Intl. Conf. on Systems Science.*

Mahlknecht, S. (2002). Virtual Wired Control Networks: A Wireless Approach with Bluetooth. In: *2002 IEEE africon.* Vol. 1. IEEE. pp. 269–272.

Rauchhaupt, L. and J. Hähniche (1999). Opportunities and problems of wireless fieldbus extensions. In: *Fieldbus Conference FeT'99.* pp. 48–52.

Stein, G. and K. Kabitzsch (2002). Concept for an architecture of a wireless building automation. In: *2002 IEEE africon.* Vol. 1. IEEE. pp. 139–142.

Tanenbaum, A. S. (2003). *Computer Networks.* 4 ed.. Prentice Hall.

12. ACKNOWLEDGEMENT

This work was supported by Asahi Kasei Electronics Co. Ltd., Tokyo, Japan.

PROPAGATION DELAYS IN SELF-ORGANIZED WIRELESS SENSOR NETWORKS

Jean-Dominique Decotignie

Centre Suisse d'Electronique et de Microtechnique (CSEM)
Jaquet-Droz, 1, CH-2007 Neuchâtel Switzerland

jean-dominique.decotignie@csem.ch

Abstract : self organized wireless sensor networks exhibit interesting properties from battery consumption and reliability viewpoints. The present study deals with the data propagation delay from the sensors to the information consumers. It appears that, compared to conventional networks in which all sensors are in direct visibility to the consumers, the propagation delay can be lower in a number of cases. *Copyright © 2003 IFAC*

Keywords : wireless networks, sensors, fieldbus, real-time constraints..

1. INTRODUCTION

Since a few years ago, self organized wireless sensor networks have attracted a lot of interest (Rabaey 2000, Heinzelman 1999). In such networks, information is transported from node to node between the information sensors to the consumer. Using intermediate nodes offers clear advantages such as a lower power consumption and a potentially better immunity to wireless transmission perturbations. This mechanism may seem inefficient in terms of the time necessary to transport the information from the sensor to the consumers because of the multiple transmissions and the time to relay the information. Our objective here is to model these networks in terms of transmission delay, evaluate the influence of various parameters and compare the results with a system in which all sensors are visible from the consumers.

The paper is organized as follows. The first part describes the features and the benefits of self organized sensor networks. The second presents the transmission time model that will be used. Different configurations are then analyzed with regards to the time it takes for the information produced by a sensor to reach the consumer of

the information. These results are compared with the direct visibility case and the influence of some transmission parameters and the medium access control protocol are evaluated.

2. SELF ORGANIZED WIRELESS SENSOR NETWORKS

Sensor networks have emerged in the 80s (Pleinevaux 1988) as a response to the increasing cabling costs and as a result of the development of new networking technologies. In the 90s (Roberts 1993, Morel 1996), thanks to the blossoming of radio transmission, wireless nodes have been integrated into such networks. What may be considered as the next step comes from a different world, the improvement of radio-based networks and the quest for robust and low consumption solutions (Sohrabi 2000, Rabaey 2000a), what is now called self organized sensor networks. These networks exhibit a few common features:

- The network organizes itself. There is no need for a initial configuration. There is not node such as a base station that plays a special role in the network.

- There is no infrastructure contrary to conventional mobile communication networks.

- Nodes are battery powered. Power consumption should hence be reduced to a minimum.

- All nodes do not see the other ones. A given node only sees a reduced set of the other nodes. Transmitting information from one node to another node of the network most of the case implies using intermediate nodes as relays.

- Each node is rather small (objective is < 1cm^3). The target cost per node is 0.5$.

- Information throughput is below 10Kbit/s.

As they are another kind of fieldbus, these networks have to fulfill the same constraints among which the most important are:

- Being able to transmit small packets of information in timely manner;

- Support on demand and cyclic or periodic transfers;

- Support producer consumer communication relationships (Thomesse 1993) in addition to client server, for instance for configuration purposes;

- Provide some indication of temporal consistency (Decotignie 1994) for data produced by different nodes. Temporal relationship between successive data should be supported.

Self organized wireless sensor networks offer an attractive alternative to conventional wireless fieldbusses with direct visibility for a number of reasons:

- The transmission distance may be much smaller. The energy used for transmission may thus be greatly reduced because attenuation is at best proportional to the square of the distance and at worst to the power 4 of the distance;

- The network is mode resilient to fading because the information may use different paths. Between any two nodes of the network, it is likely that more than route exists that use different relay nodes. There is thus a larger probability that one of the routes will not suffer from fading;

- The network is not dependent on a special node that may be a hot spot for reliability;

- Reliability is also improved thanks to multiple routes. A node breakdown will thus have a reduced impact.

Despite these advantages, using relay nodes to transmit the information from one point to another in the network may introduce additional delays. Such delays may cause a large prejudice to applications that require low latency. Let us thus try to evaluate the impact of relay nodes on the network latency.

3. SYSTEM MODEL

3.1 Message transmission delay

To assess the time required to transport the information from one node to another one, we will assume that the transmission time for a message exchanged between two adjacent nodes is given by :

$$D = T_{sync} + T_{tx} = T_{sync} + (L_O + L_M)T_{bit}$$

$$(1)$$

where T_{sync} is the time that two nodes will take to synchronize before they can exchange a message. This delay depends on the medium access control protocol.. T_{bit} is the transmission time of one bit L_M is the number of useful bits in the message and L_O is the number of bits due to the overhead caused by the protocols.

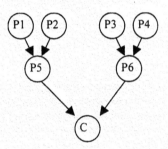

Figure 1 – Network logical topology

T_{sync} may greatly vary from one medium access control (MAC) protocol to another. If the nodes are able to synchronize by some ideal means and some form of TDMA (Time Domain Multiple Access) is used, T_{sync} becomes negligible. Conversely, if a technique such as « preamble scanning » (Mangione-Smith 1995) is employed, its value may become important. The advantage of this parameter is to offer an analysis that is independent of the MAC protocol.

3.2 Network Model

We will here assume that the network has a number of sensor nodes and one consumer node. The study may be generalized to less constrained architectures which may include several consumers. What we want to evaluate is the time it takes for the consumer node to gather all the information coming from the sensor nodes.

We will also assume that the route that the information uses is obtained from a suitable routing protocol and that this route remain valid long enough to neglect the route discovery time in the analysis. The assumption may seem excessive in conventional wireless networks. However, wireless sensors are often hooked in fixed locations. This is the case in a building where wireless communication is used to simplify cabling, installation and commissioning.

Finally, the processing time between the reception of a message and its transmission on the next hop will be assumed negligible. This is reasonable as the bit rate on wireless sensor networks is rather low and thus most processing can be performed while the message is being received.

3.3 Direct visibility

In a conventional wireless fieldbus, the consumer node can see all the sensor nodes. The transmission delay is thus the sum of the elementary delays:

$$D_{vd} = N_n D \qquad (2)$$

where N_n is the number of sensor nodes.

4. DELAY IN CASE OF MULTIPLE HOPS

4.1 Principles

In the case of multiple hop networks, information coming from one sensor node will reach the consumer node through other nodes that act as relay. We will assume that these nodes are also producing information (sensor nodes). Potentially, information goes through multiple relays before reaching the consumer. We will not deal with the route establishment and assume that it is stable for sufficiently long time to be neglected in the analysis.

A relay node may retransmit the information to the next node in the route as soon as it receives it. However, the number of useful bits is low in the case of sensors (L_M bits) and the other parts of equation 1 introduce an important overhead. Such a policy is thus not very efficient. A better policy may be to use the following algorithm. Each relay node only transmits a single message to the next relay node on the route. This message contains the useful bits of all the messages the relay node has received from the sensor nodes that are before in the route. This single message will also carry the data produced by the relay node (it is also a sensor node).

The different routes used to transmit the information from the sensors to the consumer node can be represented as a tree. Generally speaking, this tree is arbitrary. A general analysis is thus difficult to assess analytically. However, in the case of a building, there exists a spatial organization made of rooms, sides, floors, etc. The tree may easily be mapped as n-ary tree where the sensors in a room transmit to a relay node close to the room door. Several door nodes may transmit to a node in a corridor and so on.

Following this idea, we will assume a binary tree (**Error! Reference source not found.**). This would correspond to a building in which each room would be equipped with 3 sensors, two sensors inside the room and one at the door. Let N be the number of levels in the tree (N=3 in the example depicted on Figure 1). The number of sensors in the network is given as:

$$N_n = \sum_{k=1}^{N-1} 2^k = 2^N - 2 \qquad (3)$$

4.2 Sequential Transmission

Let us assume that the messages between the nodes are transmitted one after the other (no parallelism). The total time required to carry the information from the sensor nodes to the consumer node (without dead time between successive transmissions) is given by:

$$D_{sm} = (2^N - 2)D + \{(N-3)2^N + 4\}L_M T_{bit}$$
$$= N_n D + \{(N-3)2^N + 4\}L_M T_{bit} \qquad (4)$$

The first part of the equation gives the time necessary to transmit a single message from each sensor node to the next hop. The second part represents the additional load due to information retransmission (L_M bits by sensor) in the relay nodes. It may be understood as follows. The information coming from the 2^{N-1} sensors of the first level is retransmitted N-2 times. The information from the 2^{N-2} level 2 sensors is transmitted by relay nodes N-3 times and so on. The total number of retransmission is thus given as:

$$N_r = \sum_{k=0}^{N-2} (N-k-2)2^{N-1-k}$$
$$= \sum_{k=1}^{N-1} (k-1)2^k = (N-3)2^N + 4 \qquad (5)$$

which corresponds to the second par of Equation 4.

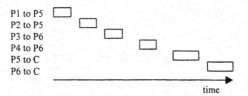

Figure 2 – Sequential tramsmission

Obviously compared to the transmission time experienced in direct visibility (Equation 2), the latency is here larger even using data aggregation in the relay nodes. The difference comes from the fact that the information needs be retransmitted.

4.3 Parallel transmission

It is possible to use the fact that only a few nodes of the network can "see" a given node to conduct the transmissions in parallel at the same instant. In Figure 1, transmitting from P1 to P5 and P3 to P6 may be performed simultaneously. Similarly, it is possible to transmit from P2 to P5 and from P3 to P6 at the same time. Conversely, transmitting from P1 to P5 and from P2 to P5 may not be handled at the same instant and should take place one after the other. The same applies to message transmission from P3 to P6 and from P4 to P6. The time necessary to transmit all the data from the sensors to the consumer becomes:

$$D_{smp} = 2(N-1)D + 4\{2^{N-1} - N\}L_M T_{bit} \quad (6)$$

The first term of the right part is the time necessary to transmit one message from each sensor node. It is linear as a function of the number of levels in the tree because message transmissions at each level are handled by groups of 2 at the same time. In each group, the two messages transmissions take place one after the other. We thus have a delay of 2D for each level.

The second term in equation 6 is due to retransmissions by relay nodes. After all the first level sensor nodes have transmitted their messages, the sensor nodes of the next level in the tree (they are relay nodes) will retransmit the data content of the received messages. At this level, the load due to retransmission corresponds to the useful bits of 4 sensors (two adjacent relay nodes cannot transmit at the same time). At the next level, this load is doubled and the load due to the 4 level 2 relay nodes is added. This process is iterated until the root of the tree (consumer node) is reached. At each level i, retransmitted (relayed) information amounts to:

$$D_i = L_M T_{bit} \sum_{k=1}^{i} 2^{k+1} = 2(2^{i+1} - 2) L_M T_{bit} \quad (7)$$

Adding the retransmitted information at all levels from 1 to N-2 gives the second term of the right part of Equation 6.

Taking profit of possible parallelism drastically reduces the latency. It is however not easy to directly compare the result given by Equation 6 with what was obtained using direct visibility (Equation 2). Let us assume that the number of useful bits is equal to the protocol overhead ($L_M = L_o$). This assumption is realistic for sensor networks as the number of useful bits is low. With this, the difference between Equation 2 and Equation 6 becomes:

$$D_{vd} - D_{smp} = 2T_{sync} \left\{ 2^{N-1} - N \right\} \quad (8)$$

It is always positive or zero (N=2). As a first conclusion, we can state that, in addition to the advantages enumerated in the introduction, multihop transmission is also interesting when the latency is considered. The binary tree is however an ideal case and, in most of the cases, the latency will be between the results given by Equations 4 and 6.

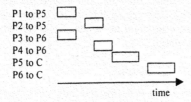

P1 to P5
P2 to P5
P3 to P6
P4 to P6
P5 to C
P6 to C

time

Figure 3 – Parallel tramsmission

5. SPECIAL CASES

Depending on the medium access control scheme, a few special cases may be analyzed:

- perfect synchronization - all the nodes are perfectly synchronized using a common clock. Medium access uses predetermined time slots for each transmission.

- partial synchronization - the first transmission in a cell is not synchronized. After this first transfer, all transmissions are synchronized.

In the first case, the synchronization time T_{sync} is equal to zero. Direct and multihop transmissions give the same latency (see Equation 8).

In the second case, T_{sync} only influences the first transmission and disappears from the subsequent ones. We will however assume a guard time T_{gap} between two

consecutive transmissions. Latency for the direct visibility case becomes:

$$D_{vd} = T_{sync} + N_n T_{tx} + (N_n - 1) T_{gap} \quad (9)$$

For multihop parallel transmissions (binary tree with $L_M = L_o$), latency is given as:

$$D_{smp} = (N-1)(T_{sync} + T_{gap}) + 2T_{tx}(2^{N-1} - 1) \quad (10)$$

because at each level, there are two transmissions and only one is impacted by T_{sync}.

Using the result of Equation 3, the last term of Equation 10 is equal to the second term of the right side of Equation 9. The latency difference between the two cases is given by:

$$D_{vd} - D_{smp} = (2 - N) T_{sync} + (2^N - N - 2) T_{gap} \quad (11)$$

This difference is highly dependent on the ration between the synchronization and the guard times. In practice, the latter is much smaller and the direct visibility case is more favorable for a small number of levels in the tree.

6. ANALYSIS

This study raises a number of questions:

- is the binary case, the best case ? In fact, preliminary studies have shown that ternary trees give even better results.

- synchronizing the nodes is another important issue. A relay node cannot retransmit to the next level before it has received all the messages from the sensor nodes from which it receives information. The results given in Equations 4, 6, 9 and 10 assume that there is no dead time (other than T_{sync}) between two successive transmissions. This means that all nodes must start to produce approximately at the same instant (T_{sync} accounts for some variations). Different solutions are available for this synchronization that need not be very precise.

7. CONCLUSION

Self organized wireless sensor networks are based on multihop transmission. They exhibit a number of appealing properties in terms of reliability and power consumption. They are thus of strong interest for industrial applications. In this paper, we have shown that this type of network may also be better when considering the time necessary to transport the information from the sensors to the consumers. This result which contradicts the intuition that multihop transmission takes longer than direct transmission calls for further study.

This study considers a network topology that can be represented as a binary tree with the consumer node being the root of the tree. Other topologies may be and should be considered. The impact of route discovery has been neglected in this study. This will obviously handicap multihop networks and should be introduced in the study.

Finally, if this new kind of wireless networks potentially offers a better reliability using route redundancy, this reliability and the influence of redundancy still need be studied.

8. REFERENCES

Decotignie J.-D., Prasad P. (1994), Spatio-Temporal Constraints in Fieldbus: Requirements and Current Solutions, *Proceedings of the 19th IFAC/IFIP Workshop on Real-Time Programming*, Isle of Reichnau, June 22-24, pp.9-14.

Morel Ph., Croisier A (1995)., "A wireless gateway for fieldbus", *Sixth IEEE International Symposium on Personal, Indoor and Mobile Radio Communications PIMRC'95*, pp. 105-109 vol.1.

CENELEC EN 50170 (1996), *General Purpose Field Communication System*, vol.3/3 (WorldFIP).

Pleinevaux P. and Decotignie J.D. (1988), Time Critical Communication Networks: Field Busses, *IEEE Network Magazine*, **vol. 2**, pp. 55-63.

Mangione-Smith, B (1995)., Low power communications protocols: paging and beyond, *Proc. IEEE Symposium on Low Power Electronics*, 1995, pp. 8 -11.

Rabaey, J. (2000), PicoRadio: Ultra-Low Energy Wireless Sensor and Monitor Networks, Invited Presentation, *CBS-ETAPS Workshop*, Berlin, April 2000.

Rabaey, J.; Ammer, J.; da Silva, J.L., Jr.; Patel, D. (2000a), PicoRadio: Ad-hoc wireless networking of ubiquitous low-energy sensor/monitor nodes, *Proceedings IEEE Computer Society Workshop on VLSI*, pp.9 –12.

Roberts D. (1993), "'OLCHFA' a distributed time-critical fieldbus", *IEE Colloquium on Safety Critical Distributed Systems*, pp. 6/1-6/3.

Sohrabi, K.; Gao, J.; Ailawadhi, V.; Pottie, G.J (2000), Protocols for self-organization of a wireless sensor network, *IEEE Personal Communications*, **volume 7** (5), Oct. 2000, Page(s): 16 -27

Thomesse J.-P.(1993), Time and Industrial Local Area Networks, *Proc. COMPEURO'93*, Paris, May 24-27, 1993, pp.365-374.

Wendi Heinzelman, Joanna Kulik, and Hari Balakrishnan, Adaptive Protocols for Information Dissemination in Wireless Sensor Networks, Proc. *5th ACM/IEEE Mobicom Conference*, Seattle, WA, August 1999.

ELSEVIER

IFAC

PUBLICATIONS
www.elsevier.com/locate/ifac

SIMULATION OF LOW POWER MAC PROTOCOLS FOR WIRELESS SENSOR NETWORKS

Amre El-Hoiydi, Jean-Dominique Decotignie

CSEM - Swiss Center for Electronics and Microtechnology
Rue Jaquet-Droz 1, 2007 Neuchâtel, Switzerland

Abstract: Wireless sensor networks have very different requirements than wireless LANs. The throughput and fairness requirements are left behind the low power consumption. Novel MAC protocols that reduce the energy spent in listening to an idle medium or in overhearing are needed. This paper presents a simulation model of a radio layer suited for the evaluation of such low power protocols. The model of a proposed low power MAC layer is described as well. Simulation results are presented and compared with theoretical expressions. *Copyright © 2003 IFAC*

Keywords: simulation, protocols, communication, networks, radio, energy, power, microsensors.

1 INTRODUCTION

Wireless sensor networks are resulting from the application of the knowledge gained in mobile ad hoc wireless networks (MANETs), to the field of distributed sensor networks (DSN). The wireless sensor networks are usually meant as a low power and low traffic variant of MANETs. Such sensors are meant to organize themselves into a network automatically after deployment and be able to remain active for years without battery replacement. Measurement data is typically forwarded in a multi-hop fashion from the sources to one sink (Figure 1). Of course, data can also need to be sent from the sink to the end-nodes, either for configuration or to perform some action, in which case the end-nodes are actuators. Applications of such networks are wide. They comprise cases where the installation of wires is not possible and where a regular battery replacement is not practical (outdoor environmental measurement, in building temperature control, etc). The main

characteristic of sensor networks is their mandatory low power consumption. As a consequence of this low power consumption, they must remain idle most of the time. The amount of traffic must hence be kept very small.

These networks are appealing for industrial use because of their resistance to interferences. In addition, the redundancy brought by a self-organizing multihop network is inherently a very good manner to counter fading, which is one of the main source of problems in industrial environments.

A large research effort has been devoted to the design of medium access protocols (MAC) for wireless LANs (Kleinrock, 1975; Bharghavan *et al.*, 1994), yielding to today's standards (IEEE 802.11, 1999; ETSI Hiperlan2, 2003). The main requirements of such networks are the fairness and the throughput. The optional support of stations in low power mode

The work presented in this paper was supported in part by the National Competence Center in Research on Mobile Information and Communication Systems (NCCR-MICS), a center supported by the Swiss National Science Foundation under grant number 5005-67322.

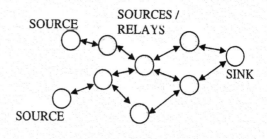

Fig. 1. Ad hoc wireless sensor network

has been added to wireless LAN protocols, trading the low power consumption against a higher latency. In wireless LANs, the low power mode is supported by a kind of TDMA, where stations in low power mode wake-up regularly to check for potential incoming traffic. This scheme requires the usage of a central node (the access point) that stays awake, buffers the packets, and sends them to stations in low power mode when they wake-up.

The requirements of wireless sensor networks are different than those of wireless LANs. The throughput and fairness requirements are left behind the low power consumption. Also, because such networks must be able to organize themselves in an ad hoc and multihop fashion, the usage of a central node that buffers and distribute packets to other nodes in low power mode is not possible. The development of a MAC protocol that is specifically designed to meet the requirements of wireless sensor networks is a need. The main goals of such protocols must be the reduction of the energy consumed for listening to an idle medium (idle listening) and the energy consumed for receiving packet destined to another node (overhearing). A number of proposals for low power MAC protocols have appeared within the last few years. As examples, Ye et al. (2002) have proposed S-MAC, a protocol where nodes are periodically active or asleep. Neighbor nodes synchronize among them to be active at the same time. The data is exchanged during those active periods using a regular 802.11 like protocol. Sohrabi et al. (2000) have proposed schemes based on TDMA, using some additional contention protocol during the synchronization phase. We have as well contributed with a proposal, consisting in a low power version of CSMA exploiting the preamble sampling technique (El-Hoiydi, 2002).

Because of the complex nature of multi-hop ad hoc networks, there is a strong need to perform simulations of a newly designed protocol, before to attempt implementation and real world testing. There exist a few discrete event network simulators that can be used for wireless MAC protocol simulation: OPNET (Chang 1999), NS-2 (Fall and Varadhan, 2003) and Glomosim (Zeng et al., 1998). All of these simulators were initially conceived for the simulation of wireless LANs, and they currently do not include the modeling of low power functions. The models must be completed by adding an idle state, where packets cannot be received, as well as the setup and turn-around delays between the states idle, receive and transmit.

A very detailed modeling of the temporal behavior of the radio transceiver is a fundamental step for the precise evaluation of the consumed power and for the correct choice of time parameters in the MAC protocol. In addition, a precise temporal behavior of the low layers is a prerequisite to any evaluation of synchronization protocols, which will for sure be needed in many sensor networks to provide time stamping.

OPNET is a very popular commercial tool, where protocols are implemented in pseudo-C language inside finite state machines. NS-2 is widely used tool among routing and transport protocols researchers, mainly in the field of TCP/IP wired networks. A wireless component has been added to enable research on adaptive routing for wireless ad hoc networks. Protocols are implemented in C++ and simulations scenarios are defined with Object Tcl scripts. Glomosim is a simulator dedicated to the simulation of wireless networks. It aims at providing a short execution time of simulation involving tens of wireless nodes. Glomosim has been developed on top of the discrete event simulation language "Parsec". This language is composed of a very limited number of commands (mainly send, send after some delay and hold). The core of the software is the scheduler, which is passing messages among nodes and continuously reordering the list of future events. Building upon simple constructs, the Glomosim software has evolved towards a complex simulator offering an OSI layered structure, with a number of different ad hoc routing and wireless medium access control protocols. Because of its processing efficiency, its clean software structure and its open source character, we have decided to use and extend this simulator. Another advantage of Glomosim is that it is written in C, a language that is likely to be used for implementations on embedded sensors, thereby allowing a potential reuse of parts of the MAC layer model source code for the implementation.

The following sections of this paper will describe the model of the transceiver (radio model), the low power MAC protocol and its model, as well as initial simulation results.

2   RADIO LAYER MODEL

For the purpose of simulating the low power wireless medium access protocol developed at CSEM for

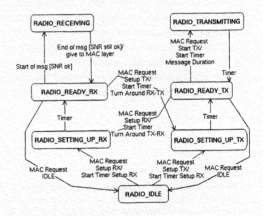

Fig. 2. Radio layer delays model

wireless sensor networks, an extended radio physical layer has been designed and implemented. The behavior and the programmability of real radio transceiver chip was modeled, including the setup and turn-around delays as well as an Idle state. This model is applicable to any half-duplex transceivers. The finite state machine of the model is shown in Figure 2. The syntax of the finite state machine is following the UML standard: *Event*[*Condition*]/*Action*. A transition is made if the *Event* happens and if the *Condition* is met. During the transition the *Action* is performed.

Transitions between states can be caused either by commands from the MAC layer (MAC Request), by timers internal to the radio layer or by the start or the end of message reception on the radio medium. In the IDLE state, the radio cannot send, nor receive. It consumes very little energy. On request from the MAC layer, the radio layer goes into the READY_TX or READY_RX states, after a waiting delay that depends from the originating state. These delays, as defined in Table 1, are typically longer from the IDLE state than from the respective opposite state (READY_RX and READY_TX). The indicated values are target values of the transceiver designed within the CSEM WiseNET project (Giroud *et al.*, 2001).

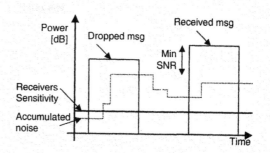

Fig. 3. Radio accumulated noise interference model

Fig. 4. Finite state machine of the accumulated noise interference model

Table 1 Transceiver delays

Symbol	Description	Values
T_{SeRx}	Settling time into RX mode.	800 μs
T_{SeTx}	Settling time into TX mode	800 μs
T_{TaRxTx}	Turn around time from RX into TX mode	400 μs
T_{TaTxRx}	Turn around time from TX into RX mode	400 μs

The transition to the TRANSMITTING state from the READY_TX state is completely controlled by the MAC layer. The transition from the READY_RX to the RECEIVING state is triggered by the beginning of the reception of a message on the radio medium (typically the detection of the start sequence). For the message to be locked with success, it must be received with a power above the receiver's sensitivity, and large enough to present a given signal to noise ratio. A message is received with success if it presents the wanted signal to noise ratio during its whole duration, as illustrated in Figure 3.

The algorithm to compute the accumulated noise and the capture behavior are illustrated in the finite state machine shown in Figure 4. This algorithm was not modified from the one in the original Glomosim accumulated noise radio model: A packet is transmitted to all neighbors, with a different power attenuation and time delay as a function of the distance. At the receiving side, in the READY_RX state, the new message is either added to the accumulated noise (if it is below the receiver sensitivity or if the SNR is below the threshold) or locked. Once a message is locked, the radio goes in the RECEIVING state and continues to compute the noise curve, to check whether the SNR is still large enough. If the power of a new message reduces the SNR of the locked message below the wanted threshold, the locked message is dropped and its receive power is added to the noise. If the power of the new message is above the needed SNR, it is captured. Otherwise, it is added to the noise as well, and the radio goes into the READY_RX state. Whenever the "End of msg" event is received, the message power is subtracted from the accumulated noise. If the message is still locked when the "End of msg" event is received, it is considered received without interferences and given to the MAC layer. In other states than READY_RX and RECEIVING, the "Start of msg" event only implies adding the message power to the accumulated noise. This message is never locked, whatever the SNR.

3   MAC LAYER MODEL

In our research on low power medium access control protocols, we focused initially on the combination of classical carrier sense multiple access protocols with the preamble sampling technique. This technique, used for example in analog paging systems (Mangione-Smith, 1995), consists in sending a preamble of size T_P in the front of every message. A receiver will sleep, and wake up every T_P to check whether the channel is idle or busy. As illustrated in Figure 5, when a preamble is detected, the receiver continues to listen until the start of the DATA section is found and the message received. This allows a node to sleep most of the time when the channel is idle.

Combined with non-persistent CSMA, this preamble sampling technique yields a protocol that is exhibiting very low power consumption when the traffic is low (El-Hoiydi, 2002). Such a protocol can be useful, for example, for the initial exchange of synchronization messages needed to setup a TDMA schedule, or for the transmission of sporadic data traffic.

The complete finite state machine of this MAC protocol being rather complex, it will be explained in the three basic procedures: Sampling, Transmission and Reception.

A regular timer, as illustrated in Figure 6, initiates the sampling procedure. The radio is instructed to be ready to sense the medium. Once the radio is in the READY_RX state, the medium is sensed and the MAC layer either goes back to sleep (if the medium was found idle) or to the LISTENING state.

From the LISTENING state (see Figure 7), the MAC layer goes into the SLEEPING state if the medium returns to the idle state without the reception of any message. If a message is received without interference, the MAC layer requests from the radio to be ready to transmit and then sends an acknowledgement.

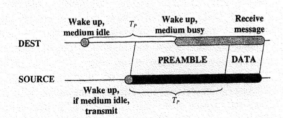

Fig. 5. Low power medium access control (MAC) protocols with preamble sampling

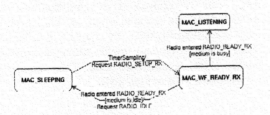

Fig. 6. Finite state machine of the MAC preamble sampling procedure

Special cases of the receive procedure include the reception of broadcast messages, in which case no acknowledgement is sent, as well as overheard unicast messages (i.e. message received but not destined to the receiving node), which are simply dropped.

The unicast transmission procedure is illustrated in Figure 8. A message transmission is initiated either by the arrival of a message from the network layer in an empty input queue, or by the backoff mechanism if the input queue contains one or more messages. After sensing the channel, the MAC either returns to the sleeping state if the medium was found busy, or initiates the transmission if the medium was found idle.

4   MAC LAYER PERFORMANCE EVALUATION

The layout of the simulated network is illustrated in Figure 9. The node 0 is the destination of all messages, sent by the nodes 1 to 10 located around it. The transmission range is selected to be larger than the radius of the circle. A transmission by a node will be sensed by the node located on the opposite side of the circle. This layout, although not representative of practical multi-hop sensor networks topologies, has been chosen for its clarity. As every node hears every message, the obtained results are, up to the hidden node effect, applicable to an ad-hoc network

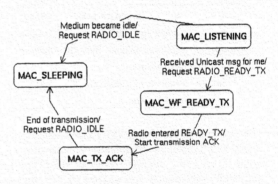

Fig. 7. Finite state machine of the MAC unicast receive procedure

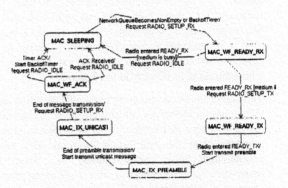

Fig. 8. Finite state machine of the MAC unicast transmit procedure

topology.

Every traffic-generating node produces packets following a Poisson process of rate g. The MAC layer attempts to transmit these packets following a Poisson process of rate g as well. If the medium is found busy, the transmission is re-attempted after a random waiting time, exponentially distributed with mean $1/g$. As a consequence, when the medium becomes busy, all packets generated by the application cannot be transmitted anymore. The queue between the network and the MAC layer reaches its maximum and packets are dropped. The performance evaluation will not consider such dropped packets. We use the Poissonian traffic generator to model an infinite packet queue at the input of the MAC layer. We are interested in the performances of the MAC protocol under maximal traffic load. These performances will be low bounds to what will be seen with real applications having less traffic to offer to the MAC layer.

We will consider the following performance parameters:

*S: **Throughput**:* the percentage of time when the channel carries data that is successfully acknowledged. This does not include header overheads, nor the preambles or the acknowledgements, but only the payload useful for the application layer.

*P: **Power**:* the consumed electrical power. The mean consumed power is computed by measuring the time spent by the radio layer in the different states. The power consumption in each state (idle, setup, ready_rx, receive, ready_tx, transmit), as given by the transceiver datasheet, is then used to obtain the average power.

P_S: ***Probability of successful transmission***. Once a node has sensed the medium idle, it initiates the transmission procedure. There remain a residual risk of collision with another transmission, resulting from the turn-around delay between the moment when the medium was sensed idle, and the moment when the transmission has started. Other nodes may have sensed the medium idle in this time interval and have initiated a transmission as well. The probability of sending one packet without interference depends only on the turn around time, on the number of traffic generating nodes and on the intensity of the Poisson traffic generated by each node. For a low range transmission (10-100 meters) the propagation and the sensing time are neglected, because they are orders of magnitude smaller than the turn-around time. This probability of successful transmission is valid for both the DATA and the ACK. The ACK is guaranteed not to suffer from interferences, because the turn around time is smaller than the setup time. All nodes that are not the destination of some unicast packet will go to the sleep state once the medium becomes idle. From there, even if they attempt a transmission right away, they will need more time to

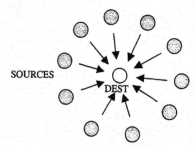

Fig. 9. Layout of simulated sensor network

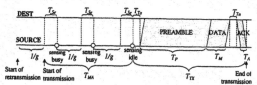

Fig. 10. Transmission delay without collisions

access to the medium than the destination node, which only needs to turn-around its transceiver.

*D: **Delay**:* the time between the start of the first transmission attempt and the successful reception of the acknowledgement.

Analytical expressions to compute the throughput, the power consumption, and the probability of successful transmission have been derived and are presented in (El-Hoiydi, 2002).

The transmission delay is one of the most relevant parameter to engineer time sensitive applications. One can compute its mean value by considering the Figure 10: If the medium was idle and if there could be no collisions at all, the transmission time would be given by:

$$T_{Se} + T_{TX} , \text{ where } T_{TX} = T_{Ta} + T_P + T_M + T_{Ta} + T_A .$$

After having setup the transceiver into READY_RX mode (delay T_{Se}), we would need T_{Ta} seconds to reverse the transceiver into READY_TX mode, send the preamble, the message, reverse the transceiver into READY_RX mode and receive the ACK.

With a medium that is busy with probability b, we may have to try several times before to find the medium idle. The probability to see the medium idle is $(1-b)$. If we are lucky at the first attempt, we have to wait only T_{Se}, otherwise, we have to backoff for an average duration $1/g$ and try again. Let $T_{MA}(k)$ be the time needed to acquire the medium if we have to try k times. The probability to wait for exactly $T_{MA}(k)$ is given by:

$$T_{MA}(1) = T_{Se} \text{ with proba. } (1-b)$$
$$T_{MA}(2) = T_{Se} + 1/g + T_{Se} \text{ with proba. } b(1-b)$$
$$T_{MA}(3) = T_{Se} + 2(1/g + T_{Se}) \text{ with proba. } b^2(1-b)$$
$$T_{MA}(k) = T_{Se} + (k-1)(1/g + T_{Se}) \text{ with proba. } b^{k-1}(1-b)$$

The average value of the waiting time is given by

$$T_{MA} = E[T_{MA}(k)] = \sum_{k=1}^{\infty} T_{MA}(k)P(K=k)$$

$$= \sum_{k=1}^{\infty} \left(T_{Se} + (k-1)(1/g + T_{Se})\right)(1-b)b^{k-1}$$

$$= T_{Se} + (1/g + T_{Se})b/(1-b)$$

From the moment when the medium is found idle, we need T_{TX} to transmit the preamble, the data and to receive the acknowledgement.

Because there is a probability $1-P_S$ that the data packet suffers from a collision, we may have to backoff for a duration $1/g$ and try again. The time $D(k)$ needed to transmit the packet collision-free at the kth transmission, and the corresponding probability are given by

$$D(1) = T_{MA} + T_{TX} \text{ with proba. } P_S$$
$$D(2) = T_{MA} + T_{TX} + 1/g + T_{MA} + T_{TX} \text{ with proba. } (1-P_S)P_S$$
$$D(k) = T_{MA} + T_{TX} + (k-1)\left(1/g + T_{MA} + T_{TX}\right)$$

$$\text{with proba. } (1-P_S)^k P_S$$

The average delay between the beginning of the first transmission attempt and the reception of the ACK message is hence given by

$$D = E[D(k)] = \sum_{k=1}^{\infty} D(k)P(K=k)$$

$$= \sum_{k=1}^{\infty} \left(T_{MA} + T_{TX} + (k-1)\left(1/g + T_{MA} + T_{TX}\right)\right)(1-P_S)^k P_S$$

$$= T_{MA} + T_{TX} + \left(1/g + T_{MA} + T_{TX}\right) P_S \sum_{k=1}^{\infty}(k-1)(1-P_S)^k$$

$$= T_{MA} + T_{TX} + \left(1/g + T_{MA} + T_{TX}\right) P_S \frac{1-P_S}{P_S^2}$$

$$= \frac{(T_{MA} + T_{TX}) + (1-P_S)/g}{P_S} \qquad (1)$$

5   SIMULATION VERSUS THEORETICAL RESULTS

This section will present the results of three sets of simulations, done with three different sizes for the preamble (50, 100 and 150 ms). The size of the preamble must be chosen as a trade-off between the energy consumed to transmit and the energy consumed to monitor the medium. The optimum length depends on the foreseen traffic. For each preamble size, the attempt rate has been varied from 10^{-2} (one attempt every 100 seconds) to 10^3 (one attempt every 0.001 second). All plots will be drawn with, on the X-axis, the transmission attempt rate at each node. Simulation results on the graphs are marked as circles. In these simulations, the setup delay of the transceiver is $T_{Se} = 0.8$ ms and the turn-around delay $T_{Ta} = 0.4$ ms (see Table 1).

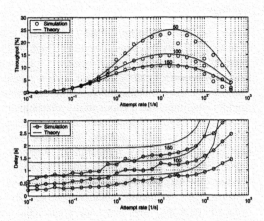

Fig. 11. Throughput and delay performances of NP-CSMA-PS for different length of the preamble (50, 100 and 150 ms), as a function of the transmission attempt rate at each node.

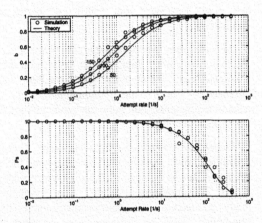

Fig. 12. Medium business and probability of collision for NP-CSMA-PS for different length of the preamble (50, 100 and 150 ms), as a function of the transmission attempt rate at each node.

The first plot in the upper part of Figure 11 shows the classical throughput curve of CSMA, with a maximum for some offered load and a performance decrease for higher loads. In sensor networks, we will not work around the throughput-optimal point. We will remain in the left part of the curve, where low traffic can permit a low duty cycling. We can observe a good match between the theory (see El-Hoiydi, 2002) and the simulation results, giving confidence both in the theoretical model and in the simulation model.

The plot of the theoretical value of the transmission delay (expression (1)) and of the simulation results can be seen in the lower part of Figure 11. We see that the theoretical curves for the delay are quickly converging to a constant value when the attempt rate decreases. For low attempts rate, the medium being quite idle, one would imaging the transmission delay to be small, as it is unlikely to need to backoff. But because the average backoff waiting time is very large when the attempt rate is small, both compensate themselves. The waiting time remains constant.

The simulations show waiting times that are up to three times lower than the theoretical values. This is caused by the way the delay is computed in the simulator: between the first transmission attempt and the reception of the acknowledgement. If the backoff expires while a transmitter isn't in the sleeping state, it will be postponed. This postponed backoff is not counted for in mean delay in the simulation, reducing this mean delay.

In any case, we see that the average measured delay is close to 500 ms, even for small attempt rates, which is still rather high. Having a large backoff window is certainly a mean to guarantee a low average traffic. However, to reduce this delay, it is preferable to use a smaller backoff window, and to give to the application the responsibility of limiting the traffic. Best results will be obtained by replacing the simple backoff mechanism of non-persistent CSMA with more elaborated persistent algorithms like the one used in 802.11.

An important parameter for the analysis of a low power contention protocol is the medium business. With the CSMA protocol, every node receives every message on the medium. A receiver is hence activated whenever the medium is busy. To be able to operate on a single 2.6 Ah alkaline battery for years, the traffic on the medium must be kept small. The upper part of Figure 12 shows the medium business b (i.e. the probability to find the medium busy at some instant). The lower part of Figure 12 shows the probability that a packet will not suffer from a collision, once the transmission procedure has been started. For both curves, the simulation results (circles) are drawn together with theoretical results from (El-Hoiydi, 2002).

The consumed power has been computed from the simulations by counting the fraction of the time spent by the radio model in the various states. It is assumed that the transceiver consumes nothing in the IDLE state (it consumes a negligible current compared to the natural loss of an alkaline battery). In the TRANSMITTING state, it consumes $P_{TX} = 9$ mW. In the RECEIVING state, as well as the READY_RX, READY_TX, SETTING_UP_RX and SETTING_UP_TX state, we assume a power consumption of $P_{RX} = 1.8$ mW. The argument behind this assumption is that in all but the TRANSMITTING state, all the electronics of the radio must be operating. In the TRANSMITTING state, we have, in addition, the final stage power amplifier. These numbers are target values for the low-power wireless transceiver integrated circuit developed at CSEM (Giroud *et al.*, 2001).

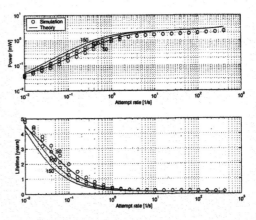

Fig. 13. Power consumed by NP-CSMA-PS for different length of the preamble (50, 100 and 150 ms), as a function of the transmission attempt rate at each node, and resulting expected lifetime using a single 2.6Ah alkaline battery.

The measured mean power consumption for the various preamble lengths are shown in the upper part of Figure 13, together with the theoretical curves introduced in (El-Hoiydi, 2002). These power consumption curves can be translated into lifetime curves shown in the lower part of Figure 13. We have assumed the usage of a single 2.6 Ah alkaline battery, with a constant energy loss every year equal to loosing 10% of its initial capacity in a year. We see that lifetimes of years can only be achieved for very low values of the attempt rate, for which the medium business is very low. We see that a preamble of 50 ms is optimum for higher values of the attempt rate, but that a larger preamble becomes more energy efficient for lower values of the attempt rate.

6   CONCLUSION

A new radio layer model has been defined and included in the Glomosim discrete event simulator, for the purpose of simulating low power wireless sensor networks. This radio layer model includes the state change delays as well as the idle state, elements that are essential to low power protocols modeling. On top of this radio layer, a custom low power MAC protocol has been defined and modeled. Simulations results have shown in most cases a good match with theoretical results presented in (El-Hoiydi, 2002). The measure and computation of the transmission delay have proven to be delicate. Further work is needed on that topic.

This simulation model will be used in the future to better understand, enhance and optimize low power MAC protocols for wireless sensor networks. The model of the MAC layer will then be usable for research on adaptive routing or synchronization in multi-hop low power networks.

REFERENCES

Bharghavan, V., Demers, A., Shenker, S., and Zhang, L., "MACAW: A Media Access Protocol for Wireless LAN's". In *Proceedings ACM SIGCOMM*, pages 212–25, London, UK, August 1994.

Chang, X., "Network simulations with OPNET", in Proceedings *Simulation Conference*, 1999. **Vol. 1**, pp. 307-314.

El-Hoiydi, A., " Spatial TDMA and CSMA with Preamble Sampling for Low Power Ad Hoc Wireless Sensor Networks", in Proc. *IEEE International Conference on Computers and Communications* (ISCC 2002), Taormina, Italy, pp. 685-692, July 2002.

ETSI Hiperlan2, http://www.hiperlan2.com/

Fall, K., Varadhan, K., "The ns Manual (formerly ns Notes and Documentation)", 2003, http://www.isi.edu/nsnam/ns/

Giroud, F., Le Roux, E., Melly, T., Pengg, F., Peiris, V., Raemy, N., Ruffieux, D., "WiseNET®: Design of a Low-Power RF CMOS Receiver Chip for Wireless Applications", *CSEM Scientific and Technical Report* 2001, p. 22.

IEEE 802.11, 1999 Edition, Wireless LAN Medium Access Control (MAC) and Physical Layer (PHY) Specifications.

Kleinrock, L., Tobagi. F. A., "Packet Switching in Radio Channels: Part I - Carrier Sense Multiple-Access Modes and Their Throughput-Delay Characteristics", in *IEEE Transactions on Communications*, **Vol. COM-23**, No. 12, December 1975, pp. 1400-1416.

Mangione-Smith, B., "Low power communications protocols: paging and beyond" in Proc. *IEEE Symposium on Low Power Electronics*, 1995, pp. 8 -11.

Sohrabi, K.; Gao, J.; Ailawadhi, V.; Pottie, G.J, "Protocols for self-organization of a wireless sensor network", in *IEEE Personal Communications*, **Vol. 7**, Issue 5, Oct. 2000, pp. 16 -27.

Ye, W., Heidemann, J., Estrin, D., "An Energy-Efficient MAC protocol for Wireless Sensor Networks", In Proceedings of the 21st *International Annual Joint Conference of the IEEE Computer and Communications Societies* (INFOCOM 2002), New York, NY, USA, June, 2002.

Zeng, X., Bagrodia, R., Gerla, M., "GloMoSim: a Library for Parallel Simulation of Large-scale Wireless Networks", Proceedings of the *12th Workshop on Parallel and Distributed Simulations* -- PADS '98, May 26-29, 1998 in Banff, Alberta, Canada

SYSTEMS WITH NUMEROUS LOW-COST SENSORS –
NEW TASKS AND DEMANDS FOR FIELDBUSSES

Herbert Schweinzer[1], Wolfgang Kastner[2]

[1] *Institute of Electrical Measurement and Circuit Design*
Vienna University of Technology
Gusshausstr.25, A-1040 Vienna, AUSTRIA
herbert.schweinzer@tuwien.ac.at

[2] *Institute of Automation*
Vienna University of Technology
Treitlstr. 1, A-1040 Vienna, AUSTRIA
k@auto.tuwien.ac.at

Abstract: New concepts in automation on the one hand, and advances in sensor technology by improved use of microelectronics and micromechanics on the other hand, lead to the perspective of a strongly increased number of low-cost sensors for measuring numerous parameters of complex models. Transmission of sensor data and sensors' positions are essential to combine their values to a global view of the controlled situation. Discussing characteristics of such systems, some aspects are presented in this paper where communication needs extend the possibilities of present fieldbus concepts.
Copyright © 2003 IFAC

Keywords: multiple sensor networks, setup of sensor nodes, sensor position measurement, spontaneous networking

1. INTRODUCTION

Since several years, advances of microelectronics and micromechanics have led to new sensor constructions with extremely small size, new sensor technologies, and universal application range. Compact and cheap constructions allow an enormous increase of the number of sensors employed in a system which raises the need for simple and inexpensive, but nevertheless flexible and powerful communication links.

Up to now, extensive use of numerous sensors can be found in research especially dedicated to situation recognition and situation-dependent behaviour. During the past years, the work mainly focused on the field of (mobile) robotics, where a big number of sensors is necessary to allow fast recognition and reaction in complex environments (Audenaert, *et al.*, 1992; Adams, 2002). Nowadays, research is directed to integrate these capabilities in home and building automation systems (Dietrich, 2000) where additional sensors not only allow an improved

quality of the automation but also enable completely new dimensions of automation levels. However, this should contribute to extend the possibilities of current fieldbus systems and might be one of the most important challenges for them in the future. Applications employing numerous low-cost sensors can be variously: Consider for example the controlling of complicated light or temperature distributions in rooms, or the monitoring of presence and movement of persons where the movement detection can be used for guidance on specific ways. Such large and spaciously distributed systems create new tasks, especially concerning communication setup and supplying geometrical models with the necessary information of sensor locations. Moreover, intermittent sensor communication and sensor malfunctions will have to be considered as a normal case.

This paper offers an overview and basics of a concept for the connection of sensors to a multistage communication network where local sensor management is the first step before integrating a sensor into the fieldbus system. In particular, we want to point out the influence of the sensor position and sensor data processing onto higher layers of the communication. The work is concentrated on sensors because the increase of system capabilities is based on getting much more and detailed information. On the other hand, the number of actuators will mostly remain within limits. However, communication concepts useful for numerous sensors will also be applicable for actuators.

2. VARIETY OF SENSORS

In general, a raising degree of automation with improved behaviour and extensive data processing is based upon manifold and more powerful sensors. Since the sensory system is responsible for the supply of "primary information", there are several issues, which are of importance.

- Measurements have to be performed with sufficient resolution and precision within well-defined ranges.

- Certain environmental conditions which are required for appropriate measurement have to be considered.

- The measurement method has to be reliable and repeatable.

- Structure, form, size, weight, etc. of the devices have to be taken into account.

- Sometimes it is necessary to pre-process the measurement values (before transmitting them).

- Regarding communication various requirements have to be taken into account: e.g. during normal operating mode real-time constraints especially by short data transmissions; during setup and maintenance mode facilities to download software-updates, data bulk transmissions for (ex-)changing parameter values, etc.

- The overall costs have to be kept within reasonable limits.

Short examples show the importance of these different aspects for both simple and complex sensors: Regarding the latter ones, a smart video system can make use of a CCD-video camera with appropriate resolution, sufficient illumination is necessary, comparison of corresponding images is based on a good reproducibility, an integrated picture analysis systems may execute high-speed pattern recognition, and a high communication bandwidth is used for online transmission of picture characteristics and continuous changes of relevant parts of the scene. Regarding simple sensor systems, a temperature-sensing device makes use of a temperature sensor for a given range and with defined reproducibility, the sensor gets its stability by appropriate arming, pre-processed information may inform about temperature trends, and communication may transport values of several temperature sensors each of them localized in a specific position.

Corresponding to the aspects and demands mentioned above, today's sensors can roughly be divided into the following groups:

- Simple sensors without integrated data processing units and with – in general – passive connection to the communication network.

- "Intelligent" or "smart" sensors with local data processing units and powerful passive or also active connection to the communication network (Gilsinn and Lee, 2001). Additionally integrated auxiliary sensors may improve the primary measurement function especially for the correction of disturbing influences.

- Multi-sensor systems with several different types of sensors. They make use of local data processing and powerful active connection to the communication network. Because of different sensing features, they have an increased flexibility and can be used for the measurement of different physical quantities and for combining of different information. Moreover, they can offer a cost-effective alternative to parallel, but independently operating sensors. In addition they allow on-line servicing and maintenance.

This overview shows the great bandwidth from simple to complex sensors, from low-cost average to expensive specialized ones. However, huge sensor systems will employ a high percentage of simple low-cost sensors, integrated on a chip and adaptable to a range of typical applications.

3. FURTHER ASPECTS OF HUGE SENSORY SYSTEMS

As stated in the introduction, the large scope of automation and the variety of solutions tend to increase the number of sensors. Thus, sooner or later solutions based on traditional fieldbus systems with their limited addressing schemes will fail. Consider that up to date fieldbusses which allow to integrate 10.000s of nodes will be ruled out by new networking facilities that allow to integrate 100.000s of nodes.

However, new approaches will show deep impacts on hardware *and* software techniques. Regarding hardware, interconnection and power supply are the most important challenges to be solved. Communication using wires is cheap to use but expensive to install. Therefore, wireless communication (point-to-point connection with, for example, frequencies of 433 MHz, 868 MHz and 2.45 GHz, Bluetooth, Wireless Lan) becomes more and more important (Brooks, 2001; Haehniche and Rauchhaupt, 2000; Townsend, *et al*., 2002). However, since they can mostly be used only for short distances they need a multistage communication network.

Power supply is an even more difficult problem though (Nordman and Kozlowski, 2001). Most of the time, wires are used for traditional powering, often also combined with data communication in a common cable. Alternatively, one can use non-regenerative energy sources like batteries (Sanderford, 2002). Up to now only applied for specific cases, regenerative sources like solar cells and thermocouples, or activation of electronics by vibration or electromagnetic fields (transponder) will have growing importance for numerous low cost sensors in the future.

For high performance and flexibility or fault-tolerant applications, as it is demanded for complex measurements, a far higher number of sensors than usual is required (Koushanfar, *et al.*, 2002). Because of that and because of possibly temporary missing power supplies, a mechanism for ad-hoc interconnection of the sensors seems to be very useful. That also includes the integration of mobile nodes, which are, for example, carried around with someone for controlling or monitoring purposes. However, as already described, today's methods are not able to handle very large number of devices – neither with regard to the communication, nor to the power supply, interconnection and costs. Thus, there is the need for new solutions.

Therefore, the main aspects of systems with a large number of low cost sensors are

- the selection of an appropriate sensor technology,

- the estimation of cost-effective concepts of sensor systems including power supply, casing and interconnection,

- the physical and functional interconnection of simple sensors and actuators at the bottom layer,

- the communication demanded by the numerous sensor nodes,

- the ability to setup completely scaleable sensor networks,

- the development of spontaneous network interconnections, and

- effective concepts for security (Sedov, *et al.*, 2001).

According to the number and complexity of these tasks, fieldbus systems get new functionalities. While during the past decade, research and development in the fieldbus system area has been directed to integrate fieldbus devices with higher networks, the new challenge is to find solutions to combine "smart dust" sensor systems with fieldbus systems. Thus, the role of fieldbus systems has inevitably changed. While formerly their devices have mainly been service providers for higher network services, they now will become service customers of big underlying sensor networks (Fig. 1). This new role of the fieldbus will be manifested as an overall (sensor) information backbone network.

4. INFLUENCE OF THE SENSOR POSITION

A large number of sensors increases the problem of modelling. Two aspects of modelling are of main importance:

- Modelling from the view of communication, and
- modelling from the view of data processing.

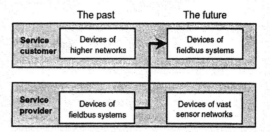

Fig. 1. The new role of fieldbus systems

Modelling the communication is a process where sensors will be located and identified corresponding to an abstract model of communication nodes. Normally this is done within the setup phase of the network which will be discussed later on.

Especially a large number of sensors of the same type will be used to gain a distribution of a certain physical quantity within the measured area. Examples for this are measurements of distributions of temperature or light intensity. In this case, information processing combines the measurement values of the sensors by use of a physical model where the geometrical positions of the sensors are parameters with primary influence. Goals of processing may be to approximate the temperature or light intensity at any defined position of the area, or to calculate dynamic effects.

Besides special cases of well-defined positions of the sensors which can be systematically included in the model (e.g. placement in a specific frame), arbitrary sensor positions are the first information which have to be get. However, a large number of sensors requires a powerful and rapid way to collect their positions.

A well-suited measurement method for the sensor positions could be an effective and fast way. A measurement scan of the sensor area has to detect the presence of each sensor which for instance has to identify itself with a unique number as reaction to the scan (Fig. 2). This process delivers both position information for geometric modelling and identification for the model of the communication. Moreover, employing a scan as a geometric addressing of the sensor may enable a bi-directional data communication with the sensor. Different ways of scanning mechanisms can be considered, mainly optical (laser scanner), inductive, or based on microwaves. The necessary speed of scanning for this purpose is slow.

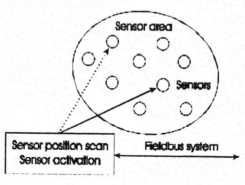

Fig. 2. Measurement scan of sensor positions

By this means, the placement of sensors can be decided primarily because of practical reasons. Scans of sensor positions can be repeated periodically which allows reconfigurations of the system: new sensors can be included in the system, missing sensors removed, and also a slow movement can be detected. However, communication of sensor's measurement values may be performed in parallel to the scanning using a different communication medium (e.g. for example a RF-link). An example of a miniature sensor device supporting optical position scanning and communication is described as "smart dust" sensor (Fig. 3).

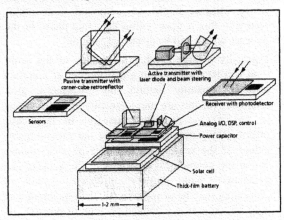

Fig. 3. Conceptual diagram of a smart dust sensor (from Warneke, *et al.*, 2001)

An important but nevertheless special case are sensors which only inform about the presence of objects where these sensors are attached. As a standard, SAW-devices (surface acoustic waves) can be used which are passive and coded to return a specific number when activated by a special microwave frequency, e.g. 2.45 MHz (Siemens, 2003). Scanning with narrow microwave beams is not the normal readout of SAW-tags but could be used to get a more precise information about the sensor position.

5. INTEGRATION OF NUMEROUS SENSORS INTO FIELDBUS SYSTEMS

With a large number of sensors a considerable expense has to be applied for the setup of the communication network. Therefore, sophisticated tools reducing initialisation time will get increasing importance.

As mentioned before, measurement of sensor position could be a very effective way for initialisation. According to the different meaning of information handled during initialisation and on the other side of transmitted sensor data, two types of nodes have to be introduced in the system:

- Sensor nodes which are able to deliver measurement data after initialisation. These nodes can be "regular" fieldbus nodes, as well as be part of an underlying vast sensor network.

- Installation or initialisation nodes which deliver position information of the sensors to according

modelling tools. Since presence and position of sensors can rapidly change in a running system, these nodes have to be permanent partners of online configuration and monitoring tools.

For the special case of scanning passive tags attached to objects (as mentioned in Section 4), these tags need not to be included as network nodes. The position scanning unit plays the role of a position measurement system which is only one special service provider for the fieldbus system.

When developing installation and initialisation nodes, a good idea might be to learn lessons from techniques already used in ubiquitous and pervasive computing (Ciarletta and Dima, 2000), where different technologies and standards for service discovery and service delivery have evolved in the last years (Helal, 2002).

For instance, in a so-called Jini federation, every device can act as a service provider as well as a service consumer. To join a federation, the service provider must locate a so-called *lookup service* by using a discovery protocol (Fig. 4). Some features of this lookup service map to issues discussed for installation and initialisation nodes: e.g. after a successful discovery and localisation, devices may upload their identification (and retrieved data values). On the other side, a service consumer, typically a fieldbus node, must obtain a reference to the lookup service, too. It can then request a particular service by sending a request to the lookup service that contains the desired service attributes. The lookup service returns a reference to the service customer for further data communication. That way, service customer and service provider may both interact directly, i.e. without the need of the lookup service.

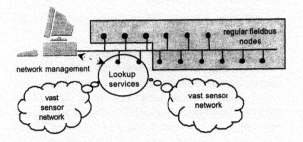

Fig. 4. Lookup service

Originally, Jini gives access to its services only on a lease basis. We could use a similar mechanism to operate the fieldbus with self-adapting data transfers. A lease is a time-based resource reservation, passed to service consumers (clients) by services. Whenever a client wants to use a service, it has to actively and periodically renew its interest. The idea behind leasing is pretty simple. A client reserves the access to the service provider only for a certain amount of time. The client has several options, before the lease period expires: First, it may try to renew its lease, before the time has elapsed, to gain more time it might need. Second, it may cancel the lease before the time period has expired, to indicate that it is no longer interested. Third, it can do nothing and wait until the time passes (thereby loosing the lease).

This easy mechanism has two benefits. On one hand, it allows to run long-living communities with minimal bus traffic and without intervention to clean up stale data, on the other hand, it allows to set up a watchdog timer for maximum response time of the service provider. In our case we have to consider sensors as service providers that may stay up for a long time but face the problem of partial failures. These failures can cause network members to hold invalid data (see also some topics discussed in the next section). Without leasing, our installation and initialisation nodes would have to scan already registered service providers and remove them "by hand". By granting the registration together with a lease for only a limited period of time, services that cease to exist, automatically stop transmitting data and thus are "plugged out" of the federation.

6. REASONS FOR INTERMITTENT COMMUNICATION AND SPONTANEOUS NETWORKING

There are several reasons why communication between partners will change dynamically:

- Power supplies are sometimes not operable which leads to intermissions of sensor data communication (i.e. sporadic communication),

- sensor failures,

- introducing new sensors in the running system, and

- mobile communication partners.

Local power supplies for sensor devices have a certain probability to fail. While non-regenerative energy sources like batteries are operating during a defined time interval which can be taken into account, regenerative energy sources are operating only under certain conditions. In this case, sensor communication will be intermittent on principle and the communication mechanism must be able to deal with sensor nodes which are not available all the time.

In periods where communication is not possible, measurement values are not available, too. This fact has to be introduced in the physical model. Problems could especially arise when integral quantities have to be calculated. However, if the sensor component locally prepares basic integral values which are not affected by a sensor breakdown caused by the loss of the power supply, also this problem can be solved. An example of such a sensor component would be a device, measuring the heating energy, powered by a peltier element, and communicating via a RF link when being powered (Scherrer and Schweinzer, 2001).

An increasing number of sensors lead to an also increasing number of sensor failures. Therefore, maintenance will become a central aspect in such a system. This will be only possible, if defect sensors can be removed from the running system and also if new sensors can be introduced in an automatic way.

Again, the role of initialisation nodes capable of features of a lookup service may help to overcome these problems.

As a consequence, two further aspects are implied: a detection method of erroneous data in case of sensor failure must be included in the system to prevent data processing from misinterpretation and disturbance. And further, the installation of a big system takes a lot of time and therefore it is likely to switch the system to the operating state although its building up is not yet finished. Also expansions of the system can be realized in this way, being the base for scalability. Thus, also corresponding algorithms have to be 'scalable' which means that they have to be able to operate on a changing base of sensor inputs.

There are some reasons why it is significant to be able to expand the system by mobile communication partners. Reasons for that are due to the primary tasks of the system. They may include a partnership for delivering information to the network or on contrary for using system information. Furthermore, the partnership may introduce the position of the mobile station as a further system parameter, or may only establish a communication link for controlling, monitoring, maintaining and servicing the system.

A crucial point for these possibilities is the system security which will influence different layers. Physically it has to be obtained by identification as a foreseen system partner. At higher levels, security has to be achieved as usual by context dependent checks of access rights of the mobile node or mechanisms for encryption.

7. DEMANDS FOR FIELDBUSSES AND FIELDBUS ORIENTED LEVELS OF INFORMATION PROCESSING

The major demands for future fieldbus systems and information processing at basic level result from the aspects discussed above:

- New network installation tools: Installation of sensor (and actuator) nodes will be strongly simplified by localisation tools that are accomplished without user interaction by dedicated initialisation nodes.

- Installation tool as a sensor node of the network: Localisation of sensors can be used as an information of a "master sensor" which is especially useful for parameterisation of physical models.

- Handling of intermittent node communication.

- Handling of spontaneous communication links.

- Bookkeeping of information of intermittent and spontaneous sensor nodes: changes of information with time have to be controlled and data consistency has to be proven.

- Detection of erroneous sensor data and elimination of faulty information, e.g. by checksum control or consistency checks.

- Security with respect to

- Attack prevention: i.e. security mechanisms that contain ways of preventing or defending against certain attacks before they can actually reach and affect the target. An important element in this category is access control, for instance via authentication for allowed (i.e., foreseen) sensor partners; identification by certified system codes; passwords dependent on the degree of system access; guest entries to the system with limited information exchange (useful for mobile partners).

- Attack avoidance: Security mechanisms in this category assume that an intruder may access the desired resource but the information is modified in a way that it makes it unusable for the attacker. The most important member of this category is cryptography, for instance via (secret) keys, or digital signatures.

8. CONCLUSION

Within this paper we discussed some of the different consequences a vast increase of the number of sensors will have for fieldbus systems and their communication. Although some aspects are far away from realisation, there are special system designs already existing showing parts of it. Especially a cost effective realisation of sensor components can be foreseen in a relatively short time. From the application point of view, systems with a great number of sensors may result in an enormous improve of the overall system capability. However, technical problems must not decrease the system reliability which makes need of a communication concept with appropriate design principle. Thus this paper is a starting point for further studies and a joint project carried out at the authors' institutes.

REFRENCES

Adams, M.D. (2002). Coaxial range measurement – current trends for mobile robotic applications, *IEEE Sensors Journal*, **Vol. 2**, No. 1, pp. 2-13.

Audenaert, K., H. Peremans, Y. Kawahara and J. Van Campenhout (1992). Accurate ranging of multiple objects using ultrasonic sensors, *Proceedings IEEE International Conference on Robotics and Autommation.*, pp.1733-1738.

Brooks, Th. (2001). Wireless technology for industrial sensor and control networks, *Proceedings Sicon/01 - Sensor for Industry Conference*, pp. 73-77.

Ciarletta, L. and A. Dima (2000). A conceptual model for pervasive computing, *Proceedings IEEE Workshop on Parallel Processing*, pp. 9-15.

Dietrich, D. (2000). Evolution potentials for fieldbus systems, *Proceedings IEEE International Workshop on Factory Communication Systems*, pp. 343-350.

Gilsinn, J.D., and K. Lee (2001). Wireless interfaces for IEEE 1451 sensor networks, *Proceedings Sicon/01 - Sensor for Industry Conference*, pp. 45-50.

Haehniche, J. and L. Rauchhaupt (2000). Radio communication in automation systems: the R-fieldbus approach, *Proceedings IEEE Factory Communication Systems*, pp. 319-326.

Helal, S. (2002). Standards for service discovery and delivery, *IEEE Pervasive Computing Journal*, **Vol. 1**, No. 3, pp. 95-100.

Koushanfar, F., M. Potkonjak and A. Sangiovanni-Vincentelli (2002). Fault tolerance techniques for wireless ad hoc sensor networks, *Proceedings IEEE Sensors*, pp. 1491-1496.

Nordman, M.M. and W.E. Kozlowski (2001). Modeling data transactions with standard protocols for low power wireless sensor links, *Proceedings Sicon/01 - Sensor for Industry Conference*, pp. 51-56.

Sanderford, B. (2002). Wireless sensor networks – battery operation of link and sensor, *Proceedings Sicon/02 – Sensors for Industry Conference*, pp. 169-171.

Scherrer, Ch. and H. Schweinzer (2001). Studie – Kostengünstiger elektronischer Heizkostenverteiler mit funkbasierter Verbrauchsdatenübertragung, *Technical Report Institute of Electrical Measurement and Circuit Design*, TU Vienna.

Sedov, I., M. Haase, C. Cap and D. Timmermann (2001). Hardware security concept for spontaneous network integration of mobile devices, *Proceedings Innovative Internet Computing Systems 2001*, pp. 175-182.

Siemens (2003). SOFIS – Automatische Fahrzeug-Identifizierung und Ortung, *Vertriebsrundschreiben*, *Siemens AG*, *Bereich Verkehrstechnik*, Berlin.

Townsend, C.P., M.J. Hamel, P. Sonntag, B. Trutor, J. Galbreath and S.W. Arms (2002). Scaleable wireless web enabled sensor networks, *Proceedings Sicon/02 – Sensors for Industry Conference*, pp. 172-178.

Warneke, B., M. Last, B. Liebowitz and K.S.J. Pister (2001). Smart dust: communicating with a cubic-millimeter computer, *IEEE Computer Journal*, **34**, No. 1, pp. 44-51.

ELSEVIER

IFAC
PUBLICATIONS
www.elsevier.com/locate/ifac

LAST-GENERATION APPLIED ARTIFICIAL INTELLIGENCE FOR ENERGY MANAGEMENT IN BUILDING AUTOMATION

Yoseba K. Penya,

yoseba@ict.tuwien.ac.at

Institute of Computer Technology,
Vienna University of Technology
Gußhausstraß6 27-29, A1040, Vienna, Austria

Abstract: Artificial intelligence has devised solutions for scheduling problems that haven't been already applied to building-automation specific issues and could be beneficial. This paper presents one of these methods, a parallel-genetic algorithm for the optimisation of a multi-objective demand-side management system, and outlines the design of the architecture that supports the use of that algorithm. *Copyright © 2003 IFAC*

Keywords: Energy management systems, genetic algorithms, parallel algorithms, scheduling algorithms, automation, artificial intelligence.

1. INTRODUCTION

Energy management is not a new application of building automation; many research works have investigated different aspects of it before. The use of last-generation artificial-intelligence tools is, however, still not frequent. This paper presents an on-going dissertation that combines the application of one of such tools with the design of an architecture that may support its work. The nature of the algorithm that issues the solution is as critical as the selection of a proper fieldbus system and the design of energy consumers that can apply such algorithm.

This paper is divided as follows: section 2 gives an overview on demand-side management, section 3 is focused on topics of building automation and energy management, section 4 outlines the genetic algorithms, section 5 introduces the architecture designed to support the optimization carried out by the algorithm of section 6. Finally, section 7 describes the scenario and section 8 draws some conclusions and further work.

2. OVERVIEW ON DEMAND-SIDE MANAGEMENT

Demand-Side Management (DSM) is a group of techniques that try to control the energy consumption of a number of devices with principally two aims: avoiding sudden load peaks (Rollet, 1993) and scheduling the energy consumption in order to do it in the cheapest possible time according to the energy tariff (Swisher *et al.*, 1997). Energy consumers may participate in a DSM system in two ways: managing and adapting their behaviour according to the global DSM-optimum plan or at least informing about their future consumption schedule (Penya *et al.*, 2003a). In a distributed DSM environment, all devices broadcast the prognosis about their energy consumption and then, they decide the new plan for each one. The problem is that this process must be done co-ordinately, since one device could decide unilaterally to change its behaviour and if other does so, the system would never reach the optimum. Therefore, they require an algorithm that searches such optimal solution and architecture to support it.

DSM can be modelled as a scheduling problem where the energy is the resource to be shared, the devices are the resource consumers and the aforementioned aims are the objectives to be satisfied. More accurately, it is a multi-objective optimization problem, since there is more than one goal (three, as explained in section 6) to be met simultaneously.

3. BUILDING AUTOMATION AND ENERGY MANAGEMENT

Traditionally, building automation has dealt with energy management and it has issued a big number of solutions. For instance, Palensky *et al.* (1997) present a number of devices connected by means of a fieldbus system and achieving a coarse DSM.

3.1 Fieldbus Systems

Fieldbus systems (or Field Area Networks, FANs) are networks or bus systems normally consisting of a high number of lightweight nodes connected by a small-bandwidth transmission medium. These drawbacks have maintained artificial intelligence far from an intensive use in FANs (Palensky, 1999). Their low cost, however, makes them a popular solution in building and industrial automation.

3.2 DSM-able Energy Consumers

There are two types of DSM-able consumers. In order to discover any other DSM-able consumer and to establish a DSM-process, they must be connected to a common network, usually a fieldbus system. On one hand, the devices that can participate in a DSM system are called active. They are able both to issue a consumption prognosis and also to adopt a new and DSM-optimal consumption plan. The prognosis consists usually on a table that represents the task that are going to need energy, the amount of this energy and the planned time for the task. Additionally, there is a list with the alternatives for tasks that may be postponed or anticipated (actions that give the DSM process its meaning).

On the other hand, informative consumers are not able to adapt their consumption to an optimal plan but they still can issue a prognosis so that active devices may take them into account. Even non-DSM-able devices may be DSM-ificated, as introduced in Penya *et al.*, (2003a) and therefore be included in the DSM calculations.

3.3 Profiling

Profiling avoids one of the biggest problems stemming from plug-and-participate systems: recognising another entity as community (or system) member (Penya *et al.* 2003b) and interacting with it in a proper way. After classifying its interlocutor, the device can decide what actions may be done, what information may be exchanged, etc. Having a number of well-known profiles, each device can declare itself as an instance of a certain profile, facilitating typical ad-hoc networks' mechanisms such as join and discovery, and more important, the interaction between devices (in this case, DSM-able devices). The profiling of energy consumers is not standardised yet. It seems, however, that the trends will follow the footsteps of transducers' profiling, where two models may be distinguished nowadays (Elmenreich 2003):

– OMG Smart Transducer standard (OMG 2003): Each device class has an unique identifier. Information and data of the profile are stored in one server (in the same network or in Internet).

– IEEE 1451 standard (IEEE 1997): Each device has the information of its profile stored locally.

4. ARTIFICIAL INTELLIGENCE AND SCHEDULING

Artificial Intelligence researchers have developed different approaches to cope with the optimization of the scheduling of a certain resource among a number of consumers. One of the results is the evolutionary algorithm family (EA), a group of computational models originally inspired by Darwin's evolution theory (Whitley 1994).

4.1 Genetic Algorithms

Genetic algorithms (GAs) form a subtype of EAs that have evolved into stochastic and heuristic-search methods. Just as each EA, GAs are based on simplifications of natural evolutionary processes, such as selection, survival-of-the-fittest, mating, mutation and extinction. A standard GA works as follows (Abramson *et al.*, 1992): After formulating the problem and modelling it as a group of solutions or genomes (that constitute the search space or population), a GA looks for the fittest one.

The model must consider the constraints of the system, which principally are:

– Hard constraints: They cannot physically be violated, including events that must not overlap in time. For instance, if scheduling John Doe's agenda, John won't be able to be in two places at the same time.

– Soft constraints: Preferences that, thought possible, imply some kind of penalisation. For example, that John Doe has a meeting with his boss on Sunday morning, which is something possible but not preferable because it is during the weekend.

The fitness function (also known as "cost function") weights the solutions according to the goal(s), evaluates penalties of the soft constraints and rules out the solutions that violate any hard constraint.

Starting from (usually) randomly selected genomes, they are mated to form a new solution. The mating process is typically implemented by crossing over genetic material from the parents to create the genetic material of the children. This process is shown in Figure 1.

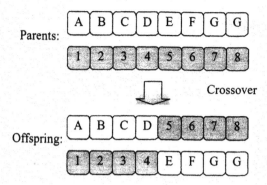

Figure 1: Uniform crossover in a genetic algorithm

Random mutation is applied periodically to promote diversity. If the new solutions are better than parents (the comparison is done using the fitness function), these are replaced and, if the stopping criteria are still not satisfied (so, the new solution is not optimal) the process starts again. Figure 2 illustrates the procedure.

Initialization, mating (normally crossover) and mutation operators are representation specific, whereas selection and replacement are independent. The representation determines the bounds of the search space, but the operators determine how the space can be traversed. Therefore, tailoring the genetic algorithm is critical to its performance (Bartschi Wall, 1996).

4.2 A Multi-Objective and Parallel GAs

Multi-objective scheduling problems must satisfy more than one goal. For instance, having the energy management as an scheduling problem, as already introduced, there are many objectives to be fulfilled, such as: consuming in the cheapest way, as smooth as possible or as soon as possible. Therefore, algorithms that optimise these scheduling problems must take

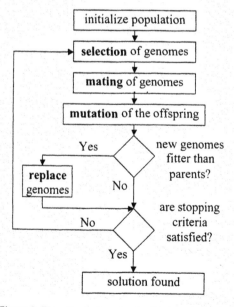

Figure 2: Flowchart of an standard GA

them all into consideration when choosing the possible solutions.

Parallel Genetic Algorithms (PGAs) are simple GAs whose operations are carried out in a distributed way. GAs are especially suited to be parallelized since the genetic operators (mating, selection, mutation) may be applied locally after receiving the individuals to operate with. Thus, they may be executed simultaneously as long as the population remains grouped (Whitley 1994).

A more sophisticated approach, called coarse grained PGA or Island Model (Eichberg *et al.* 1995), divides the population into a few demes, where they evolve separately (usually in different processors). This alternative is normally used for multi-objective optimization, since on each island the selection is made according to different criteria. Migration allows demes to receive the best individuals of the rest and this crossbreeding is supposed to mix good features and enhance the evolution.

Finally, fine-grained PGAs extend the island model to a grid one, where each subpopulation is connected to four others and the mating is only possible between neighbouring individuals. This approach is inspired by the fact that in the nature, the selection is not global but local. Compared to the Island Model, fine-grained PGAs have an easier way to disseminate good solutions across the population but it requires a higher communication effort.

5. ARCHITECTURE

The challenge is to design a plug-and-participate model where DSM regulations are performed by a number of DSM-able devices. Note that the system is distributed per se, since the devices are physically distributed and so is the knowledge as well (LeLann, 1981).

The system is a network composed of high-performance nodes, as presented in Penya *et al.* (2002b): although embedded, enough powerful to host a lightweight agent platform and a number of software agents running on it. The nodes are interconnected by a LONWorks fieldbus system and each one controls one or more devices. Some of the nodes might be not directly controlling a device, but just observing it in order to issue a prognosis and include the device into the system (the so-called Virtual Devices (VD), Penya *et al.*, 2003a).

The devices recognise themselves by using an OMG-like profiling. This allows them to include in the profile not only basic information but also to detail communication interfaces.

Each node hosts a software agent that represents it in the DSM system. Either the device itself or the agent must be able to predict the future behaviour, or at least to estimate it.

Once they have a detailed plan and the possible alternatives within the next 24 hours, they start the DSM process in a parallel way. The process finishes when an optimal solution is found and the DSM-active devices have adopted this optimal plan.

6. ENERGY CONSUMPTION OPTIMISATION

Having depicted here the scheduling of a DSM system as a multi-objective optimization problem, there two principal reasons for using a genetic algorithm:

- First, their success in this kind of problem has already been proved, in comparison to other models. For instance, in Gottlieb (2000), Zitzler (1999) or Zitzler *et al.* (2000).

- Second, GAs can be distributed easily, as explained in section 4.2. We aim to take advantage of having many nodes that could perform a simple task in a coordinated way. Therefore, the nodes of the network may be used in the execution of the GA.

Thus, a parallel genetic algorithm has been chosen for this purpose. PGAs, (and generally, all EAs) bring implicitly, however, a problem: they are blind to the constraints. In other words, the algorithm does not assure that every generation born from feasible parents will be feasible as well (Palensky, 2000). The algorithm has been adapted to the problem, so the fitness function has correctors and is able to evaluate whether new generations preserve the good characteristics of their parents, in order to eliminate them if they don't.

The system is distributed according to the principles of the Island Model PGA. Nodes (this is, consumers) that are physically close form one island. For instance, all the devices from a certain floor or section in a building. Therefore, an island may also be seen as a cluster or a sub-network. Moreover, within the island, each device is considered again to be an island. This approach enhances the scalability of the system.

Furthermore, the migration is done only between neighbouring islands (thus floors) in order to reduce the network load between islands.

The fitness function evaluates each individual according the following objectives:

- As smooth as possible: this objective tries to avoid consumption peaks by spreading it over the time.

- As cheap as possible: the fitness function calculates the prize of the individual according to the tariff.

- As soon as possible: Each task has a preferred execution time and some (or none) alternatives. This objective measures the time from the preferred to the planned execution time.

7. SCENARIO DESCRIPTION

The system issues a 24-hour prognosis. Thus, the consumers plan a one-day consumption. A leader node starts the DSM process and the active and informative devices broadcast part of their prognosis. The first generation of individuals formed with this information are delivered to the islands, so each one gets a part of the search space to explore. The islands start the algorithm themselves and spread the individuals further among their devices. Thus, every device in the system is occupied carrying out the genetic algorithms.

There are some parameters of the algorithm that must be deduced and tuned after the testing of the system, since it depends too much on the implementation. First, the number of times that the devices must broadcast parts of their prognosis in order to form a new generation of individuals (that are going to be mated, mutated, etc.). Furthermore, migrations between islands are done regularly but the frequency of these migrations is also not clear. Finally, the number of iterations that the genetic operators are applied has a very high impact on the network load. On the one hand, there are 24 hours to search for the optimal solution; on the other hand, a quick process may be preferred for whatever reason.

When the iterations planned are finished, the leader device compares all the found solutions and chooses the best one. Then, the active elements of the system adopt the new consumption and wait till the next DSM process is started.

The scope of these models includes mainly buildings, with lightning, air conditioned and heating systems that may be controlled and planned, and factories with similar appliances.

CONCLUSIONS AND FURTHER WORK

This paper presents an on-going dissertation that models the scheduling of a DSM system as a multi-objective optimization problem. Following the guideline of many successful experiments, a domain-adapted parallel genetic algorithm is used to schedule the energy consumption of the devices present in the system. The test-bed is also described, composed of some DSM-able devices connected to a fieldbus system and represented by software agents that synchronously carry out the parallel DSM-solution search algorithm. This algorithm, an Island-Model-based PGA allows to present a structure that can be more easily adapted to the physical situation of the system (floors, sections, sub-networks). It also enhances the scalability

Some of the parameter of the system, such as frequency of the migrations, number of iterations of the scheduling algorithm and number of population creating processes, must be obtained empirically, since they cannot be tuned *a priori* because they depend too much on the implementation.

Finally, there are still some interesting topics to research: as done by Palensky (2000), the comparison with a non-heuristic search algorithm would be fruitful and a the tests in real-life conditions (this is, using real fieldbus systems processors, energy consumers, etc.) should show whether developing such an intelligent-but-complicated system is worthwhile and has practical applications. Similar researchs have shown that the more complicated is the coordinated action of a multi-agent system, the higher is the network load needed (Penya *et al.*, 2002). Therefore, it seems that the scheduling algorithm should be chosen regarding its communication needs as well as its feasibility.

ABBREVIATIONS

DSM	Demand Side Management
EA	Evolutionary Algorithm
FAN	Field Area Network
GA	Genetic Algorithm
IEEE	Institute of Electrical and Electronics Engineers
OMG	Object Management Group
PGA	Parallel Genetic Algorithm

REFERENCES

Abramson, D. and J. Abela (1992). "A parallel genetic algorithm for solving the school-timetabling problem". In Proceedings of the 15th Australian Computer Science Conference.

Bartschi Wall, M. (1996). "A genetic algorithm for resource-constrained scheduling", Dissertation, Massachusetts Institute of Technology.

Eichberg D., U. Kohlmorgen and H. Schmeck (1995). "Feinkörnige parallele Varianten des Insel-Modells Genetischer Algorithmen". In Parallel-Algorithmen, -Rechnerstrukturen und -Systemsoftware, GI Mitteilungen Nr. **14**, Gesellschaft für Informatik e.V.

Elmenreich, W. and S. Pitzek (2003). "Smart Transducers – Principles, Communications, and Configuration". In Proceedings of the 7th IEEE International Conference on Intelligent Engineering Systems (INES), pg. 86. Assiut.

Gottlieb, J. (2000). *Evolutionary Algorithms for Constrained Optimization Problems*, Shaker Verlag, Aachen, ISBN 3-8265-7783-3.

Institute of Electrical and Electronics Engineers, Inc (IEEE 1997). Std 1451.2-1997, Standard for a Smart Transducer Interface for Sensors and Actuators - Transducer to Micro-processor Communication Protocols and Transducer Electronic Data Sheet (TEDS) Formats.

LeLann, G., (1981). "Motivations, objectives and characterization of distributed systems". *Lecture Notes in Computer Science* **105** *Distributed Systems – Architecture and Implementation*, Springer Verlag, Berlin.

Object Management Group (OMG 2002). Smart Transducers Interface Final Adopted Specification. Available at http://www.omg.org as document ptc/2002-10-02.

Palensky, P., D. Dietrich, R. Posta und H. Reiter (1997). „Demand Side Management by Using LonWorks", in Proceedings of the IEEE International Workshop on Factory Communication Systems (WFCS), Barcelona.

Palensky, P., (1999). "The Convergence of Intelligent Software Agents and Field Area Networks". In Proceedings of the IEEE Workshop on Emerging Technologies for Factory Automation (ETFA), Barcelona.

Palensky, P. (2000). "Distributed reactive energy management". Dissertation TU Wien.

Penya, Y.K. and T. Sauter (2002a). "Network Load Imposed by Software Agents in Distributed Plant Automation". In IEEE International Conference on Intelligent Engineering Systems, pp. 339-343, Opatija.

Penya, Y.K., S. Mahlknecht and P. Rössler (2002b). "Lightweight Agent Platform for High-Performance Fieldbus Nodes". In IEEE International Workshop on Factory Communication Systems, Work-in-Progress Proceedings pp. 9-13, Västeras.

Penya, Y.K., P. Palensky and M. Lobashov (2003a). "Requirements and Prospects for Consumers of Electrical Energy regarding Demand Side Management". In Proceedings of the International Conference on Energy Economics (IEWT), pp. 101-102, Vienna.

Penya, Y.K., S. Detter and T. Sauter (2003b). "Plug-and-Participate Functionality for Agent-Enabled Flexible Automation". In Proceedings of the 7th IEEE International Conference on Intelligent Engineering Systems (INES), pg. 136 Assiut.

Rollet, R. (1993). "Analyse der Lastgangskennlinien Wiener Haushalte nach Anwendungen". Diploma thesis, Institut für Energiewirtschaft, TU-Wien.

Swisher, J.N., de Martino Januzzi and R.Y. Redlinger (1997). "Tools and methods for integrated resource planning: improving energy efficiency and protecting the environment". UNEP Collaborating Centre on Energy and Environment/ Riso National Lab. Denmark, ISBN 87-550-2332-0.

Whitley, D. (1994). "A Genetic Algorithm Tutorial". In *Statistics and Computing* Volume **4**.

Zitzler, E., K. Deb, L. Thiele (2000). "Comparison of Multiobjective Evolutionary Algorithms: Empirical Results". In *Evolutionary Computation*, **8(2)**: pp.173-195.

Zitzler, E., (1999). "Evolutionary Algorithms for multiobjective Optimization: Methods and applications". Dissertation, Swiss Federal Institute of Technology Zurich (ETH).

WHAT IT TAKES TO MAKE A REFRIGERATOR SMART – A CASE STUDY

Georg Gaderer, Thilo Sauter, Christian Eckel

Institute of Computer Technology, Technical University Vienna
Gusshausstrasse 27/E384
A-1040 Wien, Austria
{Gaderer, Sauter}@ict.tuwien.ac.at, eckel@agcad.ict.tuwien.ac.at

Abstract: Smart devices have gained increasing importance and are meanwhile also being considered in a residential environment. For the design of such devices, cost efficiency plays an important role, although the actual demand of hard- and software resources is very difficult to determine in advance. This paper analyses the implementation effort needed for an intelligent, networked refrigerator. It gives an overview of the enhanced functionality as well as the overall usage of resources and their distribution among the different tasks. The network selected for the case study was the European Installation Bus. *Copyright © 2003 IFAC*

Keywords: Fieldbus, Connectivity, Control applications, Distributed artificial intelligence, Intelligent control

1. INTRODUCTION

It has become popular in recent years to make formerly dumb and rather simple devices smart and intelligent. Massive use of computing resources in applications that once required no microprocessor at all and a tendency to connect each and every device to a network are characteristic for this trend. What is reasonable in process and building automation (and has ultimately led to the development of various fieldbus systems) is at first sight not useful for home appliances. Indeed the first implementations of Internet Refrigerators and networked washing machines have to be seen more as marketing gags than as serious product developments.

Yet the inclusion of appliances in a probably already existing home network promises some benefits beyond having a touch screen panel in the kitchen enabling the download of recipes from the Internet. Less obvious advantages are ease of maintenance, remote control and maybe even the optimization of energy usage. All these advanced goals require a network infrastructure and consequently more intelligence than a stand-alone device. However, in a mainly academic discussion about the bright future of such smart appliances, the actual implementation efforts are often ignored.

This paper investigates what it really means – mostly in terms of software – to make an appliance smart.

The object of the study is a standard refrigerator that has been taken as starting point (Sauter and Palensky, 2001) and observes the effort necessary to connect this device to the European Installation Bus (EIB). This bus system is mainly used in building automation and therefore an interesting candidate for a network solution.

The remainder of the article is structured as follows: Section 2 starts with a collection of boundary conditions and describes the network components of the system as well as implementation details. Section 3 is concerned with the software, its realization, and a roundup of the resources needed. Finally this paper will outline some ideas for further development.

2. NETWORK COMPONENTS OF A SMART FRIDGE

Smart devices require a lot of additional functionality, and naturally the most interesting steps of the development process are focused on functionality issues and their implementation. However, if the smart device is a mass product like a refrigerator, the designer has to consider some additional design constraints:

- The network node has to be cheap. If possible, it should not interfere with the concept of the standard appliance. Thus only very cost effective, inexpensive components may be used, or the network interface must be designed as an add-on component requiring the least possible additional design effort in the standard (non-networked) version of the hard- and software.

- The assembly of a node has to be as simple as possible. For example it should be possible to program the firmware of the controller without any disassembling the controller hardware.
- Since the maintenance of high-tech, low-cost devices is rather costly, both hard- and software should simplify servicing and failure analysis. This can be achieved by implementation of remote maintenance functionality or at least a diagnosis connector.

Additional considerations especially concerning the fieldbus side are outlined by Teger and Waks (2002). The node representing the actual device is however only one component in the network, in the present case a building automation system. An equally important issue is the ways and means the information generated by the device is conveyed to the outside world, e.g., to the device manufacturer or the user/owner of the device not having a direct connection to the fieldbus. This aspect has also been addressed in the present case study.

2.1 EIB-Based Fridge Node

Basically, there are two possibilities to implement a fieldbus device and application in an EIB network (Sauter et al., 2002). The EIB Bus Coupling Unit (BCU) itself can host – beside the communication stack – an application. This possibility is used for simple devices like switches and is part of the modular hardware concept of EIB, where the BCU is universal and the additional hardware module (the switch, display, serial connector) defines the function of the complete device. The other way is to use a secondary controller. In this case an additional microcontroller implements the actual device functionality,

whereas the BCU acts only as a bus interface taking care of the correct processing of the EIB protocol. The communication between the application controller and the communication stack is done via a serial interface. This communication is standardized and known under the synonyms "Physical External Interface" (PEI) for the physical and EMI ("External Message Interface") for the logical part. For the present case study, the latter kind of implementation was chosen for two good reasons:

- Standard BCUs like the mask versions 1 and 2 offer only the computing power for very simple applications. More powerful designs, like the operating system version 70.x, are available, but due to the low demand in the installation bus domain expensive.
- With respect to the mission-critical tasks of a fridge controller, a malfunction of the EIB network has to be detected by the refrigerator. All network tasks must then be suppressed, and control must switch back into a conventional stand alone mode. It is thus not possible to implement the control application on the bus powered BCU.

The choice of the microcontroller is arbitrary. For the concrete implementation, the C161O was chosen. This low-cost 16-bit controller offers 2 kByte RAM on-chip memory, one serial interface and a bootstrap loader.

In addition to a ROM, which stores the executable code and constant data, a non-volatile memory is needed to store setup information and logging data. A very cheap way to comply with both demands is the usage of a flash memory. The chosen chip contains 128k * 8 bit reprogrammable cells. Using the bootstrap loader the executable code can be programmed into the flash memory without separating the flash memory from the board. Figure 1 shows the main components of the system.

Flash memories are typically used to store executable

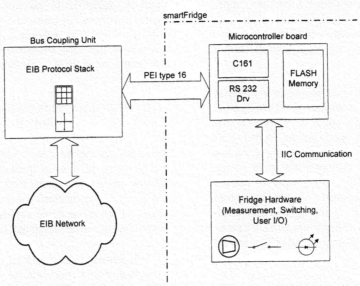

Fig. 1 Principle structure of the SmartFridge and communication with the EIB network.

program code. If it is intended to be used for setup and logging data, two restrictions apply:

- The device used in the implementation is divided into 8 sectors. One characteristic of the chip is, that only whole sectors can be erased in one time. The reprogramming can be done part-wise.
- The number of erase/program cycles is limited. For the chosen chip it is specified with 1 million erase cycles. Considering the life cycle of 25 years and the logging frequency of 1 data set per minute it turns out that sector erase/program approach is not reasonable. This estimation does not consider the setup data which would also reduce the life cycle of the system.

In the actual design this problem is bypassed with the implementation of a small database (Flash DB). Every log data set, setup information, etc. is marked with an additional timestamp and a data type information. Figure 2 illustrates the structure of such a data set.

If information is stored, the controller seeks the next free space in the flash memory. This pointer is handed over to a routine located in RAM, which commits the data to the flash memory. For reading, a subroutine also searches for the next free cell. This pointer is decremented record-set-wise until a data set with the data type of interest is found. When the memory sector of the database is full, the working sector is copied to a backup area in the flash, and afterwards the working sector is deleted. Therefore the working sector is free again to store new data sets. This makes it also easy to obtain historical information from the database simply by causing the search routine not to stop at the first occurrence of the data type of interest.

The problem of the limited programming-cycles is also significantly alleviated. Estimating 4 bytes per data set and assuming that 16 kByte of flash memory is used for the working sector, 8191 data sets can be stored in both sectors. With 1 data set per minute and 1400 data sets per day this leads to a guaranteed life time of the system of 15000 years. Figure 3 shows the described logical organization of the database.

The connection between the microcontroller and the EIB BCU is established via a serial interface, specifically the Physical External Interface (PEI) type 16. There are several PEI types (synchronous, asynchronous, soft- or hardware handshake) available which

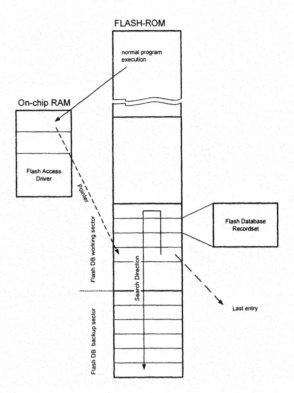

Fig. 3. Memory organization and access to the Flash database

can be configured by means of a resistor. PEI type 16 specifies a 9600 Baud asynchronous communication with hardware handshake (RTS/CTS). On top of the hardware handshake a software handshake is specified by EIBA (1999). Via the EMI (External Message Interface) information is exchanged between the bus coupling unit and the microcontroller.

An additional problem for the communication is the fact that the chosen microcontroller provides only one asynchronous interface which is used for the bootstrap loader. The connection with the BCU would require a second serial interface, which is emulated with the help of the synchronous serial interface. When reading from the serial interface the start of a transmission is detected by an additional hardware interrupt. Subsequently a write-out on the synchronous interface is started. This shifts the serial data in. The delay between the start bit and the first read operation of the shift register has to be 50 % of a bit time. The write operation works alike: to write data out a 9 bit synchronous transfer is initiated. The data, which is shifted in, is discarded, since due to the lack of synchronization reasonable data and a bidirectional communication cannot be guaranteed in this case.

2.2 Gateway

As one of the basic arguments in favor of smart appliances is the ability to allow remote management or maintenance, a typical requirement is an interconnection to an IP-based network (which need not necessarily be the Internet). Fieldbus systems like the EIB however have a limited scope and range. To allow the

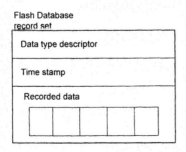

Fig. 2. Organization of a Flash DB record set.

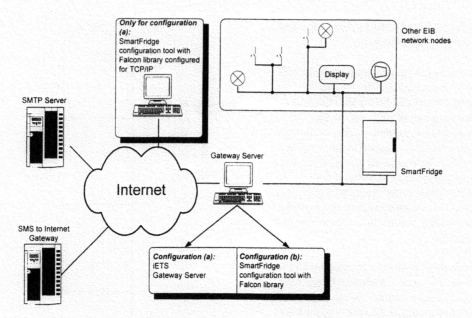

Fig 4. Network structure and interconnection between EIB and the Internet.

smart fridge to communicate with the outside world a gateway application is used. As shown in Figure 4 the intention is to use the communication infrastructure (like e-mail or SMS gateways) from the EIB network. If universality is not a primary goal, the implementation of such gateway applications is not an overly big problem. The main issue is how to access the EIB from the gateway node (which is a standard PC). Two ready-to-use libraries for EIB drivers exist for this purpose:

- *EIB4Linux* is an EIB-to-character device interface working with the versions 1.0 and 2.0 of the BCU. This GPL (General Public License) version is available for the Linux kernels 2.2.x and 2.4.x. Kastner and Thaller (2002) gives a good overview of the functionality. On top of this driver a commercial API (Application Interface) implemented in java is available.
- *Falcon* is a commercial Windows library obtainable directly from the EIB association (EIBA). It also offers a ready-to-use API. A further benefit of this driver suite is the full integration to the iETS (Internet EIB Tool Software). As shown in Figure 4 an application written with the Falcon library can be used anywhere in the Internet. In this case the iETS works as gateway server to the Internet (a). In the other case (b) the application is located directly on the gateway server. Both configurations can be made at run-time and do not need any modifications in the source code. This last argument was the reason to use the Falcon library.

In the application considered here, the main task of the gateway is to allow remote control and send messages via e-mail and SMS. After the reception of the respective data at the gateway, any message can be distributed over the Internet via an ordinary TCP/IP connection to a software enabling remote control and configuration. Further, as a reaction to special messages (like information about broken compressors,

too high temperature in the cooling compartment etc.) a notification can be send via e-mail to a predefined responsible person. Another possibility to notify the user of a critical situation is to generate a Short Message Service (SMS) telegram which is possible by the gateway tool shown in Figure 4.

3. STRUCTURE OF THE NODE SOFTWARE

Obviously the most important task of the node is the control of the cooler and the freezer with their peripherals like the temperature display, beeper and buttons for user I/O. Contrary to the standard cost-efficient hardware design, the smart fridge controller board was redesigned in a modular structure to accommodate additional hardware components (Gaderer and Sauter, 2002). The basic control algorithm of course fulfils the functional specifications and operates exactly like a standard refrigerator is supposed to do. Networking functions are designed as add-ons, so that for example the internal control algorithm can be bypassed by an external one which controls the device

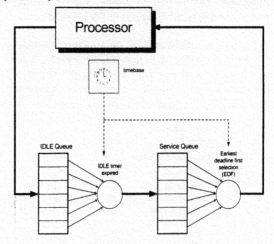

Fig. 5. Scheduling algorithm to coordinate the different task executions.

via the network (Sauter and Palensky, 2001). Apart from the basic control and the network communication facilities, there are additional utility functions. An enhancement of the original specification is the possibility to measure the energy consumption. This allows functions like current monitoring, interior light control and self-test. Special attention has to be paid to fault tolerance. A fault in the EIB network, or any peripheral has to be treated as far as possible as non-critical error and the refrigerator has to continue operation at least in an emergency mode. For remote failure analysis these functions can be triggered via the network. Likewise, status information is regularly broadcasted or can be specifically requested from remote.

For the software implementation, the multitude of individual tasks poses a problem. All tasks have to be executed independently of each other in a (pseudo) parallel way. As it is not practicable to run an operating system on the rather resource-limited controller, a simple form of multitasking has been implemented. As scheduling algorithm a non-pre-emptive timer based algorithm was selected. Two timers called *idle timer* and *service timer* are assigned to every task. After system start the idle timer of every task is started. When this timer expires, the service timer is started. Among all tasks with running service timer the one with the nearest timeout is chosen and the corresponding task is executed. After execution the task returns into the waiting queue with its idle timer restarted and service timer reset to a non-active state. Any expiration of the service timer with its corresponding task in the service queue is an overload for the system and causes a timing violation and an entry in the error log. Figure 5 illustrates this scheduling strategy useful for systems with real-time demands. With the usage of this algorithm the execution of every task within its time slice can be realized or otherwise a violation detected. The execution of each periodic task within his service-window cannot be guaranteed since no assumption about the effective execution times of the tasks can be made. Figure 6 shows the execution profile of a task in this system. Note that under normal circumstances (no timing violation) the execution frequency of a task is defined by the IDLE-Timer and the Service-Timer. The jitter of execution frequency is smaller than the service-Time. A similar implementation is described by Stallings (1998).

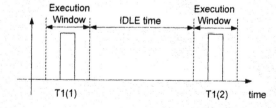

Fig. 6.Exceutuion Profile of a task.

Table 1: Memory usage in the flash memory (each sector is equally 16 kByte large)

Sector	Description
0	Program code, interrupt table
1	Program code
2	Not visible, overlapped with RAM
3	Free
4	Free
5	Flash DB working sector
6	Flash DB backup sector
7	Flash driver routines backup

Although the task of controlling a refrigerator sounds rather simple, the above listing of the various functions needed in this smart device shows that it is very difficult to estimate the effort to implement all tasks. A brief look at the memory map of the flash shows the usage (Table 1).

A special property of the flash memory affects the sector usage. When issuing sector service commands (e.g., sector erase or flash protect or the like) no reading from the flash memory is possible. The consequence is that the opcode of any flash memory manipulation command has to be executed from an other

Table 2: Categorized memory usage of the flash memory

Description	Size (Bytes)	Percent
Control	8774	31 %
EIB	6734	24 %
Utilities	4338	15 %
Drivers	4004	14 %
System	3002	11 %
Flash DB	1226	4 %
Total	**28078**	**100 %**

memory (i.e., the code must reside in the C161 on-chip RAM). At compile time this part is linked to the RAM area. To ensure the availability of the driver routines after a power-down, the very first initialization copies all these routines to a backup sector (sector 7 in Table 1). At each following boot-time the code is copied from the external flash memory to the internal RAM.

A more detailed view of the executable code is shown in Table 2. Every routine is categorized according to its functionality. The category "utilities" summarizes all commonly used routines, like floating point arithmetic, debugging extensions, as well as ANSI-C library functions (printf(), strol(), etc.) The category "system" includes the interrupt procedures (e.g., for the timebase, stack over- and underflow detection or watchdog routines), the task scheduling and all initialization routines. The sizes are taken from a compilation with debugging messages and shrink by about 3% when compilation is done in production mode depending on the number of debug statements.

Table 3: Categorized memory usage (on chip RAM). The size is given in Bytes, the fourth column (2) shows the usage without stack

Description	Size	% Total	% (2)
Control	66	5%	10 %
EIB	193	13%	28 %
Utilities	0	0%	0 %
Drivers	56	4 %	8 %
System	85	6%	12 %
Flash DB	284	20%	42 %
Stack	768	53%	-
Total	**1452**	**100%**	**100%**

Regarding the usage of the C161 RAM, Table 3 gives an overview of the requirements in the current design. In the case of RAM usage, general statements are not that easy as those about ROM utilization, since the stack oriented operation of the processor allows only case studies. The memory space which can be classified contains global variables which are mainly used for buffering and data exchange between the logically independent parts (e.g., temperature and status data between the control algorithm and the EIB Interface or interrupt-filled data queues).

The determination of stack size is very difficult since embedded applications like an intelligent EIB node use interrupt functions (Keil 1996). The working configuration uses a user stack size of 512 bytes, which did not produce an overflow in any test during the development and test phase. This value was determined by lowering the value step-by-step and running the software using a simulation tool in a way that every code line was executed until an overflow-error occurred. This stack size (340 Bytes) was increased for security reasons by approximately 40%. Analogous the same method was applied to the system stack which was set to a size of 256 Bytes.

Tables 2 and 3 can be summarized to Figure 7. It shows a comparison between the two types of usage in percent. A not so fine graded distribution of the memory utilization can be found in Flammini et al. (2002). This article describes the implementation of a web-based sensor and shows the memory usage for such a device.

The used scheduling strategy makes it easy to test the planned timing of the tasks. The complete controller software was stressed with high computing load conditions. This high load was simulated by adding a dummy task which caused a 500 ms waiting time. This task was scheduled every 1000 ms. Apparently all other tasks kept on working without any timing violation. As the obvious first source of failure in these tests it turned out, that the most critical part of the node software is the communication between the C161 and the BCU. Even if little clock differences in the asynchronous communication occur, the BCU resets the communication and restarts in a new try. Due to that, only short frames (only max. 16 bytes instead of the 32 possible) could be transferred. Any error consequently delays the ending of the EIB-communication task. The only way to solve this problem is the implementation of a real bidirectional communication between both devices. Although these timing problems occurred, a reasonable reaction time could be reached. Table 4 shows the scheduler times for each task. The timer settings are defined by the specification of the conventional refrigerator and other system-defined constraints (e.g. the current monitoring time must be at least 10 times smaller than the time constant of the hardware measurement filter). Note that not all tasks may run simultaneously (e.g., current monitoring is only needed if a compressor is switched on; in that case the freezer and cooler control task is temporarily stopped).

4. IDEAS FOR FURTHER IMPROVEMENTS

The first approach for further development can be the improvement of the communication regarding the controller and the BCU. Standard bus coupling units offer many possibilities to do so. Since mask version

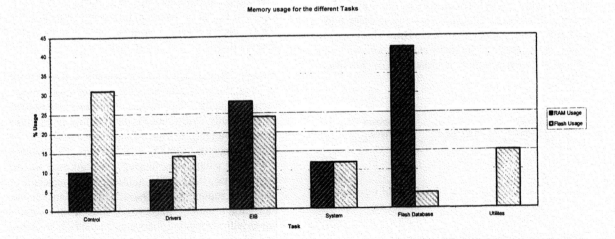

Fig 7. Comparison between flash memory and C161 on-chip RAM usage. The second series is shown without considering the stack.

Table 4: Scheduler settings for the node tasks

Description	idle time	service time
Status broadcast	1 s	50 ms
Temperature logging	5 min	50 ms
Cooler control	1 s	200 ms
Freezer control	1 s	200 ms
EIB service	500 ms	100 ms
Interior light check	200 ms	500 ms
Self check 1	20 min	200 ms
Self check 2	24 hr	500 ms
Display update	500 ms	50 ms
Current monitoring	10 ms	1 ms

2.0 a pure software handshake protocol is supported. The protocol is standardized under the synonym FT1.2 and provides a maximum connection speed of 19200 baud. Currently the EIB twisted pair standard uses a maximum speed of 9600 baud, compared to which the 19200 baud of the controller BCU-connection are sufficient for an operation without bottleneck. Another possibility is the usage of a PEI configuration which enables synchronous data exchange. The currently used software overhead to simulate an asynchronous interface would be eliminated and a real bidirectional communication possible. Also thinkable as a possibility for improvement of this area is the usage of a hardware UART. Although this modification would need an additional component, the distance between the controller and the BCU could be extended by using an UART with an integrated RS232 converter.

On the gateway side the performance of the device driver published by Kastner and Thaller (2002) could be determined. Based on the Java Bus communication architecture described in this paper the gateway side could be rewritten in Java, allowing a machine independent implementation.

5. CONCLUSION

It is certainly worthwhile to make appliances smart. However, this typically requires substantial modifications to both the hard- and software of the original device. How severe these modifications need to be of course depends on how "non-smart" the device was before. It must not be overlook that a simple enhancement of the functionality is normally not enough. Advanced functions like built-in self-test, data logging, error analysis, or remote control facilities are only part of the additional efforts. Communication is another, and the (pseudo) parallelism of all these tasks might necessitate even more sophisticated functions that come already close to implementing an operating system.

In the case of the refrigerator, the actual control task allowing for a stand-alone operation of the device and the add-on functions use less than half the memory. One quarter is needed for the communication with the network, and the rest is system software that is indispensable to make the whole thing work. The use of an inexpensive microcontroller and limitations of the peripheral hardware demand some creativity of the software designer to find solutions that are not straightforward. Examples are the flash database or the scheduler used to avoid a full-fledged operating system. How many percent of the software has to be allocated for system and communication facilities can of course vary grossly. The particular case study was maybe a worst case in that the bus control unit could not be integrated into the refrigerator controller. Therefore a part of the EIB communication had to be implemented also in the appliance controller. In other systems, where a network node has enough resources to host also the refrigerator control application, this additional task might be omitted or at least reduced in size. Such approaches will be investigated in the future.

6. ACKNOWLEDGEMENTS

The authors wish to thank Christian Ferbar for his work on the hardware and Liebherr providing the fridge itself and necessary details about its operation.

REFERENCES

EIBA (issuer) (1999) *EIBA Handbook Series: Volume 3 System Specifications,* EIBA, Brussels Belgium

Flammini, A., Ferrari, P. Sisinni, E. et. al. (2002) Sensor integration in industrial environment from fieldbus to web sensors. *Computer Standards and Interfaces.* Article in press.

Gaderer, G. (2002) In: *Entwicklung einer EIB-Anschaltung für einen intelligenten Kühlschrank,* Master Thesis, Vienna University of Technology, Vienna, Austria

Kastner, W., Thaller, B. (2002). Connecting EIB to Linux and Java. In: *Proceedings of AFRICON 2002* **Vol. 1** pp. 273.

Keil (issuer) (1996) In: *C166 Compiler Optimizing 166/167 C Compiler,* Keil Inc. Plano Texas

Sauter, T. Palensky, P. (2001) The smart Fridge - a networked appliance In: *Proceedings of FET 2001* pp. 161

Sauter, T., Kastner, W., Dietrich, D. (Eds) (2002) In: *EIB Installation Bus System,* Publicis, Heidelberg

Stallings, W. (1998) In: *Operating systems: Internals and Design Principles,* Prentice Hall, New Jersey

Teger, S. Waks, D. (2002) End-User Perspectives on Home Networking. *IEEE Communications Magazine* **Vol. 40,** pp. 114

Voit, R. (1999) *Entwicklung eines Feldbus Interfaces für eine Kühl- Gefrier- Gerätekombination* Master Thesis, Vienna University of Technology, Vienna, Austria

A NEW CAPABILITY FOR FIELDBUS SYSTEMS:
PERCEPTIVE AWARENESS

Clara Tamarit, Dietmar Dietrich and Gerhard Russ

Institute of Computer Technology
Vienna University of Technology, Austria
Gusshausstrasse 27-29 A-1040 Vienna

Abstract: Over the last years, increased efforts have been made in the area of home and building automation. The control systems have become faster and smaller, they are more useable and more efficient. However, they are still based on simple control systems supporting a pure reactive behaviour. In contrast, humans consider the consequences of their actions before embarking upon them. In this paper, extensions for fieldbus systems are described according to this insight. This paper shows how to process the information coming from different sensors to perceive the global environment, to recognise the current situation and to react in a preventive way. *Copyright © 2003 IFAC*

Keywords: Automation, Fieldbus, Situation recognition, Situation-dependent behavior.

1 INTRODUCTION

While the earliest automation systems supported individual and particular uses, new possibilities have emerged with the development of fieldbus technologies. The behaviour of these fieldbus systems is suitable for comparison to some basic actions of biological entities. These systems bear a similarity to works in neurology [Heinze, 1998 and Sacks, 1987]. Taking a step further, and as it was presented in FeT'2001 [Tamarit et al., 2001] the faculty of perceptive awareness (PA) extends automation systems by enabling them to have a measure of preventative behaviour.

Perceptive awareness concerns three mainly functions: situation perception, situation recognition and situation-dependent behaviour. Up to now researches that have applied techniques for situation perception, situation recognition, and for situation-dependent behaviour generally concentrate on the field of robotics [Ota et al., 1996, Haigh and Veloso., 1999, Coenen et al., 2001 and Ueno et al., 1999]. In contrast, this paper presents the integration of these capabilities in the area of home and building automation.

The work is structured in three main parts. The first part is an introduction to perceptive awareness. During the second part the function of perception is attended. The third part concentrates on recognition of the situation and preventive reaction of the automation system.

2 PERCEPTIVE AWARENESS

As function for extending the possibilities of automation systems the concept of perceptive awareness (PA) is presented. Based on this term, the expression perceptive awareness automation system

(PAAS) is defined to refer to automation systems that support the faculty of perceptive awareness.

2.1 WHAT IS PERCEPTIVE AWARENESS

Considering human beings, a preventative reaction takes place once a person is conscious of the current situation [Kandel et al., 2000]. Based on this statement the designation perceptive awareness is born. PA concerns the part of awareness related to data perception and processing, not considering any other of the complicated aspects that consciousness presents in human beings such as feelings [Franklin and Graesser, 1999].

As reference pattern to develop perceptive awareness automation systems a general model, the Perceptive Awareness Model (PAM) has been built as adaptation of biological systems. The first designs of the model and the perception of the environment using different kinds of sensors are already published in [Dietrich, 2000, Dietrich et al., 2001 and Tamarit et al., 2001].

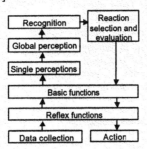

Fig. 1: Perceptive awareness model

Fig. 1 shows the structure of the PAM. The first three lower layers of PAM are defined in compatibility to fieldbus automation systems. These systems react in a predefined way that depends solely on particular inputs. Actions that consist of a direct communication between sensor and actuator are located at layer two – reflex functions, for example smoke detector and acoustic alarm. Actions that depend on more than one input are placed in layer three – basic functions. Usually, these second kind of actions are managed by an additional control unit. The service of a common light controller, which requires the input values of brightness and presence in the room, is an example of basic function.

In both cases automation systems behave without considering any side effect of the actions that they execute. In order to surpass this limitation, the upper layers of PAM are defined based on conscious human behaviour.

2.2 REQUIREMENTS OF PERCEPTIVE AWARENESS

Neural-psychological studies defend that conscious human behaviour entails three main functions [Kandel et al., 2000]:

- Perception of the global current situation

- Recognition of the global current situation

- Selection of the proper response and evaluation of the response before executing it

Since PAM is defined as a tool for developing automation systems that behave preventatively, each one of the functions previously mentioned have to be supported by the model.

3 SITUATION PERCEPTION

Humans make use of five sense organs as sources of perception. Each of these organs captures different kinds of data, which allows the person to perceive the situation from various points of view. Besides this, multi-source perception ensures the validity of the information by reducing the probability of errors concerning collected data [Kandel et al., 2000]. The analogy in automation requires working out different aspects of system's perception that are next exposed.

3.1 SOURCES OF PERCEPTION

Among other things PAM has to enable the easy integration of an extensive amount of different data collectors to get as much information as possible about the environment such as temperature, pressure and distance sensors, ammeter, cameras, microphones, etc. Until now, only a minimum number of sensors have been placed in the field, redundancy and perception from different sources was simply not cost effective. This may change in the next years with the advent of very low cost and low power wireless ad-hoc sensor networks [Sun, 2001 and Rabaey et. Al., 2000], that are capable of organising themselves aggregating data and providing much more reliable information about the environment. These sensor networks can be integrated into fieldbus existing systems [Mahlknecht 2002] and their information can be used as input for the perceptive awareness model.

Similar to the human nervous system, apart from these devices, additional equipment is required to transmit data from the sense organs to the nerves [Sacks, 1987 and Jovanov 1997]. Fieldbus technologies for HBA such as LonWorks and EIB are only a partial solution in most cases. Other higher speed technologies are needed to integrate acoustic and visual perception, e.g. FireWire [Dietrich et al., 2001].

However making use of distinct technologies results in data-format discrepancy, which makes functions such data comparison and data association more difficult. PAM faces this problematic by converting the different data representations to a unified data format.

3.2 FUNCTIONS FOR GLOBAL PERCEPTION

In order to be capable of perceiving the global situation the automation system requires the functions of data comparison and data association [Weinberg and van Wyk, 1997]. Data comparison permits to identify perceptive errors while data association leads to the perception of complexes.

The function of data comparison demands data redundancy, which is supported through the integration of various data collectors that take care of the same variable. These data collectors can either belong to the same technology or to different ones.

Also based on principles from nature, PAM integrates the function of data association. Humans do not only perceive individual data points. Though each sense organ works independently concerning the function of data collection, there is a subsequent and unconscious association of single perceptions that enables the person to perceive entities [Weinberg and van Wyk, 1997].

In PAM referring to data association the terms attribute and object are used in conceptual similitude to they use in object-oriented programming. As attribute i is to understand each one of the variables that the system detects or monitored such as room temperature.

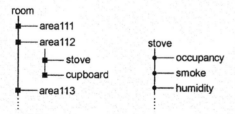

Figure 2 Objects and their hierarchy

Those variables that belong to the same unit are grouped into an object. An object can represent a person as well as an appliance or even an area. As fig. 2 shows, these objects have attributes and can contain several other objects, similar to [Royer, 1995]. The hierarchy of objects and attributes represents a detailed description of the environment including information about locations.

3.3 REQUIREMENTS OF PERCEPTION

In order to enable the automation system to easily execute the functions of data comparison and data association, PAM has to fulfil the following requirements

- enable easily data access, and

- solve the discrepancy of data formats that results from making use of different technologies.

Easy data access entails to place collected data in a unique location, so that the automation system can access data for further processing in an easier way. The integration of an OPC server (OLE for process control) and a proper client application is a possible solution when using technologies such as LonWorks.

comparison level	0	0
adaptation level	interface	interface
access level	interface	OPC
first data presentation	closet_1 open	switch_10
	↑	↑
data collector	camera	switch sensor

Figure 3 Adaptation of data formats to support comparison

As it is shown in fig. 3, above the access level a data adaptation interface level is defined. This adaptation interface level is responsible for the unification of data formats. In order to implement each adaptation interface a detailed study of the format and variable types supported by the correspondent technology is demanded. As result of the adaptation level data referring to the same monitor and/or measured parameter, but coming from different sources of perception, appears in the same format.

Besides the aspects of data access and adaptation the concept of data significance is also considered. Analysing how humans perceive temperature one observes that the sensations that temperature produces in humans are not related to particular temperature grades. In other words, humans recognise levels of temperature but not an absolute value. Since the system lives together with the user and operates for the user's benefit, it is necessary that both system and user perceive events in a similar way.

In order to achieve the exposed target the function of data significance is defined. Furthermore, this function contributes to a better understanding between system and user. Table 1 shows how significance is given to the attribute temperature by means of translating measured values to sensitive values.

In addition to the sensitive value a number is associated to each one of the temperature levels in order to make further data processing easier. Furthermore, a priority label is defined. This label allows the system to pay attention to the most important incidents first, working out the less important ones later.

Table 1 Conversion of temperature values into significant information

Temperature				
range (°C)		meaning	value	priority
from	to			
-50	-5	freezing	1	1
-5	10	cold	2	-
10	20	mild	3	-
20	32	warm	4	-
32	50	hot	5	1

4 SITUATION RECOGNITION AND REACTION HANDLING

The functions of recognition of a global situation and preventive reacting have been designed in PAM on the basis of the following statements:

- A person recognises those situations that he has previously lived and reacts to them depending on his experiences.

- The previously experienced situations and the correspondent reactions are part of the long-term memory of the person [Kandel et al., 2000].

4.1 RECOGNITION

Recognition entails knowledge of previously experienced situations. Consequently an automation system based on PAM requires the knowledge of a certain amount of situations – model scenarios. In this first model these situations have to be predefined and stored. In order to be compatible to the perceived situation the predefined ones are also described through objects and attributes.

In analogy to humans, situation recognition results from a comparison process. The system has to constantly compare the situation that is being perceived with the predefined ones. These predefined situations are part of the long-term memory of the system.

Once the situation is recognised, the next step is to find the appropriate reaction to it. At this starting point of development of PAM reactions have to be also predefined and stored - like the model scenarios. For one specific situation there are several possible reactions. To find the reaction that best fits, the system first simulates them in a virtual manner. This simulation consists on adding the reaction to the current perceived situation (Fig. 4). Again the system will search in the long-term memory for recognition of the virtual situation. The system accepts the selected reaction only if there are no negative side effects, i.e. only if the resulting virtual situation is save.

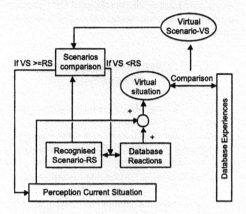

Figure 4 Process of validation of the reaction

4.2 DATA STORAGE AND DATA FLOW

For the storage of information and relations PAM uses different databases. The analogy in biology to these databases are the different kinds of memory supported by humans [Kandel et al., 2000]. PAM requires three databases:

- In moment memory are stored each one of the collected data.

- In short-term memory just significant data for a determined situation are stored.

- In long-term memory predefined information is stored such as model scenarios and reactions.

4.2.1 MOMENT MEMORY AND SHORT-TERM MEMORY

Values of the environment are stored in moment memory and whenever there is a change perceived by any sensor, the value in the database is updated. The use of symbols in the database can considerably improve the performance and usability of the system [Starks et al., 1997 and Kline et al., 1999].

As it is shown in fig.5. PAM incorporates two lists of attributes, one concerning events (event attributes) and one concerning time periods (period attributes). The list of objects is related to the focus list and to the priority list. The focus list is predefined. It contains events of special interest for the user such as smoke detection in a room.

If there is an update in an attribute, it will result in an update of the concerned object. Parallel to that, the system searches in the focus list to see if there is correlation to the current attributes. If so, it means that there is a point of attention and the attribute will be added to the priority list. The list seeks improvement of the performance: The change of an attribute is taken into account at upper layers only if it is added to that list.

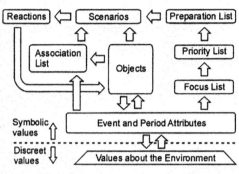

Figure 5 Data flow

4.2.2 LONG-TERM MEMORY

The scenarios and the corresponding reactions represent the long-term memory of the system. In a well-specified application, various scenarios are defined and stored. If there is a new item in the priority list, the system searches for all scenarios using that item and puts them into the preparation list. Thus every scenario in that list presents at least one of the items of the priority list. Consequently the system is "prepared" for recognising the situation by means of searching in the preparation list.

Additionally, the system holds an association list, a kind of dictionary, comparable to biological association cortices [Kandel et at., 2000]. Here numerous terms (including all objects and attributes used by the system) and the semantic relations between them are stored. If there are changes e.g. in objects, but there are no predefined scenarios related to them, the system will search in the association list for related terms to that object.

5 CONCLUSIONS AND FURTHER WORKS

This paper has presented a general framework for the perception and recognition of situations in rooms and buildings and how to behave in front of them. The Perceptive Awareness Model shows a possible way of extending and tuning custom automation systems in buildings. Depending on the definitions of the scenarios, the system can provide improvements in areas like security, safety, energy management or comfort.

PAM has been presented in context of the area of building automation. However, it is applicable in any domain where it is necessary to recognise the current situation and react to it.

This work should not be considered completed. On the contrary it has opened up several new directions for future research. The comparison of predefined scenarios with current objects would be a valuable research area. The exposed method is based on rule-based systems. However, it is possible to apply other techniques for recognising the current situation. Another interesting research direction would be to explore different methods for mapping data values onto symbols. That mapping is a decisive factor for the performance and usability of the system. The functions of learning and forgetting are also interesting for future researches.

Another area of possible research is the building up of a decentralised system by using several PAAS and by connecting them. The idea of the hierarchy of objects should rather be extended in the way that even the whole system could be seen as an object. E.g., one system (a building) contains several other systems (floors), which contain other systems (rooms) and so on.

REFERENCES

Ayddede, M. and Güzeldere, G.: "Consciousess, Intentionality, and intelligence: Some Foundational Issues for Artificial Intelligent" Journal of experimental & Theoretical Artificial Intelligence (JETAI), December 2000, pp. 263-277

Coenen, O., Arnold, M., Sejnowski, T., and Jabri M.: "Parallel Fiber Coding in the Cerebellum for Life-Long Learning", Autonomous Robots, Vol. 11, (2001) 293-299

Dietrich, D.: "Evolution potentials for fieldbus systems", IEEE International Workshop on Factory Communication Systems WFCS 2000; Instituto Superior de Engenharia do Porto, Portugal (2000)

Dietrich, D., Russ, G., Tamarit, C., Koller, G., Ponweiser, W. and Vinczen, M.: "Modellierung des technischen Wahrnehmungsbewusstsein für den Bereich Home Automation" e&I, Vol.11, November 2001, pp. 545-555

Franklin, S. and Graesser, A.: "A Software Agent Model of Consciousness". Consciousness and Cognition 8, pp. 285–305

Haigh, K. and Veloso, M.: "Learning situation-dependent costs: improving planning from probabilistic robot execution", Robotics and Autonomous Systems, Vol. 29, 145-174; 1999

Heinze, H.J.: "Bewußtsein und Gehirn aus Gene, Neurone" Qubits & Co, Gesellschaft deutscher Naturforscher und Ärzte. Hrsg. Detlev Ganten, Berlin 1998

Jovanov, E.: "A Model of Consciousness: an Engineering Aproach" Brain and Consciousness, Proceedings of the ECPD Symposium (Belgrade, Yugoslavia); pp. 291-295; 1997

Kandel, E.R., Schwarz, J.H. and Jessell, T.M.: "Principles of Neural Science", McGraw-Hill, 2000

Kline, K., Gould, L., and Zanevsky, A.Transact-SQL : "Programming. Chapter 20", O'Reilly & Associates; 1999

Mahlknecht, S.: "Virtual Wired Control Networks: A Wireless Approach with Bluetooth", Proc. Of IEEE Africon, pp. 269-272 Vol1, September 2002

Ota, J., Arai, T., Yoshida, E., Kurabayashi, D., and Sasaki, J.: "Motion Skills in Multiple Mobile Robot System", Robotics and Autonomous Systems, Vol. 19, 57-65; 1996

Rabaey, J., M. Ammer, M. J., da Silva, J. L. J., Patel, D., and Roundy, S.: "Picoradio supports ad hoc ultra-low power wireless networking". IEEE Computer Magazine, pages 42--48, July 2000

Royer, V.: "Hierarchical Correspondance between Physical Situations and Action Models", The 1995 International Joint Conference on AI, Quebec, Canada; 1995

Sacks, O.: "The man who mistook his wife for a hat" Summit Books, Simon & Schuster, New York 1987Haakenstad, L.K; 1999

Starks, S. A., Kreinovich, V., and Meystel, A.: "Multi-resolution data processing: It is necessary, It is possible, It is fundamental", Proc. of 1997 Int'l. Conf. on Intelligent Systems and Semiotics, Gaithersburg, MD 145-150: 1997

Sun, JZ.: "Mobile ad hoc networking: an essential technology for pervasive computing" Proceedings. ICII, International Conferences on Info-tech and Info-net, Beijing, vol.3 pg: 316 – 321; 2001

Tamarit, C., Dietrich, D., Dimond, K. and Russ, G.: "A definition and a model of a Perceptive Awarness System (PAS)" FeT'2001, Loria, INPL, Nancy

Ueno, A., Takeda, H., and Nishida, T.: "Cooperation of cognitive learning and behavior learning", In Proceedings of the 1999 IEEE/RSJ International Conference on Intelligent Robots and Systems (IROS 99), Vol. 1 387-392

Weinberg, I.R. and van Wyk, C.P.: "An integrative Neurological Model of Consciousness: the case for quantum-determinism". Transformation Strategies; Articles, 1997

ELSEVIER

IFAC

PUBLICATIONS
www.elsevier.com/locate/ifac

APPLICATION OF DISTRIBUTED EXPERT SYSTEMS TO THE ADVANCED CONTROL OF SMART HOUSES

Juan V. Capella, Alberto Bonastre and Rafael Ors

Department of Computer Engineering
Polytechnical University of Valencia
Camino de Vera, S/N - 46022 Valencia, Spain
{jcapella, bonastre, rors}@disca.upv.es

Abstract: A distributed expert system has been proposed for the smart house control. Its configuration is mainly based on the use of distributed nodes connected by means of a CAN network. This kind of systems allow the implementation of highly flexible control systems capable of adapting themselves to different situations. The expert system used is based on Rule Nets (RN), which are a formalism that seeks to express an automatism in a similar way to as would make it a human being: "IF antecedents THEN consequents". But at the same time Rule nets are a tool for the design, analysis and implementation of rule based systems (RBS), and consist on a mathematic-logical structure which analytically reflects the set of rules that the human expert has designed. Additionally, the RN is able to take decisions concerning possible malfunctions and has bounded response time. *Copyright © 2003 IFAC*

Keywords: Distributed and intelligent control, fieldbus based system, CAN, home automation, artificial intelligence, expert systems, smart house.

1. INTRODUCTION

Control systems are evolving in a quick way. The current challenge in this field is centered in three main aspects: In the first place, it is looked for that the control systems are flexible, programmable, so that they can be adapted easily to the process modifications. This would allow, in turn, the definition of generic modules that could be adapted to concrete problems. In second place, the modern control systems can extend in big surfaces. It seems, because, indispensable, the application of control systems distributed in the space, so that several nodes collaborates to carry out a concrete task, that they are connected by means of industrial local area networks. In third place, a makers tendency to build intelligent and standardized elements is observed (Yen, *et al.*, 1995), so that it is simple its interconnection with other makers elements. In front of these new challenges the traditional control systems should evolve. As answer to the necessity of facilitating the programming, a marked tendency is appreciated to the application of expert systems to the automated control, because it offers some interesting advantages (Gupta and Sinha, 1998). In this field, some of the techniques that probably offer the most interesting perspectives are those coming from Artificial Intelligence, and within them the so-called Expert Systems, which attempt to simulate the reasoning of a human technician. Most of them consist on the denominated rule-based systems,

since they express the desired system behavior by means of simple production rules. In this case we can add the flexibility and easiness of programming, so that an user does not require a solid computer background to specify the desired knowledge base. This is why the expert systems, simulating the knowledge of a human operator, are considered one of the best bets when implementing these control systems.

On the other hand, the most used devices in those systems have an ever-increasing capability of processing data and transmitting them through communication networks. In this way, apart from solving the problem associated to the distance among the system elements, it is also possible to build generic modules that will be assembled to meet specific requirements. It also allows the implementation of fault-tolerant systems by means of replicated nodes.

Apart from the system physical elements being distributed, it is also possible to do the same with the own control system. This permits the configuration of a control system as a distributed one. Then, a distributed system (DS) can be defined as a group of physically distributed nodes (each one of them being an element of the analysis system, e.g. an actuator, sensor or a PC) that collaborate with each other to carry out a common task (the successful system control), communicated by means of a local area network.

The application of distributed control systems to industrial systems has grown in a spectacular way. The advantage of the DS on the networked conventional systems consists on the transparency. This way, the distributed system user should not know the physical elements distribution, but rather interact as if it was a conventional centralized system, being hidden the space distribution of the elements.

As answer to the communications interfaces homogenization problem, diverse focuses have appeared. In this work it has been opted for an abstraction system that represents any device of a system as a group of input and output lines, so much analog as digital. To each one a variable will be assigned, being the value of the variable bound with the physical value of the line.

The confluence of the aforementioned trends has given rise to the distributed expert systems (DES) development as a new technique which, by combining the advantages of both approaches, offers very interesting possibilities in the future for the automated processes monitoring and control (Peris, et al., 1998).

2. DISTRIBUTED EXPERT SYSTEMS

We can define an expert system as a computer program that behaves, as it a human expert would do in certain circumstances (Liebowitz, 1988).

Another focus that we could really call distributed expert system, would be that in the one that the nodes that are part of the system are not limited to the acquisition of data, but rather they participate in the expert system execution. Evidently, this introduces a bigger complexity in the nodes design and in the communication protocols, but it offers significant advantages as the flexibility, the modularity, that allows to enlarge the system in an almost infinite way, and the hardware fault tolerance.

It can be seen how the expert system information it is distributed among the system nodes. Each one executes its corresponding part of the global reasoning, and diffuses the obtained results through the network. These data will be employed for some other nodes to continue its reasoning. This scheme corresponds to a truly distributed expert system.

Once mentioned the advantages of this system, let us pass to enumerate their inconveniences. Besides the biggest complexity in the nodes, already mentioned, serious problems arise, such as the nodes programming. This way, the programming easiness been able when using artificial intelligence techniques.

The true distributed system functionality would be obtained by means of a centralized programming system, where the expert system to use would be defined and tested in a single node. Later on, an automatic distribution of the rules among the nodes that belong to the system is required. This distribution would be carried out through the communications network, so that each node would obtain its part of the expert system to execute. It is possible to conceive several politicians of tasks distribution, as the traffic minimization or the rules redundancy in more than a node to offer fault tolerance.

A system complying with the features described above is currently being developed by the authors, and successful results having already been obtained.

3. RULE NETS THEORY

Basically, Rule Nets (RN) are a symbiosis between Expert Systems (based on rules) and Petri Nets (PN) (Silva, 1985), in such a way that facts resemble places and rules are close to transitions. Similarly, a fact may be true or false; on the other hand, since a place may be marked or not, a rule (like a transition) may be sensitized or not. In the first case, it can be fired, thus changing the state of the system.

Like Petri Nets, RN admit a graphic representation as well as a matricial one; additionally, it also accepts a grammatical notation as a production rule, which drastically simplifies the design, thus avoiding the typical problems associated with PN.

3.1 Introduction to RN

Every control system works with input and output variables, among other internal variables. In the case of RN, each variable has associated facts (corresponding to their possible status). For each associated fact, the set of complementary facts is defined as the one formed by the rest of facts associated with that variable.

Mathematically, a RN is a pair $RN = <F, R>$, where:
F is the set constituted by all the facts associated to all system variables (the ordinal of F will be f)
R is the set of rules. Each rule has one or several antecedents as well as one or several consequent facts (the ordinal of R is assumed to be r)

In order to represent a RN in matricial form, the following structures are defined:
Matrix A of antecedents. It is an $r \cdot f$ matrix where element A_{ij} is 1 if fact f_j is antecedent in rule r_i, or 0 if it is not.
Matrix C of consequents. It is a $r \cdot f$ matrix where element C_{ij} is 1 if fact f_j is consequent in rule r_i, or 0 if it is not.
Matrix D of dismarking. It is a $r \cdot f$ matrix where element D_{ij} is 0 if element C_{ik} is 1 and fact f_j is complementary of fact f_k; if it is not so, D_{ij} is 1.
Vector S of states. It is a vector with f components, so that each component corresponds to a fact of the system. In this vector, component S_i is 1 if fact f_i is true and 0 if it is false.
Rule r_j is sensitized in a determined states S if the two following conditions are met:

$$\bar{S} \text{ AND } A^j = 0 \quad (1)$$
$$\bar{S} \text{ OR } C^j \neq 0 \quad (2)$$

(where from now on X^j will denote the jth row of matrix X).

94

The firing of rule r_j sensitized in the E_m state yields a change in the state vector into a new one, E_n, which may be calculated from the equation (3):

$$\overline{S_n} = (\overline{S_m} \text{ AND } C^j) \text{ OR } D^j \quad (3)$$

It can be easily demonstrated that all operations necessaries to run the RN are implemented by means of easy logical operations. It should also be remarked that there are other figures dealt with in the RN that are useful in the automation of processes, such as timers, etc., as well as the possibility of obtaining the dynamic properties (liveness, cyclicity, etc.) owned by a system.

3.2 Distribution of RN

Given a distributed system formed by n nodes, it is possible to implement a control system through a Distributed RN (DRN), so that each one posses a part of the global reasoning, *i.e.*, a part of the variables and the rules of the system. For this purpose, each node N_i will handle a sub-set of variables and rules of the system, represented by a sub-vector S_i of the states vector S and three sub-matrices A_i, C_i and D_i of matrices A, C and D, respectively.

After defining the set of all system variables, two cases may arise. Some variables will be used in more than one node and others only in one. Those variables required in more than one node are global variables (they will need a global variable identifier), whereas if they only affect one node will be known as local variables.

Obviously, a variable will be local if all the rules referring to it are in the corresponding node, whilst those appearing in rules of different nodes will necessarily be defined as global.

By minimizing the number of global variables, network traffic will be reduced. Besides, the variables referring to a physical input or output will necessarily be located in the node of the corresponding port. Taking this into account, there is an algorithm, explained in (Bonastre, 2001), that permits the optimum assignment of variables and rules to each node (*i.e.* sub-matrices S_i, A_i, C_i and D_i). Optimal versions of this complex algorithm have been implemented successfully. It is possible to distribute the RN applying other criteria, such as fault tolerance approaches that allow several nodes to update any global variable.

Write-Through propagation mechanism guarantees that if the rule net meets a set of properties, once it is distributed and working on a broadcast network, the information will be coherent and the whole system will run in a coordinated way.

4. HOUSE AUTOMATION

One of the lines that are considered more appropriate for the application of the distributed expert systems it is the market of the smart house

control systems. The simplicity required in the used devices, their low cost, flexibility and power, besides the necessity of allowing that non expert programming users could specify the system behavior, make this proposal very interesting in this field.

The proposed system consists on a distributed system placed on a CAN network (Bosch, 1991) to which a group of nodes are connected. One of them is denominated Programming and Supervision Node (PSN). The rest corresponds to the so-called control nodes (CN), which are connected by means of the corresponding inputs and outputs to the process to be controlled.

The implementation of simple generic nodes just as it appears in the figure 1a is proposed. Each node consists of two modules, implemented each one of them in a small card.

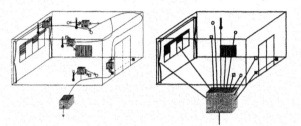

Fig 1. Distributed vs. centralized smart house control

A. The communications module, that consists on a system with a low power microcontroller with CAN (for example the T89C51CC01 of Atmel, the Siemens 515, etc.). It is the one in charge of the communications and the execution of the proposed protocol. This circuit also would contain the CAN driver and the rest of necessary basic elements for the system operation (quartz clock, reset circuit, CAN connector, etc.). This module would exist in all the system control nodes, being able to produce in mass, reducing the production costs considerably.

B. The interface module, with as many variants as different node types required in the system, it would be constituted by the necessary input/output elements and it would provide the interface with the process. It is possible to implement modules with an only one digital input and/or output, even modules that offers a group of several digital and/or analog inputs and/or outputs.

It is sought with this design that the nodes finally result very simple, reducing the functions from the interface module to the strictly necessary ones. For example, a node in charge of a switch control only requires a digital input. Another example constitutes a node for the control of a lamp that contemplates an output with a triac to activate the bulb and a digital input, connected to a light sensor to assure that the lamp has lit. Thanks to this philosophy it is also possible to install simple nodes in the registrations boxes that would control the elements to them connected, as for example appliances, heating, presence sensors, etc.

Given the system characteristics, with each one of these nodes a set of rules is given that allows the standard operation of the same ones, leaving the user the possibility to complete and to particularize the system operation according to its necessities.

For example, two types of nodes are exclusively employees for the implementation of basic illumination systems:

A. Nodes Switch: Allow the user to light or turn off the different lights. It would consist basically of a digital input, with the possibility of adding a digital output to illuminate a position led when the light is turn off.

B. Nodes Lamp: It manages only one bulb. They have a digital output that acts on a TRIAC that illuminates the bulb. Optionally, they can incorporate a digital input that, connected to a current or illumination sensor, allow to detect that the bulb is fused.
The small size of these nodes make possible its incorporation to the same carry-lamp.

For each one of these types of nodes predetermined profiles settle down, defined by a standard set of rules. The user uses this standard set of rules directly, establishing the global system behavior by means of the global variables that export each one of them, or he has the possibility to modify the set of rules to adapt the behavior in a more precise way or even to optimize the set of rules.

The node switch would have a digital input, to which will denominate *position*, with states *A* and *B*. Also, would have a digital output, *neon*, with *on* and *off* values. To communicate each switch node with the rest of the system a global switch variable will be used with two states: *stable* and *modified*. To avoid internal oscillations, the switch node has a internal variable *previous state* with two possible values *A* and *B*.

The node is accessible from the rest of the system through the global variable *switch*. The set of rules that defines the behavior of a switch node would be the following one:

> If *position* is A and *the previous state* is *B*
> then *the previous state* should be *A* and *the switch* should be *modified*.
> If *position* is *B* and *the previous state* is *A*
> then *the previous state* should be *B* and *the switch* should be *modified*.

On the other hand, a lamp node will have an digital output denominated *bulb*, with *on* and *off* values. The *sensor* variable corresponds with the digital input, being able to be in two states: *illuminated* and *not illuminated*.
The *lamp* global variable allows the access to the node functions, with three possible states: *on*, *off* and *fused*. The set of rules for the node would be the following one:

> If *the lamp* is *on* then *the bulb* should be *on*.
> If *the lamp* is *off* then *the bulb* should be *off*.
> If *the bulb* is *on* and *the sensor* is *not illuminated*
> then *the lamp* should be *fused*.

As example of this profiles capacities it is implemented two applications. The first of them consists on a typical switch - lamp system, while the second of them implements a system with a lamp governed from three switches (corresponding to a typical system with two switches and a crossing key).

Application 1: In the system they are only one bulb node and only one switch. Besides the rules foreseen in the previous profiles, it would be necessary to add the following rules:

To light the lamp the following rule will be used:
> If *the switch* is *modified* and *the lamp* is *off*
> then *the lamp* should be *on* and *the switch* should be *stable*.

To turn off it, it is necessary to indicate the following rule:
> If *the switch* is *modified* and *the lamp* is *on*
> then *the lamp* should be *off* and *the switch* should be *stable*.

Finally, to upgrade the global variable neon we will use the value of the variable lamp
> If *the lamp* is *off* then *the neon* should be *on*
> If *the lamp* is *on* then *the neon* should be *off*
> If *the lamp* is *fused* then *the neon* should be *on*

Application 2: We have three switches, denominated switch one, switch two and switch three. The internal variables of each one of them are supposed implemented in the same way. The set of rules that would implement a switch would be the following one:

To light the lamp the following rules will be used:
> If *the switch one* is *modified* and *the lamp* is *off*
> then *the lamp* should be *on* and *the switch one* should be *stable*.
> If *the switch two* is *modified* and *the lamp* is *off*
> then *the lamp* should be *on* and *the switch two* should be *stable*.
> If *the switch three* is *modified* and *the lamp* is *off*
> then *the lamp* should be *on* and *the switch three* should be *stable*.

For turn off, rules contrary to the previous ones will be used:
> If *the switch one* is *modified* and *the lamp* is *on*
> then *the lamp* should be *off* and *the switch one* should be *stable*.
> If *the switch two* is *modified* and *the lamp* is *on*
> then *the lamp* should be *off* and *the switch two* should be *stable*.
> If *the switch three* is *modified* and *the lamp* is *on*
> then *the lamp* should be *off* and *the switch three* should be *stable*.

Finally, the same as in the current case, the global variable neon is upgraded according to the value of the lamp variable:
> If *the lamp* is *off* then *the neon* should be *on*.
> If *the lamp* is *on* then *the neon* should be *off*.
> If *the lamp* is *fused* then *the neon* should be *on*.

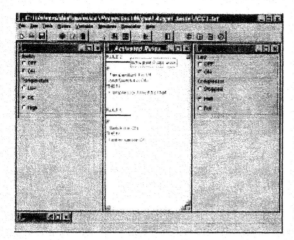

Fig. 2. Rule building in a menu oriented interface

Following this line, the design of a completely automated smart house in all the aspects can be studied and carried out. In the figure 2 can be observed the simple way of specifying the Rule Nets in the developed graphical environment.

Among others, the following aspects, features and variables should be kept in mind. The system should have an illumination control loop. Also, the presence should be detected by means of adequate sensors and optical barriers at doors, for lit or turn off the adequate lights and maintain the optimum habitability conditions together with the heating, conditioned air and fan nodes.

The node that takes charge of the heating can be implemented by means of a CANary processor with a digital output. The node that controls the conditioned air of the room can be also implemented with a CANary processor, using a digital output (ON/OFF) and an another that, by means of a triack, controls the fan speed. Finally, the node that controls external climatic variables is located outside the house, and it is necessary not only for maintain the climatic conditions of the house, but also to decide in function of the sun, wind, etc. when the windows and blinds should be closed or open.

After the design process and the specification of all the necessary rules, the protocol performance has been evaluated and was completely successful, although some design matters of the RN should need further revision. In addition, nowadays new features are being introduced.

The proposed system will execute the following protocol phases (Bonastre, *et al.*, 1999):

1. Initial phase.

The PSN diffuses a RESET message and subsequently a WHO message, to begin the protocol and to discover that nodes are in the network. Each CN responds with an IDENT message for each port (native variable) that possesses, indicating for each one of these like it is: variable type (digital input, analog input, digital output or analog output), the variable resolution and identifier.

To know in that moment all the variables have been received, after a certain prudential time the PSN sends a message with minimum priority (IDENT message), and waits until it has been sent. If this message is sent, it is because the rest of messages of the other nodes has been already sent (they have a bigger priority), and that therefore the PSN has already received all the variables.

2. System calibration.

When everything is already designed, the limit values of the analogical variables should be assigned. This task can be carried out by means of the calibration. The calibration consists on seeing physically the values that are introduced in design time. For example an output calibration consists on write physically the output (WRITE message), and an input calibration is to see in the PSN screen the logical value of the physical input that there is in the system in that moment (READ message).

3. Distribution phase.

It is one of the most important phases, since in this phase the matrices are generated, divided and distributed among the executioners nodes. Also, the result influences in a very decisive way in the response and execution time of the Distributed Rule Net (DRN).

4. Execution phase.

Once arrived to this point, the PSN has already distributed the RN, and the execution begins. This phase begins with a START message, and from this point the controller node can already retire of the protocol and to leave alone the executioners nodes to carry out its task.

This phase is the normal system operation, in which the RN is executed. The CN's begins to read their inputs, to execute their assigned rules and to write their outputs. In case of detecting a change in a global variable, will inform to the rest of nodes by means of a VAL message.

In this stage the PSN can adopt monitoring functions or be removed off the system.
If the PSN remains in the system, fault tolerance mechanisms can be implemented. It can work this way as system watch dog. In this case, PSN would send a WHO message periodically, which would be replied by means of an ALIVE message. If that message does not arrive, and after several tries, the PSN will assume a failure in the corresponding node.

More sophisticated methods are contemplated. PSN knows which rules and variables belong to each node, so PSN can watch the network for variable updates. In this case, the PSN only would transmit a WHO message when no variable of the node has been updated for a fixed amount of time. It is also possible for the PSN to keep a copy of the distributed RN, simulate it and detect if an update that should have been transmitted is lost.

When the PSN notices that a node has fallen, several actions are available. PSN always will inform the user of the failure, but is also able to restart the system with the same RN, or distribute a new RN for degraded working mode.

Also the system has bounded response time. Mathematical form of Rule Net allows the execution time evaluation of a determined RN. In this sense, and taking into account that this execution consists on simple binary operations (AND & OR) between bit matrices (usually implemented by bytes), a worst case (WCET) is defined, thus response time can be bounded.

Communication time on a CAN network can be limited, at the worst-case and in absence of transmission errors, because of the minimum latency between variable updates and deterministic collision resolution.

In this way, this system allows deterministic time responses, making it suitable for a great range of applications.

5. CONCLUSIONS

The application of Distributed Expert Systems to the smart house has been presented. The proposed system incorporates some interesting features as automatic capture of distributed nodes and their characteristics, simple programming by means of production rules and distribution of the control algorithm on the nodes in a user friendly way. Everything has been carried out using generic nodes that can be implemented with low-power microcontrollers, which means a compact, low cost, as well as powerful and flexible system.

Rule Net theory has been utilized to implement the distributed expert system, which allows the easy control of the previously mentioned system, even correcting transient faults. In the case of facing a permanent fault, the crashed subsystem is isolated and the user is notified accordingly.

Given the excellent results obtained, new approaches are being applied. One of them consists on the implementation of algorithms that take into account other user restrictions, more fault tolerance mechanisms by means of redundancy, etc.

Other interesting question to study is the impact of fault tolerance mechanisms (when a fault occurs) in the response time of the proposed system.

6. ACKNOWLEDGMENTS

Authors gratefully acknowledge financial support from the CICYT (Spanish Comisión Interministerial de Ciencia y Tecnología). Research Project No. CYGPL54X.

REFERENCES

Bonastre, A., R. Ors, and M. Peris (1999). "A CAN Application Layer Protocol for the implementation of Distributed Expert Systems" *Proceedings of the 6th International CAN Conference*. Torino (Italy).

Bonastre, A. (2001). *Industrial Local Area Network: A new architecture for control distributed systems*. Ph. D. on Computer Science. Polytechnical University of Valencia.

Bosch, R. (1991). *CAN Specification version 2.0*.

Gupta, M. and N. K. Sinha (1998). "*Intelligent control systems: theory and applications*", IEEE Trans., Piscataway, NJ.

Liebowitz, J. (1988). *Introduction to Expert Systems*, Mitchell Publishing, Inc.

Peris, M., A. Bonastre and R. Ors (1998). "Distributed Expert System for the Monitoring and Control of Chemical Processes" *Laboratory Robotics and Automation*, Volume 10, 163. Elsevier Science Publishers.

Silva, M. (1985). *The Petri Nets in Automation and in Computer science*, Editorial AC, Madrid.

Yen, J., R. Langari and A. Zadeh (1995). "*Industrial Applications of Fuzzy Logic and Intelligent Systems*", IEEE Trans., Piscataway, NJ.

RFieldbus a Survey[1]

Lutz Rauchhaupt

Institut f. Automation und Kommunikation e. V. Magdeburg

Steinfeldstr. 3

D-39179 Barleben, Germany

E-mail: lra@ifak.fhg.de

Abstract

*More flexible systems in manufacturing and process
automation systems require wireless extensions to
existing automation networks. The European RFieldbus
project was launched to develop a wireless solution for
plant floor automation. This paper describes a solution
which combines real-time communication (based on
PROFIBUS-DP) with mobility (based on radio
transmission) and multimedia transmission (based on IP
tunnelling via PROFIBUS-DP). The requirements on
such a system are listed. The industrial environmental
conditions, concerning the radio propagation, are
explained and the relevant radio technologies are
assessed. The system architecture as well as the device
architecture of RFieldbus are presented. The RFieldbus
System has been successfully tested in two pilot
applications for factory and process automation.
Copyright © 2003 IFAC*

1. Introduction

Today, fieldbus systems determine the state-of-the-art
communications infrastructure in manufacturing and
process automation systems. However, driven by the
demand on more flexible and more reliable systems,
further investigations are continuing. Wireless communication is, beyond all question, an appropriate approach
to assure flexibility and mobility in the factory. In particular, the usage of digital radio communication
promises a new quality of industrial automation systems.
Additionally to the extension of well proven real-time
control networks, wireless transmission, combined with
multimedia capabilities, can provide new opportunities
for installation and maintenance purposes e.g. the on-site
supply of documents containing execution instructions.
As another example, photos and industrial-like video
streams may be transmitted to monitor or control production processes. However, the introduction of these
new multimedia services must not impact the real-time
control traffic. This strong requirement, as well as the
special radio propagation conditions of the industrial

environment, complicate or even inhibit the usage of
even highly sophisticated telecommunication solutions
for industrial automation systems. To break the deadlock, a research project was launched with the objective
to provide mobility and multimedia services for real-
time communication systems on the industrial automation field level.

On January 2000, the IST project "High Performance
Wireless Fieldbus In Industrial Related Multi-Media
Environment (R-FIELDBUS)" has been started. The
consortium set oneself the target to develop a high-
performance wireless fieldbus architecture, providing
data rates of up to 2 Mbit/s and response times similar to
wired fieldbus solutions. To achieve this target, new
high-performance radio technologies and existing
industrial communication protocols had to be integrated
to provide a flexible wireless fieldbus architecture. This
architecture must be able to cope with the real-time
requirements of the distributed control data, to support a
user defined quality of service (QoS) concerning
industrial multimedia services, to support mobility of
devices and to support interoperability with existent
communication infrastructures (see also [13]).

The main actions which had to be undertaken for the
design, development, implementation and demonstration
of the proposed R-FIELDBUS system were as follows
(see also [10]):

- Assessment of end-user requirements and
  characterisation of the radio channel for flexible
  industrial automation systems
- Specification and development of the RFieldbus
  System
- Field trials
- Exploitation and dissemination.

The first section of this paper deals with the requirements on an industrial communication system, to be used
on the field level, supporting mobility and multimedia
access. The special environmental conditions of the field
application area are described. The decision concerning
the basic fieldbus system and the radio solution are
discussed. The system architecture is shown and the
extensions to the architecture of the devices are

---

[1] The described work is supported by the European Commission under the project name R-FIELDBUS (IST-1999-11316). The consortium
consists of 4 companies and 3 research institutes located in 4 European countries. These are ifak e. V. (project co-ordinator), Siemens, Softing, ISI,
LPC, ISEP and ST2E. For more detailed information see www.rfieldbus.de.

explained. The main results of the field trials are also briefly addressed.

2. Requirements and Environmental Conditions

The project started with a study of the end-user requirements and the characterisation of the radio transmission channel in industrial environments. For this purpose, a questionnaire has been prepared. Summarising the extraction of the questionnaire responses given by experts in the area of field technology and the results of face-to-face interviews and taking account of the knowledge of today's industrial communication structures, 18 basic requirements, to be met by RFieldbus, have been formulated [11]. The most important requirements are:

- Real-Time communication ($\geq$ 2Mbit/s)
- Integration into existing architecture
- Guarantee that no message is lost
- Mobile applications ($v \leq$ 20km/h)
- Hand-off functionality ($\leq$1 sec)
- Multi-media communication
- Low power solutions
- Mid range area (100m x100m) with $\leq$ 30 devices
- Short range area (10m x 10m) $\leq$ 10 devices
- Remote connection ($\leq$ 1000m)
- Control of harsh noise environment
- Network configuration support

In order to obtain more detailed requirements for the radio interface, the propagation path characteristics (at 1,3 and 2,45 GHz) in different industrial sites were investigated. Propagation measurements focus on identifying both long-term fading and short-term fading parameters that are: average path loss (or signal attenuation), average delay and root mean square delay spread.

Figure 1 illustrates the typical problem in industrial environments. It shows an impulse response diagram recorded in a factory hall (B:16m x L:46m x H:14m). The factory hall contains large metallic machines. The large dimension of the hall and the metallic machines cause reflections, refractions and diffractions, leading to noticeable multipath propagation. Therefore, the impulse response is characterised not only by one small impulse, but by a number of impulses, arriving one after the other (echoes). For practical use, this behaviour means an additional signal attenuation (besides the path loss due to the distance) and a spreading of the originally transmitted impulse (delay spread). This means the propagation conditions in a factory hall are worse than in other indoor areas such as offices or homes, even if there are no disturbances in the hall (see also [3]).

In the given case, the amplitude of the first path (direct path) is even smaller than the amplitudes of the echoes. Investigations of the transmission behaviour of radio modems corresponding to the IEEE 802.11 standard have shown that in such a case the receiver often is unable to synchronise [14]. The data transmission is very unreliable even if there is a line of sight path.

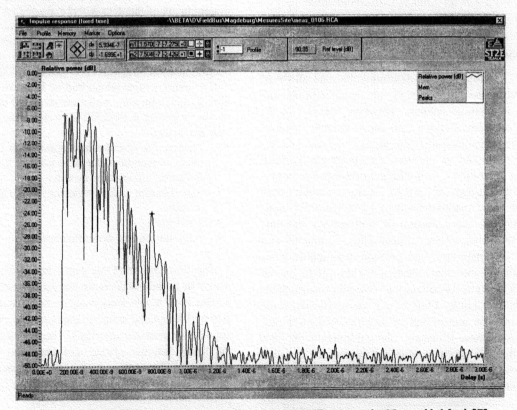

Figure 1: Impulse Response in a Factory Hall (B:16m x L:46m x H:14m) [5]

Taking into account the user requirements and the results of the radio propagation measurements, the following basic radio requirements (for the 2,4GHz band) have been deduced: The radio modem

- has to be able to transmit data with a bit rate of ≥ 2Mbit/s
- has to be able to handle a path loss of ≤ 100dB if a range of ≤70m is required
- has to be able to handle a delay spread of ≤ 200ns

3. Assessment of Radio Technologies

To be able to select a radio technology appropriate to the requirements of the RFieldbus, the following technologies, which seem to have the potential to satisfy some of these requirements, have been assessed: DECT/DPRS, Bluetooth, Hiperlan2, Hiperlan1, UMTS and IEEE 802.11b. The objective of this assessment was to give the basic characteristics for each technology, and briefly present some important features which may be useful for the RFieldbus application. Possible drawbacks were also pointed out [6].

In order to get a feeling of the existing market for the candidate technologies, a thorough investigation was undertaken in the first phase of the project in the year 2000 during which more that 40 manufacturers were contacted. From this effort it was realised that some standards like IEEE 802.11b are mature and have a large variety of products, whereas there are hardly any systems and platforms implementing more advanced technologies like Hiperlan2 and UMTS. This market research gave not only a detailed account on products in different technologies, but also a good overview on the capabilities of the respective wireless systems and the degree to which they can match RFieldbus requirements.

As a result of the technology assessment and the market scan, a comparison table was generated which, thereafter, was used to focus on the three most promising technologies, UMTS, HIPERLAN2 and IEEE 802.11b, mainly because of their superior ability of multipath reception.

A more detailed analysis proved that all three remaining candidate technologies exhibit considerable benefits and advanced radio interface aspects to support high bit rate radio communication in harsh industrial environments. At the same time, all of these standards exhibit excessive MAC-layer overhead (producing transmission delay >10 ms), which is not feasible in real-time RFieldbus applications. As a result, the physical layers Orthogonal Frequency Division Multiplex (OFDM), Wideband Code Division Multiple Access (WCDMA) and Direct Sequence Spread Spectrum (DSSS) were separately examined in order to assess their capacity for the development of the RFieldbus. Table 1 indicates the relative advantages (+) and disadvantages (-) of the three physical layer technologies. The symbol zero (0) indicates a neutral figure. It may be emphasised that this table is a summary of a physical layer (radio technology) assessment, assuming that it is possible to combine the candidates with a specific existing data link layer protocol of the RFieldbus system.

Contemplating the assessment results it has to be taken into account that the table represents a snapshot of the year 2000 when the decisions were made.

Although OFDM appears to be a very attractive technology, it was left out mainly because of its lack of maturity and its high cost which is still true in 2003. Without any doubt, WCDMA is the most powerful technology for the target bit rate of 2 MBit/s and the most promising technology for RFieldbus. However, there are some serious drawbacks such as expected high costs, availability problems and licensing requirements. On the other hand, the MAC-less IEEE 802.11b, using direct sequence spread spectrum, is a mature technology and has also many attractive features that fit well into the RFieldbus requirements.

Measurements with different radio-modems confirmed that DSSS systems with an antimultipath Rake receiver can accommodate for delay spreads and path losses met in the target environments. Whereas, DSSS chip sets commonly running in available IEEE 802.11 implementations are very sensitive to amplitude variation on the paths, and multipath effects, which can decrease its performance [6].

From the present point of view it must be said that the situation has not changed significantly. Even if the emerging radio standards and solutions are more sophisticated today (e.g. IEEE 802.11g or ZigBee) they are still being developed for another application area than the industrial automation. The higher layer protocols do not fit the requirements of industrial field level communication but are closely connected to the hardware implementations. Therefore, separated physical layer hardware which can be used is hardly available.

Table 1: Degree of suitability of candidate radio technologies

	Physical Layer Requirements	Delay Performance	Degree of Matching the Application	Degree of Maturity/ Availability	Licensing/ Frequency Band	Complexity/ Cost
OFDM	Satisfied	+	-	-	++	-
WCDMA	Satisfied	++	++	0	-	-
DSSS	Satisfied	0	++	++	+	++

4. System Architecture

The first activities during the specification of the system architecture concerned the assessment and validation of existing data communication systems to find the basic fieldbus system. From the European perspective, the RFieldbus project focused on the design and implementation of a wireless integrated fieldbus system based on the European Fieldbus Standard CENELEC EN50170 which consists of three parts: PROFIBUS, WorldFIP and P-NET systems.

Therefore, the aim was that the RFieldbus architecture bases on existing data link and application layers of one of these systems, extended by the new radio physical layer and additional necessary protocol functionality (protocol extensions) to support the mobility and multimedia services.

In order to meet the user requirements, the following communication requirements were outlined for the RFieldbus system, considering the necessity to combine wired and wireless segments/domains:

- Ability to support both periodic and sporadic traffic, with bounded response times (control-related traffic)
- Ability to support multimedia traffic with an adjustable Quality-of-Service (varying communication requirements)
- Non-interference between different traffic classes (such as multimedia traffic, control-related traffic or management traffic)
- Adequate data rate to support the envisaged RFieldbus applications
- Ability to easily handle insertion/removal of communication devices

- Integrated error detection/recovery mechanisms to provide the adequate level of reliability to the overall communication platform

After the comparison of the 3 profiles of the EN 50170 standard it became clear that the PROFIBUS profile is the most appropriate basis for the communication infrastructure of the RFieldbus system [7]. PROFIBUS is the world's leading vendor-independent open fieldbus standard for use in manufacturing automation and process control.

After the decision was taken to choose PROFIBUS, different radio related system architectures were discussed within the consortium. It was figured out that an architecture based on gateways between PROFIBUS network(s) and radio network(s), using independent network access schemes for each wireless or wired network, can not fulfil the RFieldbus real-time requirements. Therefore, it was specified that there shall be only one common access control for the combined wired/wireless network, independent of the number of wired or wireless domains. In particular, this means that there is only one token rotating between the master stations, in the complete network.

This approach has without a doubt, an impact on the real-time behaviour of the overall system. For one and the same message, the frame structure as well as the transmission rate is different between a wired and a wireless domain. The frame in a wireless domain is longer because of the additional required synchronisation fields and the necessary header fields. On the other hand, the transmission rate is 2 MBit/s in the wireless domain versa 1,5MBit/s in the PROFIBUS wired domain. Therefore the number of different domains respectively the number of repeaters (wired to wireless and vice versa) influence the overall timing behaviour (see [7] and [15]).

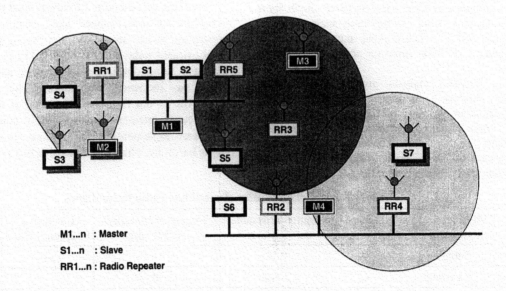

M1...n : Master
S1...n : Slave
RR1...n : Radio Repeater

Figure 2: RFieldbus System Architecture

The specified RFieldbus system architecture in principle allows two basic wireless network types, the direct link network and the base station network. In Figure 2 the master station M2, the slave stations S3 and S4 as well as the radio repeater RR1 belong to a direct link network. This type of network is characterised by:

- No central network element
- Wireless nodes are communicating directly
- Wired segments connected by Link Stations (RR1)
- Distances 30-60m (indoors)
- Single- or multi-cell network
- Intra-cell mobility
- Minimum transmission delay
- The master M3, the slaves S5 and S7 as well as the radio repeaters RR2 to RR5 belong to base station networks. The main characteristics are:Central network element (base station)
- Stations communicate via base station (RR3 and RR4)
- Link base stations (RR2 and RR5) combine base station and link station functionality
- Distances 60-120m (indoors)
- Single- or multi-cell network
- Intra-cell and inter-cell mobility
- Seamless *milli*second hand-off

The main difference in the characteristics of the two types of wireless networks is the type of mobility. Intra-cell mobility means a station must remain within the coverage area of all stations of a network to assure uninterrupted communication. The inter-cell mobility includes the possibility to extent the operational sphere by moving from one radio cell to another. This presupposes a central network element (base station) in each cell, which assures that the moving station may keep in contact to the common network. Moreover, the hand-off from one base station (cell) to another base station (cell) must proceed seamlessly. This means the communication in the common network must not be influenced by the moving station.

A notable advantage of the RFieldbus concept is that original functions of the PROFIBUS protocol are used to implement the hand-off mechanism [19]. A special PROFIBUS master station, called mobility master, is introduced which cyclically transmits one special defined PROFIBUS message, the so called beacon trigger. The beacon trigger period depends on the maximum specified velocity of the mobile stations in the system. In industrial applications this period is in the range of some hundred milliseconds. The beacon trigger forces the base stations to transmit a message. This message is not a PROFIBUS message and will only be transmitted in the wireless domains. The mobile stations use this message to assess the radio signal strength in different cells. The radio channel is switched if the signal of another channel is stronger than the signal of the current channel. This procedure (assessment and switching) lasts only about 1,2 ms [19].

In the following chapter the architecture of the RFieldbus devices is explained.

5. Device Architecture

Two main extensions characterise the RFieldbus device structure in contrast to that of a conventional PROFIBUS device. These extensions regard the new, wireless physical layer and the simultaneously real time control data and multimedia data transmission. In [7] the reasons are explained to use the internet protocol (IP) for multimedia traffic in addition to the real time data of the automation application.

The RFieldbus specification consists of four main documents: the physical layer specification, the data link layer specification, the higher layer specification and the system management specification. Each document is divided in a service part and a protocol part as usual in communication standards. The service part defines the data objects and the interfaces used by other layers of the communication stack. The protocol part defines the protocol data units and the behaviour of the specified part.

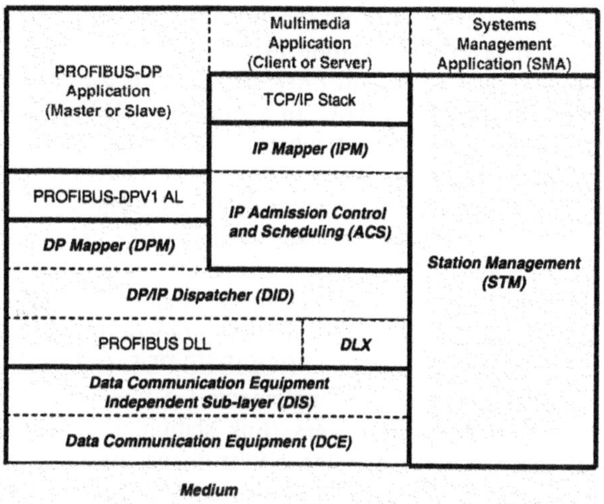

Figure 3: RFieldbus device architecture

The generic device architecture of an RFieldbus device is depicted in Figure 3. The application, consisting of a PROFIBUS-DP part, a multimedia part and a system management part, will be implemented in accordance to a real automation function. The implementations of the PROFIBUS-DPV1 application layer, the TCP/IP stack and the PROFIBUS data link layer as well as the implementations of the wired physical layer (RS485) remain unchanged.

Four modules of the RFieldbus device architecture deal with the **radio** extension of an RFieldbus device: the station management (STM), the data link layer extensions (DLX), the data communication equipment independent sub-layer (DIS) and the data communication equipment (DCE).

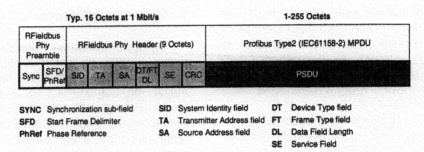

Typ. 16 Octets at 1 Mbit/s		1-255 Octets	

RFieldbus Phy Preamble	RFieldbus Phy Header (9 Octets)	Profibus Type2 (IEC61158-2) MPDU
Sync	SFD/PhRef	SID	TA	SA	DT/FT DL	SE	CRC	PSDU

SYNC	Synchronization sub-field	SID	System Identity field	DT	Device Type field
SFD	Start Frame Delimiter	TA	Transmitter Address field	FT	Frame Type field
PhRef	Phase Reference	SA	Source Address field	DL	Data Field Length
				SE	Service Field

Figure 4: RFieldbus Physical Layer PDU

In fact, the STM provides the configuration services of all newly specified extensions, not only the radio part. It is based on a common management model as described in [12] and supports local configuration as well as remote configuration. In addition, the mobility master functionality is specified as a part of the station management.

The DIS is responsible for the physical layer PDU format. It is derived from the model of the IEEE 802.11 physical layer specification [4]. However, the medium access control (MAC) of *PROFIBUS* [17] is used instead of the IEEE 802.11 MAC. In addition, the IEEE 802.11 PLCP header is eliminated, and the receiver synchronisation is accomplished by detecting one symbol lock in a pseudo noise (PN) code. This PN code is different from the PN code which will be used in the transmission/reception of the rest of the frame. This two PN code operation results in a total of 16 Octets overhead due to the physical layer protocol operation in comparison to the original PROFIBUS PDU (figure 4).

In addition, the DIS implements the functionality of the different repeater types (link station, base station, link base station) and the power management of the radio part. The DCE includes the radio transmission and reception functions in the 2,4GHz band using the direct sequence spread spectrum (DSSS) code and the Rake receiver technology, which is able to deal with the above described multipath propagation.

Four modules are specified in order to support the transmission of **multimedia** data: the IP mapper (IPM), the IP admission control and scheduling (ACS), the data link layer DP mapper (DPM) and DP/IP dispatcher (DID). These modules are responsible that the timing behaviour of the PROFIBUS-DP control traffic is not influenced by the IP traffic.

The IPM is responsible for the conversion of IP packets into/from PROFIBUS data link layer frames. The ACS is responsible for the control/limitation of the network resources used by the TCP/IP applications. This limitation is done by using specific traffic scheduling policies, capable of distinguishing the traffic generated by different TCP/IP applications [16]. In order to integrate the specification into the PROFIBUS application layer standardisation [18], special application layer service elements (ASE) were specified [20]. E.g. the IP mapper ASE is an object specification (figure 5), representing an IP packet and its attributes. The attributes determine the behaviour of the IP transmission

within the PROFIBUS fieldbus application layer (FAL). The ASEs are part of the ACS module.

From the PROFIBUS-DP application layer point of view, the DPM has a behaviour equivalent to the PROFIBUS data link layer [17]. However, it has to provide the quality of service (QoS) of a standard PROFIBUS network, but should also give the possibility to provide a certain band width to the IP high priority traffic. This is done by identifying and classifying the PROFIBUS-DP traffic. The DID provides 5 queues concerning the DP high priority requests, the DP low priority requests, the IP high priority requests (IP QoS traffic), the DP best effort requests and the IP best effort requests. The DID transfers requests from these queues to the PROFIBUS DLL, limited by an overall time.

FAL ASE:			IP Mapper ASE
CLASS:			Simple IP Mapper
CLASS ID:			not used
PARENT CLASS:			TOP
ATTRIBUTES:			
1.	(m)	Key Attribute:	Packet Index
2.	(m)	Attribute:	Data Type
3.	(m)	Attribute:	Packet Length
4.	(m)	Attribute:	State
5.	(m)	Attribute:	Destination DL Address
6.	(m)	Attribute:	Maximum Length Data Unit
7.	(m)	Attribute:	Fragment Number
8.	(m)	Attribute:	Fragment Length
9.	(m)	Attribute:	Packet Position
10.	(m)	Attribute:	Sending Mode
11.	(m)	Attribute:	Sending
12.	(c)	Constraint:	Data Type = Ethernet Packet Type
13.	(c)	Constraint:	Data Type = ARP Packet Type
SERVICES:			
1	(m)	OpsService:	Unconfirmed Packet Delivery

Figure 5: Object Specification of the IP Mapper ASE

Depending on the device type, the above described modules may be implemented or not. In Table 2, the RFieldbus device types are listed and the relevant modules are marked.

Table 2: RFieldbus devices with modules to be implemented

	Link Application	DPV1 AL	DPM/DID	DLL	DLX	IPM/ACS	STM	SMA
Wired master		X		X				
Wired slave		X		X				
Wired multimedia master		X	X	X		X	X	X
Wired multimedia slave		X	X	X		X	X	X
Wireless master		X		X	X		X	X
Wireless slave		X		X	X		X	X
Wireless multimedia master		X	X	X	X		X	X
Wireless multimedia slave		X	X	X	X		X	X
Mobility master					X	X		X	X
Link station	X				X		X	
Base station					X		X	
Link base station	X				X		X	

The mobility master is the device which transmits the beacon trigger telegram in order to force the base stations to transmit beacon telegrams. An application layer implementation is not necessary as the beacon trigger functionality is included in the station management.

The link station is a radio repeater which transparently transfers the telegrams from the wired domain to the wireless domain and vice versa. The link application implements a so called cut through repeater functionality. The link station is not a PROFIBUS device. Besides the link application it contains only the data link layer extensions and the station management in order to configure the device.

The base station is also not a PROFIBUS device. It implements a so called store and forward functionality. This means a telegram is received entirely and is transmitted afterwards. A base station does not have a connection to a wired domain.

The link base station includes both of the above mentioned functions. Therefore, it is also not a PROFIBUS device. The base station is also a store and forward repeater. Basically, the transmission into the wired domain starts before the end of a telegram due to the higher data transmission rate in the wireless domain.

It must be pointed out that all three repeater types (link station, base station, link base station) are based on one basic functionality. That is why there is only one repeater device type necessary in RFieldbus. The different functionality (link station, base station, link base station) is accomplished by implementing/activating different modules in one and the same device.

6. System Evaluation

The RFieldbus prototypes, containing the PROFIBUS, the multimedia and the station management functionality, were implemented based on an available PROFIBUS software [1], [2]. In addition, a radio physical layer hardware, consisting of a radio front-end, a baseband controller and a lower layer service board, were developed representing the radio functionality including hand-off.

After the implementation of RFieldbus prototypes, both field trials and special tests of the communication characteristics were carried out.

The limits of the key parameters as well as the key characteristics of the RFieldbus communication infrastructure were an important input to the preparation of the field trials. Therefore, these values were investigated independently of the applications and before starting the field trials. The basic scenario for these investigations was the communication architecture of one of the field trial applications. This concerns mainly the number and type of RFieldbus device and its location within the communication infrastructure (see Figure 6). In addition, 5 standard PROFIBUS devices were integrated to show the interoperability to devices of the shelf.

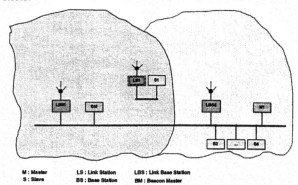

M : Master   LS : Link Station   LBS : Link Base Station
S : Slave   BS : Base Station   BM : Beacon Master

Figure 6: RFieldbus test scenario

Starting from the above shown communication structure, 11 different scenarios were investigated. The scenarios were different concerning the number of segments (wired, wireless), the number and type of network elements (link station, base station, beacon master) and the location of the master and slave devices (stationary or mobile segment). However, the number of

PROFIBUS/RFieldbus devices were the same in each scenario.

The minimum possible cycle times of the PROFIBUS-DP traffic and the Throughput of the IP traffic were the most important timing parameters which were measured. The results can be summarised as follows:

- The influence of the wireless connections to the cycle time of the PROFIBUS-DP traffic is marginal between the best case (about 11.5ms) and the worst case (about 14ms).
- The worst case scenario is if there is more than one master and they are situated in different segments connected by wireless connections.
- The IP traffic has less influence on the average cycle time of the PROFIBUS-DP traffic, but results in greater deviation.
- Simultaneously to the PROFIBUS-DP traffic, a throughput of about 4kByte/s could be achieved for the IP traffic.
- The throughput values of the IP traffic in the direction from Master to Slave is greater than in the direction from Slave to Master. The reason is that the IP traffic of the slave has to be requested by the master. The value of the request cycle time was set to 10ms.

7. Field Trails

The main scope of the field trials was to demonstrate the advanced technical characteristics of the PROFIBUS based radio system, mainly its real-time response, as well as additional operational features and services like wireless communications, mobility, and industrial related multimedia data transmission in the field. RFieldbus pilot application scenarios were defined for both process automation and manufacturing automation [8].

The process automation field trial is the so called warehouse application. It involves warehouse operation, where products are loaded/unloaded by moving forklifts. The goal is to collect data from the moving forklifts (loading-unloading vehicles) to the inventory system. The use of wires is not applicable, because the forklifts continuously move across the warehouse. Each time products are moved in or out of the warehouse the forklift's operator has to inform the person in charge in order to update the inventory database. The forklifts move across corridors and always have to be connected to the inventory system.

In the framework of the RFieldbus project, an operator panel with additional input/output capabilities (barcode scanner, alpha numeric display, numeric keypad) was installed on the moving forklift for the automatic recording of lot movements. The operator panel is equipped with a standard PROFIBUS RS-485 fieldbus interface and can be considered as an RFieldbus

slave device. The wireless connection between a warehouse and forklift operator panel is performed by an RFieldbus Link Station, installed on the forklift as well.

The operator is responsible for data entry using the input features of the operator panel (barcode scanner, numeric keypad) installed on his forklift, which transmit the data via the RFieldbus Link Station through the wireless part of the RFieldbus system. This data is repeated through Link Base Stations installed around the facility, which are connected to the central warehouseman's PC (RFieldbus Master) by the wired RFieldbus part of the network (see also [9]).

The manufacturing automation field trial involves the use of traditional distributed computer control (DCCS) and 'factory-floor-oriented' multimedia (e.g. voice, video) application services, supporting both wired and wireless/mobile communicating nodes (mobile vehicles, for example). It was also a major goal that the manufacturing automation field trial provides a suitable platform for RFieldbus timing (e.g. guaranteeing deadlines for time-critical tasks) and dependability (e.g. reliability) requirements to be tested and assessed.

RFieldbus mobility requirements impose the use of wireless nodes such as transportation vehicles and handheld terminals for supervision and maintenance. The manufacturing automation field trial also involves the use of wired segments, i.e. a hybrid wired/wireless fieldbus network. The interconnection of the different wired and wireless domains is achieved through the use of Link Stations, Base Stations and/or Link Base stations.

One very important issue to be addressed in the manufacturing automation field trial is whether to bring multimedia applications into the factory floor. Applications such as (mobile) on-line help for maintenance purposes and hazardous or inaccessible location monitoring are examples. The manufacturing automation field trial intends to be an adequate test-bed to assess the adequateness of the RFieldbus system to support both real-time control data and multimedia data in the same transmission medium (see also [9]).

The results of the investigations in both field trials can be summarised as follows:

The specified and verified RFieldbus is a communication system for automation purposes based on the most important fieldbus system PROFIBUS. It is a transparent extension of that system and is fully compatible with it. The compatibility is given for all PROFIBUS protocol versions (FMS, DP, DP V1, DP V2) including the safety related PROFIsafe.

The specified RFieldbus radio front end, working in the 2,4GHz ISM band with a maximum transmission power of 100mW, can cover indoor distances between 40m and 100m and outdoor distances > 100m. There are several options specified to extend the area covered by the wireless network, whereby the building multi-cell networks is the most important.

The RFieldbus system enables moving nodes with a speed of up to 20 km/h (6 m/s). The mobility is supported by a special handoff algorithm, based on PROFIBUS protocol elements. This algorithm needs a time slot in the order of some ms.

The PROFIBUS communication protocol, which is the basis of the RFieldbus communication protocol, includes acknowledgement services, which guarantee that requests are repeated if a message cycle could not been finished successfully. So it is guaranteed that no message is lost when a temporary disconnection occurs.

The extension of the PROFIBUS protocol makes it possible to guarantee quality of service to both, PROFIBUS-DP related real-time traffic as well as to multi-media related real-time traffic.

8. Conclusions

Within the R-FIELDBUS project, a number of key issues of an advanced communication infrastructure were addressed. The requirements analysis has shown that there is an interest in services which is provided by RFieldbus. However, the required characteristics of such a system and the special environmental conditions of industrial applications exclude the usage of telecommunication solutions as they are. This was circumstantiated by measurements in industrial environments and by the analysis of available radio solutions. The decision concerning the radio technology ensures a short-term solution which provides the required characteristics and deals with the multipath phenomena. However, the specification of these extensions to the fieldbus standard is open to new radio solutions which will be developed in the future e. g. with other modulation/coding schemes or for higher frequency bands. Furthermore, the specifications are in line with the style of international standards. This opens the opportunity to integrate the specifications in existing fieldbus standards.

The intensive discussions within the consortium lead to a solution which allows adaptable communication architectures: from simple ad-hoc radio networks to complex hybrid (combined wired and wireless) networks with a number of radio cells and an appropriate hand-off mechanism. Remarkable is that the hand-off procedure uses fundamental functions of the basic fieldbus system, PROFIBUS.

Several investigations were made to determine the influence of the system architecture to the real-time behaviour of the communication system. In the same sense, timing analyses were performed to ensure the real-time behaviour of the control traffic and to provide real-time and non-real-time multimedia traffic at the same time.

The field trials demonstrated that RFieldbus systems are successfully applicable in process and factory communication.

The next steps include the optimisation concerning the size, the power consumption and the costs of the radio physical layer implementation. Furthermore, some effort is being spent on developing the acceptance of the RFieldbus solution in the automation society. In Germany for instance a technical working group was founded to bundle all activities in the area of radio based communication in industrial automation. This working group copes with subjects which are related to radio application in automation systems such as licensing and registration, requirements and solutions concerning safety and security, radio transmission in explosive surroundings and the usage of radio frequency bands for industrial applications. The RFieldbus project provided helpful contributions to the work of this working group.

9. References

[1]   J. Hähniche, E. Hintze, M. Langer, A. Pösch-mann „Portierbare Schicht 7 Implementierung für den PROFIBUS als Sensorbus" („Portable layer 7 imple-mentation for PROFIBUS"). *DFAM Research Report* Nr. 2/94, Frankfurt (Germany), 1994.

[2]   P. Deike, J. Hähniche, E. Hintze, A Pöschmann, "Entwicklung einer portierbaren PROFIBUS-DP Proto-kollsoftware" ("Development of a portable PROFIBUS protocol software"). *DFAM Research Report* Nr. 9/96, Frankfurt (Germany), 1996.

[3]   D. Hampicke, A. Richter, A. Schneider, G. Sommerkorn, R. Thomä, U. Trautwein, "Characterization of the Direc-tional Mobile Radio Channel in Industrial Scenarios, Based on Wide-Band Propagation Measurements". in *Proc. IEEE Vehicular Technology Conf.*, Vol. 4, Amster-dam, The Netherlands, Sept. 1999, pp. 3358-2262

[4]   *ISO/IEC 802-11:1999(E):* "Information technology – Telecommunications and information exchange between systems – Local and metropolitan area networks – Specific requirements" – Part 11: Wireless LAN Medium Access Control (MAC) and Physical Layer (PHY) specifications, 1999 edition

[5]   P. Coston, C. Guegan, "Channel Propagation Measurement" in *Magdeburg Site. internal R-FIELDBUS Report*, 26.05.2000

[6]   *R-FIELDBUS Deliverable* D1.2: "Assessment and selec-tion of the radio technology". 26.9.2000

[7]   *R-FIELDBUS Deliverable* D1.3: "General System Architecture of the Rfieldbus". 26.09.2000

[8]   *R-FIELDBUS Deliverable* D5.1: "Specification of pilot applications and field trials". 08.03.2002

[9]   R-FIELDBUS Deliverable D5.2: "Report on field trials". 10.03.2003

[10] J. Haehniche, L. Rauchhaupt, "Radio Communication in Automation Systems: the R-Fieldbus Approach", in *Proc. of the 2000 IEEE International Workshop on Factory Communication Systems*, Porto (Portugal), pp. 319-326, September 2000

[11] *R-FIELDBUS Deliverable* D1.1: "Requirements for the R-FIELDBUS system". 24.11.2000

[12] A. Di Stefano, L. Lo Bello, T. Bangemann, "Harmonized and Consistent Data Management in Distributed Automation Systems: the NOAH Approach". Pueblo (Mexico), 06.12.2000. Proceedings.

[13] J. Hähniche, G. Hammer, R. Heidel, „Funk statt Draht - Kabellose Kommunikation auch in der Feldtechnik?"

("Radio instead of wire - wireless communication in the field automation?") *atp* 6/2001/pages 28-32; ISSN 0178-2320 Oldenburg

[14] L. Rauchhaupt, H. Adamczyk, "WLAN Systeme in industrieller Umgebung - Funktioniert´s? (WLAN systems in industrial environments - does it work?)". Mobile Communication over wireless LAN: *Research and applications* Wien (Austria); 26.09.2001; Vol 1, pp 530-536

[15] M. Alves, E. Tovar, F. Vasques, "On the Adaptation of Broadcast Transactions in Token-Passing Fieldbus Networks with Heterogeneous Transmission Media", 4th *IFAC Conference on Fieldbus Technology* (FET'01), Nancy (France), November 15-16, 2001

[16] F. Pacheco, E. Tovar, A. Kalogeras, N. Pereira, "Supporting Internet Protocols in Master-Slave Fieldbus Networks". 4th *IFAC Conference on Fieldbus Technology* (FET'01); Nancy, France; November 15-16, 2001; 278-284

[17] *IEC 61158*: "Digital data communication for measurement and control - Fieldbus for use in industrial control systems" Part 3 (Data link service definition) and Part 4 (Data link protocol specification), IS, IEC 2001

[18] *IEC 61158*: "Digital data communication for measurement and control - Fieldbus for use in industrial control systems" Part 5 (Application layer service definition) and Part 6 (Application layer protocol specification), IS, IEC 2001

[19] M. Alves, E. Tovar, F. Vasques, G. Hammer, K. Röther, "Mobility Management in Hybrid Wired/Wireless Token-Passing Fieldbus Networks". *2nd IFIP Networking* (Networking 2002), May, 2002, Pisa (Italy).

[20] P. Krogel, A. Pöschmann, L. Rauchhaupt, "Open Internet Protocol Fieldbus System Tunneling Specification". Addendum to "Offene Internet-API unter Berücksichtigung der Echtzeitbedingungen in der Feldtechnik (Open Internet API considering the real-time conditions of field devices)". *DFAM Research Report* Nr. 18/2002, Frankfurt (Germany), 2002

ELSEVIER

IFAC

PUBLICATIONS

www.elsevier.com/locate/ifac

Simulation of RFieldbus Networks

Elke Hintze [1], Pavel Kucera [2]

[1] Institut f. Automation und Kommunikation
e. V. Magdeburg (ifak)

Steinfeldstr. 3
D-39179 Barleben, Germany
E-mail: ehi@ifak.fhg.de

[2] Brno University of Technology
Faculty of Electrical Engineering and Communication
Department of Control and Instrumentation
Bozetechova 2
612 66 Brno, Czech Republic
E-mail: kucera@feec.vutbr.cz

Abstract

Driven by the demand on more flexibility in manufacturing and process automation systems, investigations are made in the field of wireless communication. In the R-FIELDBUS project a new wireless physical layer type was added to an existing fieldbus system. The resulting R-FIELDBUS systems allow adaptive architectures with mixed wired and wireless segments. To be able to analyse the functional and time behaviour of different configurations, a simulation model of the fieldbus system was also extended by the wireless behaviour.

This paper describes the used simulation method, the models of the communication protocol as well as the new added wireless medium type as well as providing some first simulation results. Copyright © 2003 IFAC

1. Motivation

Industrial communication systems play a decisive role in many application areas of automation. Regarding their functionalities and real time capabilities, high demands are made. Driven by the demand on more flexible and more reliable systems, investigations are made to improve the existing fieldbus solutions. One aspect is the use of wireless communication techniques to assure flexibility and mobility in the factory.

In the scope of the IST project "High Performance Wireless Fieldbus In Industrial Related Multi-Media Environment (R-FIELDBUS)" [1] the technology for a radio based fieldbus system was specified and prototypically implemented. RFieldbus is based on the PROFIBUS protocol, which was extended by a new wireless medium type as well as the tunnelling of the IP protocol for transmission of multimedia data. The wireless medium type is marked by

- Data rates of up to 2 MBit/s
- Response times similar to wired fieldbus solutions
- Support of mobility of devices
- Interoperability with existent fieldbus architecture

Figure 1 shows the system architecture of an RFieldbus network.

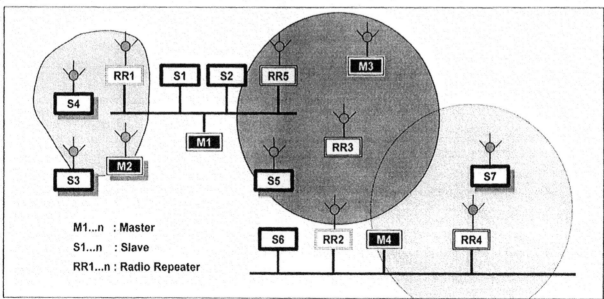

M1...n : Master
S1...n : Slave
RR1...n : Radio Repeater

Figure 1   RFieldbus System Architecture

The RFieldbus system architecture [12] allows heterogeneous networks with wired and wireless segments. The media access covers the network i.e., only one rotating token is in the system, independent of the number of wired and wireless segments. There are two wireless network types: Direct Link and Base Station Networks.

In Figure 1, the master station M2, the slave stations S3 and S4 as well as the radio repeater RR1 belong to a Direct Link Network. This network has no central network element i.e., the wireless stations communicate directly with each other. The interconnection with a wired segment is done via a Link Station (RR1).

The master station M3, the slaves stations S5 and S7 as well as the radio repeaters RR2 to RR5 belong to a Base Station Network. Each cell of this network has a central network element i.e., the wireless stations communicate via these base stations (RR3). Link Base Stations (RR4) combine base station and link station functionality.

The main distinction between both network types is the type of mobility. In Base Station Networks the stations can move between the radio cells without interruption of communication because the Base Stations assure a so-called Seamless Hand-off.

The operability of RFieldbus systems was verified in the project by prototype implementations and field trials. Time measuring was done to determine first properties. However, not all wireless network topologies and parameterisation scenarios can be built with the prototypes to investigate the resulting protocol and time behaviour (achievable cycle times, hand-off times for n radio cells, etc.). Furthermore, there is a big complexity in such networks if wired and wireless segments are combined using the whole spectrum of PROFIBUS functionality. Such systems have the following different characteristics in the individual segments as well as different processing times in the segment couplers:

- Telegram structures
- Media access controls (token)
- Data transfer rates
- Communication cycles
- Application cycles

Thereby the network segments may be unsynchronised. This can cause the following:

- Unpredictable time behaviour
- Poor system throughput because of untuned throughput in the individual segments
- Misbehaviour because of inconsistent datasets

Simulating the new designed communication functions is a method to get information about the system behaviour and achievable reactions times in an early phase. In this paper, the on-going work to simulate RFieldbus networks is presented. Section 2 provides a brief presentation of the simulation method. Section 3 and 4 present the simulation model and first simulation results. In section 5 future work is discussed.

2. Simulation Method

Basis for the specification of simulation models are Formal Description Techniques (FDT). They are used to describe structure and behaviour of transmission systems. The use of FDT offers various advantages e.g.,

- Gathering of complete and consistent requirements by means of suitable abstractions, structuring concepts as well as construction and analysis tools
- Possibility of tool-supported verification/validation of the specified behaviour already in early design phases (Rapid Prototyping)
- Possibility of code generation for prototypes and implementations
- Precondition for evidence of correctness and quality-oriented development methodology
- Improvement of documentation

There are a lot of different FDT's and simulation tools. To choose the best suited one is difficult because of complexity of communication systems and variety of sub functions that can be modelled in different FDT's with different comfort. For example, transmission media and medium-related sub layers that usually are realised in hardware can be better modelled in FDT's with synchronous events (e.g. VHDL). On the other hand application-related higher protocol layers that usually are realised in software need FDT's with asynchronous communication mechanisms and queues (e.g. ESTELLE, SDL). Furthermore, big differences can be found in distribution, tool support for design, verification and validation, simulation and analysis as soon as availability of model libraries. And finally models of communication protocols and applications in different FDT's already exist and shall be reused if possible.

For the simulation of complex, heterogeneous communication networks a method was developed at ifak [2] that considers these demands and allows the combination of models from different FDT's. It uses existing tools and technologies like code generators (e.g. for C code) and foreign language interface. Therefore, each new sub model can be specified in the best-suited language and existing sub models can be reused.

2.1. System model und tool chain

VHDL [4] and ESTELLE [6] were used to model the RFieldbus system. A certain system configuration can be built in VHDL by instantiation and composition of the sub models for applications, communication protocols and network components like media and couplers. New VHDL models were specified for the network components; existing ESTELLE models of the DP protocol were integrated via the foreign language interface of the used VHDL simulator. Telegram logging functions according to industrial bus monitors were implemented in C and integrated in the VHDL simulator too. Figure 2 shows the principle of model composition.

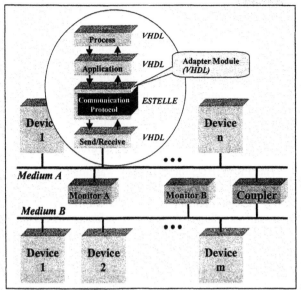

Figure 2   Modelling of a communication system with VHDL and ESTELLE

Figure 3 shows the tool chain. The NBS prototype compiler [10] is used to generate C code from ESTELLE models. This code is compiled and linked together with the execution framework and supplied as a model library. ModelSim [9] is used as the simulation environment. It allows the effective simulation of large systems. The model libraries can be integrated into ModelSim via its Foreign Language Interface.

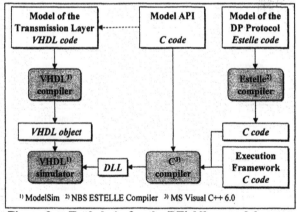

¹⁾ ModelSim ²⁾ NBS ESTELLE Compiler ³⁾ MS Visual C++ 6.0

Figure 3   Tool chain for the RFieldbus models

The execution framework of the ESTELLE compiler was modified for interconnection with ModelSim because the superior simulator must control instantiation of models, activation of model instances as soon as progress of simulation time. Each generated C model will be supplemented with a uniform function interface for interactions with the simulator. The integration in the system model is done via a VHDL adapter module that activates the interface functions of the integrated model.

2.2.   Parameterisation of the model instances

The parameterisation of the communication protocol is done by the applications via management services. For

PROFIBUS DP such a start-up contains e.g. loading of bus parameters, loading of parameter and configuration data for each slave, initialisation of process and diagnostic data as well as starting of the communication cycle. The parametering potentialities of the models should not be restricted but the specification of very simple application models should be possible. That is why a default parameterisation via parameter file was integrated in the model API. Figure 4 shows an example.

```
[M10]
;Master with FDL address 10
FDL_Add=10
EST_DLL=DpDll
Debug=6
;Communication Parameter:
Baud_Rate=6      ;1,5 MBit/s
Tsl=5000         ;[Bittime]
Min_Tsdr=1000 ;[Bittime]
Max_Tsdr=2000 ;[Bittime]
Tqui=0           ;[Bittime]
Tset=1           ;[Bittime]
Ttr=50000        ;[Bittime]
G=10
HSA=20
Max_retry_limit=1
Bp_Flag=0x00
Min_Slave_Interval=100  ;[100µs]
Poll_Timeout=60000      ;[1ms]
Data_Control_Time=100   ;[10ms]
;Slaves that shall be controlled
NoOfSlaves=3
Slaves=S20,S30,S40
;Application Parameter:
...

[S20]
;Slave with FDL address 20
FDL_Add=20
EST_DLL=DpDll
Debug=4
;Communication Parameter:
Baud_Rate=6             ;1,5 MBit/s
Default_Min_Tsdr=11 ;[Bittime]
Ident_Number=0x0000
No_Add_Chg=1
Sync_Supported=1
Freeze_Supported=1
Sl_Flag=0x80
Sl_Type=0x00
Prm_Data_Len=7
Prm_Data=0x80,0x00,0x00,0x37,0x00,0x00,0x00
Cfg_Data_Len=1
Cfg_Data=0x31 ;2 Byte Input, 2 Byte Output
Ext_Diag=0
Stat_Diag=0
Ext_Diag_Overflow=0
Ext_Diag_Data_Len=0
Ext_Diag_Data=
Outp_Data_Len=2
Outp_Data=22,22
Inp_Data_Len=2
Inp_Data=22,22
;Application Parameter:
...
```

Figure 4   Example for a parameter file

The init function of a model gets the parameter file and an instance identifier. It generates an input event with the related management service request for each parameter that can be found in the instance section. Therefore it is not necessary to model the start-up e.g. for investigations of the stationary behaviour. On the other hand, the management services can be used additionally in complex applications e.g. to investigate run up, shut down or switching behaviour. In future it shall also be possible to use the output of an industrial configuration tool instead of the shown file format.

2.3. Representation of simulation results

Besides the methods of the simulation tool event tracks shall be logged at several points of observation. The points of observations were chosen in such a way that they are also observable in real systems. Therefore a comparison between simulated and real behaviour can be done. The measurements in real systems can record the telegram traffic on the bus as well as reaction times on application level. That is why during the simulation, log files of telegrams (bus interface) as well as services and process data (application interface) will be generated.

The bus log files can be transferred into a special format to use industrial bus analysers (bus monitor) for representation and filtering of simulated telegram sequences. This increases the acceptance and practicalness in companies. Furthermore, analysis routines shall be implemented which calculate reaction times via tracking of specific data pattern.

3. Simulation model

The simulation model is divided into sub-models for the network components (bus media, segment couplers), for the communication protocols (DP master, DP slave) and for applications. The whole model of a certain network configuration can be built by instantiating and parametering of this "component models".

3.1. Model of the PROFIBUS DP protocol

The PROFIBUS protocol [3] was modelled in the FDT ESTELLE [6] that was developed by ISO especially for communication protocols with layer architecture corresponding to the ISO/OSI-reference model [5].

ESTELLE is based on asynchronous message exchange between hierarchical state machines (modules). The messages will be saved in FIFO's at the module interfaces (interaction points) until they can be processed. The behaviour specification of a module consists of a set of transitions. Execution of a transition depends on input event and internal state of the module and can include a state change, manipulation of module variables as well as sending of events to other modules.

Many fieldbus protocols like PROFIBUS are written in an ESTELLE-like semantic. Therefore the modelling can be concentrated mainly to transfer it into ESTELLE syntax (Figure 5); this can partly be automated.

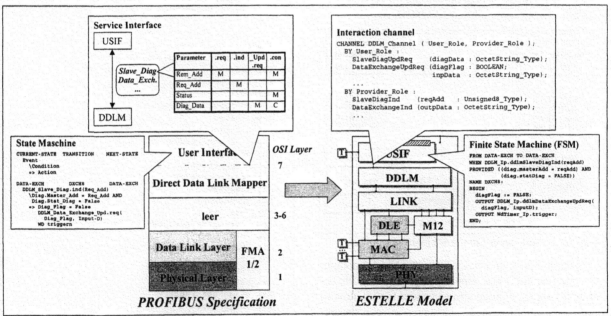

Figure 5   Modelling of the PROFIBUS protocol with ESTELLE

For PROFIBUS detailed ESTELLE models of the device types **DP Master (class 1)** and **DP slave** were created that allow e.g. generation of bus monitor files or control of a real DP slave device [11]. This model of the DPV0 protocol has about 28 000 ESTELLE code lines and allows not only simulation of stationary behaviour but also start-up, error and switching scenarios.

The NBS prototype compiler [10] was used to generate a model library in C. This code generator supports the whole language range of ESTELLE, especially for model structuring (dynamic structure, parallel and sequential modules), and generates C code in the K&R style which assures good portability. The source code of the DPV0 protocol has about 96 000 C code lines; additionally there are about 7 000 C code lines for the modified execution framework, interface implementation and default parameterisation. The model library for the VHDL simulator was built with Microsoft Visual C++ 6.0 as a DLL.

3.2. Model of the Physical Layer

The Physical Layer was modelled in VHDL [4]. VHDL is a widespread standard for the (hardware) design of digital systems. It is based on the structuring of models into hierarchical units that are communicating via signals. Concurrent signal assignment and process statements are used for the behaviour description. Several delaying models and signal attributes allow detailed and comfortable modelling of time behaviour. That is why VHDL is well suited for the modelling of hardware related units of a communication device (e.g. sender, receiver) as well as the bus medium.

For the Physical Layer (i.e. first layer of the ISO/OSI model) two medium types have been modelled: wired medium (asynchronous transmission according to TIA/EIA RS-485-A) and wireless medium (RFieldbus).

3.2.1. Wired Medium (RS-485)

Model of the communication bus is based on Recommended Standard 485 [7]. The following implementation simplifications have been realised compared to the above-mentioned standard:

- Signal on the bus is modelled as enumerate type with the four possible states (see below for details)
- Signal on the bus has infinite propagation speed, i.e. all communication partners detect on the same bus segment the identical information at the same time and no matter how long the bus is
- Mechanical parameters (wire length, wire diameter, ...) or electrical parameters (resistance, coupling capacity, ...) are not considered

Basic user-defined VHDL type that models common transfer medium (RS-485) is TxDRxD. All wired communication devices are connected together through this medium. It is based on user-defined type uart_level:

```
    type uart_level is (SPACE, MARK, EMI_SPACE,
EMI_MARK);
```

where

```
SPACE      = VOH - log. 1 (default state)
MARK       = VOL - log. 0
EMI_SPACE  = Electromagnetic Interference SPACE
EMI_MARK   = Electromagnetic Interference MARK
```

Defined states SPACE and MARK correspond to the states defined in Recommended Standard for asynchronous serial communication i.e., SPACE is idle state and MARK active state on the bus. Beside it, two other states for electromagnetic interference are defined: EMI_SPACE and EMI_MARK.

The type allows modelling 'Read' and 'Write' access to the bus for each of the communication devices. Since VHDL normally allows only one driver for a signal, it is necessary to use resolved types for signals based on uart_level type. A resolved type is based on a resolution function that uses defined access right to determine the final signal value. In our case, the resolution function is implemented as follows:

```
function resolve_uart (drivers : in uart_array)
return uart_level is
variable ret: uart_level;
begin
   ret := SPACE;
   for index in drivers'range loop
      if drivers(index) = EMI_MARK then
         return MARK;
      end if;
      if drivers(index) = EMI_SPACE then
         return SPACE;
      end if;
      if drivers(index) = MARK then
         ret := MARK;
      end if;
   end loop;
return ret;
end resolve_uart;
```

where

```
type uart_array is array (integer range <>) of
   uart_level;
function resolve_uart(drivers: in uart_array)
   return uart_level;
subtype TResolvedUARTLevel is resolve_uart
   uart_level;
```

Resolution function simply scans all possible TxDRxD signals in the model and returns the first appearance of the state EMI_MARK or EMI_SPACE or MARK in this order. If there is no such appearance on the bus (i.e. no communication device writes data), then function returns SPACE state. It is evident that EMI event has the highest priority and in every case it overrides any of the regular states (MARK or SPACE). Model of the Electromagnetic Interference will be discussed below.

3.2.2. Wireless Medium

Model of the wireless medium is based on type *SSI* (Synchronous Serial Interface) that models wireless synchronous bit-oriented communication medium. This type is based on user-defined type *ssi_level*:

```
type ssi_level is (L0, L1, EMI_L0, EMI_L1);
```

```
L0      = log. 0
L1      = log. 1
EMI_L0  = log. 0 EMI
EMI_L1  = log. 1 EMI
```

Defined states L0 and L1 correspond to the logical states use for transmitting signal by RFieldbus Radio Physical Layer Protocol [13]. Beside it, two other states for electromagnetic interference are defined: EMI_L0 and EMI_L1. Resolved function and its behaviour is formally the same as for type resolve_uart.

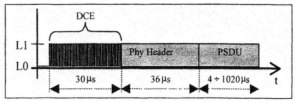

Figure 6   VHDL model of the PhLPDU

RFieldbus PhLPDU telegram format is implemented as well as it is described in [13], except RFieldbus Phy Preamble. The Preamble is generated as 30 us bit pattern of L0 and L1 states on the bus (see Figure 6) in VHDL model.

During DCE Preamble, every SSI receiver tunes its Baud Rate generator in such a way that at the end of the Preamble, the final Baud Rate (BR) value is equal to:

$$BR = \frac{59}{\sum_{n=1}^{59} DLY_n} [Baud]$$

where DLY is a time delay between the minimum state changes on the bus during synchronisation.

3.3. Segment couplers

Different types of couplers that are connecting the segments of a network have been modelled.

3.3.1. RS-485 Repeater

VHDL model of the RS-485 repeater links particular bus segments in the project, even the segments with the different communication speeds. An interface of the VHDL entity is shown in Figure 7.

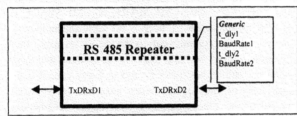

Figure 7   RS-485 Repeater Entity

Signals TxDRxD1(2) construct two bus segments that have different baud rates (user parameters BaudRate1 and 2). Propagation delay of the telegram between segments is at least t_dly1(2). It can be greater (t_gap longer) due to different communication speeds on the segments, see Figure 8 for details.

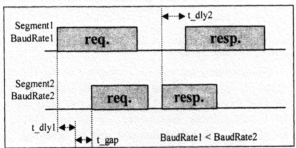

Figure 8   An example of the propagation of the telegram.

A Request telegram on segment 1 is repeated on the speedier segment 2 with the time delay t_dly1 (user parameter) and t_gap due to the fact that all data in the PROFIBUS telegram has to be transferred without gaps between the individual characters

3.3.2. Link Station

Link Station (LS) is a repeater between a wired and a wireless segment of a RFieldbus network - Figure 9.

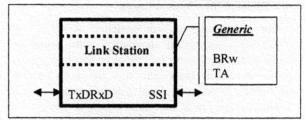

Figure 9   Link Station Entity

Time behaviour of the Link Station is very similar to the above-introduced RS-485 repeater. Wireless segment (signal SSI) has a fixed transmission rate 2Mbit/s and the speed of the wired segment (signal TxDRxD) is user defined parameter BRw. The Second parameter TA holds the physical address of the transmitting station.

3.3.3. Base Station

Base Station (BS) is a repeater between two wireless segments of a RFieldbus network, see Figure 10.

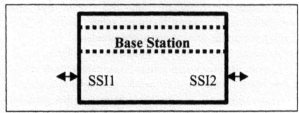

Figure 10   Base Station Entity

Function behaviour of the BS model corresponds to the [8], except the fact that only two wireless segments (SSI1 and SSI2) are possible.

3.3.4. Link Base Station

Link Base Station (LBS) is a repeater between two wireless segments and one wired segment, see Figure 11.

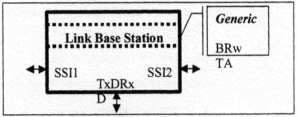

Figure 11   Link Base Station Entity

If the propagation of the PROFIBUS telegram is from the wired (signal TxDRxD) to the wireless segments (SSI1 or SSI2) then LBS behaves like a pure Link Station. If the communication path is from SSI1 to SSI2 (or vice-versa) then LBS behaves like a pure Base Station.

3.4. EMI model

Electromagnetic Interference is a very important phenomenon. Due to the fact that communication buses are frequent targets of the EMI, the possibility on how to inspect this influence was included into the VHDL model.

The basic tool is an EMI entity (Figure 12) which enables you to stimulate selected wired or wireless bus segment(s) by previous defined error patterns. This pattern is either defined implicitly by any of the distribution functions of the random variables generated during simulation or explicitly by a stimulation file (StimFile) before simulation.

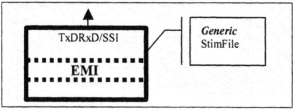

Figure 12   EMI Entity

For instance, if the StimFile includes following list:

```
14.0    0
6.0     1
8.0     x
```

EMI entity generates bus error as shown in Figure 13.

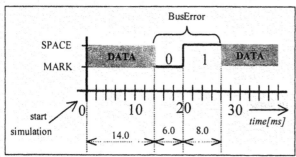

Figure 13   An example of the error on the bus

4. Simulation results

Shown is an example of a system model with one Master, one Slave and a wireless connection in Figure 14. Two Link Stations (LS1 and LS2) separate three bus segments marked by I, II and III. Segments I and III are standard RS-485 PROFIBUS segments with baud rate 1.5Mb/s and segment II is a wireless connection between Link Stations. Moreover, EMI entity stimulates wireless segment by predefined error pattern and entity UARTMonitor scans wired segment I.

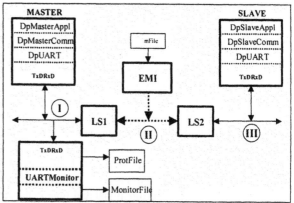

Figure 14   Simulation configuration

Figure 15a) shows global time behaviour of this heterogeneous model. DP Master on the segment I sends SD1 telegram (Request_FDL_Status) at 5.2 ms and the same telegram is repeated on the wire segment III through wireless segment II. Similarly, token telegram SD4 is distributed by the same way at 8.6 ms. The last telegram sequence (SD4, SD2, SD2 and SD1) shows how DP Slave on the segment III responds to the Master's request.

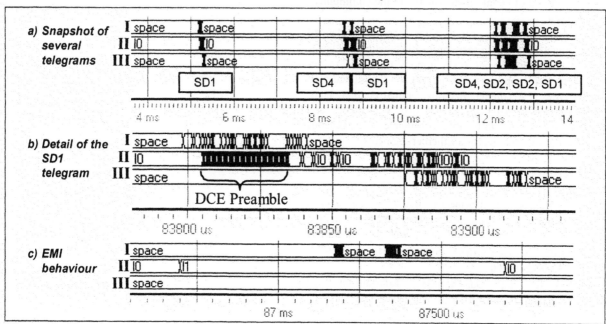

Figure 15   Waveform representation of simulated bus signal

115

Figure 15 b) shows a detail of the SD1 telegram from the Master. Synchronisation preamble mentioned above is well visible here and it is easy to measure propagation delay between telegrams on the segments I and III - in this case it is approx. 75 microseconds.

Finally, EMI behaviour is shown in Figure 15 c). 1 ms error (log. 1 stick) on the wireless segment II (between 86.7 and 87.7 ms) causes no telegram from the segment I to be repeated on the segment III (or vice-versa).

5. Summary and future work

In this contribution a method was presented that allows combining simulation models of different FDT's with the aim of timed simulation of heterogeneous industrial communication systems. This method was applied for the simulation of RFieldbus systems. An existing ESTELLE model of the PROFIBUS DP protocol was combined with new defined models of wireless medium type, related send and receive machines as soon as different types of radio repeaters. An EMI generation entity was also added to the model. First simulation results that demonstrate the functional and time behaviour at the intersections of wired and wireless segments can be shown.

The next step is to compose typical RFieldbus network configurations with more segments and devices and to find out typical parameterisation scenarios, benchmark applications and EMI models. The simulation results of these scenarios could be compared with the results of time measuring in the field trials. Furthermore, it should be possible to investigate e.g. dependency of reaction time from the number of radio cells or the influence of certain EMI patterns to the communication cycle.

Perceptively, the system model shall be extended continuously with further protocol functions (DPV1, DPV2), medium types (synchronous transmission according to [3], optical transmission) and the related coupler types (PA coupler, PA link, optical coupler, etc.). Moreover Ethernet models shall also be evaluated and integrated in order to be able to simulate heterogeneous networks with fieldbus and Ethernet segments.

References

[1] J. Haehniche, L. Rauchhaupt, "Radio Communication in Automation Systems: the R-Fieldbus Approach", in Proc. of the 2000 IEEE International Workshop on Factory Communication Systems, Porto (Portugal), pp. 319-326, September 2000

[2] Hintze, E.; Donath, U.: Kombination von Simulationsmodellen für Kommunikationssysteme auf der Basis formaler Beschreibungstechniken. VDI-Fachgespräch: Software-Engineering in der industriellen Praxis, VDI-Berichte Bd. 1666, S. 41-52, Düsseldorf, 2002.

[3] IEC 61158: Digital data communication for measurement and control - Fieldbus for use in industrial control systems, IS, IEC 2001

[4] IEEE: Standard VHDL Language Reference Manual - IEEE Std. 1076-1993. New York, 1994.

[5] ISO: ISO Reference Model of Open-System Interconnection, ISO/TC97/SC16, DP7498, 1980.

[6] ISO: Estelle: A formal description technique based on an extended state transition model. International Standard ISO, IS 9074, 1987.

[7] ISO/IEC 8482:1993, Information technology - telecommunications and information ex-change between systems twisted-pair multipoint interconnections.

[8] Koulamas C.; Koubias S.: Real-time characteristic of the RFieldbus link device bridging operation, IST 2002

[9] Model Technology Inc.: ModelSim. SE/EE User's Manual, Version 5.4 (Online-Handbuch), 2000.

[10] NBS: User guide for the NBS prototype compiler for Estelle. NBS Report No. ICST/SNA-87/3, 1987.

[11] Pöschmann, A.; Hintze, E.; Hähniche, J.: Formale Methoden in der Kommunikationstechnik. Tätigkeitsbericht 1998, Institut für Automation und Kommunikation e.V. Magdeburg, Seite 55-58, Barleben, 1999.

[12] RFieldbus Deliverable D1.3, "General System Architecture for the RFieldbus System", Technical Report, Sep. 2000.

[13] IST-1999-11316 RFieldbus, D2.1.1 - Physical Layer Specification - Part 2, Protocol & Operational Characteristic Definition.

PERFORMANCE EVALUATION OF THE RFIELDBUS SYSTEM

Weiyan Hou[1], Heiko Adamczyk[1], Christos Koulamas[2], Lutz Rauchhaupt[1]

[1]Institut f. Automation und Kommunikation
e. V. Magdeburg (ifak)
Steinfeldstr. 3
D-39179 Barleben, Germany
E-mail: ada@ifak.fhg.de

[2]Industrial Systems Institute (I.S.I.)
University of Patras

26 500 Patras, Greece
E-mail: koulamas@ee.upatras.gr

Keywords: Fieldbus, performance evaluation, throughput, cyclic time analysis, error rate
performance

Abstract

*RFieldbus is a modified kind of PROFIBUS
fieldbus system featuring the extended wireless
physical layer and the IP transmission
capability. This paper describes the results of a
number of tests which were made to assess the
RFieldbus behaviour. PROFIBUS-DP cyclic
data exchange time, TCP/IP traffic throughput
as well as error performance of the transmission
mechanisms were evaluated in laboratory as
well as in industrial environments. The results
show that in the industrial environment, an
RFieldbus system can fully meet the real-time
requirements even accompanied with the
TCP/IP traffic.* Copyright © 2003 IFAC

1. FOREWORD

Nowadays, the distributed controlling system is
more and more popular and the fieldbus
technology plays an important role in the
industry environment. This is because of its low
cost in the connection of the sensors, actuators
and equipment in the field-controlling layer.
Fieldbus systems can make the data exchange
between the nodes in a deterministic time
deadline, which means it grants the stringent
real-time property. The famous PROFIBUS
fieldbus has been installed in hundreds of
thousands of points in the manufacturing and
process control field all over the world.

However, for some scenarios such as the
turning of a remote controlling object, a wireless
connection is necessary. How to keep the real-
time property of the fieldbus system under the
wireless connection in a harsh industrial
environment is a relatively new researching
topic[1][2]. RFieldbus which was launched in
January 2000 under the support of European
Commission 5th IST-plan [3] is a new system
which integrates the PROFIBUS and wireless

LAN (WLAN) technology. It makes the
PROFIBUS-DP traffic and the TCP/IP data stream
be transmitted simultaneously on the wired and
wireless hybrid physical layer. This means that the
process controlling data and multimedia info, for
example photos and industrial-like video stream,
can go through an RFieldbus tunnel at the same
time.

This project has been in progress for 3 years.
Several prototypes have successfully run in
different field trials.

The goal of the paper is to make the performance
evaluation of this new hybrid system and to see
how well it is suitable to a real industrial controlling
environment and fulfil the real-time requirement,
how the TCP/IP and PROFIBUS-DP traffic impact
each other, how is the error performance of the
system.

This paper is organised as follows:

In the first section, a simple introduction of the
RFieldbus system is given. In the next section the
indoor experiment environment including hardware
and software is depicted in detail and the 13-
selected evaluation scenarios are put forward
according to 4 criteria. In section 3 the evaluated
performance is discussed and measuring
procedure is presented. Section 4 describes the
results of the indoor evaluation experiment and the
detailed analysing of the results on T_{cycle},
Throughput and the error properties.

In section 5 the second experiment environment
(outdoor, pilot side) is depicted and the
subsequent section contains the measurements
made on the pilot side as well as the evaluation of
the performance.

In the last section of this paper there is an overall
conclusion.

2. INTRODUCTION OF RFIELDBUS

In RFieldbus there are three kinds of network elements in addition to a conventional PROFIBUS system, as described in the following table.

- LS (link station):
  link station is a radio repeater which transparently transfers the telegrams between the wired domain and the wireless domain.
- BS (base station):
  base station implements frame store and forward functionality in the wireless domain only.
- LBS (link base station):
  link base station integrates both of the above-mentioned functions.
- Beacon Master:
  broadcasts the synchronisation signal in the whole system, as the coordinator for the mobile handoff management.

	Wireless Function	PROFIBUS Function	Bridge Function
Repeater (LS, LBS, BS)	Yes	No	Yes
Beacon Station	No	Yes	No
DP/IP-Master/ Slave	No	Yes	No

table 1: RFieldbus devices

In addition, there are DP/IP-Master and DP/IP-Slave stations, which integrate the TCP/IP and DP transmission functionality.

In RFieldbus system, the data transmission speed (baud rate) is fixed at 1.5 Mbit/s in the wired domain, and at 2Mbit/s in the wireless domain with the DSSS technology which is the same as in WLAN. In order to dampen down the Multipath interference, RAKE receiving technique is adopted. The radius of a single cell can be extended as wide as 60-100 meters in an indoor environment. The following figure shows an architecture covert by RFieldbus:

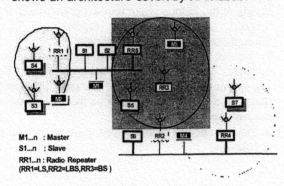

M1...n : Master
S1...n : Slave
RR1...n : Radio Repeater
(RR1=LS,RR2=LBS,RR3=BS)

Figure 1: RFieldbus infrastructure [7]

3. INDOOR EXPERIMENTAL ENVIRONMENT

3.1 Description of the testing campaign

Testing campaign The first measurement campaign is conducted in a PROFIBUS products testing room on the 1st floor, in a concrete-steel building in ifak, Magdeburg. It is depicted in figure 2. The C and D sides of the wall have been furnished with over 20 PROFIBUS-enabled metal devices, e.g. valves, motors, and pipes which were made by different vendors. On the 2nd floor there is a WLAN, which runs regularly in another channel while our testing carrys on. On the ground floor just under the testing room, there is a big machinery workshop with all the common electrical power machinery. Its purpose is to set the testing campaign in such an environment where the electromagnetic interference is strong. Although this measurement field is not a real industrial environment, most of all the industrial environment characteristics are still captured.

The second measurement campaign was made within one of the RFieldbus pilot sides which acts as a real industrial environment (see chapter 5).

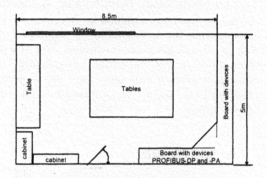

figure 2: testing campaign

Testing scenarios Depend upon the following infrastructure showed in Figure 3. 4 criteria as follows were adopted to determine 13 measuring scenarios.

- With or without wireless connection
- Intracellular or intercellular mobility
- Single Master or Multimaster (with or without Beacon master)
- One point, or a segment mobility

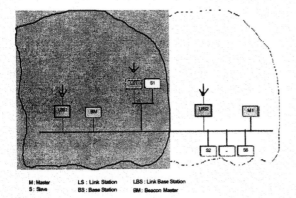

M : Master   LS : Link Station   LBS : Link Base Station
S : Slave   BS : Base Station   BM : Beacon Master

Figure 3 : Primary scenario

1. Scenario 1: Sole wired connection, as the reference standard for the rests
2. Scenario 2: Scenario1 plus Beacon Master (multimaster)
3. Scenario 3: Direct wireless connection between two LS
4. Scenario 4: Wireless connection between LS and LBS
5. Scenario 5: Scenario 4 plus Beacon master
6. Scenario 6: One slave makes intercellular mobile through 2 LBS, with Beacon master
7. Scenario 7: A segment with 5 slaves make intercellular mobile through 2LBS
8. Scenario 8: One master makes intercellular mobile
9. Scenario 9: One master makes intracellular mobile
10. Scenario10: Scenario 9 without Beacon master
11. Scenario11: Mobile segment and mobile master intercellular on direct wireless conn.
12. Scenario12: scenario11 plus one LBS
13. Scenario13: scenario11 plus one BS
*details can be retrieved in [4]

Description of the used devices
The prototype implementation of the RFieldbus Higher Layer was made using a PROFIBUS PC card. This concerns the TCP/IP integration, the station management and the mobility support (Beacon Master).
The RFieldbus Lower Layer implementation was made using a Tricore evaluation board in conjunction with the radio front end.
In addition to the RFieldbus devices a number of conventional PROFIBUS devices were used in the testing scenarios. The tests include following devices:
1. DP/IP Master:
   one ifak master ISA card with RFieldbus prototype (SW based) firmware, host PC equipped with Pentium-200 CPU, with ISA slot, run on WinNT 4.0., controlled by DP-Testapplication

2. Slaves:
   a) DP/IP Slave: ifak PCMCIA card with the RFieldbus prototype (SW based) firmware for Master, host PC is a Notebook equipped with Pentium-III 800 CPU, operates on WinNT4.0.m, I/O data length = 10 Byte
   b) DP-devices slaves with DP-V1 (HW based):
      Buerkert Ventilinsel 1470/1475+ 8640, I/O data length = 12 Byte;
      Beckhoff BK3100, I/O data length = 15 Byte;
      ET200X, I/O data length = 3 Byte;
      ET200M, I/O data length = 3 Byte;
      ET200S, I/O data length = 67 Byte
4. Beacon Master: one ifak master ISA card with RFieldbus prototype (SW based) firmware for Beacon Master, in ISA slot. Host PC is mounted Pentium 200 MHz, run on WinNT4.0.
5. LS, LBS, BS:
   complete prototype device with built-in RAKE receiver-technique modem+ Antenna.
6. PROFIBUS Monitor:
   made by Softing AG

Description of the software tools
1. DP-TestApplication:
   TestApplication is a customised software based on the PROFIBUS-DP standard interface-specification which was developed by ifak. It is oriented to the new application of the RFieldbus. It has four main functionalities:
   a) Initialisation of all the device in RFieldbus system (DP-Master, DP-Slave, IP/DP-Master, IP/DP-Slave, Beacon Master)
   b) Configuration of the Master with Bus-Parameter-Set, Slave with Slave-Parameter-Set
   c) Communication Setup (e.g. Start/ Stop of cyclic and/or acyclic data exchange, etc.)
   d) Management Application (Read and Write of RFieldbus parameters, read of the Physical layer error codes)
2. Winttcp:
   Winttcp is a windows-version software of the popular benchmark tool TTCP which has been used under Unix, Linux environment for the Throughput measure of TCP/IP traffic. It can be used in the command line in WinNT.
3. Ethereal:
   Ethereal is a free network protocol analyser for Unix and Windows. It allows you to examine data from a live network or from a capture file on disk. Testers can interactively browse the capture data, viewing summary and detail information for each packet of 355 kinds of protocols on Ethernet, FDDI, PPP, Token-Ring, IEEE 802.11, etc.

4. THE EVALUATED PERFORMANCES

In RFieldbus there are two kinds of data streams, process controlling data and multimedia data, from the application viewpoint. Therefore, two groups of performance parameters should be measured in the tests. Thirdly, the error performance evaluation of RFieldbus is also an important property for the utility in the industrial environment.

4.1 PROFIBUS-DP Performance

In RFieldbus the length of user data packets varies from 1 bytes (min) to 244 bytes (max). The user data packets that contain the controlling data are relatively short. This data is cyclic exchanged between the master station and the slave stations. From the control application viewpoint the system data message cycle time Tcycle is the most important parameter, which benchmarks the system real-time characteristics [5].

4.2 Multimedia traffic performance

To evaluate the behaviour of the multimedia data traffic, the important performance parameters are: throughput, transmission delay, reliability, effectiveness, capacity utilisation and resource utilisation [6]. The multimedia streams in RFieldbus are shown as IP data traffic. The most crucial performance properties are Throughput and the Frame loss-rate, which were focused on in our testing.

4.3 Failure analysis

RFieldbus has the hybrid wired/wireless connection that was used in the industrial environment full of noise and interferences. Especially, the wireless channel is error prone. Compared with the PROFIBUS system, the error performance is an important factor which affects the performance of the whole system.

Upon the RFieldbus protocol architecture in figure 4, the error mainly happens on wireless physical layer (PHL). To evaluate the number of PHL errors, some measurements on one of the RFieldbus pilots took place (chapter 5) due to the need of a real industrial environment.

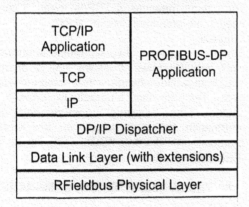

figure 4: RFieldbus protocol structure

4.4 Measuring procedure

As showed in figure 5, the DP/IP Master and DP/IP Slave run respectively as Sender and Receiver. It means that two sub-programs in Test Application have been installed on the two sides simultaneously, but will be started separately. When one side runs as the Sender, the other side runs as the Receiver, or vice-versa.

The 5 DP-V1 devices run as the loading for the bus. The bus-link in RFieldbus has the different topologies structure, which was constructed as different testing scenarios (referred to section 2.2). The two measured points: Sender and Receiver will be fxed or make the mobile, according to the scenarios respectively.

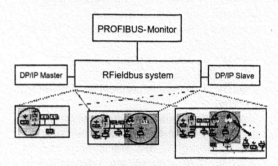

figure 5: Measuring depiction

The Bus-Monitor was connected to the bus of the RFieldbus system. It was used to record the DP-data stream on the bus near the Master side. The Tcycle can then be calculated from the recorded data easily. Concurrently, the Winttcp will be started separately as the client on Sender and as the server on Receiver to test the Throughput. This is because the TCP/IP implemented mechanism in RFieldbus is direction oriented [3]. This Test will be performed bi-directionally, e.g. Master to Slave (M->S) and Slave to Master (S->M).

In our test, the number of sent packet was fixed and the length of the sent packet will be varied. Ethereal running on the Master side will test the number of received frames (packet), which were sent out from the Slave side.

5. MEASURING ANALYSING

5.1 Measuring of Tcycle time and Throughput

In 1st test the default settings of DP-Bus parameters defined in TestApplication are used, as following Setting 1st:

- min.Tsdr = 1 000 bittimes
- Tslot = 5 000 bittimes
- TTR = 50 000 bittimes
- min. Slave. Interval = 1 ms
- Datarate = 1.5 Mbits/s (bittime = 0.667 ns)

		Tcycle time (ms)		Throughput (KB/s)	
		No TCP	WithTCP	M->S	S->M
Scenario 1	Min.	10.2	9.4		
	Ave.	11.1	11.3	4.99	3.72
	Max.	11.8	12.8		
	JS:	0	1		
Scenario 2	Min.	10.2	10.3		
	Ave.	12.7	12.8	4.12	3.84
	Max.	14.3	15.6		
	JS:	0	1		
Scenario 3	Min.	10.7	9.9		
	Ave.	11.2	11.6	4.06	3.68
	Max.	12.0	15.1		
	JS:	0	1		
Scenario 4	Min.	10.6	9.9		
	Ave.	11.2	11.6	3.87	3.52
	Max.	12.0	15.1		
	JS:	0	1		
Scenario 5	Min.	10.2	10.9		
	Ave.	11.8	11.8	4.03	2.98
	Max.	13.4	14.3		
	JS:	0	0		
Scenario 6	Min.	11.8	10.3		
	Ave.	11.9	12.0	3.79	3.23
	Max.	11.9	14.3		
	JS:	0	2		
Scenario 7	Min.	12.5	11.6		
	Ave.	12.6	13.6	6.45	4.50
	Max.	14.0	18.7		
	JS:	0	1		
Scenario 8	Min.	12.8	12.9		
	Ave.	14.5	17.07	6.16	3.77
	Max.	16.7	45.4		
	JS:	0	1		
Scenario 9	Min.	13.6	12.2		
	Ave.	13.8	14.1	4.16	3.79
	Max.	15.2	16.2		
	JS:	0	2		
Scenario 10	Min.	12.3	12.2		
	Ave.	12.8	13.6	4.07	4.85
	Max.	13.5	19.4		
	JS:	0	1		
Scenario 11	Min.	12.2	12.2		
	Ave.	12.6	13.9	4.02	4.11
	Max.	12.9	16.3		
	JS:	0	2		
Scenario 12	Min.	14.3	15.2		
	Ave.	14.9	17.4	4.64	4.13
	Max.	15.7	24.0		
	JS:	0	2		
Scenario 13	Min.	14.1	15.0		
	Ave.	14.6	16.9	3.99	3.83
	Max.	15.5	20.3		
	JS:	0	2		

Table 2: Tcycle time and Throughput

5.2 Analysing

1. Tcycle keeps stable when there is no TCP/IP traffic. Otherwise, jitters will occur while there is TCP/IP traffic. This means that TCP/IP traffic has a very small impact on the average DP-Tcycle time, it only makes Tcycle a little greater than that without TCP/IP traffic.

However, it results in the jitter of the Tcycle. In order to depict this phenomenon, two new parameters were defined.

a) Average of Tcycle

In the measured samples trace a sampling-window sized with 37 Tcycles long is chosen stochastically. So the average value is calculated as:

$$\overline{x} = \sum_{i=1}^{37} Tcycle(i) / 37$$

b) JS (jitter strength of Tcycle). To depict the jitter strength, a new parameter was defined.

$$s = stdv(x) \Big/ n*\overline{x} = \sum_{1}^{37} \sqrt{(x_i - \overline{x})^2} \ / \ n*\overline{x}$$

JS is defined as:
→ if **s** > 30% then **JS = 2**
→ if **s** > 10% then **JS = 1**
→ if **s** < 10% then **JS = 0**
// 2,1,0 mean the jitter is Big, Moderate, Small

2. The Throughput measured in the direction: Master to Slave is greater than in the direction: Slave to Master.

3. The performance difference between the wired and wireless connection is not big.

4. The Beacon Master and DP-application Master station should keep connected within one wired segment when the master makes a intercell-mobile roaming. If they are linked through the wireless connection, the max. Tcycle will get not be predicted greater when there is TCP/IP traffic (see scenario 8). It seems that the TCP/IP traffic impacts the Beacon transmission while handling on the Hand-off, when the Master roams intercell.

5.3 More Tests

After the optimising of DP-BUS parameters another two tests were further introduced. The difference to the 1st Setting is:
- Setting 2nd : min. Tsdr = 1.000 bittimes
- Setting 3rd : min. Tsdr = 11 bittimes

	Ave. Tcycle(2nd)		Ave.Tcycle(3rd)	
	No TCP	With TCP	No TCP	With TCP
Scenario 1	11.2	11.5	54.3	52.3
Scenario 2	11.6	12.4	12.1	12.1
Scenario 3	11.1	11.9	50.7	40.7
Scenario 4	11.1	11.7	11.25	11.4
Scenario 5	12.4	12.2	11.9	11.9
Scenario 6	11.9	12.2	10.9	12.1
Scenario 7	12.6	13.4	10.9	11.3
Scenario 8	18.5	20.8	18.2	14.1
Scenario 9	18.5	14.2	10.3	14.1
Scenario 10	12.8	13.9	12.9	13.7
Scenario 11	12.4	12.6	12.5	12.5
Scenario 12	15.0	73.3	>100	>100
Scenario 13	14.7	76.3	>100	>100

Table 3: Average Tcycle in 2nd, 3rd test

Explaining:

1. Since the value of Throughput in the 2nd and 3rd test are similar to the 1st test, they vary between 4 KB/s to 6 KB/s, they were not presented in Table 4.

2. By optimising the BUS-parameters the shortest Tcycle can reach up to ca.10 ms in all the tested scenarios. It can meet the time requirement of almost all the industrial applications.

3. In a single master scenario, the Tcycle seems to have related with the min. Tsdr closely. At the moment, the RFieldbus firmware was realised in software and it runs on the CPU of the DP-Master/Slave card, without a special ASIC hardware solution. When min. Tsdr is getting smaller, host CPU must take more time-slices to perform the management of the PROFIBUS state-machine. Thus, it does not have too much time to process Tasks in polling queue. As the result, the Tcycle will increase reversely.
   → in the future this point will be improved

4. The min. Tsdr was set big enough in the default parameters group. When there are multimaster, no such regulation exists, min. Tsdr can be set as 11 bittimes that was proposed in PROFIBUS standard.

6. OUTDOOR EXPERIMENTAL ENVIRONMENT

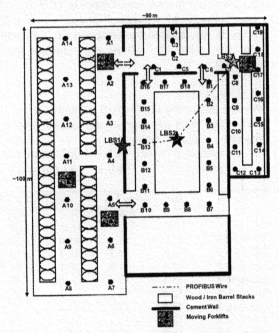

Figure 6: CYCLON Refinery plan

In the context of the RFieldbus IST project, the CYCLON HELLAS S.A. refinery plan, situated in Greece, was used as the industrial site in which a pilot RFieldbus system was installed and evaluated. An outline of the site is depicted in Figure 6., where the 'A' points correspond to the outdoor area, and 'B', 'C' points to the indoor area part of the site. The indoor part is a storage warehouse with roof and piles made of metal, surrounded by cement walls, as shown in the figure. With respect to Figure 6, the grey areas correspond to storage metal and wood racks containing mainly cardboard, plastic or metal objects. The outdoor part is a storage area for metal drums on two levels, with heights such as the forklifts moving along the corridors which do not always have a line-of-sight with the roof of the indoor part, where the three Link Base Stations (LBS) are situated. As a result, the CYCLON site encompasses the various types of environments classified in radio propagation, that is:

- Line-of-sight with Light Surrounding clutter
- Line-of-sight with Heavy Surrounding clutter
- Obstructed Path with Light Surrounding clutter
- Obstructed Path with Heavy Surrounding clutter

Each of the three LBS devices operates on its own frequency channel, while the selected channels are obviously those belonging to the only one set of non-overlapping channels of the FCC 802.11 channel set (i.e. channels 1, 6 & 11). All LBSs are connected to a 1.5 Mbps wireline (RS485) PROFIBUS system, which terminates to two PROFIBUS masters (PC stations, not presented in Figure 6). One of these masters is the application master, with an OPC Server over PROFIBUS DP, gathering one 50-byte variable from each mobile forklift. The second master plays the role of the 'Beacon Master', coordinating the channel assessment period for the mobiles, which adds a 1msec overhead every 100 msec during the system operation. Finally, on each of the forklifts there is a separate 1.5 Mbps wireline PROFIBUS segment which connects a Barcode / Numeric Pad & Text Display standard PROFIBUS slave with an RFieldbus Link Station (LS). The latter is the device which provides the radio linkage with the fixed LBSs, and therefore, the hybrid connection between the PROFIBUS slave on the forklift and the PROFIBUS masters in the warehouse.

measurement of the reception quality

Using the RFieldbus system described above, two major rounds of measurements were carried out. The first was to switch on each one of the three LBS while the others were off, in order to identify the reception quality in the moving forklift, all over the indoor and outdoor area, and for each LBS alone (that is, a total of three sub-rounds of measurements). The results are presented in Figure 7, where it is shown that in the areas reasonably close to each LBS the resulting PER was better than 10^{-4} (up to measurable zero), while, by moving away, the points where the PER was increased were identified.

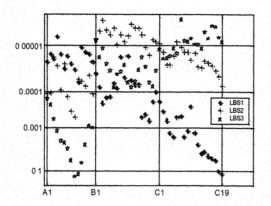

Figure 7: Measured PER in the mobile, each LBS

test of the mobility mechanism
The second round of measurements targeted to the mobility mechanism testing, as this mechanism is described in [8]. This was accomplished by switching on all LBS devices and setting up the LS in the forklift to operate in the so called 'Mobile Mode', that is, in a mode in which the mobile LS is continuously assessing its reception quality from all LBSs and selecting the best one, during its movement. As expected, in the areas where the mobile had shown high PER (due to low reception quality) while connected to a specific LBS, now the PER is much better, since the mobility mechanism forces a channel selection which corresponds to the closest LBS. Measurements with the mobility mechanism enabled showed a worst case PER of 10^{-4}, while in a complete circle of the mobile all over the A, B & C areas, 20 handovers (i.e. channel changes) were recorded.

Regarding the real-time performance of the described installation, this was measured by the usage of a PROFIBUS Network analyser, which showed a peak capability (i.e. in time windows with zero PER and not including a channel assessment phase) of 265 message cycles per second, where, at the application level, the OPC quality of the variables with a refresh period of 0.5 sec was never changed to 'BAD'.

7. CONCLUSIONS

The tests show that RFieldbus has almost the same transmission properties on the wired and wireless segment. Of course, the values of the PROFIBUS parameter set are larger due to the introduction of repeaters, the link and base stations. Apart from that, the wireless connection has no further influence on the Tcycle. It can fulfil the time demands of the processing control in the industrial environment.

The average Tcycle of the PROFIBUS-DP traffic is almost not influenced by the IP traffic. However, in the current prototype implementation the jitter of the DP traffic was considerable when the IP traffic was transmitted simultaneously. The reason is that the token is passed just after the queued traffic is transmitted. This means the Tcycle is shorter if there is no IP traffic to transmit.

The test has also shown that the packet loss rate in RFieldbus is small, also if wireless connections are used. Process control data transmission can be guarantied.

RFieldbus meet all the designed specifications.

8. ACKNOWLEDGMENTS

This evaluation measurement was performed as part of the project RFieldbus which was supported by the European Commission (IST-1999-11316) under the 5th IST Program.

9. REFERENCES

[1] Adamczyk, H., Rauchhaupt, L., "WLAN systems in industrial environments - does it work?", Mobilecommunication over wireless LAN, 09.2001, Wien/ Austria

[2] Rauchhaupt, L., "Radio in industrial communications", Wireless Communication , 05.2001, Munich /Germany

[3] General structure of the RFieldbus, White Paper, D1.3 of the IST Report, 2000

[4] Hou, W.Y., Adamczyk, H., Rauchhaupt, L. "The evaluation of RFieldbus", inner Technical Report of Ifak, Feb. 2003.

[5] Bender, K., "PROFIBUS -the field for Industrial automation", 1991

[6] RFC 2544 Benchmark Methodology for Network Interconnection Devices

[7] Rauchhaupt, L., "High Performance wireless fieldbus in Industrial Related Multimedia environment", Proceedings of 4th IFAC Conference, Sweden, 08.2002,pp.184-192

[8] D2.1.1 - Physical Layer Specification - Part 1, Service Definition, 2001

ELSEVIER

IFAC

PUBLICATIONS
www.elsevier.com/locate/ifac

BRINGING INDUSTRIAL MULTIMEDIA TO THE FACTORY-FLOOR: WHAT IS AT STAKE WITH RFIELDBUS?

Eduardo Tovar, Luís Miguel Pinho, Filipe Pacheco, Mário Alves

IPP-HURRAY Research Group
Polytechnic Institute of Porto, School of Engineering (ISEP-IPP)
Rua Dr. António Bernardino de Almeida, 431
4200-072 Porto, Portugal
http://www.hurray.isep.ipp.pt

Abstract: The support of new information technologies and applications in the factory-floor, such as multimedia applications, is moving from eagerness to a real need. In this paper we will discuss the motivation for embodying industrial multimedia application support into existent fieldbus communication systems. The RFieldbus project is one example of an initiative where this objective has been successfully carried out. Without going into details neither in the RFieldbus specification nor in the internals of the multimedia applications related with the discrete parts manufacturing field trial (a pilot application developed in the framework of the RFieldbus project), we want to put forward some functionalities which are possible to implement with RFieldbus. The field trial itself is a complex distributed computer control system with a wide set of software applications, ranging from small software components, to large (distributed) applications. This pilot implementation provides a suitable framework and test-bed for the assessment of integrated multimedia and real-time control traffic in the same network, and the use of hybrid wired/wireless networks for Distributed Computer-Control Systems. *Copyright © 2003 IFAC*

Keywords: fieldbus networks, real-time, multimedia, wireless

1. MOTIVATION: THE E-MANUFACTURING

The most strategic advantage of any manufacturing enterprise today is information. Whether the challenge is faster time-to-market, improved process yield, non-stop operations or tighter supply-chain coupling, information is the key. The plant is crucial to every manufacturing enterprise. It is the place where value is created. And technology is not new to the plant floor. Highly sophisticated combinations of relays and switches, programmable controllers, drives and sensors have long operated the lines, conveyors, and machinery required to manufacture the goods.

As industries consolidate and restructure, some companies are choosing to remain as the "producers" of goods while others are positioning themselves downstream in the supply chain to be the "marketers" of those same goods. In almost every industry, tightly integrated supply chain models are emerging. In these new models, connection of the plant floor to the broader supply chain is essential, and information access is more critical than ever. The Internet and e-Commerce have simply accelerated this trend towards the e-Manufacturing (Rockwell, 2000; Tovar, *et al.*, 2002).

The factory-floor is the starting point for greater information connectivity. Computer-based factory-floor controls for manufacturing machinery, materials handling systems and related equipment generate a wealth of information about productivity, product design, quality and delivery. Factory-floor networking arises as the key to unleashing this information in a cost-effective manner.

Over the last decade, the evolution of the manufacturing industry has assimilated "contemporary" information technologies, such as the PC and open standards for the factory-floor networking. In fact, the application of information technologies has evolved from relatively passive data collection and reporting roles to feedback control and diagnostics applications. This context unveils the role played by factory-floor networking in modern industrial automation systems.

However, it is interesting to notice that the use of communication networks at the factory-floor is more recent than at the office environment. One of the reasons for this delay was that manufacturing systems usually depend on being able to sample input data at equally spaced points in time (Skeie, et al., 2002) and this feature was not easily fulfilled using early office-room networks. This lead to the fieldbus concept, the first step on the road to networked industrial automation systems.

Fieldbus networks aim at the interconnection of sensors, actuators and controllers, being generally adapted to convey periodic and aperiodic, real-time and non-real time traffic (Thomesse, 1993; Decotignie and Prasad, 1994).

Different technologies are available in the market, being PROFIBUS (Tovar and Vasques, 1999), DeviceNet (Rockwell, 1997), WorldFIP (Almeida, et al., 2002) and TTP (Kopetz and Grünsteidl, 1994) just a few but representative examples. This diversity is partially justified by the diversity in the requirements imposed by the different targeted applications.

In the era of the Internet, however, factory-floor communication systems must also better explore commercial information technologies. This includes COTS operating systems, TCP/UDP/IP based applications (XML, Java, etc.) and general-purpose communication networks. The reason is not only to integrate factory-floor operations to work seamlessly throughout a typical plant environment, but to also make that control information transparent throughout the overall enterprise. These "open" technologies present a strategic advantage to address the control needs of the plant and the connectivity to achieve that transparency.

In this trend, networks must deliver an efficient means to exchange data for precise control, while supporting non-critical systems and device configuration at start-up and monitoring during run-time. Networks must also provide the critical link for collecting data at regular intervals for analysis and feedback control. An integrated architecture benefits from a common set of advanced network services and interfaces optimised for timeliness control, configuration and collection of data, with seamless communications up and down the architecture, allowing access to any part of a system from any location.

Advances in networking and information technologies are transforming factory-floor communication systems into a mainstream activity within industrial automation. It is now recognised that future industrial computer systems will be intimately tied to real-time computing and to communication technologies. For this vision to succeed, complex heterogeneous factory-floor communication networks (including mobile/wireless components (Alves, et al., 2002)) need to function in a predictable, flawless, efficient and interoperable way.

2. MOTIVATION: THE "CYBER-FACTORY"

In the past few years, the so-called gadgets like cellular phones, personal data assistants and digital cameras are more widespread even with less technologically aware users. However, for several reasons, the factory-floor itself seems to be hermetic to these changes ... (Pacheco and Tovar, 2002).

After the fieldbus revolution, the factory-floor has seen an increased use of more and more powerful programmable logic controllers and user interfaces, but the way they are used remains almost unchanged.

It is a common believe that new user-computer interaction techniques including multimedia and augmented reality, combined with currently affordable technologies like wearable computers and wireless networks, can change the way the factory personal works together with the machines and the information system on the factory-floor.

But, which applications are we talking about?

Some applications may seem a little awkward in the industrial environment, but can give significant productivity gains and enable new working solutions.

For instance, a data network can be used to send real-time data not only from traditional sensors like position detectors or level meters, but also from more demanding ones like microphones or video cameras. Apart from the interest in gathering information in a more intuitive manner (with more simplified data input and new user interfaces) when workers are considered as an integral part of the process, it also allows new applications to be used for control. However, this means that these new applications and communication services may need to be inside the control loop, thus still with stringent timing requirements to be meet.

A portable mobile device can be used to access tutorial databases, presentations and interactive applications. Given the appropriate bandwidth, a small, lightweight, low-power device can gather or present information, exchanging it (using a wireless link) with a more powerful processing unit located nearby. Two examples of these applications are wearable computers for data collection, and ubiquitous surveillance cameras and sensors. An industrial network with wireless extensions can be a cost-effective solution to make this kind of information available where it is needed.

The possibility of using audio or video streams to connect several users is even more interesting if wireless communications are available for the industrial environments. By definition, they are interactive applications, so requiring rather small round-trip delays, and most of the data is time-sensitive and should be discarded if out of date.

Location awareness applications rely on information from the system on the current user location and are also related to mobility issues. This capability can enable innovative ways to deal with faults and maintenance in the factory-floor.

3. TECHNOLOGICAL CHALLENGE: EXTENDING FIELDBUS FUNCTIONALITIES

A point to be noted is that most of these applications are already available and supported by the TCP/IP suite of protocols. These are widely used, vendor independent, standardised, and interoperable with almost every operating system and system architecture.

TCP/IP stacks over Ethernet networks are widely available, allowing the use of standard applications.

3.1 What about using Ethernet Technologies?

Although Ethernet has been used primarily as an information network, there is a strong belief that some very recent technological advances will enable its use in dependable applications with real-time requirements (Decotignie, 2001). First of all, Ethernet has evolved to operate at 100Mbps and 1Gbps (with 10Gbps in development). Therefore, it provides a high bandwidth, when compared to that of CAN (1Mbps) or of PROFIBUS (12Mbps), just to give a few examples.

Also, the mechanisms associated with Switched-Ethernet (such as priorities, flow control, spanning tree, port trunking, virtual LANs, etc.) seem to be promising to enable the support of flexible distributed applications (Alves, et al., 2000), whilst guaranteeing the determinism and reliability requirements of real-time applications.

Moreover, the recent standardisation of wireless Ethernet (ISO-IEC, 1999) introduces the support of time-bounded delivery services, allowing the integration of wireless links in the communication infrastructure.

3.2 Why Other than Ethernet-based Solutions?

The fieldbus fundamentalists argue that Ethernet itself does not include any features above data link layer. TCP/IP protocols can, of course, be used to fill up some of the layers above Ethernet. However, what about layers above the transport layer? Moreover, which performance characteristics will be attained with the overall ensemble?

There have been some efforts to offer overall industrial communication solutions for Ethernet-based systems. Some of the approaches are based on "encapsulation technologies" like Ethernet/IP (Moldovansly, 2002). The term encapsulation is used to describe the embedding of a frame into a TCP or UDP container as "user data". The packet is then sent over Ethernet.

Even given that this type of solutions may satisfy the typical application requirements of fieldbuses, a strong argument persists in favour of using "common" fieldbus-based communication infrastructure solutions to convey standard multimedia applications. In fact, industrial applications are often applications planned to have a relatively long life. On the other hand, there are millions of already installed standard fieldbus nodes worldwide. Therefore, compatibility and interoperability with legacy systems is still indeed a very strong requirement in industrial automation.

4. RFIELDBUS: ENABLING MULTIMEDIA IN PROFIBUS NETWORKS

One example of putting this eagerness into practice is the RFieldbus European IST Project, where PROFIBUS was extended to support tunnelling of TCP/IP traffic together with the "native" PROFIBUS-DP traffic.

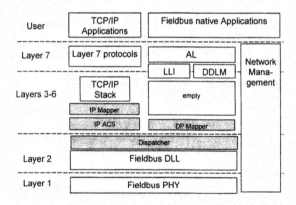

Fig. 1. PROFIBUS and TCP/IP integration

The main objective was to provide the adequate Quality of Service to the supported TCP/IP applications while, at the same time, guaranteeing the timing requirements of the PROFIBUS control-related traffic. The proposed approach was built upon previous relevant work on timing analysis of PROFIBUS networks (Tovar and Vasques, 1999). Based on that work, and in particular on the approach which considered that the PROFIBUS-DP low priority traffic was constrained in each token visit, a multi-traffic allocation approach was proposed (Tovar, et al., 2001). This is a crucial aspect of the architecture, providing the required timing guarantees for the different types of traffic. This ability, together with the IP scheduling functionality (Ferreira, *et al.*, 2001) allows providing different levels of QoS for simultaneous IP data flows handled in a PROFIBUS master, while maintaining the required timeliness for the "native" PROFIBUS-DP traffic (usually time-critical distributed control).

To implement this solution, a dual stack architecture was developed (Fig. 1), with a Dispatcher sub-layer connecting the TCP/IP and the PROFIBUS stacks over the PROFIBUS Data Link Layer (DLL). This architecture adds extra sub-layers (in addition to the above mentioned dispatcher) to the standard TCP/IP and PROFIBUS stacks.

The IP-Mapper sub-layer resides directly below the TCP/IP Protocol Stack mapping the TCP/IP services into the PROFIBUS DLL services. It performs the identification, fragmentation and re-assembly of the IP packets to/from PROFIBUS DLL frames. This layer is also responsible for the integration of the client-server model of the IP protocol into the fieldbus communication model. Since the communication model of PROFIBUS is master-slave, it is required to provide extra functionalities to compensate the lack of initiative in slave stations (Pacheco, *et al.*, 2001; Pereira, *et al.*, 2002).

The Admission Control and Scheduling (ACS) sub-layer is responsible for the control/limitation of the network resources usage by the TCP/IP applications. Each IP packet is classified examining the IP Header fields such as destination address and port. Given this classification, the corresponding fragments are placed in a specific queue. Moreover, this sub-layer implements the appropriate scheduling policies, in order to provide the desired Quality of Service for the multimedia applications (Ferreira, *et al.*, 2001).

5. MULTIMEDIA IN THE RFIELDBUS MANUFACTURING AUTOMATION PILOT

In this section we will describe the set of multimedia applications demonstrated in the framework of the RFieldbus project, in the Discrete Parts Manufacturing field trial (Alves, *et al.*, 2001).

Some of the described applications have timing requirements, for which details on how the RFieldbus communication system was parameterised is provided in (Alves, *et al.*, 2003).

One important issue addressed in this manufacturing automation field trial is the possibility of integrating multimedia applications into the factory-floor. Applications such as (mobile) system monitoring, image processing, video and audio streaming, are examples. The manufacturing automation field trial was designed to be an adequate test-bed to assess the suitability of the RFieldbus system not only to support

both real-time control data and multimedia data in the same transmission medium, but also to include multimedia applications within control loops.

The layout of the manufacturing application is presented in Fig. 2. When a new part arrives (is transported to this subsystem), it must be classified according to a certain criteria and must be distributed to storage buffers or to the next stage of the manufacturing process. This next stage could be further processing (cutting, drilling, etc.) or just transporting a storage buffer to a warehouse.

Roller belts and different pneumatic equipment are used to transport and distribute parts to output buffers, according to their type. When output buffers are full, they are moved (either by an automatic vehicle, a robot arm, or an operator) to the respective unload station, in order to be emptied. Considering the classification criteria, currently each part is distinguished by its colour.

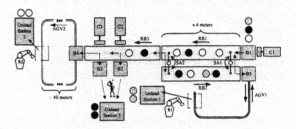

Fig.2. Mechanical system layout

5.1 Multimedia Devices

In order to exploit the multimedia characteristics of the field trial, several multimedia devices are integrated with the industrial system. Some of these devices are not common in the factory-floor, but there is a clear eagerness for their use, with clear benefits.

Head Mounted Displays (HMD). The Head Mounted Display (HMD) technology opens a new level in the way information is presented to the user. The display is in front of the user's eye, giving (due to the lenses used) the sensation of a big monitor several meters away. A headphone-like structure, special glasses or even a construction helmet are used in order to keep the monitor in position. Some HMD are opaque, while others are "see-through" (i.e. the user sees a translucid image in front of the real world). Some devices have 2 displays to provide a more real experience or even to support real 3D presentations (different images are projected in each display). HMDs can range from a monocular low-resolution (320x240) grey-scale monitors up to full colour, high-resolution (1024x768), high-luminosity and fast refresh rate binocular systems. Since the light from the HMD is focused on a small area (the eye of the viewer), it results in very low power consumption compared to other display technologies.

Video Conferencing (one-to-one). Simple Video Conferencing is becoming more and more accessible. The only additional hardware needed is a so-called "webcam". With this hardware configuration, there are several software solutions and communication standards. One of the most popular (it is freely available for several Microsoft operating systems) is Microsoft NetMeeting, that supports one-to-one video conferencing and is compatible with the H.323 video conferencing standard. Being a "symmetric" application (heavy traffic flows in both directions) and, moreover, interactive (where small delays are imperative), it is clear that this application borders the limits of the RFieldbus TCP/IP implementation and a high-quality conference cannot be expectable.

Video Capture. Nowadays the simplest and cheapest method to capture real word video sequences into a computer is to use an USB webcam. Low-resolution (320x240) USB webcams are available at a reduced cost (even for home users' standards). However, due to development constrains, the field trial is restricted to Windows NT platforms, which, unfortunately, do not support this type of cameras. Thus, a more "traditional" video camera plus video capture board combination was chosen.

Personal Data Assistant (PDA). Available for more than a decade, Personal Data Assistants (PDAs) have been used almost exclusively for their main purpose: as an electronic version of the traditional pocket agenda. However, in the last few years, new applications have been appearing ranging from recreational games to complex management applications. The latest generation of PDAs have advanced features like fast processors, full-colour displays, 64MB of RAM, TCP/IP and WWW support, connectivity (IrDA, 802.11b, Bluetooth, GSM/GPRS), expandability (PCMCIA, Compact Flash, Secure Digital), and more. Since there are no RFieldbus drivers for PDAs, the integration in the field trial is done using a standard TCP/IP network and a gateway.

Connection with a 802.11b Wireless Network. IEEE 802.11b (ISO-IEC, 1999) is a wireless network standard with a maximum speed of 11Mbps and a range up to 500 meters outdoors (at 1Mbps). It was designed to be the wireless equivalent of 10Mbps Ethernet. There are different 802.11b interfaces available (PCMCIA, ISA, PCI, etc) for several operating systems, as well as "stand-alone" bridges to Ethernet networks. It is important to note that the RFieldbus wireless physical layer is based on the physical layer of 802.11b, so the two wireless networks must coexist in the same environment.

5.2 Computing and Communication Infrastructure

The computing and communication infrastructure is based on a hybrid (wired/wireless) topology with mobile nodes. There are 2 masters in the system: PC1 is responsible for the overall control of the system and the mobility master (MobM) is responsible for triggering the mobility management procedure (Alves, *et al.*, 2002). In order to test, validate and demonstrate the technical capabilities of the RFieldbus approach, a network infrastructure including a wired segment (operating @ 1.5Mbps) and two radio cells was implemented, requiring communication between wired and wireless nodes and the handoff between radio cells. In order to have a structured wireless network supporting mobility, the RFieldbus network infrastructure is composed of two link base stations (LBS1, LBS2) that interconnect the two wireless domains (WL1, WL2) and the wired segment (WR). All nodes are PROFIBUS slaves (PC2-6, I/O1-2 and Drive 1-2), except PC1 and MobM (Mobility Master). Further details on the communication infrastructure and components are available in (Alves, *et al.*, 2003).

5.3 Multimedia Communication Flows

Within the field trial, several multimedia applications are used for control, monitoring and interpersonal communication. The multimedia message flows on the RFieldbus network are presented in Fig. 3. All the bandwidth figures presented in the remainder of this section are for raw data; that is, without TCP/IP and/or RFieldbus overhead.

Note again that all these TCP/IP flows are conveyed in the same RFieldbus network (as PROFIBUS frames), interleaving with the PROFIBUS "native" DP traffic.

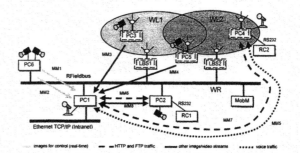

Fig. 3. Multimedia flows

TCP/IP Remote Part Classification (MM1, MM2). Two cameras in PC6 acquire images of the moving parts. These images must be transmitted to PC1 for classification within a time bound that permits proper control actions to be taken. On the capture side (PC6), images are captured at a predefined rate, down-sampled and processed to check if a part is in the active area or not. If a part is detected, the image is compressed to JPEG format and sent using a TCP connection to the remote machine (PC1), tunneled in PROFIBUS frames. On the monitoring side (PC1), each received image is decompressed and processed to classify the part. Both the image and the result of the processing are presented to the user. After the part classification, the control system reacts accordingly (e.g. issuing PROFIBUS DP commands for part transfer: B in Fig. 4). When the part is not within the camera's field of view, a small message is sent using the TCP connection to PC1 (again, tunneled in PROFIBUS frames). This message is used only for user interface purposes. This multimedia application is the most stringent in terms of real-time performance, since it interfaces directly with the control system. To guarantee correct delivery of images from Slave App to Master App (A in Fig. 4) the system was configured to guarantee QoS of this stream (see details in (Alves, et al., 2003)).

TCP/IP Remote Video Monitoring (MM3, MM4). The Image Centre application is used by the operator in the central control PC (PC1) to visually monitor the area in the trajectory of the AGVs (AGV1 and AGV2). On the capture side (PC3, PC5), images are captured at a predefined rate (1/sec by default), compressed to a JPEG stream and sent using UDP to the remote machine (again, tunnelled in PROFIBUS frames). On the monitoring side (PC1), the Image Centre application decompresses the JPEG images and presents them to the user. There are basic control facilities to start and stop the multimedia stream in the client. The used RFieldbus bandwidth is around 24 kbps. The last image from each camera (including images from auxiliary applications like Remote Part Detection) is stored in the memory of the Image Center application and forwarded using UDP fragments to the Intranet every second (thus providing this information to PDAs or HMDs connected through IEEE802.11 access points to the Intranet).

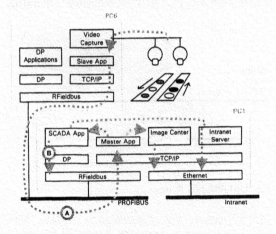

Fig. 4. The Remote Part Classification System

TCP/IP Voice Connection (MM5). This is a simple point-to-point TCP/IP bi-directional voice application connecting PC1 and PC4. The only control available to the user is a button to "dial" or "hang-up". Incoming calls are signalled to the operator and accepted automatically. This was implemented using Microsoft NetMeeting ActiveX control and Visual Basic. The used bandwidth depends on the desired sound quality, but a typical figure is around 4 kbps.

TCP/IP Remote Position Detection (MM6). The autonomous vehicle (AGV1) may slightly deviate

from the ideal position. Therefore, a visual position detection mechanism was implemented in order to make the appropriate position corrections for the robot arm to manipulate the buffer. On the capture side (PC2), an image of the AGV is captured by request of the monitoring machine (PC1), at 640x480 and compressed to JPEG format. The data is then sent using TCP to the remote machine. On the monitoring side, the image is decompressed and processed to identify the presence and location the buffer. The image is also made available to the user (down-sampled and forwarded using UDP to the Image Centre). The processing result is used to program the robot movement to pick up the buffer. This control is done remotely using the TCP connection. This application has no stringent timing requirements but demonstrates de transmission of large data blocks (14Kbyte) over the RFieldbus network.

Remote Robot Control Services (MM7, MM8). In order to be able to remotely control the two robots of the field trial, support of FTP and HTTP is provided. The FTP servers (in PC2, PC4) are configured to enable the transference of program files using the standard FTP protocol to a specific directory on the computer. The WWW application enables the transfer of these files to the robot. This application uses a standard WWW Server with specific scripts and application modules to enable communication and control of the Robots. This enables any user on the network with a WWW Browser to download a program file from the server to the robot controller, to run a particular program on the robot, stop the program on the robot and to check current robot status (including, in the future, an image for the case of Unload Subsystem 1). The program files can be transferred from a remote machine to the WWW server using the FTP service or using a WWW form. In the future, a simple username/password authentication/access control scheme will be implemented.

Intranet Interface Services. Several services are available for system monitoring and control, using standard TCP/IP stations in the Intranet attached to PC1. There are two ways to access this information. a) WWW Server. This server provides several HTML pages and forms that the user can browse to check the current system status and interact (given the proper credentials) with the system. Any WWW browser can access this information. b) UDP Server. The efficient broadcasting of information (including video streams) to several stations on an Intranet is supported by using the UDP service. This service is used by client applications developed for standard Windows PCs and for Pocket PC devices (see next section).

5.4 Some Other Details

In (Pacheco, et al., 2003), implementation details on how the diverse application components related to the multimedia issues were implemented are described.

Here, we are going to describe just a few of those. The reason for this choice is just to give further intuition on the multimedia possibilities that are opened by the RFieldbus.

Intranet Server Application. This is the main interface between the field trial manufacturing system and the Intranet clients. A GUI interface was developed in Visual Basic where the operator can change the text information and access/change/define the parameters of each of the PCs in the communication infrastructure. It is also possible to send alarm messages to the Intranet and to configure the messages to be accepted by the information devices (HMD and PocketPCs) and the SMS gateway (a functionality not referred before). On normal system operation, however, these fields will be usually changed automatically using the control system interfaces (via Windows Messages) already implemented. All the information in the Intranet is exchanged using UDP packets. This solution was selected to enable the broadcast of information to all the devices connected to the Intranet, including the Augmented Reality Client (HMD) and the SMS Gateway. Text information and parameter information is sent in a single packet. Alarm messages have 5 classes and several type identifiers. Classes are used to filter the presentation of alarm messages in the Intranet devices. For example, the HMD may be configured to present only some classes of alarms. At this moment the class identifiers are used to identify the area of the alarm: A-Manual buffers, B-Roller belts and AGVs, C-Remote Unload Station, D-System wide messages, E-Reserved/User Messages. Type identifiers are only used to illustrate the alarm using images and/or sounds. The alarm message itself is a text string. Output-only devices like HMD and SMS gateway are configured using Intranet commands. For the HMD and output-only clients it is possible to select a video stream to be displayed as well as the alarm classes. For the SMS gateway it is possible to configure the destination cellular phone number and the alarm classes. This configuration commands may be issued using the Intranet Server GUI or using one of the Intranet Clients.

Intranet Clients Applications. There are two versions of the Intranet Client Application: one for Pocket PC and the other for desktop Windows. Both offer the same services and user interface, and were developed in Microsoft Visual Basic (Embedded version for the Pocket PC clients). The current version of the application (Fig. 5) provides access to images from cameras, text information from PCs, read/write access to selected parameters in PCs, alarm messages and SMS/HMD configuration. The Pocket PC client was developed in a way that it can be used with a finger (no need to use the pen for all operations except changing parameters) or using the hardware 4-way cursor. The device used is a Compaq iPaq 3870 PocketPC with a PCMCIA Expansion Jacket and a WL 110 Wireless PC Card (802.11b). This results in a completely mobile and autonomous system that can be

used for more than 8 hours without recharging. The connection to the Intranet is done using a Compaq WL410 Access Point (i.e. an 802.11b to Ethernet bridge). Since the Microsoft Winsock ActiveX component for Windows CE does not support UDP communications, an ActiveX component was developed to perform this task for Embedded Visual Basic. Additionally, it also enables the creation and writing to binary files using strings (something that the Microsoft standard components also lack). This is a rather simple component but it provides great benefits: the component converts the UNICODE Visual Basic strings (Windows CE is a UNICODE system) to ASCII and so all file and network data is in ASCII format; it supports strings with null characters and supports broadcast UDP packets.

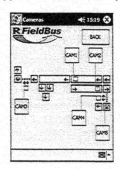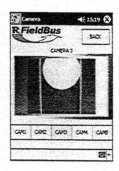

Fig. 5. Pocket PC Client Application snapshots: field trial schematics (left) and camera image (right)

The HMD application. It provides access to a video stream and/or alarm messages. The HMD used is a low-cost, monochrome 320x240 TekGear M1 (Fig. 6) connected to a PC running a Visual Basic application. The type of information that the HMD presents is selected from the Intranet Client Applications (in Pocket PC or desktop PC). This information includes image streams and alarm messages. Alarm messages have several pictures depending on the type of alarm (like a warning sign in Fig. 6), and feature animations so the user will notice the new message even without looking directly at the HMD.

Fig. 6. The TekGear M1 HMD (left). Views at the HMD: camera view (centre); information message (right)

SMS Gateway. In the field trial, a mobile phone is used, connected to a PC using a RS232 data cable. The SMS are sent to the network using standard GSM commands. Unfortunately "standard" should be used with care in this area: each phone manufacturer has implemented several nuances of the standard and a COM port sniffer had to be used in order to identify the exact format of the messages. The application forwards messages received from the UDP Intranet

Server. The application also checks if the device is connected, the GSM network signal and battery power.

6. CONCLUSIONS

The support of new information technologies and applications in the factory-floor, such as multimedia applications, is moving from eagerness to a real need. In this paper we have discussed the motivation for embodying industrial multimedia application support into existent fieldbus communication systems. The RFieldbus project is one example of an initiative where this objective has been successfully carried out. Without going into details neither in the RFieldbus specification nor in the internals of the multimedia applications related with the discrete parts manufacturing field trial, we wanted to bring into the light some functionalities which are possible to implement with RFieldbus. The field trial itself is a complex distributed computer control system, with a wide set of software applications, ranging from small software components, to large (distributed) applications. The implementation provides a suitable framework for the assessment of integrated multimedia and real-time control traffic in the same network, and the use of a hybrid wired/wireless network for Distributed Computer-Control Systems.

The interested reader may find additional information available at http://www.hurray.isep.ipp.pt/rfpilot.

ACKOWLEDGEMENTS

This work was partially supported by the European Commission (project RFieldbus IST-1999-11316).

REFERENCES

Almeida, L., Tovar, E., Fonseca, J. and F. Vasques (2002). Schedulability Analysis of Real-Time Traffic in WorldFIP Networks: an Integrated Approach. IEEE Transactions on Industrial Electronics, Vol. 49, No. 5, pp. 1165-1174.

Alves, M., Brandão, V., Tovar, E., Pacheco, F. and L. M. Pinho (2001). Specification of the Manufacturing Automation Field Trial. IPP-HURRAY Technical Report (HURRAY-BTR-0131).

Alves, M., Pinho, L.M., Tovar, E. and S. Machado (2003). Engineering Real-Time Distributed Applications with RFieldbus. IPP-HURRAY Technical Report.

Alves, M, Tovar, E., Fohler, G. and G. Buttazzo (2000). CIDER: Envisaging a COTS Communication Infrastructure for Evolutionary Dependable Real-Time Systems. In WIP Proceedings of the 12th IEEE Euromicro Conference on Real-Time Systems (ECRTS2002), pp. 19-22.

Alves, M., Tovar, E., Vasques, F., Hammer, G. and K. Roether (2002). Real-Time Communications over Hybrid Wired/Wireless PROFIBUS-based Networks. In Proceedings of the 14th Euromicro Conference on Real-Time Systems (ECRTS'02), pp. 142-150.

Decotignie, J.-D. and P. Prasad (1994). Spatio-temporal Constraints in Fieldbus: Requirements and Current Solutions. In Proceedings of 19th IFAC/IFIP Workshop on Real-Time Programming, pp. 9-14.

Decotignie, J.-D. (2001). A Perspective on Ethernet as a Fieldbus. Proceedings of the 4th International Conference on Fieldbus Systems and their Applications (FET'01), pp. 138-143.

Ferreira, L., Tovar, E. and S. Machado (2001). Scheduling IP Traffic in Multimedia Enabled PROFIBUS Networks. In proceedings of the 8th IEEE International Conference on Emerging Technologies and Factory Automation (ETFA'2001), pp. 169-176.

ISO/IEC 8802-11 (1999). IEEE Standard for Information technology—Telecommunications and information exchange between systems—Local and metropolitan area networks. ISO/IEC 8802-11: 1999.

Kopetz, H. and G. Grünsteidl (1994). TTP - A Protocol for Fault-Tolerant Real-Time Systems. IEEE Computer, 27(1).

Moldovansky, A. (2002). Utilisation of Modern Switching Technology in EtherNet/IP Networks. In Proceedings of 1st Workshop on Real-Time LANs in the Internet Age (RTLIA2002), pp.25-27.

Pacheco, F. and E. Tovar (2002). User-interface Technologies for the Industrial Environment: Towards the Cyber-factory. In the Proceedings of the 6th IST CaberNet Radicals Workshop, Funchal, Madeira Island.

Pacheco, F., Pereira, N., Marques, B., Machado, S., Pinho, L.M. and E.Tovar (2003). Industrial Multimedia put into Practice: A Manufacturing Field Trial. IPP-HURRAY Technical Report.

Pacheco, F., Tovar, E., Kalogeras, A and N. Pereira (2001). Supporting Internet Protocols in Master-Slave Fieldbus Networks. In proceedings of the 4th IFAC International Conference on Fieldbus Systems and Their Applications (FET'2001), pp. 260-266.

Pereira, N., Pacheco, F., Pinho, L. M., Prayati, A., Nikoloutsos, E., Kalogeras, A., Hintze, E., Adamczyk, H. and L. Rauchhaupt (2002). Integration of TCP/IP and PROFIBUS Protocols. In WIP Proceedings of the 4th IEEE International Workshop on Factory Communication Systems.

Rockwell Automation (1997). DeviceNet Product Overview. Publication DN-2.5.

Rockwell Automation (2000). Making Sense of e-Manufacturing: a Roadmap for Manufacturers. Rockwell Automation White Paper.

Skeie, T., Johannessen, S and O. Holmeide (2002). The Road to and End-to-End Deterministic Ethernet. In Proceedings of the 4th IEEE International on Factory Communication Systems (WFCS'2002), pp. 3-9.

Thomesse, J.-P. (1993). Time and Industrial Local Area Networks. In Proceedings of COMPEURO'93.

Tovar, E., Pinho, L. and L. Almeida (2002). Position Paper on Time and Event-triggered Communication Services in the Context of e-Manufacturing. In Proceedings of the IEEE Workshop on Large Scale Real-Time and Embedded Systems (LARTES 2002).

Tovar, E. and F. Vasques (1999). Real-Time Fieldbus Communications Using Profibus Networks. IEEE Transactions on Industrial Electronics, Vol. 46, No. 6, pp. 1241-1251.

Tovar, E., Ferreira, L., Vasques, F. and F. Pacheco (2001). Industrial Multimedia over Factory-Floor Information Networks. In proceedings of the 10th IFAC Symposium on Information Control Problems in Manufacturing (INCOM '01).

ELSEVIER

IFAC
PUBLICATIONS
www.elsevier.com/locate/ifac

QUALITY OF SERVICE ARCHITECTURE FOR WORLDFIP

Miguel Angel León Chávez

Benemérita Universidad Autónoma de Puebla
14 Sur y Av. San Claudio, CP 72570, Puebla, México
Tel. (52) 222 229 55 00 ext. 7213   Fax (52) 222 229 56 72
Email: mleon@cs.buap.mx

Abstract: This paper presents an analysis of the Quality of Service (QoS) Architecture of
WorldFIP, which is a Fieldbus used to connect all kind of devices into a factory. The
analysis shows that this Fieldbus makes use of a static resource reservation mechanism,
i.e. at network configuration time the human operator configures the nodes and their
periodic traffic in order to meet and guarantee the application requirements, at application
run-time. Nevertheless, in manufacturing applications there exist also aperiodic traffic
(events and alarms) for which the Fieldbus must provide guaranteed levels of QoS at
application run-time. The analysis shows that WorldFIP provides only the best-effort
level of QoS for a kind of aperiodic traffic, called aperiodic messages; therefore new QoS
architectures are required. Finally, this paper presents the specification of both a new
WorldFIP QoS architecture and a WorldFIP Resource ReSerVation Protocol using the
Specification and Description Language (SDL). *Copyright © 2003 IFAC*

Keywords: Fieldbus, Quality, Protocols, Specification, Verification.

1. INTRODUCTION

Fieldbuses are special purpose Local Area Networks
used to connect all kinds of devices into a factory,
such as sensors, actuators, programmable controllers,
(C)NC machines, processors, and so on (Thomesse,
1998). These networks are usually seen as a three
layers architecture, which includes the physical layer,
the data link layer and the application layer.

Typically, the distributed manufacturing applications
make use of the fieldbuses in order to monitor and
control the processes taking place in the application.
Examples of such applications are: factory
automation, automotive industry, textile machinery,
electronics manufacturing, food and beverage,
chemical processing, and so on.

With regard to the communications, the main
temporal QoS requirements of these applications are
as follows (León and Thomesse, 2000): a bounded
end-to-end delay and the periodic transmission. The
first requirement means that a message produced by
an application task, at a source node, must be
received at the destination node and delivered to the
application task, within a given time interval. The
second requirement takes into account that most of
the messages exchanged between the application
tasks are composed of periodic data that must be
transmitted before the beginning of their next period.
Nevertheless, it should be noted that there are also
aperiodic data in these applications, i.e. randomly
produced such as events and alarms, some of which
must be delivered within a given time interval.

These QoS requirements are two quantified temporal
user requirements, i.e. QoS parameters. A Fieldbus
and its users must negotiate at least the two following
levels of agreement for any QoS requirement:
guaranteed level and best-effort level. In the first
level the Fieldbus guarantees that the QoS
requirements are met, using different mechanisms
that are integrated into its QoS architecture. In the
second level the Fieldbus does its best to meet the
user requirements but there is no assurance that the
QoS will be provided.

Nowadays, there is one approach to define the Fieldbus QoS architecture, i.e. the IEC 61158 QoS architecture (IEC 61158), which is based on both the Data Link Services (DLS) and the QoS attributes which are common among the different types of DLS. In fact, the IEC 61158 can be seen as a collection of Fieldbus Standards because it defines eight types of DLS, numbered from 1 to 8, each corresponding to a Fieldbus Standard as follows: TS 61158, ControlNet, Profibus, P-Net, Foundation Fieldbus, SwiftNet, WorldFIP and Interbus.

In addition to the basic classes of DLS (i.e. connection-mode and connectionless-mode) the IEC 61158 defines the two following classes of DLS: a DL(SAP)-address, queue and buffer management service and a time and transaction scheduling service. In this approach, a DLS-user may select, directly or indirectly, the parameters of the QoS attributes (i.e. priority, maximum confirm delay, authentication, scheduling-policy and timeliness) in order to determine the quality of the DLS.

The IEC 61158 QoS architecture is well adapted to the type 1 of DLS, i.e. the TS 61158 Fieldbus, simply because the DL-entities can control the QoS attributes. Nevertheless, it is not always the case in the other types of DLS. A possible solution is mapping the QoS attributes and then to define the DL-entities to control them, as in the SwiftNet case. Another solution is to define new Fieldbus QoS architectures. This paper discusses the WorldFIP architecture and presents a new WorldFIP QoS architecture based on the Client-Server model (León, 2001). In addition, the paper presents the specification of a WorldFIP Resource ReSerVation Protocol (WorldFIP-RSVP) using the Specification and Description Language (SDL) (ITU-T Z.100), which is an object-oriented language with concepts for describing the logical structure, data and behavior aspects of the systems.

The remaining of this paper is structured as follows: Section 2 discusses the WorldFIP architecture, section 3 presents the QoS architecture based on the Client-Server model, section 4 presents the SDL specification of the WorldFIP QoS architecture and its WorldFIP-RSVP, finally section 5 presents some conclusions.

2. WORLDFIP ARCHITECTURE

World Factory Instrumentation Protocol (WorldFIP) (CENELEC EN 50170-3) is a centralized Fieldbus, which defines a Bus Arbitrator (BA), which "gives permission to speak" to each information producer. At network configuration time, the human user invokes the application layer services to send a set of service request primitives to the BA, at the data link layer. These primitives describe the execution of both the basic cycle and the macro cycles (the sequential execution of one or more basic cycles).

A basic cycle is composed of at least one window, called the periodic window, and at up to four windows, as follows: a periodic window, an aperiodic variables window, an aperiodic messages window and possibly a synchronization window to adjust the constant duration of the basic cycle, as is shown in Fig. 1.

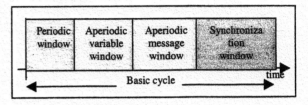

Fig. 1. WorldFIP basic cycle.

A unique identifier identifies each application's variable and each node can be a producer and/or consumer of one or more variables.

At application run-time, during the periodic window, BA reads a table of periodic variables and injects each variable identifier onto the network; each variable has only one producer. Consumers needing to utilize the variable, alerted by the identifier, store and use the value broadcasted by the producer.

Whenever an aperiodic variable or an aperiodic message is produced, the producer utilizes the response, of a received periodic variable identifier, to request its transmission. The BA stores the identifier, which is carried by the request, into the appropriate queue. After completing the current periodic window, the requested aperiodic variables or the requested aperiodic messages are then handled in the same way that the periodic variables within the proper window, where the relevant producers reply current values.

It is interesting to note that these windows provide two levels of QoS. On one hand, the periodic window provides a cyclic guaranteed service. In fact, for each variable the BA knows its periodicity and its applicative type (e.g. 8-bit integer, chain of 32 characters, etc.). Using the transmission time and the turnaround time, the BA can calculate the time required for an elementary periodic transaction made up of the transmission time for a question frame, followed by the transmission time for the associated response frame. On the other hand, the aperiodic variable window and the aperiodic message window provide a best-effort level of QoS, since if the duration of the synchronization window is zero in a current basic cycle new requested aperiodic variables or new requested aperiodic messages will be transmitted in one of the next basic cycles. Therefore, the BA cannot guarantee the delay for an aperiodic transaction.

Note also that WorldFIP uses a centralized and static resource reservation mechanism, i.e. there is one node, the BA, having a table and different queues of identifiers that define at any moment the communication requirements of all the nodes. The periodic variables are known a priori and will not change during the application run-time. Hence, it is possible to assign an identifier to each variable. However, if the number of nodes increases or if the number of periodic variables change in each node it is necessary to stop the application in order to determine whether the new set of requirements can be met and guaranteed. If yes, the new requirements are configured and the application is restarted. Nevertheless, many applications cannot be stopped without great losses.

2. CLIENT-SERVER QoS ARCHITECTURE

This section presents the Client-Server QoS architecture, in which is based the WorldFIP QoS architecture described in the next section.

The OSI Reference Model makes use of the Client-Server model to establish the relation between the adjacent layers. The OSI Reference Model provides a framework to define the structure of interconnected entities in terms of a layered architecture comprising services and protocol entities. The services in each layer are provided by the means of one or more protocols, i.e. the procedure to communicate between two protocol entities using the services of the underlying layer. Thus, each OSI layer is a client of its underlying layer, and each layer is a server of its upper layer.

The client-server model promotes modular, flexible and extensible system design (Adler, 1995). Nevertheless, the server only provides best-effort services (Thomesse, et al. 1995), i.e. the server performs the operation indicated by the service request, but no guarantee is given in what concerns the response time. For example, if a client requests a timing constraint service, there is not way to guarantee a bounded response time except if some assumptions are made on the availability of the server and its resources.

A Client-Server QoS architecture has been proposed in (León, 2001), which makes use of the resource reservation approach, i.e. the server resources are apportioned to the client requests, and subject to a management policy. Before any use of a service with some guaranteed level of QoS, the client must first reserve the server resources. Thus, in addition to the normal entities of the client-server model, i.e. the Client-Entity (CE) and the Server-Entity (SE), the following components are required, shown in Fig. 2:

Resource Allocator (RA), which stores a state structure about allocation of resources on the server and performs Admission Control (AC) tests.

Scheduler (SC), which schedules the accepted service requests towards the SE accordingly to meet the QoS requirements.

Requester Module (RM), which resides in both the client (RMC) and the server (RMS). This module is responsible for negotiating the QoS parameters between the client and the server using a resource reservation protocol.

Client-Server Resource ReSerVation Protocol (CS-RSVP), which is a reservation protocol between the client and the server.

Before any use of a service with a level of QoS, the following steps must be executed: (1) The client specifies the required QoS; (2) The QoS parameters are conveyed to the server; (3) The QoS parameters must be mapped to the required resources; (4) The required resources may be either admitted, reserved and allocated or rejected and possibly negotiated; (5) The close-down procedure concerns resource deallocation.

These steps are as follows for the Client-Server model, shown in Fig. 2: The CE passes its QoS parameters to the RMC, which creates a Resource ReSerVation (RSV) message sending it to the server.

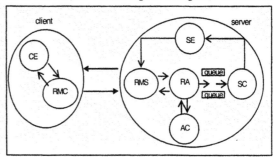

Fig. 2. Client-Server QoS Architecture.

The Requester Module of the Server (RMS) receives the RSVmessage, passing it to the Resource Allocator (RA), which invokes the Admission Control (AC) mechanism; it performs admission tests to determine whether the QoS parameters are met. If yes, RA creates a request identifier (requestID), stocks the QoS parameters into the state structure, puts the requestID into the appropriate queue, and sends the requestID to the client, which will use this identifier in future service requests.

If the QoS parameters cannot be met and guaranteed, RA creates a reject response sending it to the client. If the QoS parameters are negotiable the response contains the QoS parameters that could be met. Upon reception, the client may initiate a negotiating process. Whereas the client decides to accept or reject the proposed QoS parameters, their values are locked by the AC mechanism waiting for either a reservation message or the lock timer expiration. Finally, SE performs the requested services and sends the service response through the RMS.

It can be noted that the scheduler (SC) may use any scheduling algorithm according to the OSI layer. Note also that the schedulability test is performed by the AC mechanism and the scheduling policy is performed by the SC mechanism. Note that this description uses the term RSVmessage, nevertheless the service primitives are specified in the next section.

3. SDL SPECIFICATION OF THE WORLDFIP QoS ARCHITECTURE AND THE WORLDFIP-RSVP

This section analyzes the main design goals of the WorldFIP-Resource ReSerVation Protocol, presents its specification in Specification and Description Language (SDL) and its verification using the ObjectGEODE tool.

3.1 WorldFIP-RSVP Design goals.

This protocol will must take into account the existence of two main end-users: the human user and the field device users. The first user is the people who are in charge of the control, of the maintenance, of the production scheduling, of the technical management and so on. The field device users are the devices interconnected by WorldFIP. The human user can send service requests to the Bus Arbitrator (BA) in order to configure dynamically a variable, but also to watch the periodic table and the queues for aperiodic variables and messages. On the other hand, the device users can only produce and consume variable values, but also to configure dynamically new variables.

A key point in the WorldFIP-RSVP specification is the regularity of the variable's production, i.e. if the variable will be either periodic or aperiodic. Furthermore, in the case of periodic variables, the reservation protocol must also allow to specify its validity, called *reservation validity*, i.e. the time interval during which the variable will be produced, e.g. one day, one month, one year, etc. This approach is called reservation in advance.

In addition to the regularity, WorldFIP-RSVP must allow to specify the variable's *name* and its *starting time*. Finally, WorldFIP-RSVP must allow specifying whether the QoS parameters are *negotiable* or not.

WorldFIP-RSVP is a signaling protocol to reserve resources, i.e. to configure dynamically new variables. It is not an application layer protocol, not even a MAC protocol, and should avoid replicating any access control or application function. The WorldFIP-RSVP's task is to establish and maintain resource reservation between the end-users and the BA.

In summary, WorldFIP-RSVP is a means used by the end-users to communicate their QoS requirements to the BA in an efficient way, independent of the specific QoS requirements.

3.2 Specification and validation of WorldFIP-RSVP.

WorldFIP-RSVP is specified in the Specification and Description Language (SDL) because it is a language that is intelligible to human beings, formal enough to support analysis and comparison of behaviors (Bræk, 1996).

For validation purposes, the term validation model is utilized by SDL. A validation model is a description of the system, which is suitable for validation, i.e. to apply validation techniques such as testing the formal model, exhaustive validation (reachability analysis), non-exhaustive validation (analysis of a random subset of the reachable states), simulation and informal validation techniques (checklist) (Hogrefe, 1996).

A validation model is always executable, then in order to construct the validation model of the WorldFIP QoS architecture as well as the WorldFIP-RSVP service primitives, the toolset ObjectGEODE is utilized. It includes the following tools: SDL editor, MSC editor, SDL & MSC checker, SDL & MSC interactive simulator, SDL & MSC exhaustive simulator, C & C++ code generator, C run-time library and design tracer (Cheng, 1996).

The WorldFIP-RSVP service primitives are as follows:

- *ShowListVarPeriod*, which is a request to show the periodic variable list to the human user.
- *ShowListVarAperiod*, which is a request to show the aperiodic variable list to the human user.
- *ShowListMsgAperiod*, which is a request to show the aperiodic message list to the human user.
- *ShowVar*, which is a request to show a variable to the human user.
- *CfgVar*, which is a human-user's request to configure dynamically a variable.
- *ListVarPeriod*, which is a response to show the periodic variable list to the human user.
- *ListVarAperiod*, which is a response to show the aperiodic variable list to the human user.
- *ListMsgAperiod*, which is a response to show the aperiodic message list to the human user.
- *Var*, which is a response to show a variable to the human user.
- *CfgAccept*, which is a response to indicate that a CfgVar is accepted.
- *CfgReject*, which is a response to indicate that a CfgVar is not accepted.
- *CfgVarPC*, which is a device-user's request to configure dynamically a variable.
- *CfgAcceptPC*, which is a response to indicate that a CfgVarPC is accepted.
- *CfgRejectPC*, which is a response to indicate that a CfgVarPC is not accepted.
- *VarModify*, which is a request to modify a previously configured variable. This primitive is not specified in the current version of the protocol.
- *VarDeallocate*, which is a request to deallocate a previously configured variable. This primitive is not specified in the current version of the protocol.

In SDL, behavior is always performed in the context of a *system*, beginning with a top-level description of the system. Fig. 3 shows the newWorldFIP system, which is composed of the following *blocks*: the application and data link layers of the bus arbitrator (ApplicaLayerBA and DataLinkLayerBA), the data link layer of the producer and/or consumer nodes (DataLinkLayerPC1,...,N) and the physical layer. These blocks are interconnected through *channels*, named c0, c1, c2 and c3. The channel ca connects the BA with the environment, by which the human user configures the network.

In addition, Fig. 3 shows the *signals* passed in each direction over the channels, as indicated by the arrows on the channels.

Fig. 3 shows also the declaration of the signals as well as the declaration of the following data structures: RSV, ListVP, ListVA and ListMsgA. The RSV structure defines the QoS parameters used by the service primitives, and the lists define the identifiers for the periodic variables (ListVP), the aperiodic variables (ListVA) and the aperiodic messages (ListMsgA). This last list is for simulation purpose since the messages do not have assigned an identifier.

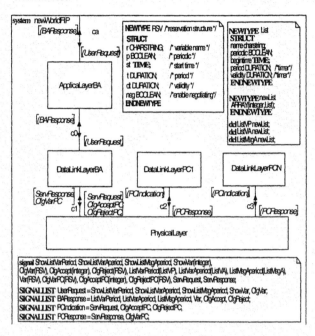

Fig. 3. WorldFIP system.

In SDL, a system occurs only at the top level, while blocks only occur inside. The system is decomposed into blocks and channels recursively over as many levels as desired until the basic components, called *processes*, are reached.

Fig. 4 and 5 present the application and data link blocks of the BA, which are composed of processes. It can be noted the Client-Server QoS architecture between these blocks.

Fig. 6 shows the data link block of the producer/consumer nodes. A reliable data transfer subsystem is supposed, i.e. the physical layer block, shown in Fig. 7.

One *SDL process* is a concurrent object with its own control flow, described by an Extended Communicating Finite State Machine (FSM), which is composed of the following four main parts: input port, FSM, timers and variables.

The input port contains an unbounded queue of incoming signals. Signals arriving at the process will be merged into the input port in the order they arrive, conflicts are resolved by selecting an arbitrary sequential order. Signals will remain in the input port until they are consumed by the FSM, which performs one transition from one state to another state for each consumed signal. This transition takes a short but undefined time. If there are not signals in the input port, FSM remains in the same state until a signal arrives.

On each transition, FSM may generate output signals; perform operations on the variables and timer operations. This FSM state-transition behavior

is expressed in terms of a *process diagram*. The process diagrams are not presented in this paper for lack of space.

In order to validate the WorldFIP QoS architecture and its RSV Protocol, the ObjectGEODE tool SDL simulator was used for interactive and exhaustive validation (reachability analysis).

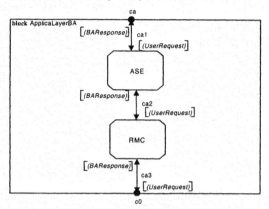

Fig. 4. Bus Arbitrator Application Block.

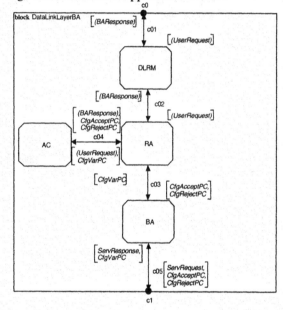

Fig. 5. Bus Arbitrator Data Link Block.

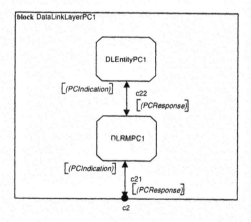

Fig. 6. Producer/Consumer Data Link Block.

A key point of the WorldFIP behavior is the timing scheduling of the traffic performed by the BA at the proper window. For this reason, the dynamic configuration of new variables will only be able to be carried out during the synchronization window. In this window the BA transmits padding identifiers, until the end of the basic cycle, to indicate to the other nodes connected to the bus that it is still functioning. A padding identifier is an identifier not produced by any node. Therefore, WorldFIP-RSVP should be invoked during the synchronization window signalled now by two padding identifiers, at the start and the end of this window.

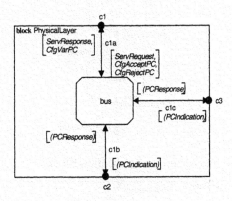

Fig. 7. Physical Link Block.

5. CONCLUSION

This paper has presented an analysis of the WorldFIP QoS architecture, which is used by the distributed manufacturing applications where most of the traffic is composed of periodic messages, which are known a priori, at network configuration time.

Therefore, WorldFIP can guarantee both the bounded end-to-end delay and the timing scheduling of the periodic traffic using a static resource reservation mechanism. Nevertheless, there are also aperiodic messages, which are known at application run-time, and for which the Fieldbus must provide guaranteed levels of QoS.

This paper has also presented the Client-Server QoS architecture, which allows guaranteeing some services between the adjacent OSI layers and between an OSI layer and its peer. Hence it can be used in centralized fieldbuses, such as WorldFIP. This architecture is based on the resource reservation approach, which is used by the Integrated Service model of the Internet profile.

In addition, this paper has presented the specification of a new WorldFIP QoS architecture and the Resource ReSerVation Protocol (WorldFIP-RSVP) between the Producer/Consumer nodes and the Bus Arbitrator node as well as between the human-user and the Bus Arbitrator. The specification is made in Specification and Description Language (SDL) and is validated by simulation with the aid of the toolset ObjectGEODE. This tool has a C & C++ code generator for different targets; a future research work is to measure the execution time of the WorldFIP-RSVP for a given target.

WorldFIP-RSVP allows configuring dynamically, i.e. at application run-time, new periodic and aperiodic variables involving a negotiation process. This reservation protocol is a signaling protocol and therefore the original service request primitives of WorldFIP are not taken into account in Fig. 3, such as those to configure the micro cycle and the macro cycle.

In order to meet the previously configured requirements, WorldFIP-RSVP can be invoked during the synchronization window in each basic cycle. For this goal it is required that the BA only points out the beginning and end of this window with a padding identifier.

The Client-Server QoS architecture will be used to define a new QoS architecture for the distributed Fieldbus architectures, such as ControlNet, and this is our future research work.

REFERENCES

Adler, R.M. (1995). Distributed coordination models for client/server computing. *Computer - IEEE Computer Magazine*, **vol. 28**, no. 4, pp. 14-22.

Bræk, R. (1996). SDL Basics, *Computer Networks and ISDN Systems*, **vol. 28**, no. 12, pp. 1585-1602.

CENELEC EN 50170-3 (1995). WorldFIP, General Purpose Field Communication System. *EN 50170-3*.

Cheng, K.E. (1996). A requirements definition and assessment framework for SDL tools. *Computer Networks and ISDN Systems*, **vol. 28**, no. 12, pp. 1703-1716.

Hogrefe, D. (1996). Validation of SDL systems. *Computer Networks and ISDN Systems*, **vol. 28**, no. 12, pp. 1659-1668.

International Electrotechnical Commission, IEC 61158. (1999). Digital Data Communications for Measurement and Control– Fieldbus for use in Industrial Control Systems.

ITU-T, ITU Recommendation Z.100 (2000). *The Specification and Description Language (SDL)*; ITU, Geneva.

León, M. and J.P. Thomesse (2000). Fieldbuses and Real-Time MAC Protocols. In *Proceedings of the 4th IFAC International Symposium on Intelligent Components and Instruments for Control Applications (SICICA)*, Buenos Aires, Argentina, pp. 51-56.

León, M. (2001). Quality of Service and Fieldbuses. In *Proceedings of the 2001 Fieldbus Technology (FeT)*, Nancy, France, pp. 160-164.

Thomesse, J.P., Z. Mammeri and L. Vega. (1995). Time in distributed systems: cooperation and communication models. In *Proceedings of the 5th IEEE Workshop on Future Trends of Distributed Computing Systems, IEEE Computer Society Press*, pp. 41-49.

Thomesse, J.P. (1998). A Review of the FieldBuses. *Annual Reviews in Control*, **vol. 22**, pp. 35-45.

CRITICAL MESSAGE INTEGRITY OVER A SHARED NETWORK

Jennifer Morris and Philip Koopman

*ECE Department
Carnegie Mellon University
Pittsburgh, PA, USA
jenmorris@cmu.edu, koopman@cmu.edu*

Abstract: Cost and efficiency concerns can force distributed embedded systems to use a single network for both critical and non-critical messages. Such designs must protect against masquerading faults caused by defects in and failures of non-critical network processes. Cyclic Redundancy Codes (CRCs) offer protection against random bit errors caused by environmental interference and some hardware faults, but typically do not defend against most design defects. A way to protect against such arbitrary, non-malicious faults is to make critical messages cryptographically secure. An alternative to expensive, full-strength cryptographic security is the use of lightweight digital signatures based on CRCs for critical processes. Both symmetric and asymmetric key digital signatures based on CRCs form parts of the cost/performance tradeoff space to improve critical message integrity. *Copyright © 2003 IFAC*

Key Words: safety-critical, error detection codes, embedded systems, fault detection, networks

1 INTRODUCTION

Distributed embedded systems often contain a mixture of critical and non-critical software processes that need to communicate with each other. Critical software is "software whose failure could have an impact on safety, or could cause large financial or social loss" (IEEE, 1990). Because of the high cost of failure, techniques such as those for software quality assurance described in IEEE Std 730-1998 (1998) are used to assure that such software is sufficiently defect free that it can be relied upon to be safe. However, such a process is expensive, and generally is not applied to non-critical system components.

In a typical safety-critical transportation system, such as in the train or automotive industries, it is generally assumed that critical components will work correctly,

but that non-critical components are likely to have defects. Defects in non-critical components could, if not isolated, compromise the ability of critical components to function. Thus, the simplest way to assure system safety is to isolate critical and non-critical components to prevent defects in non-critical components from undermining system safety. Such separation typically involves using separate processors, separate memory, and separate networks.

Separation of critical and non-critical networked messages (i.e., through the use of separate buses) can double the required network costs, both in cabling and in network interface hardware. There is strong financial incentive to share a single network between critical and non-critical message traffic. But, when such sharing occurs, it is crucial that there be assurance that non-critical network traffic cannot disrupt critical network traffic. This paper assumes that non-critical network hardware is designed to be fail-safe. For example, in railroad signaling lack of message delivery leads to a safety shutdown, so safe operation is viable with off-the-shelf networking hardware. But other challenges remain to implementing such systems.

A significant challenge in mixed critical and non-critical networks is ensuring that a non-critical process is unable to *masquerade* as a critical message sender by sending a fraudulent critical message. Masquerading is considered malicious if an internal or external attacker intentionally represents itself as a different entity within the system; however, masquerading may also occur due to non-malicious transient faults and design errors that inadvertently cause one node or process to send a message that is incorrectly attributed to another node or process.

A *software defect masquerade fault* occurs when a software defect causes one node or process to masquerade as another (Morris and Koopman, 2003). One example is a software defect that causes one process to send a message with the header identification field of a different process. Another example is a software defect that causes one node to send a message in another node's time slot on a TDMA network. A software defect masquerade fault is not caused by transient anomalies such as random bit flips, but rather is the result of design defects (e.g., the software sends the message with the incorrect header x instead of the correct header y). Fault tolerance methods designed to catch random bit flips may not sufficiently detect software defect masquerade faults.

This paper describes six successively more expensive levels of protection that can be used to guard against masquerade faults. Rather than limiting the analysis to malicious faults, the gradations presented recognize that many embedded systems have reasonable physical security. Therefore it is useful to have design options available that present tradeoff points between the strength of assurance against masquerading faults and the cost of providing that assurance.

2 FAULT MODEL

Network fault detection techniques vary from system to system. Some rely solely on a network-provided message Cyclic Redundancy Code (CRC) or other message digest such as a checksum for error detection. Some critical applications add an additional application-generated CRC to enhance error detection. These techniques can be effective at detecting random bit errors within messages, but they might not detect erroneous messages caused by software defects that result in masquerading. This is especially true in broadcast-oriented fieldbuses in which applications

have control over message ID fields and can send incorrect IDs due to component defects. In the worst case, these erroneous messages could lead to masquerading of critical messages by non-critical processes or by failed critical hardware nodes.

In order to determine the safeguards necessary to ensure correct behavior of critical components over a shared network, we must first understand the types of failures that can occur, as well as their causes. The strongest fault model, which includes malicious and intentional faults, assumes that an intruder is intentionally falsifying message traffic and has significant analytic abilities available to apply to the attack. Such malicious faults can only be detected by application of rigorous cryptographic techniques. Because a malicious attack is the most severe class of fault, such measures would also provide a high degree of fault tolerance for software defect masquerade faults. But full-strength cryptographic techniques are cost-prohibitive in many embedded systems. In systems for which malicious attacks are not a primary concern, lighter-weight techniques that protect against accidental (non-malicious) faults due to environmental interference, hardware defects, and software defects are highly desirable.

The simplest accidental failures come from random bit errors during transmission. In general, these errors are easy to detect using standard error detecting codes such as CRCs that already exist on most networks.

Errors due to software defects or hardware faults in transmitters are more difficult to guard against. They can result in undetectable failures in message content unless application-level error detection techniques are used because faults occur before the message is presented to the network for computation of a CRC. A particularly dangerous type of error that could occur is an incorrect message identifier or application-level message source field, which would result in a masquerading fault.

In an embedded system with both critical and non-critical processes, masquerading faults can occur in three different scenarios: (1) a critical process might be sending a critical message to another critical process; (2) a critical process might be sending a critical message to a non-critical process; and (3) a non-critical process might be sending a non-critical message to a critical process. Because critical processes are trusted to work correctly, critical messages are assumed to have correct information, and messages from non-critical processes are suspect. Safety problems due to masquerading can thus occur if a non-critical process sends one of the first two types of messages (i.e., if a non-critical process sends a message falsified to appear to be a critical message). An additional situation that is sometimes of concern is if a critical node suffers a hardware defect that causes it to masquerade as a different critical node, resulting in a failure of fault containment strategies (designs typically assume not only that faults are detected in critical nodes, but also that they are attributed to the correct critical node).

3 PROTECTION LEVELS

Once the fault model has been defined, an appropriate level of masquerading fault detection can be implemented based on the needs and constraints of the application. As with any engineering design, this requires tradeoffs in cost, complexity and benefits.

3.1 Level 0 - Network protection only

The first, baseline, level of protection is to rely solely on network error checking. Most networks provide some mechanism to detect network errors. Ethernet and the Controller Area Network (CAN), for example, both use CRCs.

The problem with relying on the network-level data integrity checks is that they only check for errors that occur at the network link level. Errors due to software defects or some hardware defects in the network interface are not detected. In addition, these detection techniques may not be very effective at detecting errors caused by routers and other networking equipment on multi-hop networks. For example, Stone and Partridge (2000) found high failure rates for the TCP checksum, even when combined with the Ethernet CRC.

Though relatively effective at preventing random bit errors, Level 0 remains vulnerable to defects in networking hardware that cause undetected message errors, software defects that result in masquerading by critical and non-critical processes, and malicious attacks. In terms of bandwidth and processing resources, this level requires no additional cost because it is already built into most networks.

3.2 Level 1 - Application CRC

The next step in assuring message integrity is to apply an application-level CRC to the data and to include it in the message body that is transmitted on the network. Stone and Partridge (2000) strongly recommend using an application-level CRC to help detect transmission errors missed by the network checks due to defects in routers and other networking equipment.

It might be the case that some or all of the processes within a system use the same application-level CRC. If non-critical processes use the same application-level CRC as critical processes, then the application CRC provides no protection against masquerading by non-critical processes.

Level 1 provides additional protection against defects in networking hardware that cause undetected message errors. However, it does not protect against critical message sources that falsify message source information due to faults, defects that result in masquerading by critical and non-critical processes, and malicious attacks.

With respect to resource costs, Level 1 requires some additional bandwidth and processing resources, but not many. For example, on a CAN network a 16-bit application CRC requires two of the eight available data bytes and an additional few instructions per data bit of CRC. In critical systems application-level CRCs are not uncommon.

3.3 Level 2 - Application CRC with secret polynomial/seed (symmetric)

The application-level CRC in Level 1 may be converted from a simple data integrity check into a lightweight digital signature by using different CRC polynomials for different classes of messages. In this scheme, there are three separate CRC polynomials used: one for critical messages sent between critical processes, one for non-critical messages sent by the critical processes to non-critical processes, and one for messages sent by the non-critical processes. It is important to select "good" polynomials with an appropriate Hamming Distance for the lengths of messages being sent, of course (Siewiorek and Swarz, 1992).

The Level 2 approach is a "lightweight," symmetric digital signature in which the secret key is the CRC polynomial. It is symmetric because both the sender and receiver of a message need to know the same key, and must use it to sign messages (by adding an application-level CRC using a specific polynomial), and verify signatures (by computing the application-level CRC using an appropriate polynomial based on the purported message source and comparing it to the frame check sequence (FCS) field of the message actually sent).

A straightforward implementation involves using a different secret polynomial for each class of message: CRC_1 for critical to critical messages; CRC_2 for critical to non-critical messages; and CRC_3 for non-critical message senders. (Note that the case where non-critical processes omit an application-level CRC is equivalent to using a null CRC for situation CRC_3).

Use of three CRCs is required because this is a symmetric system. Thus, it is possible that any process possessing a CRC polynomial might send a message using that polynomial due to a software defect. If CRC_1 is only known to critical processes, that means it is impossible (or at least probabilistically unlikely) that a non-critical process can falsify a message that will be accepted by a critical process as having come from another critical process. In other words, CRC_1 is a secret symmetric key, and only key-holders can generate signed messages. CRC_2 is used to provide assurance that critical messages are being sent either from critical processes or non-critical processes (with software defects) that are receivers of critical messages. CRC_3 is simply an application-level CRC for non-critical messages. It might be the case that there is no point in distinguishing CRC_2 from CRC_3 depending on failure mode design assumptions, because in either case at least one non-critical process would have access to the secret key CRC_2 for generating critical-process-originated messages.

With this scheme there is still a critical assumption being made about non-critical code. However, it is a much narrower assumption than with the Level 1 approach, and is probably justifiable for many situations. The assumption is that CRC_1 has been selected from a pool of candidate CRCs at random, and is unlikely to be used by non-critical processes on a statistical basis. (One assumes that "well known" published CRCs are omitted from the potential selection pool, of course.) For 24-bit or 32-bit CRCs this assumption is probably a good one, but there is still a finite number of "good" CRC polynomials that are significantly fewer than all possible 24-bit or 32-bit integers.

A solution that is even better for these purposes is to use a "secret seed" for a given polynomial. Conventional CRC calculations use a standardized starting value in the CRC accumulator, typically either 0 or -1. A secret seed approach uses some different starting, or "seed" value for computation of the application-level CRC that varies with the class of message. So instead of CRC_1, CRC_2 and CRC_3 for the previous discussion, the technique would involve using the same CRC with $Seed_1$, $Seed_2$, or $Seed_3$, with each seed being a different secret number. Thus, the seed value becomes the secret key for a digital signature.

Thus, the FCS of a message with a level of criticality i would be computed as follows. If CRC(M,S) takes a message M with an initial CRC seed value S to compute a FCS, then:

$$FCS_i = CRC(M, S_i) \tag{1}$$

Critical to critical process messages would be authenticated by having critical processes use S_1 to compute and compare the FCS field. Since no non-critical process would have knowledge of S_1, it would be, for practical purposes, impossible for non-critical processes to forge a correct FCS value corresponding to a critical message. There would still be a chance of an accidental "collision" between the

FCS values for two CRCs, but this is true of cryptographically secure digital signatures as well, and can be managed by increasing the size of the FCS as required.

Combining a secret polynomial with a secret seed is possible as well, of course, but does not provide a fundamentally different capability. It is important to note that CRC-based digital signatures are readily attacked by cryptanalytic methods and are *not secure against malicious attacks.* However, in a cost-constrained system it might well be reasonable to assume that non-critical components will lack cryptanalytic attack capabilities, and that software defects will not result in the equivalent of cryptanalytic attacks on secret CRC polynomials or secret seeds.

Symmetric-key CRC lightweight digital signatures of Level 2 provide the same benefits as application-level CRCs of Level 1. In addition, they provide protection against non-malicious masquerading by non-critical processes that results in acceptance of fraudulent critical messages. However, Level 2 does not protect against non-malicious masquerading of critical message sources by other critical message sources due to faults, and is inadequate protection against malicious attacks. The benefit of Level 2 is that it requires no additional processing or bandwidth to upgrade from Level 1.

3.4 Level 3 -Application CRC with secret polynomial/secret seed (asymmetric)

Symmetric CRC-based signatures ensure that non-critical processes cannot send critical messages to critical processes by accident. However, a software defect could still cause a non-critical process to masquerade as a critical process sending a non-critical message. (This is true because all noncritical processes possess the symmetric key information for receiving such messages). Additionally, Level 2 assumes that all critical processes are defect-free, providing no protection against masquerading by a critical process in the event of a hardware failure or software defect.

A further level of protection can be gained by using asymmetric, lightweight authentication. In this approach every process has a secret sending key and a public receiving key. The public receiving key is known by all processes, but only the sending process knows the secret sending key. In such a scheme every process retains the public receiving keys of all processes from which it receives messages (in general, meaning it has the public keys of all the processes). But because each process keeps its transmission key secret, it is impossible for one process to masquerade as another.

Because embedded systems tend to use broadcast messages heavily, an implementation of full public-key encryption is impractical, so the method proposed here is tailored to a broadcast environment. Additionally, CRC-based authentication is used which is of course *not secure against a cryptanalytic attack*.

One way to implement a private/public signature scheme is the following, using secret polynomials. This method may also be used in addition to the use of distinct CRCs or seeds for FCS computation as outlined in Level 2. If desired for cost and simplicity reasons, all non-critical to non-critical messaging can use a single standard polynomial, and only critical message sources need use the private/public key approach.

Each critical process has two CRC polynomials: CRC_1 and CRC_2. CRC_1 is a publicly known polynomial, whereas CRC_2 is a secret private polynomial. Every CRC_1 in the system is distinct per process. Every CRC_2 is the inverse of the corresponding CRC_1. Thus, the secrecy of CRC_2 depends on there being no code to compute an inverse polynomial in the system. Because computing inverse polynomials is performed using a bit-reverse operation (with adjustments to account for an implicit 1 bit within the polynomial in most representations), the validity of the assumption of secrecy is one that must be made in the context of a particular system design. However, computing inverse polynomials off-line and putting them in as constants within the system code avoids the presence of inverse polynomial code, and might well be a reasonable approach for systems that cannot afford the cost of full-strength cryptography. (The creation of stronger, but efficient, methods for asymmetric signatures is an open area for future research.)

A sending process S appends a signature X to a critical message M and its FCS field (X is not included in the FCS computation), where "|" denotes concatenation:

$$M \mid FCS \mid X \tag{2}$$

where:

$$X = CRC_2(FCS) \tag{3}$$

Receiving processes then verify the authenticity of the transmission by ensuring that:

$$FCS = CRC_1(X) \tag{4}$$

What this is doing is "rolling back" the FCS using an inverse CRC, CRC_2, to compute a signature that, when rolled forward through CRC_1, will yield the FCS. Because only the sending process knows the inverse CRC for its public CRC, no other process can forge messages.

This method protects against software and hardware defects that cause a process to send a message that should not be sent (e.g., forged source field or incorrect message identifier/type information).

This method is vulnerable to the following: malicious attacks using cryptanalysis (even without knowledge of the public CRC polynomial); software defects involving CRC code that computes CRCs "backwards" from the critical CRC computation (e.g., right-to-left CRC computations when the critical code is using a left-to-right shift-and-xor computation); and software defects in critical or non-critical software that compute the bit-reverse of a public polynomial and then use that as the basis for signing a message. While some of these defects could probably happen in real systems, the specificity of the defects required would seem to provide a higher degree of assurance than not using such a technique. As stated previously, this is an example of a simple lightweight signature technique; it is possible that future research will yield even better approaches to fill this niche in the design space.

If the system is originally designed at Level 1 or Level 2 with an application CRC, then there is an additional cost to compute and transmit the signature X. In a CAN network with a 16-bit signature X, this would be an additional handful of instructions per CRC bit and two bytes of the remaining six available data bytes.

3.5 Level 4: Symmetric cryptography

Levels 1 through 3 all use some form of CRC to detect masquerading errors due to defects in the non-critical software, and provide no credible protection against malicious faults. The next higher level of protection can be achieved through the use of cryptographically secure digital signatures. Although designed primarily for malicious attacks, such digital signatures can also prevent defective non-critical software components from forging critical messages. This can be accomplished via use of a Message Authentication Code (MAC), which is a keyed one-way hash function. A detailed description of MACs appears in Section 18.14 of Schneier (1996).

Symmetric digital signatures must be sufficiently long to preclude successful malicious attacks via cryptanalysis or brute force guessing. Additionally, they take significant computational capability beyond the means of many embedded systems. However, a symmetric key approach is secure against malicious attacks unless the attacker compromises a node possessing a secret key. In the case that the attacker compromised a critical code, it would be possible to maliciously forge a message that apparently originated in any node in the system. Malicious attacks aside, a Level 4 approach has the same strengths and weaknesses as a Level 2 approach in that it is a similar general signature method, but using strong cryptography.

3.6 Level 5: Public-key digital signatures

Level 4 protection was analogous to Level 2 protection, but used cryptographically secure symmetric digital signatures. Level 5 is, in turn, generally similar to Level 3 CRC lightweight public key signatures, but uses cryptographically secure signature algorithms. Various public-key digital signature algorithms are described in Section 20 of Schneier (1996).

A Level 5 approach provides protection from forgery of message sources to the limits of the cryptographic strength of the digital signature scheme used. Moreover, if a node is compromised by malicious attack, forgery of messages can only be accomplished with compromised node(s) as originators, because each node has its own distinct secret signature key. However, public-key methods have longer signatures and are much slower than symmetric cryptography (Menezes *et al.*, 1997).

3.7 Tradeoffs

Each of these methods provides a certain level of fault protection; however, they each have a commensurate cost. The developers must decide what protection is required to attain safe operation, and adjust system design decisions on how much safety critical operation to delegate to computers based on budget available to provide protection against realistic masquerading threats. For example, a system with a fault model that includes software defect masquerade faults but excludes malicious attacks might chose Level 2 or Level 3.

An additional burden that must be assumed when using any masquerading detection technique is that of cryptographic key management. Any technique discussed assumes that only a certain set of nodes have access to secret keys. This restricted access results in significant complications in configuration management, deployment, and maintenance, especially when insider attacks are considered a possibility. (As a trivial example, every time a disgruntled employee leaves a company, it is advisable to change all cryptographic keys that the employee might have had access to if attacks by that employee are a substantive threat.)

Figure 1 shows all of the levels, in order of effectiveness. In general, the stronger the protection, the more expensive the method. Levels 3 and 4 have a partial ordering, because the protection of Level 3 (asymmetric secret CRC) might be more useful than the protection afforded by Level 4 (symmetric secure

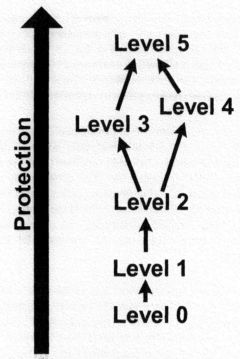

Figure 1. Masquerading fault protection levels

digital signature), depending on whether malicious attacks are a part of anticipated threats. However, it is expected that CRC-based signatures will be substantially less expensive to implement than cryptographically secure digital signatures.

4 CONCLUSIONS

This paper presents six levels of fault detection techniques that can be deployed against the possibility of masquerading faults on shared critical/non-critical fieldbuses. Level 0 (network-provided protection) provides no protection beyond what is included in the network protocol. For level 1 (published CRC), the application must be modified to apply the application-level CRC before sending messages on the network, and after messages have been received. Once an application-level CRC is present in the code, the polynomial or seed value used in the calculation can be changed to achieve Level 2 (symmetric secret polynomial/seed) protection. A novel Level 3 (asymmetric secret polynomial/seed) approach is proposed to provide very lightweight digital signatures with a public key flavor that are suitable for broadcast bus applications, but that further assume malicious faults are not a threat. Levels 4 and 5 complete the taxonomy and consist of using well known cryptographically secure approaches to guard against malicious masquerading faults.

Typical fieldbus systems today operate at Levels 0 and 1, and are not secure against masquerading faults. It might be attractive in some applications to upgrade to a Level 2 or Level 3 capability to improve resistance to non-malicious software defect masquerade faults without having to resort to the complexity and expense of cryptographically secure Level 4 or Level 5 approach.

5 ACKNOWLEDGMENTS

This work is supported in part by the General Motors Collaborative Research Laboratory at Carnegie Mellon University, Bombardier Transportation, and by the Pennsylvania Infrastructure Technology Alliance. The authors would also like to thank Bob DiSilvestro for his loyal support.

6 REFERENCES

IEEE (1990). IEEE Standard Glossary of Software Engineering Terminology, IEEE Std 610.12-1990.

IEEE (1998). IEEE Standard for Software Quality Assurance Plans, IEEE Std 730-1998.

Menezes, A. J., P.C. Van Oorschot, and Scott A. Vanstone (1997). *Handbook of Applied Cryptography*. CRC Press LLC, Boca Raton.

Morris, J. and P. Koopman (June 2003). Software Defect Masquerade Faults in Distributed Embedded Systems. *IEEE Proceedings of the International Conference on Dependable Systems and Networks Fast Abs*.

Schneier, B. (1996). *Applied Cryptography*. second edition, John Wiley & Sons, New York.

Siewiorek, D. P. and R. S. Swarz (1992). *Reliable Computer Systems Design and Evaluation*. Second Edition. Digital Press, Bedford, MA.

Stone, J. and C. Partridge (2000). When the CRC and TCP Checksum Disagree. *ACM SIGCOMM Computer Communication Review: Proc. of the Conference on Applications, Technologies, Architectures, and Protocols for Computer Communication*, 309-319.

ANALYZING ATOMIC BROADCAST IN TTCAN NETWORKS *

Guillermo Rodríguez-Navas and Julián Proenza

Departament de Matemàtiques i Informàtica
Universitat de les Illes Balears, Palma de Mallorca, Spain
email: vdmigrg0@uib.es, dmijpa0@uib.es

Abstract: TTCAN networks are considered as being suitable for safety-critical systems because of the deterministic communication they provide. However, we show that inconsistent reception of messages are very likely to occur in TTCAN networks. Despite of the high probability of such inconsistency scenarios, they have not been addressed by the TTCAN specification. Therefore, it must be studied if the solutions which solve these inconsistency scenarios in CAN networks can be also applied to TTCAN networks. A first analysis shows that many problems arise when trying to integrate such solutions with the TTCAN protocol. Copyright ©2003 IFAC.

Keywords: Distributed control, Embedded systems, Real-time systems, Dependability, Atomic broadcast, Fieldbus, Controller Area Network

1. INTRODUCTION

Controller Area Network protocol (ISO, 1993) has been proved to be suitable for many real-time systems, since the response time of any message in a CAN network is bounded and can be calculated (Tindell *et al.*, 1995). Nevertheless, due to the event-triggered nature of CAN networks, the response time of a CAN message presents a high variability, which basically depends on the bus load and the channel errors. Because of this high variability, CAN networks are not suitable for being used in real-time distributed systems which require a deterministic communication. That is the case of many real-time distributed control systems which are used in safety-critical applications (for instance in *x-by-wire systems*).

The Time-Triggered CAN protocol is a higher-layer extension of the CAN protocol which aims at providing the deterministic communication that those systems require (Hartwich *et al.*, 2000; Führer *et al.*,

2000). TTCAN relies on the use of the CAN protocol at the physical and data link layer, but implements a TDMA mechanism to access the medium. This mechanism makes the jitter of each message independent of the bus load and, therefore, significantly improves the determinism of the communication.

Nevertheless, communication determinism is not the only requirement that the real-time distributed control systems which are used in safety-critical applications must fulfill. Dependability is an important requirement as well.

In principle, the CAN protocol is adequate for being used in dependable systems. Firt, because of the powerful error-detection and error-signaling capabilities which CAN controllers implement. And second, because of the automatic frame retransmission which a sender CAN controller performs whenever a frame is rejected due to errors. This functionality allows CAN protocol to guarantee that any message issued to a CAN network is consistently received by all the non-faulty nodes. This property would correspond to *atomic broadcast*. However, as remarked in (Rufino *et al.*, 1998), some error scenarios exist in which

* This work has been partially supported by the Spanish MCYT grant DPI2001-2311-C03-02, which is partially funded by the European Union FEDER programme

CAN protocol does not provide atomic broadcast. This scenarios are called *inconsistency scenarios* as they result in the inconsistent reception of a message. Due to the relevance that atomic broadcast has in dependable distributed systems (Mullender, 1993), various solutions have been proposed to guarantee atomic broadcast even in such inconsistency scenarios (Rufino *et al.*, 1998; Livani, 1999; Proenza and Miro-Julia, 2000; Pinho and Vasques, 2001).

In spite of the fact that TTCAN is based on the CAN protocol, and therefore it has inherited the inconsistency scenarios that CAN presents, the TTCAN specification does not manage to solve them. In this work, we show that such inconsistency scenarios turn out to be much more relevant in TTCAN networks than in CAN networks, as they are much more likely to occur. Furthermore, as TTCAN is intended to be used in safety-critical systems, it is important to study whether it is possible to actually provide atomic broadcast by means of the solutions already proposed for CAN networks. A first analysis shows that many problems arise when trying to integrate such solutions with the TTCAN protocol.

2. COMMUNICATION SCHEME IN TTCAN

In order to increase the determinism of the communication, TTCAN implements a TDMA mechanism to access the medium. Due to space limitation, only the basic characteristics of this scheme are discussed. Readers are suggested to search in (Führer *et al.*, 2000; Hartwich *et al.*, 2000) a complete description of these mechanisms.

In TTCAN, each node is allowed to transmit only at predefined time intervals, which correspond to a cyclic pattern of transmission. This pattern is named the *basic cycle*, and is divided into a number of temporal segments, which are named *temporal windows*. During every one of these temporal windows, only one node is allowed to transmit. Furthermore, TTCAN controllers are not allowed to automatically retransmit erroneous frames, which is a basic feature of the CAN protocol. In this way, TTCAN prevents transmitters from interfering the following temporal windows. Thus, the determinism of the communication is ensured as the delay of any message does not depend on the bus load.

3. INCONSISTENCY SCENARIOS IN TTCAN NETWORKS

TTCAN protocol is a higher-layer extension of the CAN protocol. Due to this, TTCAN inherits the inconsistency scenarios which CAN presents. In this section, after discussing the inconsistency scenarios of CAN, we show that such inconsistency scenarios are more likely to occur in TTCAN networks than in CAN networks, basically because TTCAN controllers are

not allowed to automatically retransmit the erroneous frames. Readers are assumed to be familiar with the format of the CAN frames as well as with the error management functionality which CAN controllers perform (ISO, 1993; Etschberger, 2001).

3.1 Inconsistency scenarios in CAN networks

In the presence of errors in the last bit of the EOF, the behaviour of the CAN controllers is special so as to be able to cope with specific error situations. If a transmitter detects an error in the last bit of the EOF, it handles it in the usual way: an error flag is started in the next bit, frame transmission is considered as being erroneous and the frame is retransmitted. In contrast, if a receiver detects an error in the last bit of the EOF, it accepts the frame as being correct and, instead of an error flag, it generates an overload flag. The reason of this behaviour is illustrated by the scenario in Fig. 1a. A set of receiving nodes, called the X set, detect an incorrect dominant value in the last bit of the EOF, whereas the transmitter and another set of receiving nodes, called the Y set, see a correct recessive bit. The nodes of X start the transmission of an overload flag in the bit after the error. The rest of nodes see the first dominant bit of the overload flag in the first bit of the interframe space and then start the transmission of their overload flags as well. Therefore, the transmitter as well as the nodes belonging to Y consider the frame as correctly transmitted. Thanks to the last-bit rule, the nodes belonging to X also accept the frame and consistency is achieved.

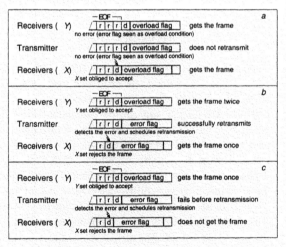

Fig. 1. Error scenarios in CAN networks

Unfortunately, this last-bit rule also causes that nodes may receive the same frame twice. Fig. 1b shows a scenario in which this situation happens. A disturbance corrupts the last but one bit of the EOF of the nodes belonging to X, so that in the next bit, these receivers start the transmission of an error frame. The first dominant bit of this error flag is seen by the transmitter and the nodes belonging to Y as an error in the last bit of their EOF. Then, the nodes belonging to

X reject the frame, the transmitter retransmits it, and the nodes belonging to Y accept the frame because of the last-bit rule. Therefore, the nodes belonging to Y receive the frame twice.

The double reception of frames is a well-known phenomenon that has lead to a set of common recommendations (Zeltwanger, 1998), *e.g.* not to transmit messages that toggle the state of the receivers. Besides the double reception, Rufino *et al.* (1998) have identified new error scenarios in which the last bit behaviour produces *inconsistent message omissions* (IMO): some nodes receive a frame and others never do. This is illustrated in Fig. 1c. This case is similar to the one depicted in Fig. 1b, the difference being that in Fig. 1c after the first transmission of the frame, the transmitter suffers a failure which avoids the retransmission of the frame (a *crash failure*). Therefore, the nodes belonging to Y receive the frame whereas those of X do not.

In (Rufino *et al.*, 1998), the probability of the inconsistency scenarios of CAN is calculated as a function of the channel bit error rate (*ber*). The way in which this analysis is performed is briefly introduced next. Equation 1 corresponds to the probability of having an error in the last but one bit of the EOF, assuming that the number of bits of a frame is equal to τ_{data} and that the probability of having an error in one particular bit follows a geometric distribution. On the other hand, the probability of a node crash failure obeys a Poisson distribution with a failure rate λ, as shown in Equation 2. Finally, the probability of an inconsistent message duplicate (IMD) is defined as p_{imd} in Equation 3, whereas the probability of an inconsistent message omission (IMO) is defined as p_{imo} in Equation 4.

$$p_{ifo} = (1 - ber)^{\tau_{data}-2} * ber \qquad (1)$$

$$p_{fail} = 1 - exp^{-\lambda*\Delta t} \qquad (2)$$

$$p_{imd} = p_{ifo} * (1 - p_{fail}) \qquad (3)$$

$$p_{imo} = p_{ifo} * p_{fail} \qquad (4)$$

Note that p_{imd} and p_{imo} are complementary, since every error in the last but one bit of the EOF drives either to an IMD or to an IMO; if the transmitter remains non-faulty then an IMD occurs whereas if the transmitter crashes then an IMO occurs. In (Rufino *et al.*, 1998), these probabilities are calculated and used in order to estimate the number of inconsistencies that occur during the operation of a CAN network.

Table 1 shows some of the results which were obtained assuming a CAN network made up of 32 nodes, with a bit rate of 1Mbps, an overall load of 90% and an average frame length (τ_{data}) of 110 bits . Note that the average number of IMOs is much lower than the average number of IMDs as the former require the crash of a node to happen, and it is a very unlikely failure. Nevertheless, as remarked in (Rufino *et al.*, 1998), the number of inconsistent omissions is still

greater than the values which are recommended for safety-critical systems (e.g. 10^{-9} incidents/hour in aerospace industry), and therefore solutions which actually achieve atomic broadcast are required for CAN networks. Solutions which solve this problem have been already suggested in (Rufino *et al.*, 1998; Livani, 1999; Proenza and Miro-Julia, 2000; Pinho and Vasques, 2001).

Table 1. CAN inconsistent errors per hour as estimated in (Rufino *et al.*, 1998)

Bit Error Rate (*ber*)	Node failures per hour (λ)	IMD/hour	IMO/hour
10^{-4}	10^{-4}	2.84×10^3	3.94×10^{-7}
10^{-5}	10^{-4}	2.86×10^2	3.98×10^{-8}
10^{-6}	10^{-4}	2.87×10^1	3.98×10^{-9}

3.2 Inconsistent message omissions in TTCAN

As remarked above, in a CAN network the probability of having an IMD is much higher than the probability of having an IMO. Nevertheless, this relation does not hold for TTCAN networks, due to the fact that TTCAN controllers are not allowed to retransmit erroneous frames. In particular, the value of p_{fail} which is used in the analysis of (Rufino *et al.*, 1998) must be equated to 1 for TTCAN networks, as the effect of not retransmitting the frames is equivalent to the effect of a crash failure in the sender. Thus, Equations 3 and 4 result in the following equations:

$$p_{imd}^{ttcan} = 0 \qquad (5)$$

$$p_{imo}^{ttcan} = p_{ifo} \qquad (6)$$

This means that in TTCAN networks, the probability of suffering an IMO increases dramatically whereas IMDs are not possible, since every error in the last but one bit of the EOF can only cause an IMO. Following the same calculations which where discussed in the previous section, we can infer that in a TTCAN network, and assuming a *ber* of 10^{-4}, an average number in the order of 10^3 IMOs occur every hour (see Table 2). This value is so high that implementing some solution to solve these inconsistency scenarios seems to be mandatory in TTCAN networks. However, to the authors' best knowledge, this issue has not been addressed by the TTCAN specification.

Table 2. Inconsistent errors in TTCAN

Bit Error Rate (*ber*)	IMO/hour
10^{-4}	2.84×10^3
10^{-5}	2.86×10^2
10^{-6}	2.87×10^1

A first step to solve the problem of data consistency in TTCAN should be to analyze whether the solutions already suggested for atomic broadcast in CAN networks can be applied to TTCAN networks or not. This analysis is the subject of the next section.

4. SOLUTIONS TO ACHIEVE DATA CONSISTENCY IN TTCAN NETWORKS

When attempting to provide atomic broadcast for CAN networks, two different approaches exist. The first one is followed in (Rufino *et al.*, 1998; Livani, 1999; Pinho and Vasques, 2001), and consists in implementing an algorithm based on message retransmissions. The second approach is the one followed in (Proenza and Miro-Julia, 2000), which consists in solving the inconsistencies at the frame level, without sending further messages.

The solutions which follow the first approach have been specifically designed for event-triggered systems and, therefore, are not suitable for TTCAN networks. On the one hand, such solutions require reservation of a significant amount of bandwidth in order to retransmit the frames and this causes a low utilization of the channel. On the other hand, those solutions are in conflict with the determinism which is supposed for time-triggered communication, as the high number of messages that must be exchanged implies an increment in the variability of the response time of the messages. However, although these solutions cannot be directly applied to TTCAN networks, it must be studied if they can be modified so as to be suitable for TTCAN networks.

In contrast, the solution suggested in (Proenza and Miro-Julia, 2000) seems to be directly applicable to TTCAN networks, as it solves the problem of inconsistency in a frame by frame basis. In this way, the fact of not having the possibility to retransmit any frame does not cause any inconsistency. However, this solution does not follow the standard CAN protocol exactly, and therefore it is not compatible with current CAN controllers. Nevertheless, further research is being carried out in order to determine in which way this solution can be used in TTCAN networks.

5. CONCLUSIONS

In this paper, it has been shown that inconsistent reception of messages are very likely to occur in TTCAN networks. Despite of the fact that TTCAN protocol is intended to be used in safety-critical system, these scenarios of inconsistency are not addressed in the TTCAN specification. A first step to overcome this problem must be to analyze if the solutions previously suggested to achieve atomic broadcast in CAN networks can be also applied to TTCAN networks. A first analysis has shown that the solutions for atomic broadcast which are based on message exchanges are not suitable for TTCAN. In contrast, a solution which implements a frame-wise mechanism to achieve consistency seems to be very adequate for TTCAN networks, as it does not require transmission of any other message and therefore it is compatible with the time-triggered paradigm which TTCAN uses.

REFERENCES

Etschberger, K. (2001). *Controller Area Network*. IXXAT Press. Weingarten.

Führer, T., B. Müller, W. Dieterle, F. Hartwich, R. Hugel, M. Walther and Robert Bosch GmbH (2000). Time Triggered Communication on CAN. *Proceedings of the 7th International CAN Conference, Amsterdam, The Netherlands*.

Hartwich, F., B. Müller, T. Führer, R. Hugel and Robert Bosch GmbH (2000). CAN network with Time Triggered Communication. *Proceedings of the 7th International CAN Conference, Amsterdam, The Netherlands*.

ISO (1993). ISO11898. Road vehicles - Interchange of digital information - Controller area network (CAN) for high-speed communication.

Livani, M.A. (1999). SHARE: A Transparent Approach to Fault-tolerant Broadcast in CAN. *Proceedings of the 6th International CAN Conference, Torino, Italy*.

Mullender, S.J., Ed.) (1993). *Distributed Systems, 2nd edition*. ACM Press, Addison Wesley.

Pinho, L. and F. Vasques (2001). Improved Fault-tolerant Broadcasts in CAN. *Proceedings of the 8th IEEE International Conference on Emerging Technologies and Factory Automation (ETFA'01), Antibes, France*.

Proenza, J. and J. Miro-Julia (2000). MajorCAN: A modification to the Controller Area Network to achieve Atomic Broadcast. *IEEE Int. Workshop on Group Communication and Computations. Taipei, Taiwan*.

Rufino, J., P. Veríssimo, G. Arroz, C. Almeida and L. Rodrigues (1998). Fault-tolerant broadcasts in CAN. *Digest of papers, The 28th IEEE International Symposium on Fault-Tolerant Computing, Munich, Germany*.

Tindell, K., A. Burns and A. J. Wellings (1995). Calculating controller area network (CAN) message response time. *Control Engineering Practice* **3(8)**, 1163–1169.

Zeltwanger, Holger (1998). Failure Detection and Error Handling in CAN-Based Networks. *Seminario Anual de Automática, Electrónica Industrial e Instrumentación. SAAEI'98. Pamplona, Spain*.

ELSEVIER

IFAC
PUBLICATIONS
www.elsevier.com/locate/ifac

THE NECESSITY OF AN UPGRADE IN INDUSTRIAL COMMUNICATIONS

Roland Heidel

Siemens AG Karlsruhe, Rheinbrueckenstr.50, Germany

Abstract: Communication systems in industrial automation have become an integral part of automation systems since years. As a consequence of upcoming new requirements an upgrade of industrial communications systems seems to be necessary forming an infrastructure for distributable applications. The usage of product data management will be one of the prerequisites for the engineering of distributable applications. The related electronic engineering process has to be defined taking into account the enhanced communication infrastructure extended i.e. by wireless communications. The overall security issue in automation has also to be addressed in detail in the near future. *Copyright © 2003 IFAC*

Keywords: Communications Systems, Industrial Control, Engineering, Product Data Management

1. INTRODUCTION

After more than ten years successful and useful application of Fieldbus Systems it is time to take a stock of the past and to open up the perspective for further development. The first generation of Fieldbusses is actually the second generation of industrial communication systems introduced in the late 1970s. Now a certain kind of consolidation of Fieldbus Communication Systems can be observed. One reason is the recently published international Fieldbus standard IEC 61158. The other one is the fact that it is now time to reconsider industrial communications and to identify areas of concentration in order to prepare the third generation of industrial networks solutions.

Considering the major trend towards Ethernet based LANs in Industrial Automation and concluding the result of many initiatives in the area of automation there are some main requirements to be covered by industrial communication systems in the future.

The most important one is a consistent support of "distribut*able* applications". In today's solutions there is some support already for distribut*ed* applications but neither in a satisfying manner by the operating systems nor by appropriate engineering systems. The operating system functions for the coordination of distributed application programs have to be improved and the distribution of application programs into different automation components have to be facilitated by appropriate algorithms. There is not to much work on distribution algorithms today. But without them no customer will be able to move application tasks into the different automation components depending of its requirements such as minimal number of components, minimum number of vendors or optimisation of performance and flexibility.

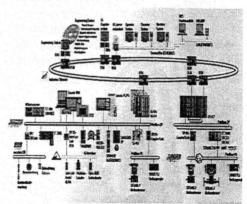

Fig. 1. A typical automation plant

Another strong requirement is the extension of wired networks by wireless ones forming an new architecture consisting of any combination of wired and wireless network segments. This new architecture supports as well "legacy" (wired) applications as new ones like combinations of fixed and mobile automation tasks but also "small" applications like slip ring replacement etc.

Fail Safe network solutions are going to guarantee protection of human life (Emergency Shut down) and of any equipment not only in wired but also in wireless applications.

Finally there is the overall Information Security issue in both wired and wireless segments of a network. Separate security solutions for wired and wireless segments must be avoided. There has to be an overall integrated solution for all network segments. Otherwise there will be no guarantee for complete and consistent security rules in both wired and wireless networks and security gaps may occur just because the segment security rules in wired and the wireless networks do not fit together.

2. STATE OF THE ART

Figure 1 shows the typical architecture of today's automation systems. It is characterized by two or three different network types. The properties of these network types are determined by their area of application for which they are optimised for. Intercommunication between the different network types is mostly done by data/memory coupling inside of one of the automation components. This is very often a PLC or an automation PC. Thus the architecture supports today "*distributed* applications". These are characterized by intelligent automation components sharing the automation task by each other and interacting in a smoothly manner. However no common and standardized application object model exists which could "hide" the different data transport mechanisms of the network types.

One of the new requirements for automation systems solutions is to provide the infrastructure for free "*distributable* applications". Such applications are characterized by the option to move parts of the application program from one component to an other one in a transparent manner. Figure 2 shows the structure of the former typical centralized automation tasks. One automation component collects values from transmitters of the plant, runs the application program completely and controls the actuators in the plant with the results of the calculations.

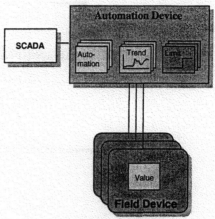

Fig. 2. Centralized Automation

Figure 3 shows how things have changed in the last 10 years. Now each of these islands is organized in a decentralized manner. The automation components share the automation task and interact in a decentralized manner realizing the specific distributed application. This is state of the art.

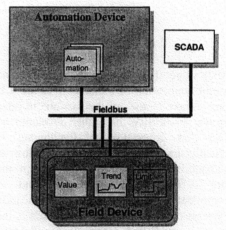

Fig. 3. Decentralized Automation

3. INFRASTRUCTURE FOR DISTRIBUTABLE AUTOMATION APPLICATIONS

Figure 4 finally shows the future architecture for automation systems providing the complete infrastructure for moving application programs from one component (resource) to another one. This means that (ideally) *any* component is able to run one or more parts of the complete application program. Usually such parts are called (moveable) Function Blocks. Whereas Function Blocks up to now mostly run in the automation computer environment such as PLCs and Automation PCs, the new automation system architecture represents the functionality of a common distributed operating system.

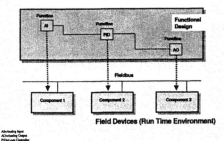

Fig. 4. Distributable Automation Applications (Example: PID Loop Controller)

Any of the moveable function blocks finds the same run time environment in any of the appropriate automation components. However there will be restrictions. Therefore there will be classes of the run time environment of components characterizing their ability to support certain kinds of Function Blocks. Two main automation component classes are obvious:

- The fixed function class
- The free programmable class.

Components of the fixed function class contain function Blocks with a certain functionality which can be enabled or disabled but not moved. They are more or less similar to them used today in the distributed environment but are characterized by one important difference: they contain Function Blocks compliant to a common object model and to the common (kernel) profile definitions of the open distributable system architecture. Today's components contain functions conformant with common profile definitions either open or vendor specific but do not follow common application object definitions and are therefore not so easy integratable.

The components of the free programmable class are very interesting in the future: they allow the free shift of functionalities from one component to another one installed in the plant. This means that a Function Block "A" running in component "A" can freely be shifted into a component "B" without any explicit reconfiguration in an engineering process. The distributed operating system recognizes by itself just the new path to the already known Function Block and adapts all associated data paths appropriately.

In order to achieve this three things are necessary at least:

- The Distributed Operating System must support the (ideally online) linking and locating

- The run time properties of the components (resources) to which the Function Block has been shifted cover the needs of this Function Block
- The Engineering System supports the (re-)engineering of the automation task appropriately.

4. EXTENDED AUTOMATION SYSTEMS WITH PRODUCT DATA MANAGEMENT

Whereas the first requirement has to be covered by the Operating System Functionality, the second one is a typical Product Data Management application as part of the engineering support. Any automation component candidate must present its abilities and properties as a machine readable data set in the future. Such a data set can be compared with a significantly extended component data sheet which is machine readable. If all components provide information about their abilities in a standardized way the operating system is able to decide whether a function block is able to run in a certain component or not. Even if the decision would not be made by the operating system automatically, the designer at the engineering system console should be supported automatically in order to avoid configurations which are not runnable per se. Such support is necessary at least during the planning and configuration phase of a plant but may be also needed at any offline phase in the life time of a plant. As a consequence future Engineering Systems have to consist of a Product Data Management System as an *integral* part of the future Plant Management System.

This brings up the question which product data may be part of a device descriptive data set. Today this question cannot be answered completely. There is a lot of activities in this area mainly motivated by eCommerce issues such as procurement and (internet mall) eMarketing. It is obvious that requirements brought up for these applications are sub quantities of the overall Plant Engineering and Maintenance System requirements. In any case the need for Product Data Descriptions in this area pushes the whole product data management activities forth to an overall solution for Plant Management Systems managing all phases of the life cycle of a plant. Very interesting in this context are the Electronic Device Description Language EDDL according to

CENELEC EN 50391 and the recently published NAMUR Guideline NE 100 which is a detailed preference list for plant equipment with a detailed formal machine readable product data format specification for each attribute of a component (Figure 5).

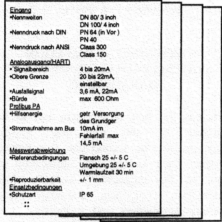

Fig. 5. NAMUR Guideline NE 100 (in German only at this time)

A real challenge is to design a tool chain for creating, changing and reducing product data during the whole product data flow in the whole life cycle of these components respective the plant in which they are installed as shown in Figure 6.

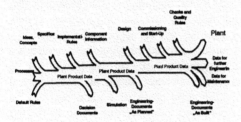

Fig. 6. Product Data Flow in the future Plant Engineering eProcess

5. A PRODUCT DATA MANAGEMENT MODEL

The prerequisite to a future engineering tool chain with product data management is to define a method to split product data in their basics. There is a concept on the table which comes originally from work in the EU project NOAH proposed by Doebrich, U. (2003). It splits the product data in 5 basic elements as shown in Figure 7.

Each of these elements is specified by the appropriate description method in terms of the model and the description language. The Construct Element may be represented to an engineering tool by ISO 10303 product descriptions specifying the relationships of mechanical parts of a component. The

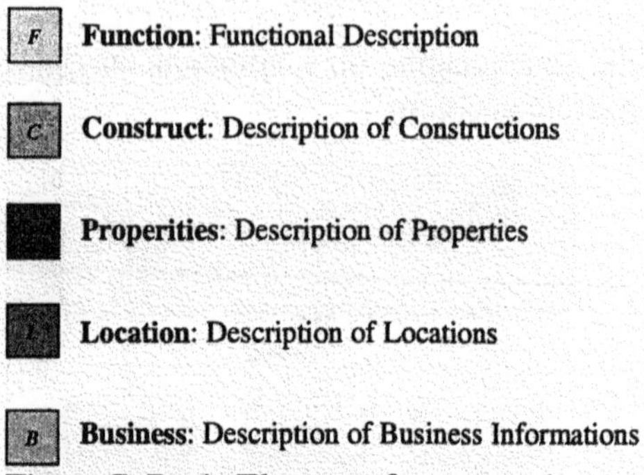

F **Function**: Functional Description

C **Construct**: Description of Constructions

Properties: Description of Properties

L **Location**: Description of Locations

B **Business**: Description of Business Informations

Figure 7. Basic Elements for structuring Product Data

function may be Function Blocks written in IEC 61131 language(s). The Location Element describes the relative or absolute location of any mechanical part of a component in an appropriate granularity. The Business Element may be catalogue description methods such as BMEcat including customer specific rebates etc. Finally the Property Element represents all properties of a component resulting of the combination of the other Basic Elements. Figure 8 shows how any component can be described by the relationships between the basic elements using this model. The advantage of this BCFLP model is its recursivity. As shown in Figure 9 it may be used to describe as well sub components of a component as the component itself, a system or a plant composed by the components etc.

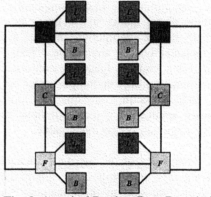

Fig. 8. A typical Product Data Description of a component by using Basic Elements

One of the benefits of this model is allowing diversity in description methods depending on the product data *usage*. Whereas the mechanical part of a component, represented by the Construct Element, may be described completely in STEP EXPRESS language following i.e ISO 10303-AP 214 or using another method. The automation function may be described in a PLC language selected from IEC 61131 or in "C" or anything else. The same is true for the Business Element which may be described using BMEcat specification or "OTD" of UNSPSC. The only prerequisite is that the engineering for a certain plant is done with a tool chain supporting the Basic Element Engineering Profile which specifies just one selected method for each Basic Element.

Process Control/ Manufacturing Plant									
Plant	F_P		C_P		■				B_P
System	F_S		C_S		■				B_S
Device	F_D		C_D		■				B_D
Component	F_C		C_C		■		L_C		B_C
Circuit									

Fig. 9. The recursive Model can be used on all levels of a plant Elements

6. CRITERIA CONTROLLED DESIGN PROCESS FOR DISTRIBUTABLE AUTOMATION APPLICATIONS

It can be expected that the method proposed by the EU project ACORN 1479 and shown in Figure 10 will lead to one or more open frameworks in the context of the overall engineering process. The concept is called "criteria controlled engineering process" (Brownlie, I. et al (1997)). The main idea is the design of the automation task independent of any (hardware) component just by combining Function Blocks according to the automation task in conjunction with already known global plant constraints (criteria) in the first step (Functional Application Designer). In the second step the created automation program is mapped on to appropriate (real) components by the Configuration Composer generating a plant proposal. In order to generate the appropriate plant proposal the Configuration Composer needs the Function Block Design (automation program) and the product data of the components managed by the Product Data Management System consisting of the machine readable product data (library) and the Product (Criteria) Selector. Each generated plant composition created by the Configuration Composer is a variant representing certain criteria entered into the system during the engineering process. It may be an optimum or not depending on the specific requirements. In order to check the usefulness and the "quality" of the composition an iterative process may be helpful to simulate parts of the application or even the whole plant in a computer. The result may require to vary

criteria. For example people may change from "chose cheapest components only" to "chose component manufacturers as less as possible" in order to minimize the number of suppliers for the plant (i.e. for simplifying maintenance).

Engineering Functionalities as described are significant extensions to these ones used today. However they are prerequisites for the engineering of automation systems for distributable applications.

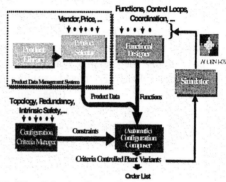

Fig. 10. The Criteria Controlled eEngineering Process of an Automation Application

7. FUTURE NETWORKS IN DIFFERENT APPLICATIONS

Up to now the pure data transport was implicitly assumed. In fact there is a major trend away from network types specially designed for some applications. The trend is to cover more applications with one network type. So it is consequent to ask which percentage the most commonly used Ethernet may cover in automation applications in the future. Due to the significant increase of speed in Ethernet Systems special real time protocols compensating the lack of bandwidth may not be necessary any longer in many real time applications. But whereas today real time networks like Fieldbusses cover most of the automation application requirements implicitly it will be necessary in the future to consider these issues more in detail. Higher bandwidth is not in all cases equivalent to "better" real time behaviour. The inherent properties of the originally not for real time purposes designed *internet* protocols may cause problems. This is also more true for wireless networks with their inherently unstable reliability. It may make sense to split applications in the following communication classes:

- Non real time applications
- Soft real time applications
- Hard real time applications

155

The PROFINET solution initiated by the PROFIBUS User Organisation reflects these communication classes already.

Non real time automation applications such as storage management or data acquisition are covered by nearly any of the available network types including the general purpose ones. Maintenance applications may also be candidates for general purpose networks depending on bandwidth needs (i.e. electronic handbooks etc). For soft real time applications the general purpose networks may also be a good choice in the future depending of the detailed requirements whereas hard real time solutions will be covered by the existing real time networks and new Ethernet based solutions (i.e. PROFINET). A real time automation application is, for example, the safety application "Emergency Switch Off" which will be desired to be even wirelessly by some parties.

8. WIRED AND WIRELESS NETWORKS

The real time classification above is independent from the used physical layer technology either it is wired or wireless. Wireless networks will come up in two flavors at least. There will be solutions adapted from wireless office LANs such as IEEE 802.11b/a or others for the non and soft realtime automation applications and especially designed wireless solutions for real time purposes.

It is expected that the general purpose network for automation will be based on Ethernet (wired) and any of the IEEE 802.11x wireless solutions. The wireless real time networks will be extensions of the existing real time networks in order to avoid real time drawbacks by protocol conversions in any device linking wired to wireless segments and vice versa.

The challenge in wireless applications will be:

- Planning and management of wireless networks
- Closing the security gap

Radio Frequencies are managed all over the world because they are limited. Therefore it is a must to coordinate each wireless cell in order to optimize the number of networks in a certain area. Today's wireless networks use the ISM bands mostly because they are free of use. But this is also their drawback. If everybody can use these bands they become more and more unreliable the more applications use them. The time is foreseeable, when these bands are so unreliable that they cannot be used for automation purposes any longer. Therefore according to the results of the RFieldbus Project (Haehniche, J., Hammer, G., Heidel, R. (2001)), a recently finished EU funded real time wireless network project, special frequencies for automation are required. It is up to the automation community to care for this worldwide automation frequency band.

9. SECURITY

There is one thing left: security. As described above a common application object model should "hide" data transport mechanisms below the application layer interface. The more open the systems will be the more security issues come up. In addition any wireless environment has to be considered as inherently unsafe. Therefore security has to be an integral part of the overall system architecture for distributable applications. First of all Security in an automation plant means adopting the common security solutions for business applications as much as possible. But "Security in Automation" is more. Some application related security holes do not simply exist in business applications such as cyclic data traffic with small periods and others. The circumstances are different. Whilst business applications represent mostly a balance between *Security and Privacy* automation applications do not have any Privacy at all. On the contrary: Privacy is explicitly not desired because any Privacy reduces transparency of a plant. As a consequence automation system designers must check the privacy part of any business security application and decide whether he tolerates the potential security hole, creates workarounds or designs its own solution. The situation is worse as in the regular business. A big number of *different* devices have to be managed in one plant which means typically some hundred types of components and some thousand devices of up to several hundred manufacturers in one installation.

As a conclusion it would be helpful to create an "Open Framework for Security in Automation" as proposed by the EU funded project "RFieldbus" (Haehniche, J., Rauchhaupt, L. (2000)). Figure 11 shows the components of an overall security architecture in automation. The Security Architecture consists of four components:

- Security Philosophy
- Protection Needs Analysis Method
- Security Specific Quality Check Procedure
- Open Smooth Security Standard

Fig. 11. Security Elements in a Plant

The Security Philosophy defines the way security will be provided. Major items are the specification how to manage keys and the definition of Security Levels which the Protection Needs Analysis Method may be based on. The selected level has to be built in the real plant. In order to validate the real security installation in the plant a check procedure has to be defined which contains also rules for change management. Finally the Open Smooth Security Standard has to be defined in a formal manner rapidly changeable for closing upcoming security gaps and open to be used in all devices of a plant installation.

Thus based on such a framework for functional security processes can be established allowing customers to determine their security *needs* including test procedures for *conformity* of a commissioned ("built") plant with the planned/designed security functions.

10. CONCLUSION

We are on the way to automation architectures supporting "distributable automation functions". The key component is the extended communication system forming a distributed operating system. One prerequisite is a common application object model and the appropriate Engineering Platform with its integrated Product Data Management. Due to the open platform and especially in wireless environments security issues come up which may not be solved by just adopting regular business solutions. From the user's perspective there will be much more flexibility and transparency in the plant as today, assumed the appropriate tool chains exist. The open environment will offer a market of automation functions (Function Blocks) selected from products by competing manufacturers.

REFERENCES

Brownlie, I. et al (1997). ACORN 1479 - Design Tools for Fieldbus, Feldbustechnik in Forschung, Entwicklung und Anwendung. *Proceedings of FeT '97, ISBN 3-211-83062-6.* Springer, Wien/New York.

Doebrich, U. (2003)
Sachsenmeier
Challenges Between Competition and Collaboration
Springer, Wien/New York.

Haehniche, J., Hammer, G., Heidel, R. (2001). Funk statt Draht - Kabellose Kommunikation auch in der Feldtechnik ?". *atp 6/2001* Oldenbourg.

Haehniche, J., Rauchhaupt, L. (2000). Radio Communication in Automation Systems : the RFieldbus Approach. *IEEE Proceedings WFCS-2000*

ELSEVIER

IFAC
PUBLICATIONS
www.elsevier.com/locate/ifac

A PERIOD-BASED GROUP MEMBERSHIP STRATEGY FOR NODES OF TDMA NETWORKS

Elizabeth Latronico, Philip Koopman

Carnegie Mellon University, ECE Department
Pittsburgh, PA, USA
beth@cmu.edu, koopman@cmu.edu

Group membership provides strong guarantees for safety-critical fieldbus systems. However, using a single group for all messages provides an unnecessarily high risk of temporary node outages in the face of transient faults. Using groups based on virtual nodes that are divided by message period can increase the availability for the most critical, high-speed network messages with potentially reasonable bandwidth cost and without giving up the assurances of strong group membership algorithms. *Copyright ©2003 IFAC*

Keywords: Automobile industry, Availability, Communication protocols, Embedded systems, Fault tolerance, Fieldbus, Group membership, Safety-critical, TTP/C

1. INTRODUCTION

In automotive embedded systems, all nodes are not created equal. Automotive embedded systems generally have a variety of nodes, sending messages with up to two orders of magnitude difference in period. For example, a brake activation message might be sent every 5 milliseconds while a battery status message might be sent at 1000 milliseconds (Tindell and Burns, 1994). Cost concerns often preclude point-to-point connections between each pair of nodes, leading to use of a broadcast bus for messages with such significantly different periods. Current automotive buses such as the Controller Area Network (CAN) use a priority-based message sending scheme (CAN, 1991). However, there is pressure to adopt a time-based sending scheme in order to provide a more predictable platform for safety assurance.

Time Division Multiple Access (TDMA) networks provide a statically-scheduled method of transmitting data on a broadcast bus. TMDA networks contain slots for each message to be transmitted. Messages are transmitted in frames, which also include overhead. Slots are defined according to their order in a round. Each node typically has at least one slot per round (Bauer and Paulitsch, 2000). Rounds are assembled into a cluster cycle (TTP/C, 2002). Typically a dual-redundant bus is used, and a node sends each message once on each of the two channels.

TDMA systems often allow nodes to share their views of the system with other nodes. These group membership services protect against a variety of faults including processor faults, link faults, and noise on the communication bus (Kim and Shokri, 1993). The faults may be permanent or transient. Group membership services form the basis for important services such as replication and clock synchronization. In automotive embedded systems, there is typically a single group, and recovery involves ensuring that all nodes eventually belong to that single group.

Unfortunately, with a single group strategy it is a difficult task to label a fault as permanent or transient. Each frame in an embedded network typically carries a Cyclic Redundancy Code (CRC) to detect if the frame has been corrupted. If a node receives a frame with an invalid CRC, it cannot tell from this information alone whether the fault is permanent or transient. The source of the fault is also unknown – the sender could be faulty, or the bus could be noisy. Therefore group membership services in state-of-the-art TDMA protocols, such as TTP/C, must take a pessimistic approach upon receiving a faulty frame. A node that receives faulty frames on both bus channels will consider the sender to be faulty. If enough nodes consider the sender to be faulty, the sender loses membership in the group and must reintegrate.

A more optimistic approach is possible if multiple, period-based groups are used. This paper demonstrates how the same group membership algorithms for a single group strategy can be used in a multiple group strategy. Greater tolerance is achieved for transient faults caused by noise on the communication bus at an acceptable bandwidth cost, without altering the group membership algorithm. This technique is demonstrated for a braking application based on SAE benchmark data.

Section 2 discusses domain characteristics and the Society of Automotive Engineers (SAE) workload from Tindell and Burns (1994) that is used as a reference example. Section 3 presents relevant concepts of group membership and explores a standard single group system. Section 4 presents our multiple group solution. Section 5 presents an availability and bandwidth analysis comparing the single group and multiple group strategies.

2. DOMAIN

A number of system constraints help structure the solution space. First, the SAE standard workload from Tindell and Burns (1994) is reviewed as a representative automotive workload. Next, other relevant constraints are discussed.

The SAE standard workload (Tindell and Burns, 1994) contains a set of periodic and sporadic messages sent in a prototype electric car with seven subsystems. These subsystems include the Batteries (Battery), Brakes (Brakes), Driver (Driver), Inverter/Motor Controller (I/M C), Instrument display panel (Ins), the Transmission control (Trans), and the Vehicle Controller (V/C). For our purposes it is assumed that each subsystem constitutes one node except the Brakes, where it is assumed that there will be one Brakes node per wheel for a total of four Brakes nodes. Actual systems might differ from this configuration; the workload in Tindell and Burns (1994) was originally designed for a point-to-point system. The SAE workload contains messages with six different periods: 5 ms, 10 ms, 20 ms, 50 ms, 100 ms, and 1000 ms. All of the 50 ms messages are sporadic messages, but are assumed to have a 50 ms period as Tindell and Burns assume (1994), in accordance with standard automotive practice.

Embedded systems are often highly constrained, and those constraints can be used to our advantage. In particular, the following properties are useful:

- *Harmonic periods*
  Messages are commonly scheduled with harmonic periods so it is easier to prove schedulability. Hence, messages can easily be grouped by period.
- *Period and deadline usually equal*
  A message's period is often the same as its deadline. If so, increasing the period is not an option.

- *Short payloads relative to overhead*
  Data payloads are often on the order of one to eight bytes long. Other common frame fields include an ID field and Cyclic Redundancy Code (CRC) field for error checking, which can consume a few bytes. A solution must be cautious about adding overhead, but overhead is acceptable in many cases.
- *Sender has "ground truth"*
  Regardless of how many other nodes disagree, the message sender has the correct state variable value of a message being sent. Therefore it is best to treat transient errors differently than permanent errors if possible, as nodes are usually not interchangeable.

3. GROUP MEMBERSHIP CONCEPTS

This paper will show that forcing all nodes to be members of a single group is a limiting restriction. Specifically, this lowers availability when transient faults are treated as permanent faults. A single faulty frame causes a sending node to lose membership even if the fault is due to noise on the communication bus. This paper shows that having multiple groups can ease the effects of this pessimistic restriction. There are several known group membership algorithms, with varying levels of guarantees. This work refers to the group membership algorithm of TTP/C.

A key advantage to this approach is that the group membership algorithm remains unchanged. This allows existing proofs to be reused. A central design problem for any group membership service is determining when a node should lose membership, if at all. Inventing a new algorithm is difficult - there are many tradeoffs and subtle points to consider. Instead, availability of the system can be increased by using multiple groups, operating by the same rules a single group would operate by. The next sections review relevant group membership concepts and constraints.

3.1. *Fault Model*

Group membership algorithms are usually designed to withstand node crashes, send faults, and receive faults. Algorithms handle both permanent and transient faults, typically with restrictions on fault interarrival rates (Kim and Shokri, 1993). Group membership algorithms cannot compensate for loss of network connectivity or semantically incorrect data that is syntactically correct. Group membership requires at least four nodes to tolerate one faulty node (Pfeifer, 2000). Faulty nodes that lose membership may reintegrate into the group, after the group has reached consensus on its members. Consensus is guaranteed to be reached within two rounds after a fault has been identified (Pfeifer, 2000). If a fault occurs in the group, additional faults are not tolerated while member nodes have inconsistent views of membership, although better fault tolerance is possible for some faults if more time is allowed (Kim and Shokri, 1993).

3.2. Clique Avoidance, Implicit Acknowledgment

Clique avoidance is one of two mechanisms employed in order to ensure that a group does not partition into two or more separate groups, called cliques (Bauer and Paulitsch, 2000). Each node maintains a list regarding who it thinks the members of its group are, sometimes called a membership vector (TTP/C, 2002). Since a node considers a frame incorrect if the sender does not have the same membership vector, nodes in separate cliques would not be able to communicate with each other. Clique avoidance is also designed to identify nodes that are receive-faulty. Clique avoidance requires a node to have received more correct frames than faulty frames in the last round in order to retain membership (not counting null frames). Clique avoidance may prohibit a node from sending a frame in its next two slots following a fault, sometimes when the node was not the source of the actual fault (Bauer and Paulitsch, 2000).

Implicit acknowledgment ensures that a faulty sender will lose membership. After sending a frame, the sending node waits to see if subsequent nodes have received its frame. Protocols use some sort of a broadcast membership vector per node (either explicit or implicit) to relate a node's opinion of who is in its group. A sender will lose membership if not enough other nodes receive the frame correctly (Pfeifer, 2000).

3.3. Performance Implications

Due to the interaction of clique avoidance and implicit acknowledgment services, group membership requires at least one round and at most two rounds to achieve consistent membership (Bouajjani and Merceron, 2000). A node may also be prohibited from sending frames during these two rounds to ensure consensus on a single group is reached.

According to the TTP/C specification and group membership proofs, each node transmits exactly once per round. Specifically, if all nodes in the system belong to a single group then:

- Each node must transmit at least once per round (Bauer and Paulitsch, 2000)

In order for the system to reach consensus, it must hear from all member nodes. Mandating that each node must transmit at least once per round allows the guarantee of a maximum of two rounds to achieve consistent membership to be made.

- Each node may transmit at most once per round (TTP/C, 2002, p. 18)

The TTP/C specification does not list a specific reason for this; however, it can be inferred that allowing a node to transmit multiple times per cycle would give this node an unequal weight in the failed slots counter that is incremented every time a faulty frame is

received. Also, a sending node must always wait for at least one subsequent valid frame to be acknowledged (Bauer and Paulitsch, 2000).

Therefore, for a system with a *single* group, each node must transmit exactly once per round (although it is not necessarily the same message that is transmitted each round). The TTP/C protocol also allows shared slots, where distinct nodes (called multiplexed nodes) may alternate sending messages in a designated slot in a round (TTP/C, 2002). Multiplexed nodes are not employed here, because the results are undesirable regardless of whether the group membership algorithm considers these nodes to be separate member nodes or a single member node. If the group membership algorithm considers the nodes sending in the shared slot to be separate member nodes, then the time to achieve consistent membership will increase as consensus requires the opinions of all member nodes. If the group membership algorithm considers the nodes as a single member node, then loss of membership for the member node implies that all of the nodes sending in the shared slot will lose membership. This work also does not consider redundant nodes with distinct sending slots due to space considerations. Redundant nodes with separate slots would consume a larger amount of bandwidth.

4. OUR SOLUTION

In order to tolerate transient faults, one can take advantage of the fact that nodes often send messages at different periods. Therefore, redundant information about the state of a node is available – a single corrupted message might be considered to be a transient failure if the next type of message from that node is correct. However, transmission of the next type of message from that node might be suppressed by the clique avoidance algorithm. Thus, the next type of message might not be sent.

In order to track different message types separately, message periods need to be the basis for group membership, not physical nodes. The obvious approach to separating messages is to try to create separate groups of physical nodes. Unfortunately, in general it is difficult to split automotive network nodes into disjoint sub-groups. For example, one cannot create two distinct groups of nodes for the SAE workload because the Vehicle Controller is either the producer or a consumer for all messages.

Table 1 shows the physical sending nodes and their sending periods in this system, and the total number of payload bits sent per period. For example, the Battery node sends 8 bits worth of data every 50 ms, 32 bits of data every 100 ms, and 17 bits of data every 1000 ms. These numbers only include payload data, not other fields in the frame, which will be discussed in the Performance Analysis section. This paper assumes four Brakes nodes instead of a single Brakes node as in Tindell and Burns (1994).

4.1. *Virtual Groups and Virtual Nodes*

Instead of anchoring our groups on physical nodes, we use *virtual groups* made up of *virtual nodes*. The algorithm for constructing virtual groups is to create one virtual group per unique message period. A virtual node is created according to the periods of messages that a physical node sends. One virtual node is created per period for each physical node. Each virtual group must have at least four members (as Pfeifer shows is required to tolerate one faulty node) (2000). If there are fewer than four distinct physical nodes sending messages at a particular period, the system designer has two choices. The designer can elect to send a message more frequently, and assign the virtual node to a smaller period virtual group. Alternatively, the designer may create additional virtual nodes by having physical nodes send placeholder messages at that period. In general, virtual nodes sending placeholder messages can be added to any virtual group if additional fault tolerance is desired.

Tables 2 through 6 show the virtual groups for the SAE benchmark system. There is one virtual group per unique message period in our system, with the exception that there is no 10 ms group. There is only one message sent at 10 ms, and a group of one node will not be fault tolerant, so this message is sent at 5 ms instead, incurring a small amount of extra bandwidth. VirtualGroup5 only had three virtual node members, so the virtual node I/MC1000 is added, sending a one bit message in this group (Table 6). There is one virtual node per message period that a physical node sends. For example, as Table 1 shows, the Battery Node sends messages at 50 ms, 100 ms, and 1000 ms. This results in three virtual nodes - Battery50, Battery100, and Battery1000 - shown in Table 4, Table 5, and Table 6.

With this strategy, nodes may not have to send once per round as Section 3.3 discussed. Instead, a virtual node need only send once during its virtual group's period. The virtual group will then be guaranteed to reach consensus within two times its associated period (not twice the round length). This strategy will tolerate one fault in two times the period of the virtual group. The round length will remain the same, and virtual nodes will send at most once per round, and at least once per period. Note that the period of the virtual group is always greater than or equal to the round length. One might expect this to negatively impact availability, but this is not the case as Section 5 will show.

A main benefit of this strategy is increased tolerance for transient faults, namely corrupted frames due to noise on the communication bus. Each virtual group keeps its own membership; therefore if a frame is

Table 1. Nodes and Sending Message Periods
From data in Tindell and Burns (1994)

Node / Period	5ms	10ms	20ms	50ms	100ms	1000ms
Battery				8	32	17
BrakesOne	16		1		8	
BrakesTwo	16		1		8	
BrakesThree	16		1		8	
BrakesFour	16		1		8	
Driver	8			13		2
I/M C	16			14		
Trans	8				8	
V/C	16	16		25		3

corrupted only the virtual node will lose membership, not the physical node. A physical node that sends messages at different periods will still be able to send some of its messages if one of its frames gets corrupted. For example, assume that a 100 ms frame from the physical Transmission node is corrupted by noise on the bus. This frame corresponds to virtual node Trans100 in VirtualGroup4. Assuming all virtual nodes in VirtualGroup4 detect a corrupted frame, virtual node Trans100 will lose membership and may not be able to send messages for the next 200ms (twice the period). However, virtual node Trans5 (which is likely to be more critical, as it has a shorter period) is unaffected. In addition to providing increased availability, this group strategy provides some protection against critical messages being prevented from sending by non-critical message failures.

Note that neither a virtual group alone, nor virtual nodes alone would solve this problem. If virtual groups were created involving physical nodes, a single faulty frame from a physical node would affect all virtual groups. The physical node would lose membership in all virtual groups; thus reintegration time would not be improved. In fact, reintegration would take longer, because some of the virtual groups have periods longer than a round. If only virtual nodes were created and a single group was used, the bandwidth cost would be prohibitive as nodes must send exactly once per round to guarantee consensus occurs in two rounds.

4.2. *Other Possible Sources of Faults*

Using multiple groups provides a fairly robust way to identify transient faults. It would be unlikely for a fault other than a transient bus error to corrupt one of the messages a node sends and not the others. Since the group membership service depends on the CRC included with the frame for error detection, a node that has sent an invalid value will be deemed correct as long as all receiving nodes correctly receive that value. A permanent fail silent processor fault would affect all messages. An outgoing link failure on a node would affect all messages. An incoming link failure would

VirtualGroup1 (5 ms period)		VirtualGroup2 (20 ms period)		VirtualGroup3 (50 ms period)		VirtualGroup4 (100 ms period)		VirtualGroup5 (1000 ms period)	
Virtual Node Name	Total Payload Bits	Virtual Node Name	Total Payload Bits	Virtual Node Name	Total Payload Bits	Virtual Node Name	Total Payload Bits	Virtual Node Name	Total Payload Bits
BrakesOne5	16	BrakesOne20	1	Battery50	8	Battery100	24	Battery1000	17
BrakesTwo5	16	BrakesTwo20	1	Driver50	13	BrakesOne100	8	Driver1000	2
BrakesThree5	16	BrakesThree20	1	I/MC50	14	BrakesTwo100	8	I/MC1000**	1
BrakesFour5	16	BrakesFour20	1	V/C50	25	BrakesThree100	8	V/C1000	3
Driver5	8					BrakesFour100	8		
I/MC5	16					Trans100	8		
Trans5	8								
V/C5	*24								

* This includes two 8-bit 10ms messages, since the V/C was the only node to send at 10 ms and it cannot be in a group by itself

** This virtual node was added so this virtual group would have 4 members

affect all messages. A faulty clock would likely affect all messages, although additional investigation is necessary for this topic.

It is important to note that our approach is amenable to services that need to treat the system as a single group. Specifically, a clock synchronization algorithm may need to consider the system as having a single group, since the 'correctness' of a frame mandates that the sender's view of members in the group matches the receiver's (TTP/C, 2002). Since each virtual node can be mapped back to a physical node, a clock synchronization algorithm will be able to determine which nodes are functional and which are not from the virtual group information.

5. PERFORMANCE ANALYSIS

This section delves into a detailed analysis of the availability provided and bandwidth required by the single group approach and the multiple group approach. First, frame overhead and payload sizes are determined. Next, the probability of all four Brakes nodes being unavailable at the same time with both membership approaches is computed. Finally, the bandwidth required by each approach is discussed.

5.1. Frame Overhead

In order to estimate bandwidth and availability, one needs to determine how many bits will be sent in a frame. In addition to the data payload, each frame

Table 7. Estimated Frame Overhead

Field Name	Estimated Bits
CRC	24
Frame Type Identifier	2
Mode Change Request	1

includes some overhead fields. The size of the overhead fields was determined according to version 1.0 of the TTP/C specification. For these estimates, the smallest size possible was used, in order to compare the new strategy to the best performance possible under the existing single group strategy.

Table 7 lists the additional fields (besides the data payload) sent with each frame in the TTP/C system (TTP/C, 2002). There is also a Schedule ID field calculated into the CRC but not sent explicitly. Note that a length field is not required for TDMA protocols, since this information may be placed in a Message Descriptor List (MEDL) deployed before startup on all nodes. The TTP/C Specification mandates a CRC with a minimum Hamming distance of 6 (TTP/C, 2002, p. 44). The maximum allowable data length (plus implicit C-state) depends on the CRC length (TTP/C, 2002, p. 24). A 24-bit CRC will adequately protect the maximum allowed data payload of 240 bytes. For the Frame Type Identifier, there are at least three types of frames: cold start, implicit C-state, and explicit C-state (TTP/C, 2002, p. 39-40). Therefore, at least two bits are needed to represent the Frame Type Identifier. For the Mode Change Request field, 'Each MEDL contains at least two modes, the startup mode and one application mode', so at minimum this field is one bit (TTP/C, 2002, p. 11). This gives a total of 27 bits of overhead per frame.

If an explicit membership vector is used, this will also incur overhead. The membership vector contains one entry for every node in the sending node's group, usually a true/false flag. In standard group membership, there will be one bit per node in the system. With virtual groups, a virtual node will send one bit for every node in its virtual group. A membership vector for a virtual group will always be equal to or smaller in size than a membership vector that treats the system as a single group, since each real node will have at most one virtual node in each virtual group. So the size of the single group membership vector is an upper bound for the size of a virtual group membership vector.

Membership vectors can also be sent implicitly (except during startup), consuming no bandwidth. Implicit membership vectors are assumed; therefore, the bandwidth calculations do not include overhead for the vectors. Explicit membership vectors are required for reintegration, however. For this analysis, it is assumed that explicit C-state frames are sent at regular intervals, but do not occur during the two rounds in question. Recall that a node that loses membership may not be able to reintegrate in the round following a fault, due to clique avoidance. Therefore, reintegration is not a factor in the performance analysis.

5.2. Payload Size

In order to determine the data payload size, recall that for single group membership a node will send exactly once per round. Table 1 lists the number of data bits a physical node has to send for each period. Because of the send-once-per-round rule, a node will need to combine payloads into one frame, or the round will need to be shorter than the minimum period. Our example assumes the payloads will be combined into a single frame, since shortening the round would result in more overhead bits total. Looking at Table 1, the shortest message period in this system is 5 ms, so the round length will be 5 ms.

The worst-case payload size for a Brakes node can be computed from the information in Table 1. For the single group strategy, a Brakes node will send out an aggregate frame exactly once per round, or every 5 ms. Each Brakes node (BrakesOne, BrakesTwo, BrakesThree, BrakesFour) has 16 bits of data at a 5 ms period. Then, each Brakes node also must send 1 bit of data every 20 ms and 8 bits of data every 1000 ms. Careful design can avoid having the 20 ms payload and the 1000 ms payload in the same aggregate frame. Therefore the worst case payload size for a Brakes node using the single group strategy is 24 bits. For our multiple group strategy, the payloads will contain data only for other messages sent at the same period. Therefore the 5 ms message will have a 16 bit payload,

the 20 ms message will have a 1 bit payload, and the 100 ms message will have an 8 bit payload.

5.3. Chance of Losing All Brake Nodes

This section explores the probability that all four Brakes nodes lose membership due to corrupted frames, given as the probability per hour. Our fault model is random, independent noise on the bus. This probability provides a conservative estimate of the chance that the brakes will be unavailable due to a transient bus error. For the multiple node strategy, it is possible that only some of the virtual nodes lose membership. Additionally, the probability of losing a particular message from the brakes is different from the probability of losing all messages from the brakes. Cases where a node loses membership due to other types of errors are not considered (for example, a node that is receive-faulty). Since the fault model is independent noise, a conditional probability analysis would produce the same results because the errors are uncorrelated. A more inclusive fault model is an avenue for future work.

Single Group Strategy
One needs to know the worst-case size possible for two frames in a row, since reintegration can take two rounds. From the Payload section, the worst-case payload size for a Brakes node with the single group strategy is 24 bits. Therefore the worst case for the first frame is a size of 24 data bits plus 27 overhead bits for a total of 51 bits. For the next frame, the worst case occurs when the 20 ms data and the 5 ms data are sent. Adding overhead gives a frame size of $16 + 1 + 27 = 44$ bits. Since the 20 ms data is sent along with every fourth 5 ms frame, and the 1000 ms data is not sent with the 20 ms data, the 20 ms data will fall either in the round after or before the 1000 ms data. The order is immaterial to the reliability equations. Therefore the largest number of bits sent by a Brakes node in 10 ms is 95 bits.

Our Multiple Group Strategy
For the virtual group strategy, recall that the consensus time will be equal to the period of that group. The Brakes nodes send messages in three different virtual groups - the 5 ms group, the 20 ms group, and the 100 ms group, as can be seen in Tables 2-6. An upper bound on the probability of losing all frames from all

Table 8. Probability Per Hour of Losing Brake Nodes

BER	Single group, any/all message(s)	Multiple group, any 5 ms message	Multiple group, any 20 ms message	Multiple group, any 100 ms message	Multiple group, all messages
10^{-4}	2.39 E-11 /hour	5.76 E-18 /hour	9.67 E-19 /hour	2.99 E-17 /hour	9.73 E-38 /hour
10^{-5}	2.39 E-19 /hour	5.76 E-26 /hour	9.67 E-27 /hour	2.99 E-25 /hour	9.73 E-62 /hour
10^{-6}	2.39 E-27 /hour	5.76 E-34 /hour	9.67 E-35 /hour	2.99 E-33 /hour	9.73 E-86 /hour
10^{-7}	2.39 E-35 /hour	5.76 E-42 /hour	9.67 E-43 /hour	2.99 E-41 /hour	9.73 E-110 /hour

Brakes nodes can be found by multiplying the probabilities of losing all 100 ms frames within twice that period (200 ms), all 20 ms frames within 40 ms, and all 5 ms frames within 10 ms. Note that each frame is sent twice, since there is a dual-redundant bus. Therefore a total of eight frames must be lost. Table 8 gives the probability per hour that a group will lose one type of message from all of the Brakes nodes at the same time. Equation 1 gives the formula for the first four columns:

$$\left(BER * \frac{\# bits}{frame} * \frac{\# frames}{two\_period} \right)^8 * \frac{two\_period}{hour} \quad (1)$$

The first portion of the equation represents the chance that one frame will be corrupted in a window of two periods. This is raised to the power of 8, since 8 frames (two per node) must be corrupted within this window of time. Then, the second portion of the equation shows how many windows occur in an hour. For example, the first entry in the 'Multiple Group Any 5 ms Message' column was calculated as:

$$\left(.0001 * (16+27) * 2 \right)^8 * \frac{1}{10ms} * \frac{3600000ms}{hour} \quad (2)$$

The BER is 0.0001. There are 16 payload bits (from Table 1) plus 27 overhead bits (from Section 5.1) in a frame. One frame is sent per 5 ms period, giving two messages per two periods. This probability of a corrupted frame (i.e., losing all messages from the physical node in a frame) is raised to the power of 8. Then, this is multiplied by how many 10 ms rounds occur per hour.

For a single group strategy, the probability of losing a particular type of message from all nodes is the same as the probability of losing all messages from all nodes. A transient error in both redundant frames a node sends will cause that node to lose membership and it will not be able to send any message in its next slot, and possibly its next two slots. However, for the multiple group strategy, it is possible that only some of the virtual nodes lose membership and not others. Therefore some of the messages may be entirely lost without affecting other messages. The final column gives the probability that all messages will be lost for the multiple group strategy, per hour. This probability is obtained by multiplying the previous three columns together. This is a pessimistic estimate, as this is the probability that all three virtual groups will fail in an hour, not necessarily at the same time.

Observations

For this data set, all of these approaches seem reasonable for automotive applications. Automotive protocols are typically designed to have lower than a 10^{-9} probability of failure per hour (PALBUS, 2001) for a network with a BER of 10^{-5} to 10^{-6}. The estimates

for all four Brakes nodes losing membership do not violate that criterion; however, designs with heavier workloads might. The results in Table 8 are not failure rate calculations and do not account for nodes that lose membership for any other reason besides a transient bus error while the node is sending a frame. These results should not be used as a sole estimate of any failure rate, but rather are intended to illustrate the possible benefits of a virtual node approach.

The multiple group strategy has a lower chance of losing a single particular message from all four Brakes nodes. This is due to the fact that the frame sent is smaller because the payload is smaller. In the multiple group strategy, less data needs to be sent in the payload, since the virtual nodes may send only once per their virtual group period instead of once per round.

The multiple group strategy has a far lower chance of losing all messages from all four Brakes nodes. Since there are virtual Brakes nodes in three virtual groups, the chance of all virtual nodes losing membership is roughly equal to the chance of the single group strategy raised to the power of three. In general, a node with an X chance of losing membership in a single group will have a X^n chance of losing membership in all n of its virtual groups. So in Table 8, the last column is roughly equal to the first column cubed. This is a significant benefit for safety-critical systems.

It is important to note that some forms of redundant components, such as shadow or backup nodes, will not improve availability in this situation that involves a transient fault. A node is not allowed to integrate into a group until the group has reached consensus on its members. So the shadow node will be prohibited from sending frames if the original node would have been prohibited from sending frames.

5.4. *Bandwidth*

This availability gain has low bandwidth cost. Determining the bandwidth needed to send frames requires computing the slot size for each node. According to the TTP/C specification, each node is assigned one slot in a round (TTP/C, 2002). The message size may vary; however, the slot size remains constant. Therefore a node's slot size must be at least as large as the largest frame the node has to send. (The actual slot size is slightly larger, but that is true for both strategies, and only serves to make this bandwidth comparison conservative.) The slot size will determine the bandwidth required – even if a portion of

Table 9. Slot Size Required Per Node (Bits)

	Battery	Brakes(4)	Driver	I/M C	Trans	V/C
Single	35	204	39	51	43	59
Multiple	35	312	66	78	70	113

the slot goes unused, other nodes are prohibited from sending until the end of the slot.

For the single-group strategy, the payloads are aggregated into a single frame because a node may only send once per round. One can do slightly better than simply adding up all payload bits given in Table 1 for a node by using the complete data listing from Tindell and Burns (1994). Since the system is statically scheduled, and some of the long period messages do not send every five milliseconds, more conservative figures can be given for the slot size. Table 9 summarizes the slot size required by each node for the single and multiple group strategies. These figures include overhead. There will be 200 of each slot in one second (since the round is 5 ms long), giving a total required bandwidth of 86,200 bits/second.

For a multiple group strategy, there is a tradeoff between error detection ability and bandwidth. If data payloads from the same physical node but different virtual nodes are sent in the same message with a single CRC, and the message is corrupted, all participating virtual nodes will lose membership in their virtual groups. For example, if the actual Transmission node sends its 5 ms and 100 ms payloads using only one CRC, both virtual nodes Trans5 and Trans100 will lose membership if the message is corrupted.

We choose to preserve the error detection ability by assuming that a physical node will send back-to-back separate, complete messages per time slot for each virtual node that needs to send. This is slightly different than the TTP/C specification approach for sending multiple messages per slot where overhead is not duplicated (TTP/C, 2002). This also represents maximum bandwidth consumption, so if the bandwidth required is too great, further points in the tradeoff space can be explored. Another option is to have some of the virtual nodes use a shared slot, since virtual nodes are not required to send a frame once per round. Both approaches incur the same bandwidth, since the same data is being sent – it is just a matter of whether a single slot is reserved, or multiple slots are reserved. It is acceptable to combine payloads for messages with the same period. These calculations assume all overhead is duplicated; it may be possible to combine some overhead fields.

The maximum slot sizes required for the multiple group virtual node strategy are listed in Table 9. There will be 200 of each slot in a second, giving a total required bandwidth of 134,800 bits/second. This represents a 1.56 times bandwidth cost compared to single group, in exchange for a dramatically reduced probability of losing group membership for the most critical messages.

6. CONCLUSION

A multiple-group, period-based group membership strategy provides an attractive way to tolerate transient errors due to bus noise in TDMA protocols. Algorithms for single group membership can be applied to virtual groups with virtual nodes without changing the existing theoretical framework. By creating a virtual group composed of virtual nodes for each period in the system, the virtual nodes are isolated from each other. Assigning a virtual node per each message period of a physical node allows the system to tolerate a corrupted message from the physical node without affecting some of the other messages the node sends.

Multiple-group, period-based group membership provides increased availability at an affordable bandwidth cost. For N virtual groups, the unavailability can be reduced by as much as a power of N over a single group strategy. The bandwidth increase is not prohibitive (1.56 times the single group bandwidth for the system studied), as the requirement that nodes send once per round can be relaxed to once per period for the virtual nodes.

ACKNOWLEDGMENTS

This work is supported in part by the General Motors Collaborative Research Laboratory at Carnegie Mellon University and by the United States Department of Defense (NDSEG/ONR).

REFERENCES

Bauer, G. and M. Paulitsch (2000). An Investigation of Membership and Clique Avoidance in TTP/C, *Proceedings of 19th IEEE Symposium on Reliable Distributed Systems SRDS-2000,*, pp. 118-24.

Bouajjani, A. and A. Merceron (2002). Parametric Verification of a Group Membership Algorithm, *Proceedings of the 7th International Symposium on Formal Techniques in Real-Time and Fault Tolerant Systems.*

Controller Area Network (CAN) Specification v. 2.0 (1991). Robert Bosch GmbH, Stuttgart.

Kim, K.H. and E. Shokri (1993). Minimal-Delay Decentralized Maintenance of Processor-Group Membership in TDMA-Bus LAN Systems, *Proceedings of the IEEE International Conference on Distributed Computing Systems - ICDCS '93*, pp. 410-19.

Analysis and Test of Bus Systems, PALBUS Task 10.2, 10.3 (2001), SP Swedish National Testing and Research Institute.

Pfeifer, H. (2000). Formal Verification of the TTP Group Membership Algorithm, *Proceedings of FORTE XIII/PSTV XX*, pp. 3-18.

Time-Triggered Protocol TTP/C High-Level Specification Document (2002), TTTech Computertechnik AG, Schoenbrunner Strasse 7, A-1040 Vienna, Austria.

Tindell, K. and A. Burns (1994). Guaranteeing Message Latencies on Control Area Network (CAN), *Proceedings of the 1st International CAN Conference.*

A DISTRIBUTED APPROACH TO ACHIEVE PREDICTABLE ETHERNET ACCESS CONTROL IN INDUSTRIAL ENVIRONMENTS

[¶]A. Bonaccorsi, [*]L. Lo Bello, [¶]O. Mirabella, [§]P. Neumann, [§]A. Pöschmann

[¶]*Dipartimento di Ingegneria Informatica e delle Telecomunicazioni*
University of Catania, Italy
abonacco@diit.unict.it, omirabel@diit.unict.it

[*]*Facoltà di Ingegneria - Sede di Enna, Cittadella Universitaria, Enna Bassa*
Dipartimento di Ingegneria Informatica e delle Telecomunicazioni
University of Catania, Italy
llobello@diit.unict.it

[§]*Institut für Automation und Kommunikation, Magdeburg, Germany*
Peter.Neumann@ifak-md.de, Axel.Poeschmann@ifak-md.de

Abstract: This paper presents an overview of a new protocol, called the Predictable Ethernet Access Control protocol (PEAC), which overcomes the unpredictable behaviour of the CSMA/CD, thus guaranteeing deterministic access to the medium to hard real-time traffic. The PEAC is a distributed protocol able to support both hard real-time and non-real-time traffic in an integrated way. The PEAC does not require any modification to the standard hardware, so COTS hardware can be used. The paper outlines the PEAC protocol, presents a timing analysis and shows experimental results obtained in a real scenario. *Copyright © 2003 IFAC*

Keywords: CSMA/CD, TDMA, Real-Time, Process Control, Timing Analysis

1. INTRODUCTION

The main obstacle to using Ethernet in real-time communication is that, due to the CSMA/CD access protocol [1], Ethernet cannot provide connected stations with deterministic channel access times and therefore guarantee that data delivery deadlines will be met.

As Ethernet technology today offers a number of appealing features, which suggest adopting it even in time-constrained environments, like distributed process control and factory automation systems, recently a lot of research addressed the problem of enforcing a predictable behaviour on Ethernet networks. Various approaches have been proposed in the literature to support real-time traffic on Ethernet networks [2-13]. They differ according to the nature (soft or hard) of real-time applications to be supported.

To support soft real-time applications, which do not require determinism and can accept a *statistical* bound on packet delivery time, traffic smoothing mechanisms on a Shared Ethernet have been proposed. Traffic smoothing, which was introduced in [5], is based on the idea that, to provide statistical guarantees on the timely delivery of Ethernet packets, it is sufficient to keep the total arrival rate for new packets generated by stations below a given *network-wide input limit*. Different approaches for *dynamic* traffic smoothing, which consists of dynamically assigning each station a *station input limit* according to the current network workload, have been proposed in the literature [7][8]. In particular, in [8] the smoothing action is dynamically gauged according to the actual workload by means of a fuzzy controller which applies rules to choose the most appropriate station input limit on a case-by-case basis.

However, traffic smoothing can only provide

statistical guarantees to soft real-time traffic, so it is not suitable to support hard-real time applications. In hard real-time environments, therefore, other solutions, like for example, topologies based on extensive use of full duplex switches to separate the collision domains have been recently investigated. Many papers thus focus on the real-time capabilities of Switched Ethernet [3-6][10][11], proposing possible solutions or addressing analysis techniques for end-to-end response time calculation.

It should be highlighted that simply adding a switch to an Ethernet network is not enough to make it able to provide hard real-time guarantees. An example is found in those scenarios where the producer/consumers communication model is adopted. As switches handle this kind of interaction as broadcast traffic, one of the major benefits deriving from the use of switches, that is, the existence of multiple simultaneous transmission paths, can be affected. Moreover, it is known that, in the absence of collisions, switches introduce an extra latency, which increases the delivery time of real-time traffic.

A recent approach to support hard real-time communication, called the Flexible Time-Triggered Ethernet protocol (FTT-Ethernet), has been proposed in [13]. This protocol offers time-triggered communication with operational flexibility, but requires a centralized traffic scheduling based on a master/multislave paradigm.

Here a distributed time-triggered approach for real-time communication over Ethernet, called the Predictable Ethernet Access Control protocol (PEAC), is presented. The PEAC overcomes the unpredictable behaviour of the CSMA/CD using a TDMA-based approach. The key concept in the PEAC protocol [14][15] is the PEAC cycle, which comprises a Control Part, an IP Part and a Safety Margin. The Control Part is ruled by a TDMA scheme which, by avoiding collisions, provides the stations with deterministic bus access. The Control Part is, therefore, reserved for transmission of hard real-time traffic, which is assumed to be a priori known and off-line scheduled. In the IP Part, non-real-time traffic is dynamically transmitted according to the CSMA/CD protocol.

The PEAC does not require any modifications to the Ethernet hardware on the network interface cards and assures total compatibility with Ethernet frames.

In the paper, the PEAC protocol is outlined and analysed. Experimental results obtained in a real scenario are also given.

2. THE PREDICTABLE ETHERNET ACCESS CONTROL PROTOCOL

The features of the PEAC protocol can be summarized as follows.

♦ Support for hard real-time communication, thanks to deterministic medium access control achieved by means of a TDMA approach;

♦ Support for non real-time traffic;
♦ Applicability to both Shared and Switched topologies;
♦ No need for tokens or administration frames;
♦ No modification to the Ethernet standard: full compatibility with COTS hardware on the network interface cards.

The above-mentioned features enable the PEAC protocol to be used in a wide range of industrial real-time applications.

2.1 Protocol Description

A key concept in the PEAC protocol is the PEAC cycle, depicted in Fig. 1.

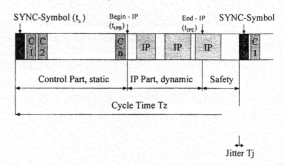

Fig. 1. The PEAC Cycle

The PEAC cycle time (Tz) is divided into three parts. The first one, called the Control Part, is ruled by a TDMA scheme and provides deterministic bus access to the communication medium. This is the part reserved for transmission of hard real-time traffic, which is assumed periodic, a priori known and off-line scheduled. [1]

The Control Part is divided into a fixed number of time slots. Each station, before running, will receive, besides all the other parameters, the offset corresponding to the beginning of its time slot, when it will be allowed to send its control packet. After this parameterisation phase, all the stations will be aware of the timing information they need to work correctly.

The second part, called the IP Part, allows non-real-time packets to be transmitted in a best effort way according to the CSMA/CD algorithm.

The third part of the PEAC cycle is called the Safety Margin. When the Safety Margin is reached, no stations are allowed to send new frames. However, the Safety Margin does not represent an idle interval, but it is exploited to complete the delivery of the

---

[1] All the hard-real-time exchanges envisaged in this work are periodic. However, occasionally, an aperiodic hard real-time message (e.g. an alarm) can be transmitted within the Control Part. In fact, it is up to each station using any slot assigned to it to transmit the most critical message in its queue. So, if it is the case, the hard real-time aperiodic message will be transmitted instead of the periodic variable.

frames sent within the IP Part and still on the way (for example, stored in the switches buffers), in order to avoid possible collisions between these frames and the first control packet of the next PEAC Cycle. This is an important point. If an IP frame were still being transmitted when a new cycle starts, there would be a collision with the frames of the control part which would prevent the system from functioning properly. All the control data packets have to be scheduled within the Control Part. The IP Part is only used to transmit non real-time frames (here called IP packets).

Figure 2 shows the Finite State Automata of the PEAC protocol. The top part of Fig.2 is relevant to the Control part, while the bottom one refers to the IP part. As the PEAC can be applied to both Shared and Switched Ethernet, the potential occurrence of collisions in a shared scenario is also depicted in Fig.2. Of course, the "Jamming" state and the two arcs related to such a state in Fig.2 can be removed in the switched scenario.

PEAC Protocol Finite State Automata

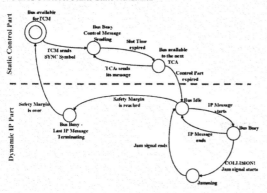

Fig. 2. PEAC Protocol Finite State Automata

When the Safety Margin is over, a new PEAC cycle starts. In the PEAC protocol, all the stations have to be equipped with the PEAC driver and are numbered from 0 to 255. For synchronization purposes, the station with the number 0 acts as a Traffic Control Manager (TCM) for the remaining stations, which are Traffic Control Agents (TCAs).

Each PEAC cycle starts with the SYNC_Symbol issued by the TCM. The Sync_Symbol is therefore a normal frame containing control data which is also used to maintain network synchronization marking the start of a new PEAC cycle.

On receipt of the SYNC_Symbol, all the TCAs will be able to check if they are still running synchronously or not. To perform the synchronization each TCA calculates the reception of two consecutive Sync_Symbols. The result is compared with the PEAC Cycle length Tz. If the distance between the two values is greater than a given Δ, corresponding to the lowest time accuracy of the system (in the current implementation, $\Delta = 35\mu s$), then a clock adjustment in the order of $\pm \Delta$ (or a multiple of it) is made. A guard time separates consecutive slots and protects the system from

misalignment between the clocks on the various stations.

3. TIMING ANALYSIS OF THE PEAC PROTOCOL

In the following, a timing analysis of the PEAC protocol aimed at setting the length of the Safety Margin for a switched topology is presented.
The PEAC cycle time can be expressed as the sum of three main components, given in formula (1) :

$$T_Z = T_{CPL} + T_{IPL} + T_{SM} = T_{CPL}(n) + T_{IPL} + T_{SM}(n) \quad (1)$$

where T_Z represents the PEAC Cycle Time, T_{CPL} the length of the Control Part, T_{IPL} the length of the IP Part, T_{SM} the Safety Margin, n the number of stations belonging to the network.
T_Z is fixed by the generation period of the fastest node of the network (i.e. the one generating the variable with the highest dynamics), so this is a known value for the scheduling phase.
The T_{CPL} and the T_{SM} are function of the number of stations in the network. T_{CPL} can be further decomposed in :

$$T_{CPL} = n*(T_S + T_G) \quad (2)$$

T_S, the Slot Time, should be set at least as the actual maximum propagation time in the network, thus it is dependent on the network topology.
T_G, the Guard Time, i.e. the time between two consecutive time slots, should be at least equal to the clock granularity of the system.
T_{IPL} has to be chosen in such a way to adapt the T_Z to the desired cycle time.
T_{SM} is related to the number of stations and network topology. T_{SM} has to be sufficiently long to avoid the eventuality of a collision of one of the last IP messages still "on the way" with the next Sync_Symbol. For a proper tuning of the T_{SM} parameter, a worst-case analysis should be done, in relation to the actual network configuration.

As an example, let us consider here a scenario of a fully switched network. As messages can be stored inside the buffers of the switches, the total delay that a frame can experience inside a switch should be predicted. This total delay can be decomposed in three components :

- Switch Latency: a fixed value depending on the hardware, usually provided by the vendor (typical values are around 10-12μsec);
- Frame Forwarding Delay: it varies with different switch forwarding mode and with respect to the frame length;
- Buffering Delay: it is related to the input traffic pattern.

Under the assumption of a periodic input traffic, a worst-case analysis can provide a maximum delay.

For periodic input pattern, in a fixed priority task scheduling, a frame of priority i experiences the worst case delay when, at its generation, all frames of higher or equal priorities arrive at the same time and must be transmitted before the tagged frame and, moreover, the longest frame of lower priority has just begun its transmission and will last $B_i = max_{j>i} (C_j)$. So, the worst-case response time of a frame of priority i is given by :

$$R_i = C_i + I_i \qquad (3)$$

$$I_i^{N+1} = B_i + \sum_{j \leq i} \left\lceil \frac{I^N}{T_j} \right\rceil * C_j \qquad (4)$$

where C represents the transmission duration and T the frame generation period. Formula (4) can be calculated using fixed-point method, starting by $I_i^0 = 0$ until the convergence $I_i^{N+1} = I_i^N$.

Here we are interested in calculating the longest time that an IP packet, sent during the IP Part, has to wait in the switch queue before being forwarded. In a fully switched network, collisions are completely avoided, but it has to be sure that the first control packet of the next cycle will not find any other packet inside the switch buffer, and will be directly delivered without any additional buffering delay.

In order to reduce the complexity of the analysis for IP traffic let us consider here a scenario where four stations are directly connected to a switch to realize a full micro-segmentation.

PEAC provides four buffers for IP packets: if such buffers are full, a negative indication is sent to the upper layer protocol. Here we assume that all the stations will transmit a burst made up of four frames of the maximum length to the same destination every cycle, namely with a generation period equal to the PEAC cycle period. This way, all the frames will be stored in the same queue within the switch and the IP traffic can be modelled as periodic traffic with a known generation period.

The worst case delay is experienced by the last frame of the last IP burst queued in the switch buffer. We need to calculate this time in order to properly adjust the T_{SM}.

As PEAC does not provide any priority mechanism, all the frames have the same probability to wait in a buffer for a time x, thus it is possible to calculate the buffering delay by using the following formula :

$$R = C + Q \qquad (5)$$

where C is the same of formula (3) and Q is given as follows

$$Q = C + \sum_{k=0}^{n-2} C_k \qquad (6)$$

As the worst case is given by the presence in the queue of frames of the maximum allowed size, it is not necessary to distinguish between frames of different dimension, so a further simplification of the last formula is

$$Q = C + (n-2)*C = (n-1)*C \qquad (7)$$

Formula (5) gives the time that the last frame out of n queued frames has to wait before being forwarded. This formula derives from the model assumed for the IP traffic and from the presence, in the PEAC driver, of four buffer for the IP frames. The differences between formula (4) and (6) lie in the following assumptions: absence of prioritization, equal size for all the frames, frame generation period equal to the PEAC cycle time.

4. PEAC PDU FRAME

The PEAC Protocol Data Unit (PDU) (shown in Fig.3) uses the Ethernet frame as a basis for the Control Part, including the destination address information (DST, Broadcast FF:FF:FF:FF:FF:FF) and the CRC sequence of the frame, but with some slight modification as regards the interpretation of some fields. The source address is modified by the PEAC driver and is used for other protocol information. The SRC field contains the address of the sending node referred to as a node number. The NMB field is the packet number that characterises the repetition of unchanged data content. The 4 octets of the *res* field are reserved for future extensions of the protocol. The TYPE field contains the Ethernet type which provides a distinction between IP and ARP packets. The PEAC header follows as the first octet of data in the Ethernet frame. The PDU structure is defined following the convention of the Ethernet frame. According to this, from 47 to 1499 data octets can be used for application data. Padding octets are used if a node sends less than 47 data octets by the Ethernet protocol. The shortest PDU has 64 octets including address information and CRC trailer.

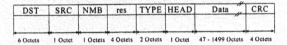

DST	SRC	NMB	res	TYPE	HEAD	Data	CRC
6 Octets	1 Octet	1 Octets	4 Octets	2 Octets	1 Octet	47 - 1499 Octets	4 Octets

Fig. 3. PEAC PDU Frame

5. EXPERIMENTAL RESULTS

In order to investigate the behaviour of the PEAC protocol an experimental testbed was realized. The experiments were carried out on a real scenario, i.e., a 100Mbit/s Ethernet network made up of four stations equipped with the PEAC driver in a Linux environment[2]. The time granularity of the system, here called the time tick, was 35μs.

The parameters chosen to evaluate the performance of the PEAC were propagation time and jitter for real-time control traffic and throughput for non real-time IP traffic, respectively.

For the sake of brevity, here we will not present the results of all the tests performed, but we will refer only to a case study.

The propagation time of a control packet and its jitter in two different scenarios, i.e. Switched and Shared Ethernet (100Mbit/s), in the presence of IP traffic were measured. On each cycle, a burst made up of three IP packets was generated by each station. In the experiments referred below, the Ethernet frame size was 1500 bytes, the real-time workload on the network was around 118 Kbyte/s, the IP workload was about 66 Mbit/s. The scheduling parameters were: T_S = 1 tick = 35μs, T_{CPL} = 8 ticks = 280μs, T_{IPL} = 10 ticks = 350μs, T_Z = 62 ticks = 2170μs. The results obtained are shown in Table 1 and Table 2.

Table 1 – Control Message Propagation Time

	Minimun	Average	Maximum
Switched	27μs	30μs	45μs
Shared	15μs	22μs	29μs

As it was expected, the real-time traffic in the Control Part was transmitted without collisions. No interference with the IP traffic occurred. The propagation time was measured by an oscilloscope starting from the sending time of the Sync_Symbol to the time at which it was received by the destination station. The Sync_Symbol, as well as all the other control packets, has a fixed dimension of 64 bytes. Since the minimum propagation time measured in the switched scenario is 24μs and the maximum is 38μs, the maximum jitter results equal to 14μs. In the same way, the maximum jitter value in the shared scenario was 10μs. The scenario with the hub shows lower values for both propagation time and maximum jitter than the configuration with the switch. This can be accounted for by the inherent switch latencies due to the store-and-forward mode. The maximum jitter obtained in both scenarios (Switched and Shared) is shown in Table 2. Both the values are satisfying for the real-time traffic.

Table 2 – Control Message Jitter

	Maximum Jitter
Switched	14μs
Shared	10μs

In the same scenario discussed before, the IP traffic throughput has been measured too. It was about 18 Mbit/s in the Shared scenario and 66 Mbit/s in the Switched one. In the Switched case, all the generated

traffic was actually sent on the network, while in the Shared case, only part of it was transmitted. The difference derives from the different features of switches and hubs. The "parallelism" introduced by the switch in the micro-segmented topology enables each station to successfully transmit all the three frames, which are first stored in the switch queue and then transmitted to destination within the T_Z, without any interference with the hard real-time traffic in the Control Part of the next PEAC cycle. On the contrary, in the Shared scenario, contention for the bus causes collisions during the IP Part. The Safety Margin is therefore exploited (partially or totally) to re-transmit any frame which experienced collisions during the IP Part.

If measures are to be taken on a network with more than one cascaded switch (or hub), it is sufficient to re-measure the propagation time between two of the most distant nodes on the network, i.e. those separated by the greatest number of active elements on the network. The propagation time thus measured will represent the Ts on the new network.

Lower throughput values will be obtained with the same PEAC cycle if the number of stations is greater than 4. The equations obtained from the time analysis will, in fact, have to be applied again to obtain a new Safety Margin that meets the constraint imposed by the variable generation period on the fastest node.

Having set T_Z and T_{CPL}, T_{IPL} needs to be suitable sized by reducing the maximum number of IP frames each node can send per cycle, and thus restricting the T_{IPL}. The T_{SM} will be calculated on the basis of these new values and will have to be sufficiently large as to ensure that all the frames queued in the switches are dealt with before the next PEAC cycle starts. If the T_{SM} thus calculated is too large and exceeds the constraint imposed by T_Z, T_{IPL} will have to be re-dimensioned.

The equations allow us to calculate the Safety Margin on a switched network in the worst case, i.e. when all the stations send their IP frames to the same address, thus overloading the output queue. By adopting the T_{SM} calculated it is guaranteed that there will not be IP frames and control packets on the network at the same time.

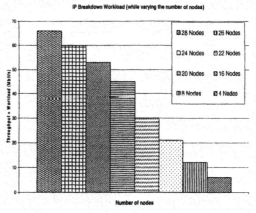

Fig. 4. IP Throughput (by varying the number of nodes)

[2] The Linux Mandrake release 7.1, kernel 2.2.18-rtl 3.0 – real-time Linux extension, was used.

Fig.4 shows the values of the maximum IP throughput that can be obtained for a fully switched network by increasing the number of nodes still maintaining a constant value of Tz. The need to end the transmission of IP frames within the T_{SM} requires, during configuration, to reduce the number of IP frames each station can send and their maximun size.

6. CONCLUSIONS AND FURTHER WORK

In this paper the PEAC protocol, a distributed approach for real-time communication over Ethernet, has been proposed. The PEAC protocol can support both periodic hard real-time traffic and non real-time traffic over an Ethernet network, both Shared and Switched. The maximum latency for real-time packets is guaranteed by the use of the TDMA approach, which avoids collisions in the Control Part, which is reserved for hard-real-time traffic.

In the paper, a timing analysis of the PEAC protocol aimed at setting the length of the Safety Margin for a switched topology has been presented. Timing analysis of the PEAC protocol for a shared topology will be addressed in further work. Experimental results obtained in a real scenario have confirmed the attractive performance in terms of latency and jitter for real-time traffic both in a Shared and in a Switched scenario. Also attractive were the results for IP traffic throughput in the micro-segmented topology, while in the Shared scenario throughput is affected by collisions. However, it should be highlighted that, in the industrial environment here considered, the main goal is providing hard real-time guarantees to control traffic, while the non-real-time IP traffic concerns sporadic activities such as, for example, software downloading, parameter configurations of intelligent devices or backup operations, which are not critical and require a limited amount of bandwidth. The above mentioned results therefore suggest that the PEAC protocol can be suitable for hard real-time communications in the process control and factory automation area. Further refinements of the PEAC protocol are currently under investigation. As an example, evaluation of different approaches for transmission scheduling both in the Control Part and in the IP Part is in progress. Support for soft-real time traffic is also currently addressed.

REFERENCES

[1] IEEE, 802.3 CSMA/CD Access Method and Physical Layer Specification, 1985.

[2] C. Venkatramani, T. Chiueh, "Supporting Real-Time Traffic on Ethernet", Proc. of 15th IEEE Real-Time Systems Symposium, pp.282-286, 1994.

[3] C. Venkatramani, T. Chiueh, "Design and Implementation of a Real-Time Switch for Segmented Ethernets", International Conference on Network Protocols, Atlanta, G.A., Oct.1997.

[4] S. Vanadarajan, T. Chiueh, "A Host-Transparent Real-Time Fast Ethernet Switch", Department of Computer Science, State University of New York, 1998

[5] S. Kweon, K. G. Shin, Q.Zheng, "Statistical Real-Time Communication over Ethernet for Manufacturing Automation Systems", Proc.of the 5th IEEE Real-Time Technology and Application Symposium, Vancouver, Canada June 1999.

[6] S. Rüping, E. Vonnhame, J. Jasperneite, "Analysis of Switched Ethernet Networks with Different Topologies Used in Automation Systems", Proc. of Fieldbus Conf. FeT'99, Magdeburg, Germany, pp.351-358, Springer-Verlag, Sept. 99.

[7] L. Lo Bello, M. Lorefice, O. Mirabella, S. Oliveri, "Performance Analysis of Ethernet Networks in the Process Control", Proc. of the 2000 IEEE International Symposium on Industrial Electronics, Puebla, Mexico, Dec. 2000.

[8] A. Carpenzano, R. Caponetto, L. Lo Bello, O. Mirabella, "Fuzzy Traffic Smoothing: an Approach for Real-time Communication over Ethernet Networks", 4th IEEE Int. Workshop on Factory Communication Systems, WFCS'02, Aug 2002, Västerås, Sweden.

[9] J.D. Decotignie, "A perspective on Ethernet-TCP/IP as a Fieldbus, Proc. of the 4th FeT'2001International Conference on Fieldbus Systems and their Applications, pp. 138-143, Nancy, France, Nov. 2001.

[10] Jasperneite, J., P. Neumann. "Switched Ethernet for Factory Communication", Proc. of ETFA2001, 8th IEEE International Conference on Emerging Technologies and Factory Automation, Antibes, France, Oct. 2001.

[11] J.Jasperneite, P. Neumann. "Performance Evaluation of Switched Ethernet in Real-Time Applications", Proc. of the 4th FeT'2001, International Conference on Fieldbus Systems and their Applications, pp.144-151,Nancy, France, Nov. 2001.

[12] M. Alves, E. Tovar, F. Vasques, "Ethernet Goes Real-Time: a Survey on Research and Technological Developments", Tech. Rep. HURRAY-TR-0001, Polytechnic Institute of Porto, Jan 2000.

[13] P. Pedreiras, L. Almeida, P. Gai, "The FTT-Ethernet protocol: Merging flexibility, timeliness and efficiency", 14th Euromicro Conference on Real-Time Systems, Vienna, Austria, June 2002.

[14] A. Pöschmann, D. Heise, "Internet Protocol and Hard Real-Time", Tech.Rep., Institut für Automation und Kommunikation, ifak, Magdeburg, 2002.

[15] A. Pöschmann, D. Heise, "Internet Protokoll und harte Echtzeit", Tagungsband 1 Embedded Intelligence 2002, pp. 79-88, Nürnberg, 2002.

OPTIMAL REPLICA ALLOCATION FOR TTP/C BASED SYSTEMS [1]

Bruno Gaujal [1] **Nicolas Navet** [2]

[1] ENS Lyon - LIP
46, Allée d'Italie
69007 Lyon - France
bgaujal@ens-lyon.fr

[2] LORIA - INPL
ENSEM - 2, Avenue de la forêt de Haye
54516 Vandoeuvre-lès-Nancy - France
nnavet@loria.fr

Abstract: In this study we show how one can use Fault-Tolerant Units (FTU) in an optimal way to make a TTP/C network robust to bursty random perturbations. We consider two possible objectives corresponding to well defined situations of the field of fault-tolerance. If one wants to minimize the probability of losing all replicas of a given message, then the optimal policy is to spread the replicas over time. On the contrary if one wants to minimize the probability of losing at least one replica, then the optimal solution is to group all replicas together.
Copyright © 2003 IFAC

Keywords: Real-Time Systems, Vehicles, Fault Tolerance, Noisy Channels, Network Reliability.

1. INTRODUCTION

Multi-access protocols based on TDMA (Time Division Multiple Access) are particularly well suited to real-time applications since they provide deterministic access to the medium and thus bounded response times. Moreover, their regular message transmissions can be used as "heartbeats" for detecting station failures. There exists several variants of the TDMA scheme. In this paper, we study the TTP/C protocol (Time Triggered Protocol - see TTTech Computertechnik GmbH (2002)) which implements the synchronous TDMA scheme: the stations (or nodes) have access to the bus in a strict deterministic sequential order and each station possesses the bus for a constant period of time called a *slot* during which it has to transmit one frame. The sequence of slots such that all stations have access once to the bus is called a *round*. The use of TTP/C is currently considered in high-dependability real-time applications where fault tolerance and guar-

anteed response times has to be provided. Examples of such applications are "brake-by-wire" and "steer-by-wire" in-vehicle applications (see Dilger et al. (1998)) or avionic applications. In such so called "X-by-wire" applications, mechanical and hydraulic components are replaced by computer control which has to be fault-tolerant. A *Fault-Tolerant Unit* (FTU) is a set of two or more nodes that performs the same function and thus may tolerate the failure of one or more of its constituent stations. Actually, the role of FTUs is two-fold considering the type of failure of the stations. They make the system resilient in the presence of *transmission errors* (some frames of the FTU may still be correct while others are corrupted). They also provide a way to fight against *measurement and computation errors* occurring before the transmission (some nodes may send the correct values while others may make errors). In the following we will see that according to which role is the most important, the optimization will lead to very different solutions.

Embedded systems may suffer from strong EMI (electro-magnetic interferences) which may repre-

---

[1] An extended version of this paper is available as INRIA Research Report RR-4614.

sent a serious threat to the correct behavior of the system. For instance, in automotive applications, the EMI (see Noble (1992); Zanoni and Pavan (1993)) can either be radiated by some in-vehicle electrical devices (switches, relays..) or come from a source outside the vehicle (radio, radars, flashes of lightning..). EMI could affect the correct functioning of all the electronic devices but the transmission support is a particularly "weak link" and the use of an all-optical network, which offers very high immunity to EMI, is not generally feasible because of the low-cost requirement imposed by the industry (see Barrenscheen and Otte (1997) for more details on the electro-magnetic sensitivity of different types of transmission support). Even with a redundant transmission support, such as in TTP/C, the network is not immune to transmission errors since a perturbation is likely to affect both channels in quite a similar manner because they are identical and very close one to each other. Unlike CAN (Controller Area Network - ISO (1994)), TTP/C do not provide automatic retransmission for corrupted frames and their data are actually lost for the application.

Goal of the paper. The system under study is an application with redundant nodes distributed over a TTP/C network. The problem we address here is to find the best allocation of the slot of each station in the round in such as a way as to maximize the robustness of the system against transmission errors. We consider two distinct objectives :

(1) **Objective 1 :** minimize, for each FTU, the probability that all frames of the FTU carrying the same information will be corrupted. In the rest of the paper, this probability will be termed the "loss probability" and denoted by $\mathbb{P}_{all}$.

(2) **Objective 2 :** maximize, for each FTU, the probability that at least one frame of each station composing the FTU is successfully transmitted during the production period of a data. For this objective, we will assume that the production period of the data is equal to the length of a round. Under this assumption, it comes to minimizing, for each FTU, the probability that one (or more) frame of the FTU will be lost during a round. The corresponding probability is denoted by $\mathbb{P}_{one}$.

As it will be further discussed in Subsection 2.3, the two objectives correspond to well-defined situations in the field of fault-tolerance that are distinguished with regard to the concept of "fail-silence".

Assumptions on the error model. In this study, we will consider an error arrival process where "bursts" of transmission errors may occur. This is very likely in the context of in-vehicle multiplexing applications.

If successive transmission errors are not correlated (i.i.d.), it is clear that the location of the slot of each station of an FTU has no influence on the loss probability since each slot has the same probability of being corrupted independently. However in practice transmission errors are highly correlated and one observes bursts or errors leading to successive transmission errors. The assumptions made for the error arrival process will thus influence the solution to the problem of locating the FTU slots.

We will consider an error model that can take into account both error frequency and error gravity as proposed in Navet et al. (2000) with the following assumptions:

(A_1) each time an EMI occurs, it will perturb the communications on the bus during a certain duration and each bit transmitted during this perturbation is corrupted with some probability π.

(A_2) the starting times of the EMI bursts are independent random variables uniformly distributed over time.

(A_3) the distribution of the size of the bursts is arbitrary provided that it is independent of the starting point of the burst.

(A_4) if a perturbation overlaps a whole slot, then we assume that the probability that the frames remains uncorrupted is neglectable (with $\pi = 0.5$ and a 100 bits frame, this probability is about 10^{-30}).

Without further knowledge on the considered application and its environment, assumptions (A_2), (A_3) and (A_4) are rather reasonable.

Related work. The TTP/C protocol, which is defined in TTTech Computertechnik GmbH (2002), is a central part of the Time-Triggered Architecture (TTA - see Kopetz (1997); Kopetz et al. (2001)) and it possesses numerous features and services related to dependability such as the bus guardian (see Temple (1998)), the group membership algorithm (see Pfeifer (2000)) and support for mode changes (see Kopetz et al. (1998)). The TTA and the TTP/C protocol have been designed and extensively studied at the Vienna University of Technology. Closely related to our proposal is the work described in Grünsteidl et al. (1991) where the reliability of the transmission on a TTP/C network is studied with the taking into account of transmission errors on the bus as well as failures in the TTP/C nodes. Under the assumption that all failures and transmission errors are statistically independent, a measure of the reliability of the transmission is given in terms of Mean Time To

Failure (MTTF) where a communication failure for an FTU is defined as the loss of all messages of an FTU sent in the same round. From the MTTF of each individual FTU, a global measure of the reliability of the system is derived.

There exist two main differences with our work. One concerns the assumptions made on the perturbations and the second the data production. In Grünsteidl et al. (1991) the errors are assumed to be independent, the location of the FTU slots has thus no influence and is not considered. Here on the contrary, we take into account the burstiness of the perturbation process. Hence the time allocations of the FTU replicas will have a big influence on the transmission error probabilities.

The second difference with Grünsteidl et al. (1991) is that we do not compute the reliability of a given system but provide a way to optimize it via time allocation of the replicas. This does not require any modification of the protocol or of the parameters of the system.

2. FRAMEWORK OF THE STUDY

In this section, we first describe the Medium Access Control (MAC) protocol, namely the synchronous TDMA scheme, then the model of the application and the notations used. We then justify the two distinct objectives that were identified with regard to the concept of "fail-silence". Finally, we describe the TTP error handling mechanisms that have to be taken into account.

2.1 MAC Protocol description

Throughout this paper, we will considered the synchronous TDMA protocol. The time needed to transmit one bit over the bus is taken as the time unit. In the following all time quantities are given using this time-bit as unit.

The number of *stations*, S, is static and the stations have access to the bus in a strict deterministic sequential order. Each station possesses the bus for a constant period of time called a *slot* during which it has to transmit one *frame*. The size of the slots is not necessarily identical for all stations but successive slots belonging to the same station are of the same size. The sequence of slots such that all stations have access once to the bus is called a *round*

2.2 Application model

To achieve fault-tolerance, that is the capacity of a system to deliver its service even in the presence of faults, some nodes are *replicated* and are clustered into *Fault-Tolerant Units* (FTUs). An FTU is a set of several stations that perform the same function and each node of an FTU possesses its own slot in the round so that the failure of one or more stations in the same FTU might be tolerated. The stations forming an FTU are called *replicas* in the following. For the sake of simplicity, a non-replicated station will also be termed an FTU (of cardinality one).

One denotes by $\mathcal{F}$ the set of FTUs : $\mathcal{F} = \{A, B, C...\}$ and C_A is the cardinality of FTU A, *i.e.* the number of stations forming FTU A. The size (in bits) of the slots of all the stations in A is the same and is denoted by h_A.

By definition, the total number of bits in a round, denoted R, is equal to:

$$R = \sum_{A \in \mathcal{F}} C_A h_A.$$

The problem consists in choosing the position of the slots of all stations forming an FTU in a round. This is done under the form of a binary vector x^A of size R (called an allocation for A) defined by

$$\forall 1 \leq i \leq R, \quad x_i^A = \begin{cases} 1 & \text{if some station in } A \\ & \text{transmits at time-bit } i \\ 0 & \text{otherwise .} \end{cases}$$

Note that the construction of x^A must follow several constraints. First the binary vector x^A must be made of C_A "blocks" of ones, each of size h_A to correspond to an allocation of all the slots of A. Second, the allocations of all the FTUs must be *compatible*, meaning that the same bit cannot be allocated to two different FTUs. Finally all bits in a round must be allocated to some FTU. In mathematical terms these compatibility constraints can be written

$$\sum_{A \in \mathcal{F}} x^A = (1, \cdots, 1).$$

Finally, the frame sent by a node contains some data whose value is periodically updated as it is generally the case in distributed control applications. For instance, in a typical car environment, a frame sent by the engine controller may contain the RPM value plus the engine temperature and a new frame is built every 10ms.

Since they are replicas, all nodes of an FTU update their data with the same period denoted by T_A and called a *production cycle*. The data sent during one production cycle is also called a *message* in the following. It is also assumed that all nodes of a FTU are synchronized using the global time service requested by the communication protocol so that at each point in time each node of an FTU sends the data corresponding to the same production cycle.

The length of the TDMA round R is a function of the number of nodes, of the maximal size of the message sent in each slot, and on some characteristics of the network and of the communication controllers. The value of R is thus not generally correlated with the production cycle of the data [2]. If $\exists A \in \mathcal{F}$ s.t. $T_A < R$ then some data may not be transmitted which is generally unacceptable. If $\forall A \in \mathcal{F}$, $T_A > R$ then the same data is transmitted in more than one round. Also, if the beginning of the production cycle does not correspond to the beginning of a round and if FTU A has more than one replica, then data corresponding to different production cycles may be transmitted in the same round as it is the case in the first and third round of the example drawn on Figure 1.

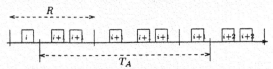

Fig. 1. Three successive rounds. Only the slots allocated to the FTU A of cardinality 3 are shown. The message corresponding to the $(i+1)^{\text{th}}$ production cycle is sent over 3 rounds.

2.3 Which objective with respect to fail-silence ?

The number of replicas per FTU which is required to tolerate k faults heavily depends on the behaviour of the individual components (see Dilger et al. (1998)). For instance, if the failure of k nodes must be tolerated, the least necessary number of replicated nodes is $k + 1$ when all nodes are *fail-silent*. A node is said fail silent if

(1) a) it sends frames at the correct point of time (correctness in the time domain) and b) the correct value is transmitted (correctness in the value domain),

(2) or it sends detectably incorrect frames (eg. wrong CRC) in its own slot or no frame at all.

TTP/C provides very good support for the requirements 1.a) and 2) (whose fulfillment provide the so-called "fail-silence in the temporal domain") especially through the bus guardian concept while the value domain is mainly the responsibility of the application. The reader is referred to Dilger et al. (1998); Temple (1998); Poledna et al. (2000) for good starting points on the problem of ensuring fail-silence.

For FTUs composed of a set of fail-silent nodes, the successful transmission of one single frame for the whole set of replicas is sufficient since the value carried by the frame is necessarily correct. In this case, the main objective to achieve with regard to the robustness against transmission errors is the minimizing of $\mathbb{P}_{all}$, that is the probability that all frames of the FTU (carrying data corresponding to the same production cycle) will be corrupted.

In practice replicated sensors may return slightly different observations and, without extra communication for an agreement, replicated nodes of a same FTU may transmit different data. If a decision, such as a majority vote, is taken by a consumer node with regard to the value of the transmitted data, the objective is to maximize the probability that at least one frame of each FTU is successfully transmitted. If the production cycle is equal to one round then it comes to minimizing $\mathbb{P}_{one}$, the probability that one or more frames of an FTU become corrupted.

2.4 TTP/C error handling mechanisms

The TTP/C protocol includes powerful but complex algorithms such as the clique avoidance and membership algorithms. In this paragraph, we give a simplified description of the functioning schemes of TTP/C version 1.0 that are related with transmission error handling and that might a priori interfere with our analysis. For instance, TTP/C defines the concept of "shadow" node. A shadow node replaces a defective node but does not possesses its own slot in the round. This redundancy scheme does not protect against transmission errors and we won't consider them in the rest of the paragraph.

A TTP/C controller is always in one of the nine states defined by the protocol (see TTTech Computertechnik GmbH (2002) page 101). Three are of particular importance in our context :

- the "active" state which is the normal functioning state,
- the "passive" state : the controller is synchronized and can receive frames but no transmission is allowed,
- the "freeze" state : the execution of the protocol is halted and the reintegration process will not be started before the controller is turned on by the application software.

The protocol distinguishes frames with and without "C-State". The C-State is a collection of control data that describes the state of the network as seen by the sending node : current time, current operating mode, membership of the stations (i.e. their operational state) . . .

The most important TTP/C functioning schemes related to transmission error handling are listed below :

[2] The version 1.0 of the TTP/C specification TTTech Computertechnik GmbH (2002) enables the designer to insert an idle time after the transmission of a frame so that the duration of a round can take an application related value. In particular it could be equal to the length of the production cycle of a data but the problem remains with data having different production cycles.

(1) Lost of membership due to a incorrect transmission : if a frame is corrupted during its transmission the sender loses its membership and enters the passive state. It waits in the passive state until it can re-acquire its slot. To re-acquire a slot the controller must have received the "minimum integration count" (MIC) correct frames (the first correct frame must contain an explicit C-state). The value of the MIC should be set at least to two.

(2) Maximum Membership Failure Count (MMFC) check : if a node do not possess its membership in MMFC successive sending slots, then the controller terminates its operation by entering the "freeze state". It is an optional feature since MMFC can be set to zero which means no verification.

(3) Re-integration of a node (transit from freeze state to passive state) : a "frozen" node must wait until the application sets the Controller On (CO) field to the value "on". Then it must listen to a valid frame containing explicit C-state before entering the passive state. Then the node has to re-acquire its slot as described in point 1.

(4) Clique avoidance algorithm : before starting to send a frame, a node must verify whether the number of frames that have been successfully sent in the last S slots (where S is the number of slots in the round so that it includes its own last transmission) is greater than the number of incorrect frames. In the latter case, the node enters the "freeze state" otherwise it transmits its frame and reset its counters. This rule will be termed the "majority rule".

3. MINIMIZING $\mathbb{P}_{ALL}$

In this section, we investigate the problem of minimizing the loss probability $\mathbb{P}_{all}$ on TTP/C. The problem has been studied in Gaujal and Navet (2002) for the general synchronous TDMA case. The TTP/C rules 1,2 and 3 actually affect the value of $\mathbb{P}_{all}$ but not which allocation scheme is optimal. However, the majority rule of TTP/C (item 4 above) changes the solution with respect to the general TDMA case. In fact, it makes it easier to reach optimal allocation for all FTUs together compared to the pure synchronous TDMA network.

One constructs two stacks S_1 and S_2 of slots. for each FTU i with C_i replicas, push $\lfloor C_i/2 \rfloor$ slots in the largest stack and $\lceil C_i/2 \rceil$ slots in the smallest stack.

The allocation x_{stack} is constructed by concatenating S_1 and S_2. The construction is illustrated by Figure 2.

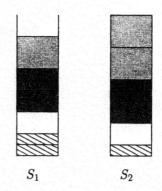

$S_1 \qquad S_2$

Allocation of a round

Fig. 2. Construction of the optimal allocation x_{stack}.

Theorem 1. Using TTP/C and under the foregoing assumptions, the allocation x_{stack} minimizes $\mathbb{P}_{all}$.

PROOF. The replicas of an FTU can be corrupted by several perturbations each touching exactly one frame. Since starting points of EMI bursts are uniformly distributed over time (assumption (A_2)), this probability is equal under all allocations. Several replicas can also be corrupted by a same perturbation with a probability decreasing when the distance between the replicas inside the round becomes larger.

The allocation x_{stack} has the following property: each FTU with more than two replicas has two replicas separated by at least $\lfloor S/2 \rfloor$ slots. Now, as soon as two replicas of the same message are allocated more that $\lfloor S/2 \rfloor$ slots apart, no single perturbation can destroy both of them without freezing all the nodes of the network. It is thus useless to consider a distance between replicas larger than $\lfloor S/2 \rfloor$. This means that x_{stack} is optimal. □

Corollary 1. If the probability to have more than one perturbation in the same round is sufficiently low, and because of the TTP/C majority rule, it is useless to have more than two replicas per FTU if the objective is to minimize the corruption of all the replicas.

4. MINIMIZING $\mathbb{P}_{ONE}$

Unlike the previous case, the technique used to find the optimal allocation of the replicas of one FTU is based on majorization and Schur convexity.

4.1 Schur convexity and majorization

Let $u = (u_1, \cdots, u_n)$ and $v = (v_1, \cdots, v_n)$ be two real vectors of size n. We denote by $(u_{[1]} \cdots, u_{[n]})$ and $(v_{[1]}, \cdots, v_{[n]})$ the permutations of u and v such that $u_{[1]} \leq \cdots \leq u_{[n]}$ and $v_{[1]} \leq \cdots \leq v_{[n]}$. The vector u *majorizes* v ($u \succ v$) if the following conditions hold:

$$\sum_{i=1}^{n} u_i = \sum_{i=1}^{n} v_i, \qquad (1)$$

$$\sum_{i=1}^{k} u_{[i]} \leq \sum_{i=1}^{k} v_{[i]} \quad k \leq n. \qquad (2)$$

A function f from $\mathbb{R}^n$ to $\mathbb{R}$ is *Schur convex* (resp. *Schur concave*) if $u \succ v$ implies $f(u) \geq f(v)$ (resp. $f(u) \leq f(v)$). For more details on these notions, the reader can refer to Marshall and Olkin (1979).

4.2 Schur concavity of $\mathbb{P}_{one}$

In this section, we will show that the probability that an error burst corrupts at least one replica within a production cycle ($\mathbb{P}_{one}$) is a Schur concave function with respect to the allocation of the replicas. Using the definition of Schur concavity, this will provide directly the best allocation minimizing $\mathbb{P}_{one}$. Note that the result will be proven for arbitrary production cycles although, in our context, $\mathbb{P}_{one}$ is only meaningful for a production cycle equal to one TTP/C round.

Let x be an allocation of the K replicas forming FTU A. We denote by $t = NK$ the number of frames (of size h) composing a message for FTU A.

The quantity $I_i(x)$ denotes the interval between the end of replica r_{i-1} and the beginning of replica r_i. We denote by $I(x)$ the sequence of intervals $(I_1, \cdots I_t)$ and by $|I(x)|$ the vector of the length of the intervals, $|I(x)| = (|I_1|, \cdots, |I_t|)$. Note that $|I_1(x)| + \cdots + |I_t(x)| = N(R - Kh)$ does not depend on the allocation x.

Lemma 1. Let us consider a single error burst starting at a random time uniformly distributed over one round. Let x and x' be two allocations of A. If $|I(x)| \prec |I(x')|$ then the probabilities of losing at least one frame satisfy $\mathbb{P}_{one}(x) \geq \mathbb{P}_{one}(x')$.

PROOF.

A replica can either be corrupted by a perturbation that starts between two replicas of the FTU or by a perturbation that starts during the transmission of a replica of the FTU. Both cases are independent and can be studied separately.

Let us first consider the first case. Note that if $t = 1$ then $|I(x)| = |I_1(x)| = N(R - Kh) = |I_1(x')| = |I(x')|$ and all allocations are equivalent since the error model is time homogeneous.

If $t \geq 2$, we renumber the intervals of x and x' such that $|I_{[1]}| \leq \cdots \leq |I_{[t]}|$ and $|I'_{[1]}| \leq \cdots \leq |I'_{[t]}|$. Using the majorization condition, one gets for all j, $\sum_{i=1}^{j} |I_{[i]}| \geq \sum_{i=1}^{j} |I'_{[i]}|$.

We now prove by induction that for all $1 \leq j \leq t$ one can construct a coupling between $I_{[1]}, \cdots, I_{[j]}$ and $I'_{[1]}, \cdots, I'_{[j]}$ such that the probability $\mathbb{P}'_j$ that an error starting in $I'_{[1]}, \cdots, I'_{[j]}$ and corrupting at least one replica is smaller than the corresponding probability $\mathbb{P}_j$ in $I_{[1]}, \cdots, I_{[j]}$. For $j = 1$, the coupling is done according to Figure 3.

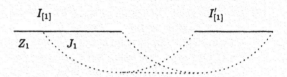

Fig. 3. Coupling for the smallest interval.

After the coupling, the interval $I_{[1]}$ is split into two intervals, Z_1 and J_1 such that $I_{[1]} = Z_1 \cup J_1$ and $|I'_{[1]}| = |J_1|$. A burst starting in J_1 has the same probability of corruption that a burst starting in $I'_{[1]}$ because

• both intervals are of the same size and both are contiguous to replicas having the same length,

• if a perturbation overlaps the whole replica then the corruption occurs with probability 1 (assumption (A_4)) under x and x' otherwise the corruption probability is also identical under x and x'.

The remaining zone (Z_1) is such that an error starting in Z_1 corrupts one replica with a non-negative probability. Therefore, $\mathbb{P}_1 \geq \mathbb{P}'_1$.

The proof continues by induction on j. The induction property is that for a given j one can construct a splitting of $I_{[1]}, \cdots, I_{[j]}$ into $(J_1, Z_1), \cdots, (J_j, Z_j)$ such that the probability that a burst starting in $J_1 \cup \cdots J_j$ is larger or equal than in $I'_{[1]} \cup \cdots \cup I'_{[j]}$ and the zone $Z_1 \cup \cdots \cup Z_j$, has a non-negative total probability of corrupting a replica.

We now add $I_{[j+1]}$ and $I'_{[j+1]}$. Two cases can occur.

1) If $I_{[j+1]} \geq I'_{[j+1]}$ then one splits $I_{[j+1]}$ as it has been done for $I_{[1]}$ and $I'_{[1]}$ in Figure 3. We get new intervals Z_{j+1} and J_{j+1} and the induction remains true by using the argument given for $j = 1$.

2) If $I_{[j+1]} \leq I'_{[j+1]}$, we couple according to the following procedure. The interval $I'_{[j+1]}$ is split

into two intervals U and V such that $|V| = |I_{[j+1]}|$, which are coupled together.

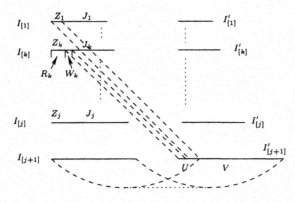

Fig. 4. Coupling when $I_{[j+1]} \leq I'_{[j+1]}$.

Note that by the majorization property, $|U| = |I'_{[j+1]}| - |I_{[j+1]}| \leq |Z_1| + \cdots + |Z_j|$. Let $k := \min\{k : |Z_1| + \cdots + |Z_k| \geq |U|\}$. We split the interval Z_k into two intervals R_k, W_k such that $|W_k| = |U| - (|Z_1| + \cdots + |Z_{k-1}|)$. The coupling is illustrated in Figure 4.

- An error starting in V has the same probability to corrupt a frame than an error starting in $I_{[j+1]}$.
- An error starting in U has a smaller probability of corruption than an error starting in $Z_1 \cup \cdots Z_{k-1} \cup W_k$ because $|V| > |J_i|$ for all $i \leq k$.
- An error starting in $I'_{[1]} \cup \cdots \cup I'_{[j]}$ has a probability of corruption smaller or equal than an error starting in $J_1 \cup \cdots \cup J_j$ by the induction hypothesis.
- An error starting in $R_k \cup Z_{k+1} \cup \cdots \cup Z_j$ has a non-negative probability of corruption.

In total, $\mathbb{P}_{j+1} \geq \mathbb{P}'_{j+1}$.

Finally, the induction assumption is carried one more step by using the new splitting of $I_{[1]}, \cdots, I_{[j+1]}$ into

$$((J_1, \emptyset), \cdots (J_{k-1}, \emptyset), (J_k, R_k), (J_{k+1}, Z_{k+1}), \cdots,$$
$$(J_j, Z_j), (I_{[j+1]} \cup Z_1 \cdots \cup Z_{k-1} \cup W_k, \emptyset)).$$

We will now consider the case where a replica is corrupted by a perturbation starting during the transmission of a replica. The perturbation might corrupt either the replica during which it occurred, with probability $\mathbb{P}_a$ under allocation x and $\mathbb{P}'_a$ under x', or the next replica (using assumption (A_4)) respectively with probability $\mathbb{P}_b$ or $\mathbb{P}'_b$. Since perturbation starting points are uniformly distributed over time and slots have the same size under all allocations, $\mathbb{P}_a = \mathbb{P}'_a$. The same proof based on the length of the intervals between replicas used for $\mathbb{P}_t$ shows that $\mathbb{P}_b \geq \mathbb{P}'_b$ since $|I(x)| \prec |I(x')|$.

The proof is concluded by noticing that $\mathbb{P}_{one}(x) = \mathbb{P}_t + \mathbb{P}_a + \mathbb{P}_b \geq \mathbb{P}_{one}(x') = \mathbb{P}'_t + \mathbb{P}'_a + \mathbb{P}'_b$.

Theorem 2. Under assumptions (A_1), (A_2) and (A_4) for each FTU A, the optimal allocation x_{one} minimizing $\mathbb{P}_{one}$ is to group together all replicas of A.

PROOF. Under (A_1), (A_2) and (A_4) each burst may corrupt a same replica independently. Therefore, $\mathbb{P}_{one}$ is a function of the probability that one burst corrupts one replica (denoted by q). By conditioning on the number of bursts, say K, one gets

$$\mathbb{P}_{one} = \sum_{i=0}^{K} q(1-q)^i = 1 - (1-q)^{K+1}.$$

This is an increasing function of q for all K. Therefore, minimizing q (*i.e.* minimizing the impact of one burst) also minimizes the combined effect of all bursts.

Finally, let x be an arbitrary allocation. The restrictions over one round R of x and x_{one} are denoted $x|_R$ and $x_{one}|_R$ respectively. They obviously satisfy $I(x|_R) \prec I(x_{one}|_R)$. By periodicity $I(x) = (I(x|_R), I(x|_R), \cdots, I(x|_R))$ (repeated N times). This implies $I(x) \prec I(x_{one})$. Finally, applying Lemma 1 concludes the proof. $\square$

The combined minimization of $\mathbb{P}_{one}$ for all FTUs is not a problem since the optimal solution is to group all replicas of each FTU together.

4.2.1. Performance comparison against random allocations To assess the robustness improvement brought by the optimal allocation for $\mathbb{P}_{one}$, simulations were performed against random allocations. A configuration is defined by a number of FTU and the cardinality of each FTU. In these experiments, the number of FTUs ranges from 3 to 12. Two hundreds configurations were randomly generated with FTUs having a cardinality between 2 and 4. For each configuration, we randomly pick up 500 hundred slots (in the 2000 first rounds) where a data is transmitted for the first time. The duration of the production cycle of the data is chosen equal to one round which length is R. Then for each selected start of transmission, 10000 bursts of errors are generated with a size exponentially distributed of mean $c \cdot R$ with $c \in \{0.5, 1, 1.5, 2, 2.5, 3\}$. The starting point of each burst is randomly chosen in the first 2000 rounds.

The event that has to be avoided is the corruption of one or more frames of the FTU by a perturbation. The results of these experiments are shown on Figure 5. The clustering of the replica significantly diminishes the total number of lost data (around 18.5% for $c \in \{2, 2.5, 3\}$) knowing that there are cases where the start of the burst

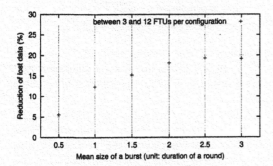

Fig. 5. Reduction of the number of lost data when the optimal allocation is used instead of a random allocation. The data being lost when at least one replica of a same FTU is corrupted. The mean burst size ranges from 0.5 to 3 times the length of a round.

and its size are such that at least one replica will be corrupted whatever the allocation. The loss of robustness with a random allocation tends to be more important when the size of the burst is becoming bigger.

5. CONCLUDING REMARKS

The position of the replicas inside a TTP/C round has an impact on the robustness of the system to transmission errors when bursts of errors may occur. The first result of this study is to give an optimal way to spread the replicas in order to minimize the loss probability of all replicas. In a second part, it has been proven that clustering together all replicas minimizes the probability to lose one or more replicas when the production cycle of a data is equal to the length TTP/C round. A first extension of this study is to consider arbitrary data production cycles.

In a future work, one may consider the case where a subset of FTUs requires the minimization of the loss probability while the rest of the FTUs need to maximize the probability that at least one replica of each FTU is successfully transmitted. This may be a situation arising on systems made of fail-silent and non fail-silent nodes.

Another future work is to consider the use of Forward Error Correction techniques (such as Reed-Salomon codes) instead of replicas in order to make the system even more robust to transmission errors. Finally, we intend to study the robustness against transmission errors of an hybrid event-triggered/time-triggered network such as Flexray which is also considered for use in X-by-Wire automotive applications.

REFERENCES

J. Barrenscheen and G. Otte. Analysis of the physical CAN bus layer. In *4th international CAN Conference, ICC'97*, pages 06.02 06.08, Octobre 1997.

E. Dilger, T. Führer, B. Müller, and S. Poledna. The x-by-wire concept: Time-triggered information exchange and fail silence support by new system services. Technical Report 7/1998, Technische Universität Wien, Institut für Technische Informatik, 1998. also available as SAE Technical Paper 98055.

B. Gaujal and N. Navet. Maximizing the robustness of tdma networks with applications to TTP/C. Technical Report RR-4614, INRIA, 2002.

G. Grünsteidl, H. Kantz, and H. Kopetz. Communication reliability in distributed real-time systems. In *10th Workshop on Distributed Computer Control Systems*, 1991.

International Standard Organization ISO. *Road Vehicles - Low Speed serial data communication - Part 2: Low Speed Controller Area Network*. ISO, 1994. ISO 11519-2.

H. Kopetz. *Real-Time Sytems : Design Principles for Distributed Embedded Applications*. Kluwer Academic Publishers, Boston, 1997.

H. Kopetz, G. Bauer, and S. Poledna. Tolerating arbitrary node failures in the time-triggered architecture. In *SAE 2001 World Congress, March 2001, Detroit, MI, USA*, Mar. 2001.

H. Kopetz, R. Nossal, R. Hexel, A. Krüger, D. Millinger, R. Pallierer, C. Temple, and M. Krug. Mode handling in the time-triggered architecture. *Control Engineering Practice*, 6 (1998):61 66, Mar. 1998.

A. W. Marshall and I. Olkin. *Inequalities: Theory of Majorization and its Applications*, volume 143 of *Mathematics in Science and Engineering*. Academic Press, 1979.

N. Navet, Y.-Q. Song, and F. Simonot. Worst-case deadline failure probability in real-time applications distributed over CAN (Controller Area Network). *Journal of Systems Architecture*, 46 (7):607 618, 2000.

I.E. Noble. EMC and the automotive industry. *Electronics & Communication Engineering Journal*, pages 263 271, Octobre 1992.

H. Pfeifer. Formal verification of the ttp group membership algorithm. In *FORTE/PSTV 2000*, 2000.

S. Poledna, P. Barrett, A. Burns, and A. Wellings. Replica determinism and flexible scheduling in hard real-time dependable systems. *IEEE Transactions on Computers*, 49(2):100 111, Feb. 2000.

C. Temple. Avoiding the babbling-idiot failure in a time-triggered communication system. In *International Symposium on Fault-Tolerant Computing (FTCS)*, pages 218 227, 1998.

TTTech Computertechnik GmbH. *Specification of the TTP/C Protocol - version 1.0*, July 2002.

E. Zanoni and P. Pavan. Improving the reliability and safety of automotive electronics. *IEEE Micro*, 13(1):30 48, 1993.

ELSEVIER

IFAC
PUBLICATIONS
www.elsevier.com/locate/ifac

FRAME PACKING UNDER REAL-TIME CONSTRAINTS

Ricardo Santos Marques - Nicolas Navet - Françoise Simonot-Lion

LORIA - TRIO
2, Avenue de la forêt de Haye
54516 Vandoeuvre-lès-Nancy
France
{rsantos,nnavet,simonot}@ensem.inpl-nancy.fr
Tel: +33.3.83.59.55.77 - Fax: +33.3.83.59.56.62

Abstract: The set of frames of an in-vehicle application must meet two constraints: it has to be feasible from a schedulability point of view and it should minimize the network bandwidth consumption. The latter point is crucial for enabling the use of low cost electronic components and for facilitating an incremental design process. This study proposes two heuristics for the NP-complete problem of generating a set of schedulable frames that minimizes the bandwidth usage. The proposed strategies are complementary. The first one can be applied to large sized problems (in the context of in-vehicle applications) while the second one, slightly more efficient in our experiments, is limited to small size problems (less than 12 signals emitted by each stations). These proposals has proved to be effective in comparison with other possible strategies. *Copyright © 2003 IFAC*

Keywords: Embedded Systems, Vehicles, Scheduling Algorithms, Heuristics, Bandwidth-Minimization Problems.

1. INTRODUCTION

The purpose of the study is a contribution to the configuration of a middleware in the context of in-vehicle embedded systems. More precisely, it proposes a method for building off-line the set of frames when the Medium Access Protocol (MAC) is a priority bus such as CAN which is a de-facto standard for automotive communications. Application processes in the different Electronic Control Units (ECUs) send and receive data, termed signals, over the network. The middleware layer, that masks the communication system services to producer and consumer processes, performs at runtime the packing of the frames and their emission according to the configuration.

In-vehicle applications are subject to real-time constraints and most signals have a limited lifetime. The freshness constraint associated to a signal is not necessarily the same for all consumers of the signal, in the following we will consider that it is the most stringent one. Knowing the set of ECUs and for each ECU the set of signals that are to be sent over the network, their size, their deadline and their production period, one must build the set of frames, the sequence of signals composing each frame as well as decide their emission period. In the case of a priority based MAC protocol, the priority of each frame has to be chosen and those priority choices will obviously have an impact on the respect of the deadlines.

In addition to schedulability, the set of frames must be constructed with the objective of minimizing the bandwidth consumption. The first reason lies on the fact that the possibility to add new functionalities in the form of one or several ECUs must be preserved and adding more ECUs implies that more signals will be exchanged on the bus. Such an incremental design is a standard procedure in the automotive industry. The second reason to minimize the bandwidth consumption is to allow the use of low power processors which are less expensive. The problem to solve is to find a configuration of frames under schedulability and bandwidth minimization constraints. This problem is NP-complete (see Norström et al. (2000)) and, as it will be shown, it cannot be solved using an exhaustive approach even for a small number of signals and/or ECUs. The solution is thus to find efficient heuristics.

A solution to the problem of frame configuration under schedulability constraints is proposed in the same applicative context by the middleware Volcano (see Casparsson et al. (1999)) but the algorithms of this commercial product are not published. Some heuristics are presented in Norström et al. (2000) to build a set of frames over CAN that minimizes the bandwidth consumption but without explicitly searching one feasible solution. Furthermore, the proposed solutions does not decide the priority of the frames. Finally, in the context of production management, several strategies were developed to place elements of different sizes inside some boxes. This problem is usually known as the 'bin-packing' problem (see Coffman et al. (1996) for a survey).

In this paper we propose two heuristics. The first one is a greedy algorithm inspired by the 'bin-packing' approaches, its complexity allows its usage on large size problems. The second heuristic searches more extensively through the solution's space than the second. It can only be applied to limited sized problems and in this case it is slightly more efficient. In this study, the performance metrics are the network bandwidth consumption and the capability to find schedulable solutions. Our proposals will be evaluated by comparison with two effective 'bin-packing' heuristics and with a naive strategy (one signal per frame).

Section 2 is devoted to a formal description of the problem, the assumptions and notations are given and the complexity of an exhaustive approach is derived. In sections 3 and 4, the proposed heuristics are described and their performances are assessed.

2. PROBLEM FORMULATION

The problem is to constitute a set of k frames $F = \{f_1, f_2, ..., f_k\}$ from a given set of n signals $S = \{s_1, s_2, ..., s_n\}$ in such a way as to minimize the bandwidth consumption while respecting the deadline constraint. The location of application processes is fixed so each signal is associated to the station that produces it. Each signal s_i is characterized by a tuple (N_i, T_i, C_i, D_i):

- the sending station denoted N_i,
- its production period T_i on this station. In practice the clocks of the stations are not synchronized but we assume that the first production of all signals located on the same ECU takes place at the same time,
- C_i is the size of s_i in bits with the assumption that C_i is always smaller or equal than the maximal possible frame size (there is no segmentation of a signal),
- D_i is the relative deadline of s_i, it is the maximum duration between the production of the signal on the sender side and its reception by all consumers. In this paper we assume that the deadline of a signal is equal to the production period but the proposed strategies are still valid for deadlines smaller than the period.

The communication network is a priority bus that might be CAN(ISO (1993)), VAN(ISO (1994)) or the J1850(SAE (1992)). The numerical results are obtained with a CAN network at 500kbits/s. Each frame contains up to 8 data bytes and we consider an overhead of 64 bits[1].

Each frame f_i resulting from the packing process is characterized by a tuple $(N_i^*, T_i^*, C_i^*, D_i^*, P_i^*)$:

- N_i^* is the station that sends the frame,
- T_i^* is the frame emission period that is the smallest period of all the signals composing the frame,
- C_i^* is the size of the frame in bits. It is composed of data plus the constant overhead fixed to 64 bits,
- D_i^* is the deadline of the frame which depends on the signals composing the frame,
- P_i^* is the priority for the MAC protocol.

It is usual on in-vehicle networks that a part of the traffic has no real-time constraint (for instance, diagnosis frames) and such frames are assigned a priority lower than all real-time frames. It is assumed that these frames have a size of 128 bits (8 bytes of data plus the overhead) which will be the blocking factor for all real-time frames (ie. the

---

[1] The exact size of the CAN frame overhead is dependent of the data because of the 'bit-stuffing' mechanism. An overhead of 64 bits by frame is a reasonable value.

maximum time interval during which one frame can be delayed by a lower priority frame).

2.1 Problem complexity

The problem of building frames from signals is similar to the 'bin-packing' problem and it was proven to be NP-complete in Norström et al. (2000). Nevertheless on small size problems an exhaustive approach might be used. One thus has to determine the exact complexity of our specific problem.

To group a set of n signals in k non-empty frames comes to create all the partitions of size k from the set of signals. The complexity of this problem is known, it is the Stirling number of the second kind (see Abramowitz and Stegun (1970) page 824):

$$\frac{1}{k!} \sum_{i=0}^{k} (-1)^{(k-i)} \binom{k}{i} i^n$$

The number k of frames per station can vary from 1 to n where n is the number of signals produced by the station. The number of frames to envisage on a station i that produces n signals is thus:

$$S_i = \sum_{k=1}^{n} \frac{1}{k!} \sum_{i=0}^{k} (-1)^{(k-i)} \binom{k}{i} i^n$$

For example, three signals a, b, and c on a same station leads to five different partitions : $[(a,b,c)]$, $[(a),(b,c)]$, $[(a,b),(c)]$, $[(a,c),(b)]$ and $[(a),(b),(c)]$.

If we consider a set of m stations, the solutions space becomes $\prod_{i=1..m} S_i$ where S_i is the number of possible frames for station i. The solutions space grows very quickly as soon as n and m increase. For instance, with 10 signals per station ($S_i = 115975$) and 5 stations, an exhaustive search would need to consider about $2 \cdot 10^{25}$ cases. Moreover, this evaluation did not take account of the determination of frame priority. One can see that for real industrial cases an exhaustive approach cannot be applied. This justifies the design of specific heuristics.

2.2 Optimal priorities allocation

Under the Fixed Priority Preemptive (FPP) policy, the response time of a task only depends on the set of higher priority tasks and not on the relative priorities between these higher priority tasks. Starting from this observation, Audsley proposed in Audsley (1991) a priority allocation algorithm that runs in $O((n^2 + n)/2)$ where n is the number of tasks. This algorithm is optimal: if

a solution exists then it will necessarily be found by the algorithm. The strategy is to start from the lowest priority level (m) and to search a task that is feasible at this priority level. In case of failure, one can conclude that the set of tasks is not schedulable. The first feasible task at level m is assigned to that priority. Once the priority level m is given to a task, the algorithm tries to find a feasible task at level $m - 1$ and so on until priority 1 which is the highest priority of the system.

In the general case this algorithm cannot be applied to message scheduling under the Non-Preemptive Fixed Priority (NPFP) policy which is the medium access strategy for priority buses. Indeed, on a network the response time of a frame not only depends on the higher priority frames but also on the set of lower priority frames because of the blocking factor (maximal time interval during which one frame can be delayed by a lower priority frame). In our particular context the blocking factor is equal for all frames since we assume the existence of a non real-time traffic (e.g. diagnostic frames). Without any other assumptions on this traffic, one must consider the blocking factor as being the size of the largest frame compliant with the communication protocol. In this case, the conditions are fulfilled to apply the Audsley algorithm in order to evaluate the schedulability of a set of frames.

3. THE "BANDWIDTH-BEST-FIT DECREASING" HEURISTIC

The name of this heuristic has been given by analogy with the terminology used in 'bin-packing' problems with off-line resolution (*Best-Fit decreasing - First-Fit decreasing*, see Coffman et al. (1996)). In 'bin-packing' problems, the objective is to minimize the number of boxes (of frames here). In our context the goal is to minimize the network bandwidth consumption without taking account of the number of frames. The underlying idea of the heuristic is to place a signal in the frame that minimizes the bandwidth consumption induced by the signal. The heuristic builds only one solution which feasibility is assessed with the Audsley algorithm. If the solution is feasible then a local optimization procedure is applied. If it fails to find a feasible priority assignment, the heuristic applies some transformations on the frames in order to shift some load to lower priority levels.

3.1 Description

The algorithmic description of the 'Bandwidth-Best-Fit decreasing' heuristic is given below:

(1) On each station, sort all signals in decreasing order of bandwidth consumption.

(2) Insert the signal s_i in a frame:

 (a) there is at least one frame in the station that can accept s_i, it means a frame with a size including s_i smaller or equal to the protocol maximum and whose deadline is strictly positive (see Appendix A). Find which is the frame that minimizes the most the bandwidth consumption with s_i as part of the data field. Compare this bandwidth consumption augmentation with the one caused by placing s_i alone in one frame. Insert the signal s_i in the frame f_k corresponding to the best solution. If f_k corresponds to a frame that was already created, then one updates the frame timing characteristics: $T_k^* = \min(T_k^*, T_i)$ and D_k^* is set according to the algorithm described in Appendix A. If f_k corresponds to a frame whose data field is only composed by s_i then one sets the frame timing characteristics to $T_k^* = T_i$ and $D_k^* = D_i$.

 (b) there is not a single frame that can accept s_i: create a new frame with timing characteristics $T_k^* = T_i$ and $D_k^* = D_i$.

(3) As long as there is a signal not inserted in a frame, return to step 2.

(4) Feasibility test of the set of frames with the Audsley algorithm:

 (a) the configuration is not feasible:

  (i) one constructs $\hat{F}$, the set of frames for which no priority has been found.

   (A) there is one frame in $\hat{F}$ that contains at least two signals. Find the frame in $\hat{F}$ with at least two signals and for which the difference between the worst-case response time and deadline is the smallest when the frame is given the lowest priority that has not been successfully assigned. This frame is denoted $\hat{f}$.

   (B) all the frames were fully decomposed: FAILURE.

  (ii) one removes from the signals composing the frame $\hat{f}$, the one with the smallest deadline and one places it in a new frame. Update $\hat{f}$'s timing characteristics (period, deadline and size).

  (iii) return to step 4 (the algorithm will stop when all non-feasible frames have been completely decomposed).

 (b) the configuration is feasible: SUCCESS

This heuristic is composed of distinct parts. The first part (steps 1 to 3) aims at constructing a solution that minimizes the bandwidth consumption. These steps are inspired from the very efficient 'Best-Fit decreasing' heuristic from the 'bin-packing' domain. However, the choice criterion is not the size of the objects but the bandwidth consumption. The second part (step 4) deals with the feasibility of the proposed solution. Should the Audsley test fail, then the heuristic tries to reduce the deadline constraints of the frames by isolating the most demanding signals. The frame initially chosen for the decomposition is the one that exceeds the least its deadline at the lowest priority level that has not been successfully assigned. It is thus the frame that is the more likely to respect its deadline if it should contain less data. To determine which frame to decompose, it is necessary to compute a response time for each frame for which a priority has not been found (one gives the considered frame the first priority not assigned by the Audsley algorithm, the higher priority frames stay unchanged and the lower priority frames have no influence).

The step 1 of the heuristic is a sorting process, its complexity can thus be reduced to $n \log(n)$ (where n is the number of signals in the station). Steps 2 and 3 require in the worst case $O((n^2 + n)/2)$ steps. During step 4, there is at most n response time computations (no priority level was already assigned) and n calls to the Audsley algorithm. In our experiments made on industrial-scale problems (one hundred signals distributed over ten ECUs), this complexity did not raise any computation time problem.

3.2 Performances evaluation

The performances of the BBFd heuristic are evaluated with regard to two metrics that are crucial in our context: the feasibility of the system and the network bandwidth consumption. The competing strategies are:

- 'One Signal per Frame' (1SpF): each frame contains only one signal,
- 'First-Fit decreasing' (FFd): the signals are sorted decreasingly according to their size and they are inserted in the first frame that can contain them,
- 'Best-Fit decreasing' (BFd): the signals are sorted decreasingly according to their size and they are inserted in the frame that will have the smallest space left after the insertion.

The heuristics were implemented in C++ and the feasibility of each solution is tested by the *rts* software (written by Jörn Migge, see Migge (2002)) that implements the Audsley algorithm (see paragraph 2.2) and response time computation on CAN.

3.2.1. Bandwidth consumption Only the results induced by configurations that are feasible with all strategies are taken into account since the use of non-feasible set of frames is ruled out in the context of in-vehicle applications. Figure 1 shows the average network load on 100 feasible configurations for an useful load varying from 15 to 35% (the useful or nominal load is the load brought by the data alone). Signals are randomly generated with a period ranges from 5 to 100 ms (uniform distribution - step of 5ms), a deadline equals to the period, a size between 1 and 8 bytes (uniform distribution - step of 1 byte). For each test, the number of generated signals is the average number of signals necessary to reach the desired nominal load level. The signals are randomly located on a set of 10 stations. The network transmission rate is 500 Kbits/s and the overhead is 8 bytes for one data frame that can contain up to 8 data bytes (CAN network).

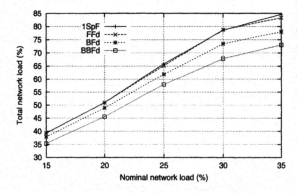

Fig. 1. Total network load for a nominal load ranging from 15 to 35%. Average results on 100 feasible configurations for the 4 competing strategies.

As it can be seen on figure 1, the relative performance of the heuristics remains identical whatever the network load. BBFd always produces the best results with a gain varying from 9.9 to 13.7% over the 1SpF approach and from 6.3 to 7.6% over BFd which is the closest heuristic. The gains in terms of bandwidth consumption do exist for all network load without being very important.

3.2.2. Configurations feasibility The conditions of the experiments are those described in paragraph 3.2.1. The results are based on 100 randomly generated tests. Table 1 shows the results obtained for the 4 competing strategies.

Up to 25% of nominal load, all the heuristics find a feasible solution. Beyond that load level, the performances strongly differ but the relative performances remains identical as for bandwidth consumption. BBFd is always the best. In particular it enables us to obtain 13 supplementary feasible configurations at the highest load level.

nominal load	15%	20%	25%	30%	35%
1SpF	100	100	100	97	36
FFd	100	100	100	97	54
BFd	100	100	100	97	76
BBFd	100	100	100	100	89

Table 1. Number of feasible configurations over 100 tests for a nominal load varying from 15 to 35%.

This suggests to us that the constraint relaxation technique used in case of non-feasibility is effective (see step 4 of the algorithm in paragraph 3.1).

3.3 Local optimization algorithm

When the BBFd heuristic returns a feasible solution, one can additionally execute a local optimization (LO) algorithm as it is classically done in optimization problems (see for instance Osogami and Okano (1999)). We propose a LO strategy that basically tries to permutes pairs of signals or to move some signals. This procedure is executed a large number of times (n) in each station ($n = 10000$ in our experiments) and the solution of the LO becomes the current best solution if it improves the bandwidth consumption and if it is feasible. The following algorithm is executed n times in each station:

(1) $\{f_a, f_b\}$ are two frames randomly chosen from the set of frames with more than one signal, bc_{before} is the bandwidth consumption of $\{f_a, f_b\}$. Choose randomly one signal from each frame: s_a from f_a and s_b from f_b and execute the following operations:
   (a) remove s_a from f_a and insert it in f_b. If the size of f_b is smaller or equal than the protocol maximum and if the deadline of both frames is strictly positive then bc_{after1} = bandwidth consumption of $\{f_a, f_b\}$ otherwise $bc_{after1} = \infty$.
   (b) same procedure as step (1a) with s_b instead of s_a. The corresponding bandwidth consumption is bc_{after2}.
   (c) remove s_a from f_a and insert it in f_b, remove s_b from f_b and insert it in f_a; if the size of both frames is smaller or equal than the protocol maximum and if their deadline is strictly positive then bc_{after3} = bandwidth consumption of $\{f_a, f_b\}$ otherwise $bc_{after3} = \infty$.
(2) if one of the bandwidth consumptions $\{bc_{after1}, bc_{after2}, bc_{after3}\}$ is lower than bc_{before} then update the set of frames of the station with the best solution otherwise go to step (1).
(3) Feasibility test of the new set of frames with the Audsley algorithm:
   (a) the new set of frames *is not* feasible, undo the changes operated made at step (2) and go to step (1).

(b) the new set of frames is feasible, go to step (1).

The possibility of improvement comes from the fact that, by swapping or moving signals, one can test a configuration with a signal located in a frame that did not exist, or not with the same content, at the time where the signal's insertion was decided (see step (2a) of the algorithm described in Section 3.1).

3.3.1. Performance evaluation The performance of the LO procedure is evaluated by the gain of network bandwidth consumption and by the percentage of cases where it improves the solution of BBFd.

The conditions of the experiments are those described in subsection 3.2.1. LO, with parameter $n = 10000$, is applied on 100 feasible solutions computed with BBFd. Table 2 shows the resulting network bandwidth usage.

nominal load	15%	20%	25%	30%	35%
BBFd	35.32	45.54	57.96	67.81	73.01
BBFd - with LO	35.21	45.36	57.63	67.47	72.73

Table 2. Average network bandwidth consumption over 100 feasible solutions for a nominal load varying from 15 to 35%.

Although the improvements are modest, BBFd with LO is clearly always better than BBFd with a gain varying between 0.3% and 0.5%. Table 3 shows the percentage of cases where LO brought a gain against BBFd alone.

nominal load	15%	20%	25%	30%	35%
%	56	64	87	94	95

Table 3. Percentage of cases where BBFd with LO decreased the bandwidth consumption. Experiments made over 100 feasible solutions with a nominal load varying from 15 to 35%.

The improvement with LO tends logically to be more important (Table 2) and more frequent (Table 3) when the load increases since BBFd is likely to be more distant from the optimal solution.

4. THE 'SEMI-EXHAUSTIVE' HEURISTIC

The term 'semi-exhaustive' is chosen because this strategy is an exhaustive search through the solution's space where one would cut some branches that are judged not promising. This algorithm starts by the construction of the set of all possible frames. In practice, the complexity of this first step, determined in paragraph 2.1, does not allow

us to deal with cases where there are more than 12 signals in one station ($S_i = 4213597$ for $n = 12$, see paragraph 2.1). With less than 12 signals per station, the generation of the set of possible frames can be done using the technique detailed in Orlov (2002).

4.1 Description

The algorithmic description of the 'semi-exhaustive' heuristic is given below:

(1) On each station i, construction of F_i^* which is the set of all possible partitions of the set of signals. A partition is 'possible' if it does not include a frame that violates the maximum data size imposed by the communication protocol or a frame having a negative deadline. The transmission period of a frame is the smallest production period of the signals it contains while the deadline of the frame is set according to Appendix A.

(2) On each station i, the set F_i^* is sorted in increasing order according to the partition bandwidth consumption (i.e. the bandwidth consumed by all the frames of a partition).

(3) $F_{i,1}^*$ is the partition that minimizes the bandwidth usage on station i. The stations are sorted in increasing order according to $F_{i,1}^*$ (e_k is the index of the k-th station in this order). For each F_i^*, one defines a depth p_i from which no further search will be performed: the partitions $F_{i,j}^*$ with $j > p_i$ won't be considered since they are the least satisfactory in terms of bandwidth consumption. The values chosen for p_i enable us to control the complexity of the heuristic. Let a_i be a temporary index variable initialized to 1 on each station i.

(4) The current set of frames is composed of the frames contained in the partitions $F_{e_1,a_{e_1}}^* \bigcup F_{e_2,a_{e_2}}^* \bigcup ... \bigcup F_{e_m,a_{e_m}}^*$ where m is the number of stations in the network.

(5) Feasibility test of the set of messages with the Audsley algorithm:
(a) the configuration is not feasible: the next solution to test is built according to the algorithm

```
for u:=a_{e_m} downto a_{e_1} do
    if (u<p_u)
    then a_u:=a_u+1; break;
    else a_u:=1; fi
```

If all the configurations were tested: FAILURE otherwise return to step 4. Figure 2 illustrates this search technique.
(b) the configuration is feasible: SUCCESS.

As soon as the first feasible solution is found, the algorithm stops. Nevertheless it is possible to continue the search for instance until a fixed number

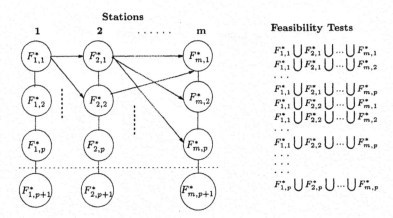

Stations

Feasibility Tests

$F_{1,1}^* \bigcup F_{2,1}^* \bigcup \cdots \bigcup F_{m,1}^*$
$F_{1,1}^* \bigcup F_{2,1}^* \bigcup \cdots \bigcup F_{m,2}^*$
$\cdots$
$F_{1,1}^* \bigcup F_{2,1}^* \bigcup \cdots \bigcup F_{m,p}^*$
$F_{1,1}^* \bigcup F_{2,2}^* \bigcup \cdots \bigcup F_{m,1}^*$
$F_{1,1}^* \bigcup F_{2,2}^* \bigcup \cdots \bigcup F_{m,2}^*$
$\cdots$
$F_{1,1}^* \bigcup F_{2,2}^* \bigcup \cdots \bigcup F_{m,p}^*$
$\cdots$
$\cdots$
$F_{1,p}^* \bigcup F_{2,p}^* \bigcup \cdots \bigcup F_{m,p}^*$

Fig. 2. Search trough the solution's space in the 'semi-exhaustive' heuristic.

of feasible solutions is found. This algorithm does at most $\prod_{i=1..m} p_i$ calls to the Audsley algorithm and the choice of the values of p_i should be done according to the available computation time. In our experiments, the values of p_i were all chosen equal to 25 but we think that it is possible to obtain better results by individualizing the values of p_i on each station according to some criterion that is to be defined. It is noteworthy that the algorithm finds the optimal solution in terms of bandwidth consumption in the particular case where the set of partitions $F_{e_1,1}^* \bigcup F_{e_2,1}^* \bigcup \cdots \bigcup F_{e_m,1}^*$ is feasible.

4.2 Performances evaluation

The experiments were conducted with the parameters described in subsection 3.2.1 except for the number of signals on each station and the number of stations which are no longer randomly chosen. Indeed, to ensure that the 'semi-exhaustive' (SE) heuristic may be applied (no more than 11 signals on each station), the number of signals is fixed to 10 per station while the number of stations varies according to the desired nominal load.

On figure 3, one observes that SE and BBFd-wlo are always significantly better than 1SpF (up to 13.46% for SE and 13.22% for BBFd with LO) which shows the effectiveness of the proposed heuristics. The performance of SE and BBFd with LO are very close one to each other (the curves are almost superposed on Figure 3) but SE remains better whatever the load. However the gain is limited since it ranges from 0.3 to 0.56%. From a schedulability point of view, SE, as well as BBFd with LO, always found a feasible solution in our experiments (made with deadline equal to period) for nominal loads lower than 35%. Beyond this load level, the computation time for SE may become problematic due to the rarefaction of feasible solutions.

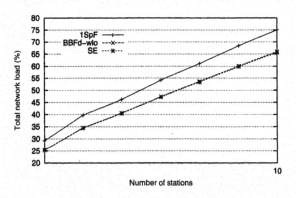

Fig. 3. Total network load for a number of stations varying from 4 to 10 with 10 signals by station. The corresponding nominal load is 11% with 4 stations, 13.75% with 5 stations, 16.5% with 6 stations, 19.25% with 7 stations, 21.5% with 8 stations, 24% with 9 stations and 27.5% with 10 stations. Average results over 100 feasible configurations with "one signal per frame" (1SpF), BBFd with LO (BBFd-wlo) and 'semi-exhaustive'(SE) heuristics.

CONCLUSION

In this study, two heuristics for building feasible sets of frames that minimize the bandwidth consumption have been proposed. The strategies are complementary, the first one (BBFd with LO) may be applied on large size problems (in the context of in-vehicle applications) while the second one (SE), slightly more efficient in the experiments, is usable only on limited size problems (less than 12 signals by ECU). These proposals have proven to behave much better than other a priori possible approaches that are 'Best-Fit decreasing', 'First-Fit decreasing' and 'One Signal per Frame'.

These heuristics can be used as a starting point for other optimization algorithms to direct the search towards promising parts of the solution's space. In particular, they might be included in the initial population of a genetic algorithm, the initial population having a strong impact on the performance of the algorithm (see for instance Westerberg and Levine (2001)).

REFERENCES

M. Abramowitz and I.A. Stegun. *Handbook of Mathematical Functions.* Dover Publications (ISBN 0-486-61272-4), 1970.

N.C. Audsley. Optimal priority assignment and feasibility of static priority tasks with arbitrary start times. Technical Report YCS164, University of York, 1991.

L. Casparsson, A. Rajnák, K. Tindell, and P. Malmberg. Volcano - a revolution in on-board communications. Technical report, Volvo Technology Report, 1999.

E.G. Coffman, M.R. Garey, and D.S. Johnson. *Approximation Algorithms for NP-Hard Problems*, chapter Approximation Algorithms for Bin Packing: a Survey. PWS Publishing Company, 1996.

ISO. ISO International Standard 11898 - Road vehicles - Interchange of digital information - Controller Area Network (CAN) for high-speed communication, 1993.

ISO. ISO International Standard 11519-3 - Road vehicles - Low-speed serial data communication - Vehicle Area Network (VAN), 1994.

J. Migge. RTS, 2002. program and manual available at http://www.loria.fr/~nnavet.

C. Norström, K. Sandström, and M. Ahlmark. Frame packing in real-time communication. Technical report, Mälardalen Real-Time Research Center, 2000.

M. Orlov. Efficient generation of set partitions, 2002. available at http://www.cs.bgu.ac.il/~orlovm/papers/partitions.pdf.

T. Osogami and H. Okano. Local search algorithms for the bin packing problem and their relationships to various construction heuristics. In *Proc. IPSJ SIGAL*, 1999. also available as IPSJ Technical Report number 99-AL-69-5.

G. Quan and X. Hu. Enhanced fixed-priority scheduling with (m,k)-firm guarantee. In *IEEE Real-Time System Symposium (RTSS)*, 2000.

SAE. SAE J1850 Class B data communication network interface, 1992.

C. H. Westerberg and J. Levine. Investigation of different seeding strategies in a genetic planner. In Springer-Verlag, editor, *Proc. EvoWorkshops2001: EvoCOP, EvoFlight, EvoIASP, EvoLearn, and EvoSTIM*, LNCS 2037, 2001.

Appendix A. DEADLINE OF A FRAME AFTER THE ADDITION OF A SIGNAL

The transmission period of a frame containing several signals is the smallest production period of the signals and the transmission of the frame will be synchronized with the production of the signal having the smallest period. On the other hand, the deadline of the frame is not the smallest deadline due to possible offsets between the production dates of the signals and the actual transmission dates of the frame. Let us for instance consider the example shown on figure A.1 with two signals s_1 and s_2 having respectively a period $T_1 = 10$ and $T_2 = 14$ and a deadline (relatively to the production date) $D_1 = 10$ and $D_2 = 14$. The signal s_2 produced at time 14 is actually sent at time 20 and the deadline of the frame must thus be equal to 8 in order to respect the timing constraint of s_2. One can also note that the transmissions made at times 10 and 40 do not include any new value for s_2. In practice, the value of s_2 might be re-transmitted or not but there will not be any timing constraint on the frame induced by s_2.

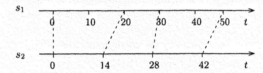

Fig. A.1. Two signals with production periods equal to 10 and 14. The signals are transmitted in a frame synchronized with the signal having the smallest period. The dotted line indicates when the signal with period 14 is actually transmitted.

To determine the deadline of the frame, one must find the largest offset between a production date and the transmission of the next frame. One wants to include signal s_i in frame f_k already containing the signals $s_1^k, s_2^k, ..s_n^k$. One denotes s_{min} the signal with the smallest period of the set $\{s_1^k, s_2^k, ..s_n^k\} \cup s_i$, the period of f_k becomes $T_k^* = T_{min}$. The relative deadline of f_k is $D_k^* = \min\{D_j - \text{Offset}(T_{min}, T_j) \mid s_j \in \{s_1^k, s_2^k, ..s_n^k\} \cup s_i\}$ where $\text{Offset}(a, b)$ returns the largest possible duration between the production date of a signal with period $b \geq a$ and the transmission of the frame of period a that contains the signal. It has been shown in Quan and Hu (2000) that $\forall k_1, k_2 \in \mathbb{N}$ $k_1 \cdot a - k_2 \cdot b = q \cdot \gcd(a, b)$ with $q \in \mathbb{Z}$. In our context, one imposes $a > k_1 \cdot a - k_2 \cdot b \geq 0$ and thus $\text{Offset}(a, b) = (\frac{a}{\gcd(a,b)} - 1) \cdot \gcd(a, b) = a - \gcd(a, b)$. In our example, the deadline must be set to 6.

ELSEVIER

IFAC
PUBLICATIONS
www.elsevier.com/locate/ifac

EXPRESSIVENESS AND ANALYSIS OF
SCHEDULING EXTENDED TIME PETRI NETS

Didier Lime* Olivier H. Roux*

*IRCCyN, UMR CNRS 6597,
1 rue de la Noë - BP92101
44321 Nantes Cedex 3 - France
{Didier.Lime |
Olivier-h.Roux}@irccyn.ec-nantes.fr

Abstract: The most widely used approach for verifying the schedulability of
a real-time system consists of using analytical methods. However, for complex
systems with interdependent tasks and variable execution times, they are not well
adapted. For those systems, an alternative approach is the formal modelisation
of the system and the use of model-checking, which also allows the verification
of more varied scheduling properties. In this paper, we show how an extension
of time Petri nets proposed in (Roux and Déplanche, 2002), scheduling extended
time Petri nets (SETPN), is especially well adapted for the modelisation of real-
time systems and particularly embedded systems and we provide a method for
computing the state space of SETPN. We first propose an exact computation
using a general polyhedron representation of time constraints, then we propose an
overapproximatiion of the polyhedra to allow the use of much more efficient data
structures, DBMs. We finally describe a particular type of observers, that gives us
a numeric result (instead of boolean) for the computation of tasks response times.
Copyright © 2003 IFAC

Keywords: Scheduling, formal methods, time Petri nets, realtime systems

1. INTRODUCTION

A real-time system is typically composed of several tasks that interact. An important problem
consists of ensuring that the tasks can be executed
in such a way that they respect the constraints
they are subject to (deadlines, periods, ...), and
that the overall system performs correctly with
respect to its specification, *i.e.* that it satisfies a
given property (for example, the duration between
two successive samples of a signal is always in
the interval of time $[a, b]$). Solving this problem
usually requires scheduling the tasks in an appropriate way.

Two approaches are considered in order to solve
the scheduling problem. The off-line approach:

given a system S and a property P, a pre-runtime
schedule (scheduler) is constructed in such a way
that the system S satisfies P. The on-line approach: a scheduling policy based on priorities,
mostly derived from the temporal parameters (e.g.
Rate Monotonic, Earliest Deadline, Least Laxity)
is selected and implemented within the scheduler
; it is then necessary to make a schedulability
analysis to verify that with this given scheduling
policy, every task meets its deadline.

Consequently, scheduling theory is studied a
lot, mainly in the form of an analytical study.
It mostly consists of determining schedulability tests: according to the complexity of the
tasks model, these schedulability tests are ex-

acts (necessary and sufficient conditions) or over-approximations (sufficient conditions).

For analysis of sets of independent tasks, exact methods have been developped that take into account both offsets and jitter (Tindell, 1994a; Palencia and Harbour, 1998; Hladik and Déplanche, 2003).

Precedence relations lead to much more complex problems. The main approach tries to fall back into an independent task model by computing an additional jitter simulating the time that the tasks submitted to precedence relations have to wait for ((Tindell, 1994b)). Some works refine this approach by reducing the jitter by a best-case response-time computing ((Henderson et al., 2001) for instance). However, the independent model with jitter is an overapproximation of the original one, which leads to pessimistic response times. Other approaches analyze the precedence graph to reduce the number of scenariis to look at ((Harbour et al., 1991; Palencia and Harbour, 1999)), but they are still based on the identification of a worst-case scenario which does not always exist. Finally, communication between tasks by messages in a distributed system has also been investigated in (Tindell and Clark, 1994), which adds the computation of the messages response times into the classical method.

These works generally consider a fixed execution time but while in independent tasks configurations, the worst case is obtained by considering the longest execution time, this is not true anymore when considering precedence relations, which leads to pessimism in the computation of response times. Moreover, they are not yet well adapted to complex synchronization schemes, such as those allowed by real-time executives services.

Related work

Consequently some work proposes to model complex behaviors of the task using a formal model. Concerning the off-line approach, works are mainly based on the controller synthesis paradigm (Altisen et al., 2000). A scheduler is considered as a controller of the processes to be scheduled, which restricts their behavior by triggering their controllable actions. The models used include Petri nets with Deadlines (Altisen et al., 1999), Petri nets with a maximal firing functioning mode ((Grolleau and Choquet-Geniet, 2002)).

Concerning the on-line approach, the modeling of pre-emptive process scheduling has been the subject of recent works. The first approach consists of modelling priorities of tasks with inhibitor arcs added to the Petri net ((Robert and Juanole,

2000)). An inhibitor arc is then placed from each transition of the pattern representing a task towards each transition of the patterns representing the other lower priority tasks. A similar approach consists of using Petri nets with priorities (Janicki and Koutny, 1999). Okawa and Yoneda ((Okawa and Yoneda, 1996)) propose an approach with time Petri nets consisting of defining groups of transitions together with rates (speeds) of execution. Transition groups correspond to transitions that model concurrent activities and that can be simultaneously ready to be fired. In this case, their rate are then divided by the sum of transition execution rates. Roux and Déplanche (Roux and Déplanche, 2002) propose an extension for time Petri nets (SETPN) that allows to take into account the way the real-time tasks of an application distributed over different processors are scheduled. Finally, Fersman, Pettersson and Yi (Fersman et al., 2002) propose extended timed automata with asynchronous processes. The main idea is to associate each location of a timed automaton with an executable program called a task and to construct the preemptive scheduler (fixed priority or EDF) with timed automata with subtraction. However, they consider the worst case execution times as a fixed time and it is easy to show that reducing the computation time of a task may surprisingly induce a decrease of timing performances for the application.

Except the work of (Fersman et al., 2002), all these models include the concept of stopwatch. The reachability analysis problem for stopwatch automata (as well as for time Petri nets with inhibitor arcs...) is undecidable. There is no guarantee for the termination of the analysis. Moreover, for these models, the state space computation is often inefficient and difficult to implement.

We propose to use the SETPN model (Roux and Déplanche, 2002) that have the advantage of being able to express both concurrency and real-time constraints in a natural way and to express the indeterminism of the task execution time by the firing intervals of transitions. We propose a method for computing the state space and an overapproximation of that method that allows a compact abstract representation of transitions firing domains with DBM (Difference Bound Matrix).

Outline of the paper

The paper is organized as follows : section 2 gives the formal definitions of the SETPN model and of its semantics. Section 3 illustrates the expressivity with regard to the classical services provided by real-time executives. Section 4 describes an exact state space computation and also a more efficient

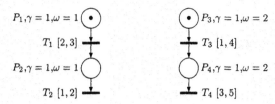

Fig. 1. SETPN of two tasks on one processeor

computation method (at the price of an overapproximation) that uses compact abstract representations of the state-space based on DBM to analyze SETPN. Finally, in section 5, we describe special type of observer that allows us to compute response time with our method and we apply this method on a set of examples for which we have exact response times computation methods.

2. SCHEDULING EXTENDED TIME PETRI NETS

This extension of time Petri nets introduced in (Roux and Déplanche, 2002) consists of mapping into the Petri net model the way the different schedulers of the system activate or suspend the tasks. For a fixed priority scheduling policy, SETPN introduce two new attributes (γ and ω) associated to each place that respectively represent allocation (processor) and priority (of the modeled task). From a marking M, a function Act (formally defined in (Roux and Déplanche, 2002)) gives the projection of the behaviour of the different scheduler in the following sense: Let us suppose that the place P models a behavior (or a state) of the task T. $M(p) > 0$ means that the task T is activable and $Act(M(p)) > 0$ means that the task T is active.

All places of a SETPN do not require such parameters. Actually when a place does not represent a true activity for a processor (for example a register or memory state), neither a processor (γ) nor a priority (ω) have to be attached to it. In this specific case ($\gamma = \phi$), the semantics remains unchanged with respect to a standard TPN [1]. One can notice that it is equivalent to attach to this place a processor for its exclusive use and any priority. An example of a SETPN is presented in figure 1. The initial marking of the net is $\{P_1, P_3\}$. However, since those two places are affected to the same processor, and that the priority of P_3 is the highest, the initial active marking is $\{P_3\}$. So the first transition fired will be T_3.

Definition 1. (Scheduling Extended Time Petri net). A scheduling extended time Petri net is a n-tuple $\mathcal{T} = (P, T, {}^\bullet(.), (.)^\bullet, \alpha, \beta, M_0, Act)$, where

- $P = \{p_1, p_2, \ldots, p_m\}$ is a non-empty finite set of *places*,
- $T = \{t_1, t_2, \ldots, t_n\}$ is a non-empty finite set of *transitions* ,
- ${}^\bullet(.) \in (\mathbb{N}^P)^T$ is the *backward incidence function*,
- $(.)^\bullet \in (\mathbb{N}^P)^T$ is the *forward incidence function*,
- $M_0 \in \mathbb{N}^P$ is the *initial marking* of the net,
- $\alpha \in (\mathbb{Q}^+)^T$ and $\beta \in (\mathbb{Q}^+ \cup \{\infty\})^T$ are functions giving for each transition respectively its *earliest* and *latest* firing times ($\alpha \leq \beta$),
- $Act \in (\mathbb{N}^P)^P$ is the active marking function. $Act(M)$ is the projection on the marking M of the scheduling strategy. In (Roux and Déplanche, 2002) $Act(M)$ is defined for a fixed priority scheduling policy, starting from three parameters :
  - $Proc\{\phi, proc_1, proc_2, \ldots, proc_l\}$ is a finite set of processors (including ϕ that is introduced to specify that a place is not assigned to an effective processor of the hardware architecture),
  - $\omega \in \mathbb{N}^P$ is the priority assignment function ,
  - $\gamma \in Proc^P$ is the allocation function.

We define the semantics of scheduling extended time Petri nets as *Timed Transition Systems* (TTS) (Larsen *et al.*, 1995). In this model, two kinds of transitions may occur : *continuous* transitions when time passes and *discrete* transitions when a transition of the net fires.

A *marking* M of the net is an element of $\mathbb{N}^P$ such that $\forall p \in P, M(p)$ is the number of tokens in the place p.

An *active marking* $Act(M)$ of the net is an element of $\mathbb{N}^P$ such that $\forall p \in P, Act(M(p)) = M(p) or Act(M(p)) = 0$.

A transition t is said to be *enabled* by the marking M if $M \geq {}^\bullet t$, (*i.e.* if the number of tokens in M in each input place of t is greater or equal to the valuation on the arc between this place and the transition). We denote it by $t \in enabled(M)$.

A transition t is said to be *active* if it is enabled by the active marking $Act(M)$. We denote it by $t \in enabled(Act(M))$.

A transition t_k is said to be *newly* enabled by the firing of the transition t_i from the marking M, and we denote it by $\uparrow enabled(t_k, M, t_i)$, if the transition is enabled by the new marking $M - {}^\bullet t_i + t_i{}^\bullet$ but was not by $M - {}^\bullet t_i$, where M is the marking of the net before the firing of t_i. Formally,

$$\uparrow enabled(t_k, M, t_i) = ({}^\bullet t_k \leq M - {}^\bullet t_i + t_i{}^\bullet) \\ \wedge ((t_k = t_i) \vee ({}^\bullet t_k > M - {}^\bullet t_i))$$

---

[1] When $\gamma = \phi$, the parameter is ommited in the figures of this paper.

By extension, we will denote by $\uparrow enabled(M, t_i)$ the set of transitions newly enabled by firing the transition t_i from the marking M.

A *valuation* is a mapping $\nu \in (\mathbb{R}^+)^T$ such that $\forall t \in T, \nu(t)$ is the time elapsed since t was last enabled. Note that $\nu(t)$ is meaningful only if t is an enabled transition. $\overline{0}$ is the *null valuation* such that $\forall k, \overline{0}_k = 0$.

Definition 2. (Semantics of a SETPN). The semantics of a scheduling extended time Petri net $\mathcal{T}$ is defined as a TTS $\mathcal{S}_{\mathcal{T}} = (Q, q_0, \rightarrow)$ such that

- $Q = \mathbb{N}^P \times (\mathbb{R}^+)^T$
- $q_0 = (M_0, \overline{0})$
- $\rightarrow \in Q \times (T \cup \mathbb{R}) \times Q$ is the transition relation including a continuous transition relation and a discrete transition relation.

  · The continuous transition relation is defined $\forall d \in \mathbb{R}^+$ by :

$$(M, \nu) \xrightarrow{d} (M, \nu') \text{ iff } \forall t_i \in T,$$
$$\begin{cases} \nu'(t_i) = \begin{cases} \nu(t_i) \text{ if } Act(M) < {}^\bullet t_i \\ \wedge M \geq {}^\bullet(t_i) \\ \nu(t_i) + d \text{ otherwise,} \end{cases} \\ M \geq {}^\bullet t_i \Rightarrow \nu'(t_i) \leq \beta(t_i) \end{cases}$$

  · The discrete transition relation is defined $\forall t_i \in T$ by :

$$(M, \nu) \xrightarrow{t_i} (M', \nu') \text{ iff },$$
$$\begin{cases} Act(M) \geq {}^\bullet t_i, \\ M' = M - {}^\bullet t_i + t_i^\bullet, \\ \alpha(t_i) \leq \nu(t_i) \leq \beta(t_i), \\ \forall t_k, \nu(t_k)' = \begin{cases} 0 \text{ if } \uparrow enabled(t_k, M, t_i), \\ \nu(t_k) \text{ otherwise} \end{cases} \end{cases}$$

3. EXPRESSIVENESS

Time Petri nets in general have already been used for a long time to model real-time systems and protocols. SETPN add expressiveness with regard to the scheduling policies provided by real-time executives like OSEK/VDX (*OSEK/VDX specification*, 2001). In this section, we show some examples of modelisation of services of these executives. Modelisation possibilities are of course not restricted to what is presented here.

3.1 Basic task model

The basic model we propose for tasks is shown on Fig. 2a. Basic patterns of that type may be concatenated to express sequentiality (Fig. 2b).

Activation of tasks may be modeled very easily. Fig. 3a shows a periodic activation and Fig. 3b a delayed periodic activation. Cyclic activations are shown on Fig. 3c.

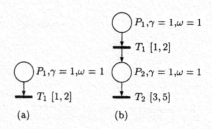

Fig. 2. Basic task model

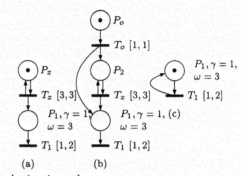

Fig. 3. Activation schemes

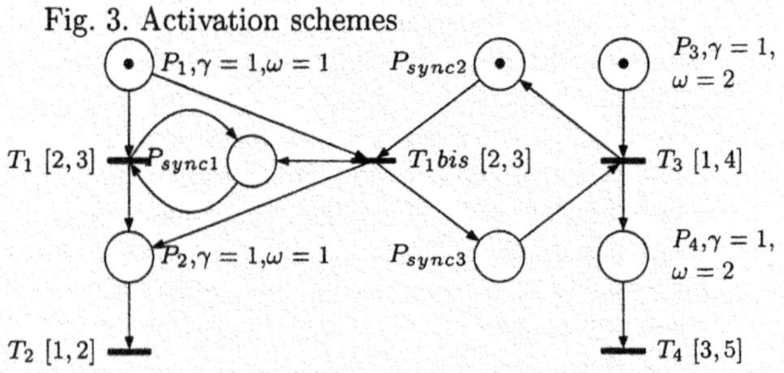

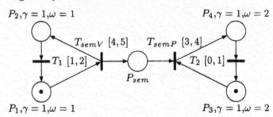

Fig. 4. Synchronization with memorized events

Fig. 5. Two cyclic tasks synchronized *via* a semaphore

3.2 Concurrent tasks

3.2.1. Basic synchronization
Real-time executives provide several services for synchronization. In particular, semaphores and events are widely used. Fig. 5 shows a model for Dijkstra's semaphores and Fig. 4 proposes a model for memorized events as found in OSEK/VDX, for instance. Higher level or other synchronization mechanisms are as easily modelisable.

3.2.2. Shared resources access
Access to critical resources may involves a mutual exclusion mechanism. This is usually done using a semaphore (Fig. 6). However, it is well-known that this policy may result in deadlocks, so real-time execu-

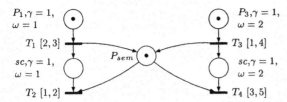

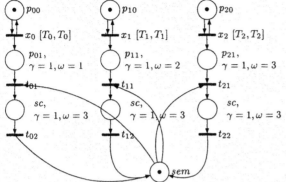

Fig. 6. Semaphore for mutual exclusion

Fig. 7. Priority Ceiling Protocol

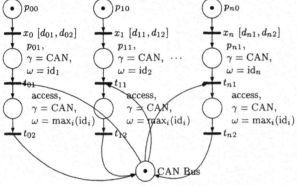

Fig. 8. CAN bus access

tives provide some higher level protocol to handle shared resources access. The most used, in OSEK/VDX ((*OSEK/VDX specification, 2001*)) for instance, is Priority Ceiling Protocol. It can be modelised by SETPN with the pattern in Fig. 7.

3.3 Messaging

We provide here, as an example, a model for the messaging component of a distributed real-time system in which tasks communicate with a CAN bus. This is a high-level modelisation of the CAN protocol, since we only consider that the bus is a shared resource, for which the access is made according to the priorities corresponding to the station number of each node. A much more precise modelisation has been done in (Juanole, 1999), which completly modelize the MAC sublayer of the CAN protocol.

4. ANALYSIS

In order to analyze a time Petri net, the computation of its reachable state space is required.

However, the reachable state space of a time Petri net is obviously infinite, so a method has been proposed to partition it in a finite set of infinite *state classes* (Berthomieu and Diaz, 1991). This method is briefly explained in the next subsection. The following paragraphs then describe its extension in order to compute the state space of SETPN.

4.1 State class graph of a time Petri net

Given a bounded time Petri net, Berthomieu and Diaz have proposed a method for computing the state space as a finite set of *state classes* (Berthomieu and Diaz, 1991). Basically a state class contains all the states of the net between the firing of two consecutive transitions.

Definition 3. (State class). A *state class* C, of a time Petri net, is a pair (M, D) where M is a marking of the net and D a set of inequalities called the *firing domain*.

The inequalities in D are of two types (Berthomieu and Diaz, 1991)

$$\begin{cases} \alpha_i \leq x_i \leq \beta_i \ (\forall i \text{ such that } t_i \text{ is enabled}), \\ -\gamma_{kj} \leq x_j - x_k \leq \gamma_{jk}, \ \forall j, k \text{ such that} \\ j \neq k \text{ and } (t_j, t_k) \in enabled(M)^2 \end{cases}$$

x_i is the firing time of the enabled transition t_i relatively to the time when the class was entered in.

Because of their particular form, the firing domains may be encoded using DBMs (Dill, 1989), which allow the use of efficient algorithms for the computation of the classes.

Given a class $C = (M, D)$ and a firable transition t_f, computing the successor class $C' = (M', D')$ obtained by firing t_f is done by:

(1) Computing the new marking as for classical Petri nets: $M' = M - {}^\bullet t_f + t_f{}^\bullet$
(2) Making variable substitutions in the domain: $\forall i \neq f, x_i \leftarrow x_i' + x_f$
(3) Eliminating x_f from the domain using for instance the Fourier-Motzkin method ((Dantzig, 1963))
(4) Computing a canonical form of the new domain using for instance the Floyd-Warshall algorithm ((Berthomieu, 2001))

With the method for obtaining the successors of a class, computation of the state space of the TPN consists merely of the classical building of the reachability graph of state classes. That is to say, that starting from the initial state class, all the successors obtained by firing firable transitions are computed iteratively until all the produced successors have already been generated.

The semantics of definition 2 obviously implies that the domain of state classes cannot be computed for SETPNs as for classical TPNs. However, minor changes are proposed in (Roux and Déplanche, 2002) to allow its construction.

Precisely, the variable substitution in the firing domain is now only done for *active transitions*. Moreover, when determining firable transitions, the *active firing domain* must be considered *i.e.* the firing domain restricted to variables corresponding to active transitions.

While we still define a state class of a SETPN as a marking and a domain, the general form of the domain is not preserved. The new general form is that of a polyhedron with constraints involving up to n variables, with n being the number of transitions enabled by the marking of the class:

$$\begin{cases} \alpha_i \leq \theta_i \leq \beta_i, \forall t_i \in enabled(M), \\ a_{i_1}\theta_{i_1} + \cdots + a_{i_n}\theta_{i_n} \leq \gamma_{i_1\ldots i_n}, \\ \forall(t_{i_1},\ldots,t_{i_n}) \in enabled(M)^n \\ \text{and with } (a_{i_0},\ldots,a_{i_n}) \in \mathbb{Z}^n. \end{cases}$$

The following paragraphs give the details of the computation of the new domain on a semi-general example. We start from a DBM-like class to show how additional constraints appear.

Let $C = (M, D)$ be a state class of a SETPN, such that

$$D = \begin{cases} \alpha_1 \leq x_1 \leq \beta_1, \\ \alpha_2 \leq x_2 \leq \beta_2, \\ \alpha_3 \leq x_3 \leq \beta_3, \\ \alpha_4 \leq x_4 \leq \beta_4, \\ -\gamma_{21} \leq x_1 - x_2 \leq \gamma_{12}, \\ -\gamma_{31} \leq x_1 - x_3 \leq \gamma_{13}, \\ -\gamma_{41} \leq x_1 - x_4 \leq \gamma_{14}, \\ -\gamma_{32} \leq x_2 - x_3 \leq \gamma_{23}, \\ -\gamma_{42} \leq x_2 - x_4 \leq \gamma_{24}, \\ -\gamma_{43} \leq x_3 - x_4 \leq \gamma_{34} \end{cases} \quad (1)$$

We also suppose that t_1, t_2 and t_3 (corresponding to variables x_1, x_2, x_3) are active, and that t_4 is not. Additionally, we suppose that t_1 is firable and we compute the domain of the class obtained by firing t_1. The first step is the variable substitution $x_i \leftarrow x_i' + x_1$ for all active transitions but t_1. The domain becomes

$$\begin{cases} \alpha_1 \leq x_1 \leq \beta_1, \\ \alpha_2 \leq x_1 + x_2' \leq \beta_2, \\ \alpha_3 \leq x_1 + x_3' \leq \beta_3, \\ \alpha_4 \leq x_4' \leq \beta_4, \\ -\gamma_{21} \leq x_1 - x_1 - x_2' \leq \gamma_{12}, \\ -\gamma_{31} \leq x_1 - x_1 - x_3' \leq \gamma_{13}, \\ -\gamma_{41} \leq x_1 - x_4 \leq \gamma_{14}, \\ -\gamma_{32} \leq x_1 + x_2' - x_1 - x_3' \leq \gamma_{23}, \\ -\gamma_{42} \leq x_1 + x_2' - x_4 \leq \gamma_{24}, \\ -\gamma_{43} \leq x_1 + x_3' - x_4 \leq \gamma_{34} \end{cases} \quad (2)$$

The next step is the elimination of the variable x_1. We use the Fourier-Motzkin method. For that, we rewrite the inequations as follows :

$$\begin{cases} \alpha_1 \leq x_1, & x_1 \leq \beta_1, \\ \alpha_2 - x_2' \leq x_1, & x_1 \leq \beta_2 - x_2', \\ \alpha_3 - x_3' \leq x_1, & x_1 \leq \beta_3 - x_3', \\ -\gamma_{41} + x_4, \leq x_1 & x_1 \leq \gamma_{14} + x_4, \\ -\gamma_{42} + x_4 - x_2' \leq x_1, & x_1 \leq \gamma_{24} + x_4 - x_2', \\ -\gamma_{43} + x_4 - x_3' \leq x_1, & x_1 \leq \gamma_{34} + x_4 - x_3', \\ \alpha_4 \leq x_4' \leq \beta_4, \\ -\gamma_{32} \leq x_2' - x_3' \leq \gamma_{23}, \\ -\gamma_{21} \leq -x_2' \leq \gamma_{12}, \\ -\gamma_{31} \leq -x_3' \leq \gamma_{13} \end{cases}$$

$$(3)$$

The Fourier-Motzkin method then consists in writing that the system has solutions if and only if the lowers bounds of x_1 are less or equal to the upper bounds. The obtained system is then equivalent to the initial one. After a few simplifications we obtain

$$\begin{cases} -\gamma_{12} \leq x_2' \leq \gamma_{21}, \\ -\gamma_{13} \leq x_3' \leq \gamma_{31}, \\ \alpha_4 \leq x_4 \leq \beta_4, \\ -\gamma_{32} \leq x_2' - x_3' \leq \gamma_{23}, \\ -\gamma_{42} - \beta_1 \leq x_2' - x_4 \leq \gamma_{24} - \alpha_1, \\ -\gamma_{43} - \beta_1 \leq x_3' - x_4 \leq \gamma_{34} - \alpha_1, \\ \alpha_2 - \gamma_{14} \leq x_2' + x_4 \leq \beta_2 + \gamma_{41}, \\ \alpha_3 - \gamma_{14} \leq x_3' + x_4 \leq \beta_3 + \gamma_{41}, \\ \alpha_2 - \gamma_{34} \leq x_2' + x_4 - x_3' \leq \beta_2 + \gamma_{43}, \\ \alpha_3 - \gamma_{24} \leq x_3' + x_4 - x_2' \leq \beta_3 + \gamma_{42} \end{cases} \quad (4)$$

The final step consists in computing a canonical form of this domain and will not be detailed here. What we can see here is that eight inequations, given on the four last lines, are generated, which cannot be expressed with a DBM. Furthermore, we can easily see that those new inequations may give even more complex inequations (*i.e.* involving more variables) when firing another transition for the obtained domain.

4.3 Overapproximation

Manipulating polhydra in the general case involves a very important computing cost. In order

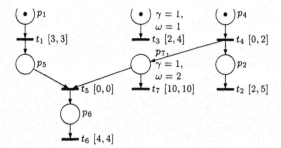

Fig. 9. A Scheduling Extended Time Petri Net

to be able to keep our algorithms efficient for SETPN, we approximate the polyhedron representing the firing domain to the smallest DBM containing it. By doing this, we clearly add states in our classes that should not be reachable and thus we do an overapproximation. This is illustrated by the net in Figure 9. After the firing sequence t_4, t_1, t_5, the transition t_6 is not firable because either t_2 or t_3 (depending on the firing time of t_4) must be fired first. The class obtained is:

$$\begin{cases} \{p_2, p_3, p_6\}, \\ \begin{cases} 0 \leq x_2 \leq 4, \\ 0 \leq x_3 \leq 4, \\ 4 \leq x_6 \leq 4, \\ -4 \leq x_2 - x_3 \leq 4, \\ -4 \leq x_2 - x_6 \leq 0, \\ -4 \leq x_3 - x_6 \leq 0, \\ 1 \leq x_2 + x_3 \leq 7 \end{cases} \end{cases}$$

We can easily see that t_6 is indeed not firable, for this implies $x_2 = x_3 = x_6 = 4$ and thus $x_2 + x_3 = 8$. But if we remove the $x_2 + x_3 \leq 7$ constraint in order to keep a DBM form, t_6 becomes firable. So we have here an overapproximation.

However, for the verification of safety properties the overapproximation is not a too big concern. Since we want to ensure that something "bad" never happens, we only need to check a set of states which contains the actual state space of the SETPN. Of course, there is still a risk of being pessimistic.

In addition, for the subclass of SETPN with fixed firing times on transitions, the following theorem holds:

Theorem 1. Let $\mathcal{T} = (P, T, {}^{\bullet}(.), (.)^{\bullet}, \alpha, \beta, M_0, Act)$ be a SETPN. If $\alpha = \beta$, then inequations in all state classes are reduced to equalities.

Proof. This will be shown by induction: since $\alpha = \beta$, the initial state class domain can be written

$$\begin{cases} \theta_{i_0} = \alpha(t_{i_0}), \\ \vdots \\ \theta_{i_n} = \alpha(t_{i_n}) \end{cases}$$

So the claim of the theorem holds for the initial state class. Let $C = (M, D)$ be a state class such

that

$$D = \begin{cases} \theta_{i_0} = \alpha_{i_0}, \\ \vdots \\ \theta_{i_n} = \alpha_{i_n} \end{cases}$$

Let us suppose that t_{i_n} is a firable transition from C and $C' = (M, D')$ the class obtained by firing t_{i_n} from C. D' is computed by the four following steps:

(1) for all active transitions, say $t_{i_0}, \ldots, t_{i_m}$ with $m < n$, for instance, the variable substitution $\theta = \theta' + \theta_{i_n}$ is made,

(2) Disabled transitions (including t_{i_n}) are eliminated from the system of equalities. The resulting system is obviously still composed of equalities,

(3) Inequalities relative to the newly enabled transitions are added. Since, $\alpha = \beta$, these inequalities are actually equalities,

(4) The canonical form is not needed since the system of equalities can only be written in one way.

$\square$

As a consequence, we never have overapproximations with that subclass of SETPN. Actually, DBMs are not even required to code that computations.

5. SCHEDULABILITY VERIFICATION

Using the classical observer notion, we can verify varied timed accessibility properties. In order to verify the schedulability, we need a special kind of observers that we formally describe in the following section. While classical observers (Toussaint *et al.*, 1997) give a boolean response for a given property, thanks to these observers we are also able to compute the response time of the tasks that we have modelised and then to compare with other approaches.

5.1 Response time observers

Definition 4. (observer). Let a scheduling extended time Petri net
$\mathcal{T} = (P, T, {}^{\bullet}(.), (.)^{\bullet}, \alpha, \beta, M_0, Act)$, and two transitions $\{in, out\} \in T$.
$\mathcal{T}_{\mathcal{O}_{in,out}} = (P_o, T, {}^{\bullet}_o(.), (.)^{\bullet}_o, \alpha, \beta, M_{o0})$ is the TPN $\mathcal{T}$ observed by the observer
$\mathcal{O}_{in,out}(\mathcal{T}) = (p_{obs}, x, m_{obs})$ where

- $P_o = P \cup \{p_{obs}\}$,
- $M_0(p_{obs}) = m_{obs}$,
- ${}^{\bullet}_o out = {}^{\bullet}out + p_{obs}$ and $\forall t \in T - \{out\}$, ${}^{\bullet}_o t = {}^{\bullet}t$,
- $in^{\bullet}_o = in^{\bullet} + p_{obs}$ and $\forall t \in T - \{in\}$, $t^{\bullet}_o = t^{\bullet}$,

- $\forall M \in \mathbb{N}^P, Act_o(M(p_{obs})) = M(p_{obs})$ and $\forall p \in P, Act_o(M(p)) = Act(M(p))$.

We note $\mathcal{Q}$ the set of state $Q = (M, \nu, \nu_x)$ of the observed TPN $\mathcal{T}_{\mathcal{O}_{in,out}}$. The semantics of an observed time Petri net $\mathcal{T}_{\mathcal{O}_{in,out}}$ is the semantics of the TPN $\mathcal{T}$ with the following modification : The continuous transition relation is defined $\forall d \in \mathbb{R}^+$ by :

$$(M, \nu, \nu_x) \xrightarrow{d} (M, \nu', \nu'_x) \text{ iff}$$
$$\begin{cases} \forall t_i \in T \begin{cases} \nu'(t_i) = \begin{cases} \nu(t_i) \text{ if } Act(M) < {}^\bullet t_i \\ \wedge M \geq {}^\bullet t_i \\ \nu(t_i) + d \text{ otherwise }, \end{cases} \\ M \geq {}^\bullet t_i \Rightarrow \nu'(t_i) \leq \beta(t_i) \end{cases} \\ M(p_{obs}) \geq 1 \Rightarrow \nu'_x = \nu_x + d \\ M(p_{obs}) = 0 \Rightarrow \nu'_x = \nu_x = 0 \end{cases}$$

As the transition *in* models the request of a task t_a and the transition *out* its termination, we can used this observer to determine the response time of t_a and to check that the task respects its deadline.

For a schedulable task such that $M(p_{obs}) \leq 1$, the response time of a task *task* is :

$$t_r(task) = \max_{\mathcal{Q}}(\nu_x)$$

One can notice that since inequations in the domain cannot be strict, a task for which a null execution time remains may be preempted while it has not done the discrete transition corresponding to its termination.

5.2 Experimental results

For testing, we have compared the results of the computation of response times given by a modelisation with HyTech (Henzinger *et al.*, 1997) for an exact result on relatively small but complex models, analytical methods for bigger but simpler models (independent tasks) and our method using the DBM overapproximation.

Experimentation on several hundreds of examples of varied models of real-time systems with non-fixed execution times gives the same results for the three (or two when HyTech or analytical method is not adapted) methods. This allows us to think that there is a relatively big subclass of SETPN for which the DBM algorithm is exact, or at least that the approximation is quite small on classical real-time applications.

6. CONCLUSION

In this paper, we have shown how to express most of the features of real-time systems with SETPN, from basic activation schemes for tasks

to complex ressource access protocols. We also provide a theoretical framework for analysis of SETPN. In particular, we have given an exact method for computing the state space of a SETPN as well as a faster method at the cost of an overapproximation. While this approximation is not quantified, experimental results allow us to think it is not very important. We have also given a formal description of a special type of observers that allow us to compute response times of tasks modeled by SETPN.

Further work includes the identification of the class of SETPN for which there is no overapproximation, extension of the modelisation for the round-robin and dynamic scheduling policies and the generation of the state-space as timed automaton.

REFERENCES

Altisen, K., G. Goßler, A. Pnueli, J. Sifakis, S. Tripakis and S. Yovine (1999). A framework for scheduler synthesis. In: *20th IEEE Real-Time Systems Symposium (RTSS'99)*. IEEE Computer Society Press. Phoenix, Arizona, USA. pp. 154–163.

Altisen, K., G. Goßler and J. Sifakis (2000). A methodology for the construction of scheduled systems. In: *6th International Symposium on Formal Techniques in Real-Time and Fault-Tolerant Systems (FTRTFT'00)*. Vol. 1926 of *Lecture Notes in Computer Science*. Springer-Verlag. Pune, India. pp. 106–120.

Berthomieu, B. (2001). La méthode des classes d'états pour l'analyse des réseaux temporels. In: *3e congrès Modlisation des Systèmes Réactifs (MSR'2001)*. Hermes. Toulouse, France. pp. 275–290.

Berthomieu, B. and M. Diaz (1991). Modeling and verification of time dependent systems using time petri nets. *IEEE transactions on software engineering* **17**(3), 259–273.

Dantzig, G. B. (1963). Linear programming and extensions. *IEICE Transactions on Information and Systems*.

Dill, D. L. (1989). Timing assumptions and verification of finite-state concurrent systems. In: *Workshop Automatic Verification Methods for Finite-State Systems*. Vol. 407. pp. 197–212.

Fersman, E., P. Petterson and W. Yi (2002). Timed automata with asynchronous processes : Schedulability and decidability. In: *8th International Conference on Tools and Algorithms for the Construction and Analysis of Systems (TACAS'02)*. Vol. 2280 of *Lecture Notes in Computer Science*. Springer-Verlag. Grenoble, France. pp. 67–82.

Grolleau, Emmanuel and Annie Choquet-Geniet (2002). Off-line computation of real-time schedules using petri nets. *Discrete Event Dynamic Systems* **12**(3), 311–333.

Harbour, M.G., M.H. Klein and J.P. Lehoczky (1991). Fixed priority scheduling of periodic tasks with varying execution priority. In: *12th IEEE Real-Time Systems Symposium (RTSS'91)*. IEEE Computer Society Press. San Antonio, USA. pp. 116–128.

Henderson, W., D. Kendall and A. Robson (2001). Improving the accuracy of scheduling analysis applied to distributed systems. *Real-Time Systems* **20**(1), 5–25.

Henzinger, T.A., P.-H. Ho and H. Wong-Toi (1997). Hytech: A model checker for hybrid systems. *Journal of Software Tools for Technology Transfer* **1**(1-2), 110–122.

Hladik, P.-E. and A.-M. Déplanche (2003). Analyse d'ordonnançabilité de tâches temps-réel avec offset et gigue. In: *11th international Conference on Real-Time Systems (RTS'03)*. Paris, France. p. to appear.

Janicki, R. and M. Koutny (1999). On causality semantics of nets with priorities. *Fundamenta Informaticae* **38**(3), 223–255.

Juanole, G. (1999). Modélisation et évaluation du protocole mac du réseau can. In: *Rapport LAAS No99303. Ecole d'Eté Applications, Réseaux et Systémes (ETR'99)*. Poitiers, France. pp. 187–200.

Larsen, K. G., P. Pettersson and W. Yi (1995). Model-checking for real-time systems. In: *Fundamentals of Computation Theory*. pp. 62–88.

Okawa, Y. and T. Yoneda (1996). Schedulability verification of real-time systems with extended time petri nets. *International Journal of Mini and Microcomputers* **18**(3), 148–156.

OSEK/VDX specification (2001). http://www.osek-vdx.org.

Palencia, J.C. and M.G. Harbour (1998). Schedulability analysis for tasks with static and dynamic offsets. In: *19th IEEE Real-Time Systems Symposium (RTSS'98)*. IEEE Computer Society Press. Madrid, Spain. pp. 26–37.

Palencia, J.C. and M.G. Harbour (1999). Exploiting precedence relations in the scheduling analysis of distributed real-time systems. In: *20th IEEE Real-Time Systems Symposium (RTSS'99)*. IEEE Computer Society Press. Phoenix, Arizona, USA. pp. 328–339.

Robert, P.H. and G. Juanole (2000). Modélisation et vérification de politiques d'ordonnancement de tâches temps-réel. In: *8th Colloque Francophone sur l'Ingénierie des Protocoles (CFIP'00)*. Hermes. Toulouse, France. pp. 167–182.

Roux, O. H. and A.-M. Déplanche (2002). A t-time petri net extension for real time-task scheduling modeling. *European Journal of Automation (JESA)*.

Tindell, K. (1994a). Adding time-offsets to schedulability analysis. Technical Report YCS-94-221. University of York, Computer Science Departement.

Tindell, K. (1994b). Fixed priority scheduling of hard real-time systems. PhD thesis. Department of Computer Science. University of New York.

Tindell, K. and J. Clark (1994). Holistic schedulability analysis for distributed hard real-time systems. *Microprocessing and Microprogramming* **40**(1-2), 117–134.

Toussaint, J., F. Simonot-Lion and Jean-Pierre Thomesse (1997). Time constraint verifications methods based time petri nets. In: *6th Workshop on Future Trends in Distributed Computing Systems (FTDCS'97)*. Tunis, Tunisia. pp. 262–267.

ELSEVIER

IFAC
PUBLICATIONS
www.elsevier.com/locate/ifac

SIMHOL – A GRAPHICAL SIMULATOR FOR THE JOINT SCHEDULING OF MESSAGES AND TASKS IN DISTRIBUTED EMBEDDED SYSTEMS

Mário J. Calha

Politechnic Institute of Castelo Branco - EST
Av. Do Empresario
6000 Castelo Branco, Portugal

José Alberto Fonseca

Dep. Electronic and Telecomunications
Of University of Aveiro
Campus Santiago
3800 Aveiro, Portugal

Abstract: In this paper a simulator to preview the timeliness of the transmission of messages and of the execution of tasks in a distributed system is presented. The simulator, called SIMHOL, builds on previous work by the authors in which a simple mechanism to dispatch tasks and messages was proposed for CAN-based distributed systems. The inputs to the simulator are the so-called data streams, which include the producer tasks, the correspondent messages and the tasks that use the transmitted data. Using the worst-case execution time and transmission time, the simulator is able to verify if deadlines are fulfilled in every node of the system and in the network. Besides discussing the simulator construction and operation, the paper presents some examples of distributed systems requirements and the results obtained by using the tool to analyze the respective timeliness. This is also used to illustrate the outputs of the simulator. One important issue also discussed is the easy way, due to the object oriented approach chosen, to extend the simulation to cover different networks and to use different scheduling techniques either at the task level or at the message level. *Copyright © 2003 IFAC*

Keywords: Simulators, Scheduling, Algorithms, Fieldbus.

1. INTRODUCTION

In previous works we've proposed a simple mechanism for distributed systems based in nodes with low computational power that allowed the dispatching of tasks and messages in a time-triggered manner (Calha and Fonseca 2002; Fonseca, *et al.*, 2002). These works led to data flow analysis and to the derivation of requirements to the tasks and messages parameters.

The aim of this work is to demonstrate the validity of the proposed analysis and requirements presenting a simulator.

The proposed simulator, SimHol, is an experimental platform for a centralized task and message scheduling. The supported architecture is based in a distributed control system with a centralized dispatcher. A central node dispatches both tasks and messages to other nodes.

For this work, other simulators were studied.
Henderson, *et al.*, (1998) have developed a design tool, Xrma, that supports the schedulability analysis of hard, uni-processor and distributed real-time systems. Xrma automates the analysis and supports performance verification of diverse real-time systems composed of tasks executing on multiple processors which communicate using the CAN fieldbus.

Vector Informatik GmbH has developed CANalyzer/DENalyzer (Vector Informatik GmbH, 2002) that is a universal development tool for bus systems. CANalyzer/DENalyzer conforms to CAN specification V 2.0 B and makes it easy to observe, analyze and add to bus traffic on as many as 32 channels.

Our approach is different because this simulator supports different architectures and scheduling algorithms. The SimHol uses the flexible time-triggered (FTT) paradigm for scheduling both tasks and messages.

This paper is organized as follows. Section 2 details some background information about the use of the FTT-CAN to the joint dispatching of tasks and messages. Section 3 explains the SimHol internals. Section 4 explains how to use the simulator and describes some experiments. In the last section some conclusions are drawn and new directions are presented.

2. USING FTT-CAN TO THE JOINT DISPATCHING OF TASKS AND MESSAGES

In a typical distributed control system a process is controlled by a closed loop, where a signal is acquired, some data is processed and an actuator is activated. Each node in a system represents a processing unit with private memory. All the nodes are connected with a common bus. At each node there is, at least, a task that produces and/or consumes one or more messages. In a real world system there are, usually, several tasks communicating. There is some probability that the producer tasks try to send a message at approximately the same time, in a way that a collision would occur. To avoid such a collision, some kind of synchronization is required. Now if the system is expanded to a higher number of nodes, each with several tasks, all communicating through the same bus, the probability of collisions is much higher. The problem is how to guarantee a global synchronization without compromising system responsiveness to external events.

In order to describe a system like this, tasks, messages and their relations have to be characterized. In a distributed system, the interaction between tasks can be classified according to 2 basic types: stand-alone and interactive. In the first type are included the tasks that perform some kind of closed-loop control and don't communicate with other tasks. While the tasks that exchange data with other tasks are included in the second type. Each interactive task communicates with other tasks in the system, using the message-passing paradigm, and can be decomposed to a simpler form where, at most, they produce and/or consume a single message.

The set of all the synchronous tasks in the system can be typically expressed as:

$$S_T = \{ ST_i : (C_i, T_i, D_i, Pr_i), i=1..N_{st} \}$$

Where Nst is the number of synchronous tasks, and each task ST_i is characterized by the following parameters: Ci – The worst-case execution time (on each release); Ti – The period; Di – The deadline measured relatively to the release instant; Pri – The priority.

In order to achieve a global synchronization, other parameters have to be defined, namely: Ni – Node where the task runs; Phi – The relative phasing, which determines the first release instant after system boot; MPi – Message produced (only for interactive tasks); MCi – Message consumed (only for interactive tasks). Considering the new parameters, the set of all the synchronous tasks in the system can now be expressed as:

$$S_T = \{ ST_i : (C_i, T_i, D_i, Pr_i, N_i, Ph_i, MP_i, MC_i), \\ i=1..N_{st} \}$$

Every interactive task uses messages to exchange data with other tasks. Particularly, the periodic message set can be typically expressed as:

$$S_M = \{ SM_m : (C_m, T_m, D_m, Pr_m), m=1..N_{sm} \}$$

Where Nsm is the number of synchronous messages, and each message SM_m is characterized by the following parameters: Cm – The maximum transmission time; Tm – The period; Dm – The deadline measured relatively to the release instant; Prm – The priority.

In order to achieve a global synchronization, other parameters have to be defined, namely: Phm – The relative phasing, which determines the first release instant, after system boot; PTm – Producer task; $CTLm,i$ – Consumer task list. Considering the new parameters, the set of all the synchronous messages in the system can now be expressed as:

$$S_M = \{ SM_m : (C_m, T_m, D_m, Pr_m, Ph_m, PT_m, \\ CTL_{m,i}), m=1..N_{sm}, i=1..N_{sct} \}$$

Where $Nsct$ is the number of synchronous consumer tasks.

A data stream is a group of interacting tasks that starts with a producer task and finishes with a consumer task. Producer/consumer tasks will appear inside the data streams. An example of data streams is shown in Fig. 1. The analysis of the data streams leads to the definition of the relative phasing, Ph, of each task and message. For this calculus the Ci, Di, Cm and Dm must be considered. For each stream these task and message parameters must be summed resulting in an accumulated execution and transmission times. This accumulated execution time is the worst-case time because the deadlines are considered.

Fig. 1. Examples of data streams.

In a data stream a task may communicate with several tasks with the same message. This means that either the message should be consumed by every consumer task, or it must be buffered in every node with a consumer task for the message.

In order to guarantee a global (tasks and messages) synchronization a reliable mechanism to dispatch both tasks and messages is required. In (Calha and Fonseca, 2002), a distributed architecture and a dispatching scheme were proposed to solve the problem.

The architecture uses a special node, known as the Master, which is added to the remaining nodes (Stations) that form the distributed control system (Fig. 2). The Master triggers the execution of tasks in the Stations and the exchange of messages in the network, in a time-triggered manner.

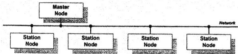

Fig. 2. Distributed system architecture showing the Master and Stations.

Each Station node acts upon a trigger event and, in accordance, dispatches any task or message. The Stations can have a variable number of tasks to be executed and can produce both synchronous and asynchronous/sporadic messages (Fig. 3). The network acts as a triggering vehicle for tasks and messages.

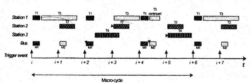

Fig. 3. Triggering of task execution and message sending.

Currently this architecture can be implemented using the Controller Area Network (CiA, 1994; Tindell, et al., 1994) and the FTT-CAN protocol (Almeida, et al., 2002) to trigger the dispatching. It is also possible to use FTT-Ethernet (Pedreiras, et al., 2002) for the same purpose.

A common property of the possible implementations is the fixed periodicity of the trigger events. The period value, usually designated by EC – Elementary cycle, imposes restrictions to some of the parameters of tasks and messages. This is the case of the periods (Ti and Tm), the deadlines (Di and Dm) and the relative phasing (Phi and Phm) which must be integer multiples of the EC.

A Simulator, SimHol, that has as input a system description using the above entities (nodes, tasks and messages) and allows the task and message scheduling is presented in the next chapter.

3. SIMHOL

The main concern for the development of this simulator was to use an object oriented approach due to its well known advantages. Two sets of requirements were identified, namely: external and internal.

The external requirements were:
- Simple interface;
- Input scenarios using plain text files;
- All simulation data should be output to plain text files;
- Fast execution time.

The internal requirements were:
- To allow the integration of different node interconnection technologies, currently Controller Area Network (CAN) is already built-in;
- To allow the integration of different algorithms for the scheduling of tasks and messages, currently Rate Monotonic (RM) and Earliest Deadline First (EDF) are already built-in;
- Function operation as close to actual working system as possible;
- Clear separation between the core elements and the user interface, allowing an easy porting to other platforms;
- Exception handling, providing a good fault coverage resulting in a smooth operation.

The static diagram is organized in four areas:
- Entities,
- Algorithms,
- Architectures and
- Simulator

The entities are: nodes, tasks and messages.
The static diagram showing simplified classes (stripped of more internal attributes and functions) is depicted in Figs. 4, 5 and 6.

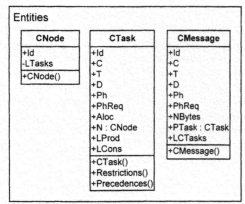

Fig. 4. Static diagram of entities.

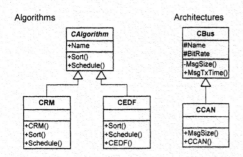

Fig. 5. Static diagram of algorithms and architectures.

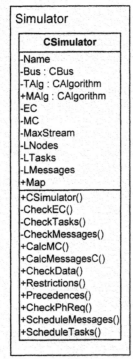

Fig. 6. Static diagram of the simulator.

2.1. Object oriented language

Due to the speed advantage, the simulator was developed using the C++ programming language instead of Java, but a future Java implementation is being considered due to its benefits in what concerns portability. All the features of exception handling were used to cover the critical areas of the code.

2.2. Scheduler

The scheduler is invoked every EC. At every invocation the scheduler considers all entities to be scheduled. This can be either the tasks of a node or the messages to be transmitted. The simulator offers the possibility of using features such as:

- To impose deadlines greater than the periods, and to set the maximum number of delayed tasks/messages;
- To set priority inversion on/off;
- To break the task execution for several ECs, that may not be adjacent;
- To break the message transmission for several ECs, that may not be adjacent.

With this simulator it is possible to test the system allowing priority inversion, or not. If priority inversion is allowed then the simulator tries to use the remaining time in each EC to fit whatever message is waiting to be transmitted, independently of its priority. If priority inversion is not allowed then if the highest priority message, which is ready, can't fit in the remaining time in the EC, this interval is left unused (inserted idle time).

If a task is allowed to be executed across several ECs, than the MayBreak parameter is set to TRUE. Because of this a task may be pre-empted due to the arrival of another task with higher priority. In this situation the context switch overhead must be considered.

If a message is allowed to be transmitted across several ECs, than the MayBreak parameter is set to TRUE. In this situation the kernel has to be aware of this possibility and break the message in smaller pieces. This overhead has to be considered. This technique may be acceptable when the majority, and most frequently used, messages are short and a few, and rarely used, are much longer. In this case the EC might be chosen to best suit the needs of the short messages but without this possibility the transmission of long messages might become impossible.

The flowcharts for the task scheduling are depicted in Fig 7.

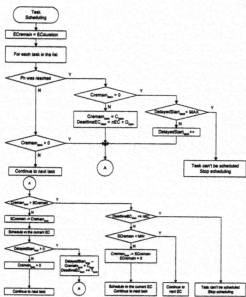

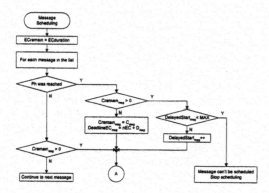

Fig. 7. Task scheduling flowcharts.

To schedule a task several parameters need to be checked, these are:

- the task's initial phase (Ph),
- the task's EC deadline (DeadlineEC),
- the task's remaining execution time (Cremain),
- the remaining time in the current EC (ECremain).

Apart from these, another important variable is the DelayedStart that is used whenever a new task is scheduled to start but a previous instance of the same task is still running. In this situation the new instance is marked for execution as soon as the previous instance finishes.

Scheduling continues until either the simulation ends or a deadline is missed. If the deadline was reached and the task is unable to finish in the current EC then the scheduler stops due to a missed deadline.

The flowcharts for the message scheduling are depicted in Fig. 8.

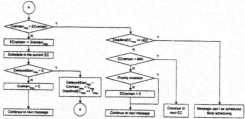

Fig. 8. Message scheduling flowcharts.

The message scheduling is very similar to the task scheduling but priority inversion is also considered.

2.3. Upgrading

Currently all upgrades, to architectures and to algorithms, are accomplished through a source file that is added to the SimHol project and to minor changes in the interface. So this means that the project needs to be re-compiled.

2.3.1 Upgrading architectures

In order to add another architecture a new class has to be defined. This class inherits the CBus class and has to implement the method *MsgSize(unsigned NBytes)* that returns the number of bits necessary to transmit a message in this architecture. The *Bus* class is shown bellow:

```
class CBus
{
  virtual unsigned MsgSize(unsigned NBytes);

protected:
  AnsiString Name;
  unsigned BitRate;

public:
  unsigned GetBitRate() { return BitRate; }
  AnsiString GetName() { return Name; }

  unsigned MsgTxTime(unsigned);
  void CheckBitRate();
};
```

The CAN architecture class is defined as:

```
class CCAN   public CBus
{
  unsigned MsgSize(unsigned);
public:
  CCAN(unsigned BR) { Name="CAN"; BitRate=BR;}
};
```

2.3.2 Upgrading scheduling algorithms

In order to add another scheduling algorithm a new class has to be defined. This class inherits the CAlgorithm class and has to implement the methods *Sort()* and *Schedule()*. The *Algorithm* class is shown bellow:

203

```
class CAlgorithm
{
protected:
  AnsiString Name;

public:
  AnsiString GetName() { return Name; }
  virtual void Sort(TList*) {};
  virtual unsigned Schedule(TList*, unsigned*,
  unsigned*, unsigned*, unsigned, unsigned, char*, char,
  char) {return 0;}
};
```

One of the algorithm classes, in this case the CRM (refers to the Class Rate Monotonic), is defined as:

```
class CRM : public CAlgorithm
{
public:
  CRM() { Name="RM"; }
  void Sort(TList*);
  unsigned Schedule(TList*, unsigned*, unsigned*,
  unsigned*, unsigned, unsigned, char*, char, char);
};
```

2.3.3 Connecting with other software

Due to the plain text and normalized data output to files, it's easy to develop tools that grab this data and make further analysis allowing a simple integration on a broader software suite. The Fig. 9 shows an extract of an output file containing simulation data.

```
--- Nodes List ---
N Id: 1
N Id: 2
N Id: 3
N Id: 4
--- Tasks List ---
T Id: 1 C:   520(  1) T:  3120(  2) D:  3120(  2) PhReq:    0(   0)
T Id: 2 C:  1040(  1) T:  3120(  2) D:  3120(  2) PhReq:    0(   0)
T Id: 3 C:  1248(  1) T:  3120(  2) D:  3120(  2) PhReq:    0(   0)
T Id: 4 C:   390(  1) T:  3120(  2) D:  3120(  2) PhReq:    0(   0)
T Id: 5 C:  2184(  2) T:  4680(  3) D:  6240(  4) PhReq:    0(   0)
T Id: 6 C:  1404(  1) T:  4680(  3) D:  4680(  3) PhReq:    0(   0)
--- Messages List ---
M Id: 1 C:   520(  1) T:  3120(  2) D:  1560(  1) PhReq:    0(   0)
M Id: 2 C:   520(  1) T:  3120(  2) D:  1560(  1) PhReq:    0(   0)
M Id: 3 C:   520(  1) T:  4680(  3) D:  1560(  1) PhReq:    0(   0)
--- Scheduling Map ---
EC   T  T  N  T  N  T  N  T  N  M  M  M Bus(%)
 0   1  5 100  0  0  0  0  0  0  0  0  0  0
 1   0  5 73   0  0  0  0  0  0  0  0  0  0
 2   1  0 33   0  0  0  0  0  0  1  0  0 33
 3   0  5 100  2 66  0  0  0  0  0  0  3 66
 4   1  5 73   0  0  0  0  0  0  0  2  0 33
 5   0  0  0  2 66  0  0  0  6 90  0  2  0 33
 6   1  5 100  0  0  3 80  4  0 25  1  0  0 33
 7   0  5 73   2 66  0  0  4  6 100  1  0  0 33
 8   1  0 33   0  0  3 80  4  0  6 15  0  2  0 33
 9   0  5 100  2 66  0  0  4  0 25  1  0  3 66
10   1  5 73   0  0  3 80  4  0 25  1  0  3 66
11   0  0  0  2 66  0  0  0  6 90  0  2  0 33
--- Execution Window Map ---
EC   T  T  N  T  N  T  N  T  N  M  M  M
 0   1  5  0  0  0  0  0  0  0  0  0  0
 1 255  5  0  0  0  0  0  0  0  0  0  0
```

Fig. 9. Example of an output file.

The output file is organized in two main areas: entities and maps. The first includes the characteristics of all nodes, tasks and messages in the system. The second includes the scheduling map and the execution window map (described in section 4). These maps are organized in the following way. The first column shows the current EC. The node

columns (N) relate the sum of the tasks execution time in the current EC to the EC time. The last column (Bus) shows the bus utilization that relates the sum of the messages transmission time in the current EC to the EC time. The task columns (T) are organized according to the node where they have been allocated and are displayed to the left of their node.

In the next chapter a simple example is used to demonstrate the SimHol operation.

4.   EXPERIMENTS

In order to begin a simulation a scenario file has to be loaded. This file includes the details of the simulation and the characteristics of every node, task and message. See an extract in Fig. 10.

```
# Bus type
# B Type BitRate(b/s)
# - Type: (CAN)ControllerAreaNetwork or ...

B CAN 1000000

# Elementary cycle duration
# E TDuration

E 1560

# Scheduling algorithm
# A Tasks Messages
# - Tasks: (RM)RateMonotonic or (EDF)EarliestDeadlineFirst
# - Messages: (RM)RateMonotonic or (EDF)EarliestDeadlineFirst

A RM RM

#
# Entity declaration
#

# The time unit is microseconds
# The greatest allowed time value is aprox. 35 minutes (2147483647)

# Node declaration
# N Id
# - Id: 0<Id<255

N 1
N 2
N 3
N 4

# Task declaration
# T Id C T D PhReq Allocation NodeId
# - Id: 0<Id<255
# - T: Only valid for independent tasks
# - PhReq: Phase Requested
# - Allocation: (F)ixed o (A)llocable

T 1 00520 0 0 0 F 1
```

Fig. 10. Extract of an input file.

The input file has two main areas: system characterization and entity declaration. The system characterization area is used to configure the simulation selecting the network architecture, the elementary cycle and the scheduling algorithms used for task and message scheduling. The entity declaration is used to define the nodes, the tasks, the messages and the transactions.

After this the simulator determines the size of the macro cycle, i.e. the interval with a pattern, of tasks and messages, that will be repeated indefinitely, and displays the simulation options, see Fig. 11.

The second step is applying all constraints.

Property	Value
Task Scheduling Algorithm	RM
Message Scheduling Algorithm	RM
Bus Type	CAN
Bit Rate	1000000
Elementary Cycle (EC)	1560
Macro Cycle	6
StartUp Interval	6

Fig. 11. Simulation characteristics.

After this the simulator checks the restrictions imposed by the messages to the tasks, builds the data streams (Fig. 12) and makes the precedence analysis. The start-up interval is also determined, i.e. the time between the first EC trigger message and the EC trigger message corresponding to the highest relative phasing of all tasks and messages.

Below is depicted an example scenario with 6 tasks, running in 4 different nodes and that exchange 3 messages.

Property	Abs. Value	Rel. Value (EC)
Node	1	-
C	520	1
T	3120	2
D	3120	2
PhMin	0	0
PhReq	0	0

Fig. 12. Data streams.

The Fig. 12 shows the two data streams. For instance *task 5* sends *message 3* to *task 6*.

The task and message parameters can be changed and a new scenario file, reflecting the changes, can be saved.

The last step is to execute the scheduling.

For the scheduling the user has two alternatives. Either message scheduling is selected, or message and task scheduling is selected. If message scheduling is selected than only messages will be scheduled to the bus, which means that each task is considered to be executing in a different node, see Fig. 13. If on the contrary, message and task scheduling is selected then messages will be scheduled to the bus and tasks will be scheduled at their nodes of execution (according to the allocation node), see Fig. 14. The scheduling takes place at an EC basis, i.e. at every EC messages and tasks are scheduled.

The scheduling is executed until either the simulation terminates or until a task, or message, misses its deadline.

Another option for the simulation is to choose between the requested initial phases (Phreq) or the minimum initial phases (Phmin) that were determined during the analysis.

The area depicted in the Fig. 13 shows the transition from the startup cycle (that finishes in the EC 5) to the first macro cycle. In this figure we can check the bus utilization (or we can check the output file).

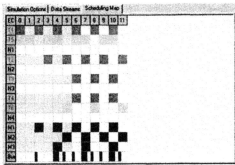

Fig. 13. Area of the message scheduling map.

The Fig. 14 also shows the nodes utilization.

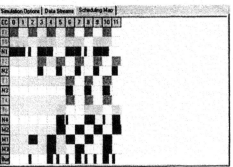

Fig. 14. Area of the task and message scheduling map.

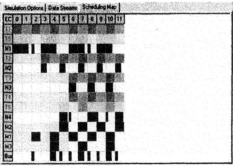

Fig. 15. Area of the task and message scheduling map with display of the execution windows.

An execution window represents the interval, in ECs, between the minimum and maximum allowed time for the task execution. Examples of execution windows can be seen in the Fig. 15.

Even for large macro cycles, the simulation is executed within a very short time frame. An example (shown in Fig. 16) with 10 nodes, 12 tasks and 8 messages, is organized in 4 data streams, and has a resultant startup time of 804 ECs and a macro cycle of 3510 ECs.

The task and message scheduling of this example stops at EC 3534 due to a missed deadline by message 4. An extract of the scheduling map is shown in the Fig. 17.

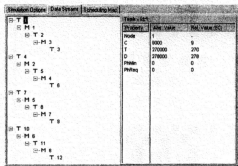

Fig. 16. Data streams.

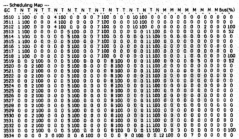

Fig. 17. Area of the output file.

5. CONCLUSIONS

The SimHol offers a suitable experimental platform for the simulation of distributed systems based on the time-triggered (FTT) paradigm. Using a simple interface it allows the simulation of various interconnection architectures, although currently only CAN is supported. Different scheduling algorithms can also be chosen. At this moment SIMHOL operates with Rate Monotonic and Earliest Deadline First. The user can start with a simple message scheduling in order to make a first essay to the system and then proceed to the full scheduling (tasks and messages) that takes into account the load of each node. The scheduling takes place at an EC basis closely resembling an actual system.

This simulator has validated the set of requirements previously derived, namely the data flow analysis and precedence requirements. The migration of the SimHol to a Java implementation is foreseeable due to a possible full integration with the execution environment. Other advantages are: improved portability and the use of standardized XML for the input/output files.

REFERENCES

Tindell, K., J. Clark (1994). *Holistic Schedulability Analysis for Distributed Hard Real-Time Systems*. In *Microprocessing & Microprogramming*, **40**, 117-134.

CiA DS 201-207 (1994). *CAN Application Layer for Industrial Applications*. CiA, CAN in Automation International Users and Manufacturers Group, 1994.

Tindell K., H. Hansson and J. Wellings (1994). *Analysing Real-Time Communication: Controller Area Network (CAN)*. Proc. of RTSS'94 (15th IEEE Real-Time Systems Symposium).

Henderson, W.D., Kendall, D., Robson, A.P. and Bradley, S.P. (1998), *Xrma: An holistic approach to performance prediction of distributed real-time CAN systems* . Proceedings of Can In Automation Conferance, San Jose.

Almeida, L., J. Fonseca and P. Fonseca (1998). *Flexible Time-Triggered Communication on a Controller Area Network*. In *Proc. of Work-In-Progress Session of RTSS'98 (19th IEEE Real-Time Systems Symposium)*. Madrid, Spain.

Almeida, L., R. Pasadas and J.A. Fonseca (1999). *Using a planning scheduler to improve flexibility in real-time fieldbus networks*. IFAC Control Engineering Practice, **7**, 101-108.

Henderson, W.D., Kendall, D. and Robson, A.P. (2000). *Improving the Accuracy of Scheduling Analysis Applied to Distributed Systems* . RTS 2000.

Vector Informatik GmbH (2002). *CANalyzer: The Tool for CAN*. Product information at http://www.vector-informatik.de

Martins, E., P. Neves and J.A. Fonseca (2002). *Architecture of a Fieldbus Message Scheduler Coprocessor Based on the Planning Paradigm*. In *Microprocessors and Microsystems*, Vol. 26, Issue 3.

Almeida, L., P. Pedreiras and J.A. Fonseca (2002). *The FTT-CAN protocol: Why and How*. To appear in IEEE Transactions on Industrial Electronics.

Calha, M.J., J.A. Fonseca (2002). *Addapting FTT-CAN for the joint dispatching of tasks and messages*. Submitted to WFCS'2002 – IEEE Workshop on Factory Communication Systems.

Pedreiras, P., L. Almeida and P. Gai (2002). *The FTT-Ethernet protocol: Merging flexibility, timeliness and efficiency*. To appear in *Proceedings of the 14th Euromicro Conference on Real-Time Systems*. Viena, Austria.

Fonseca, J.A., J. Ferreira, M. Calha, P. Pedreiras and L. Almeida (2002). *Issues on Task Dispatching and Master Replication in FTT-CAN*. Africon'2002 – IEEE International Conference, George, South Africa

ELSEVIER

IFAC
PUBLICATIONS
www.elsevier.com/locate/ifac

FLEXIBLE SCHEDULING TRAFFIC FOR VIDEO TRANSMISSION IN PROFIBUS NETWORKS

Silvestre, J.* Sempere, V.*

* Dpto. Computer Engineering. UPV. jsilves@disca.upv.es
** Dpto. Communications. UPV. vsempere@dcom.upv.es

Abstract: The mechanisms for static scheduling are inefficient and inadequate when the traffic being carried varies greatly in its bandwidth requirements, especially if there are also variations in run time. For this reason, dynamic traffic scheduling over Profibus, with the objective of developing flexible scheduling systems over this network is analyzed. This will allow us to address a scenario where the traffic load can vary over time. As a typical example of this scenario, the transmission of video sequences for monitoring applications has been selected, where the bandwidth presents important time variations. With the application of the proposed scheme, we will obtain flexibility in the network traffic scheduler, and also, the throughput of the video traffic is improved with respect to static scheduling. *Copyright © 2003 IFAC*

Keywords: Networks, Scheduling, Monitoring elements, Image flows

1. INTRODUCTION AND PREVIOUS WORK

The traffic scheduling schemes for DCCS (Distributed Computer Control Systems) applications where real-time is an important characteristic, usually employ static schemes which demand a prior knowledge of all the characteristics of the process. Although this is adequate and/or usual for hard-real time systems this is not so when speaking of soft real-time processes or when the complexity of the interactions between different nodes cannot be known in advance, or when traffic flows with large variations in bandwidth requirements are in the network. Therefore flexible scheduling systems are needed. This is becoming more and more important in industrial systems (Stankovic, 1996) (Thomesse, 1998) which need mechanisms that permit the behavior of the traffic scheduler to adapt to the traffic requirements

continuously. This demands the implementation of dynamic scheduling systems which allow the adjustment of the resources used at each moment by each of the elements implicated in a DCCS system, satisfying at all times the time restrictions of the real-time traffic.

Almeida (Almeida, 2001; Almeida *et al.*, 2002b) analyzes in detail this need, proposing several examples, also addressing the dichotomy existing in the current fieldbuses between the flexibility/timeliness behavior and the capacity to satisfy the synchronous/asynchronous traffic requirements. Another interesting aspect proposed by Almeida, although not analyzed in this paper is the definition of QoS parameters in control requirements. This allows flexibility of assignment of resources for control traffic in determined situations which thus gives space for other flows of information.

An example of this type of application with flexibility requirements and changes in run time re-

[1] supported by UPV (Polytechnic University of Valencia. Spain)

quirements is the video transmission for monitoring applications. Here the use of video codecs generates frame sequences with different sizes (relation between sizes of I frames and P and B frames), presenting different bandwidth requirements in each moment, which makes its transport complicated in timed token protocols (Kee-Yin, 1999), which is the case of Profibus.

The system proposed in this paper is also well suited to address the dynamic scheduling of image transport for control applications different to the static scheme presented by the authors in (Sempere and Silvestre, 2001; Silvestre *et al.*, 2002), in spite of the fact that the traffic load requirement for a determined application has an high stability in its traffic requirements. This greatly facilitates the installation and configuration. Also the proposed system is well suited from the point of view of QoS traffic control adjustment.

The flexible scheduling scheme is based on the carrying of synchronous communication in flexible form and with guaranteed timeliness. Almeida (Almeida *et al.*, 2002*a*) presents a WorldFIP scheduling algorithm, with the property of being usable on-line, although the objective of these algorithms is to improve the response time as much for the synchronous traffic as the asynchronous. Also analyzed is the architecture and management aspects over CAN in the system known as FTT-CAN (Almeida *et al.*, 1999) (Almeida *et al.*, 2002*b*) for applications embedded in this network. Pedreiras (Pedreiras *et al.*, 2002) presents a complete flexible and dynamic architecture for Ethernet, known as FTT Ethernet, with the objective of adequately satisfying, with flexibility, the typical fieldbus traffic requirements over Ethernet networks, avoiding some of the inconveniences of using this network as a fieldbus.

As for video transmission over Timed Token Networks, Feng (Feng *et al.*, 1996) analyzes the end-to-end delays in a local scenario (only uses the local information of a node to allocate its synchronous bandwidth) where the synchronous server manages the assigned bandwidth between the multiple flows of a node. Zheng (Zheng and Shin, 1993) proposed a local scheme for the scheduling of video traffic, taking into account variations in demand and avoiding the restriction of a deadline of the same period as other previous synchronous (Agrawal *et al.*, 1993) bandwidth assignment schemes(SBA) . However, in spite of the global schemes producing an overhead in the network, Han (Han *et al.*, 2001) showed that optimum local SBA schemes over timed token protocols don't exist. Moreover, given the dynamic nature of multimedia traffic, the parameters of this traffic in execution can be different from

those used in its design, and so a management system capable of dynamic adjustment to the requirements of this traffic is necessary (Chan *et al.*, 1995). Kee-Yin (Kee-Yin, 1999) proposes, for a timed token control network FDDI, a regulation scheme to distribute the request in the most suitable way in the bus, avoiding the superimposition of type I frames between different video sources.

The objective of this paper is to analyze the use of a flexible scheduling scheme over a Timed Token network such as Profibus for the transport of real-time control traffic and video traffic, analyzing the different aspects of the viability of its use, as such evaluating the advantages that this scheme can give to a video transmission system for monitoring applications.

Before analyzing the model, and since we are working with video, it is necessary to review some basic concepts. The convenience given by the use of MJPEG codecs for image compression when they are going to be used in artificial vision processes has been analyzed by the authors in (Sempere and Silvestre, 2003). However, when the images are going to be used for monitoring, it is possible to take advantage of the greater compression capacity which video codecs offer which make use of temporal redundancy of the images (Wu and Irwin, 1998).

Within these codecs three types of frames usually exist. The I frames or intra-frames are coded as still images, not using a past history, and thus can be decoded without the need for any other frame. The compression ratio for these frames is the lowest of all frame types. The P frame or predicted frames are predicted from the most recently constructed I or P frame. By exploiting temporal redundancies, these frames achieve higher compression ratios than the I frames. Finally there are B frames, or bidirectional frames. These are predicted from the nearest two I or P frames; one in the past and one in the future. They have the highest compression ratio among all frame types. The order of the frames obtained with the codec follows usually the pattern IBBPPBBPBB... and repeats itself. This sequence is called Group of Pictures (GOP).

2. MODEL

The proposed system to obtain a flexible scheduling scheme over Profibus, has the objective of realizing an optimum [2] (Han *et al.*, 2001) and centralized (see Fig.1) SBA. The stations make their traffic requirement request to the central

---

[2] SBA scheme that finds a feasible set of bandwidth allocation subject to the protocol and deadline constraints whenever such a set exists

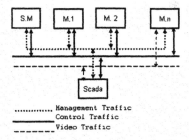

Fig. 1. Architecture of proposed system

station know as Scheduler Master SM (as a LAS in Foundation Fieldbus H1). The SM allocates access rights according to the request received and the requirements of the other network stations ($M_1 to M_n$) existing a station (SCADA) where the video flows are visualized and where the human operator acts according to Chan's model (Chan et al., 1995) being responsible for achieving management objectives. That is, providing network services that meet the need of customer applications and defining resource allocation strategies that provide benefits for the service provider.

2.1 Cycles considered

Elementary Cycles

In this type of scheduling the Elementary Cycle (EC) is seen as a bus time window used to exchange the traffic associated with each EC (see Fig.2a). The duration of the EC is typically set to the highest common factor (HCF) of the periods of synchronous traffic flows. In this type of scheduling the macrocycle (MC) is defined as the period of time in which the synchronous requests are repeated, derived therefore from the cyclical orders and with a period equal to the least common multiple (LCM) of the periods of traffic flow (Fig.2a).

Often many of the proposed scheduling schemes, work in a static way with two windows of fixed size existing for all the EC's, one for the synchronous traffic and another for the asynchronous traffic (HS and HA in Fig.2a). In these cases, within the window size for synchronous traffic, the peak is used in order to guarantee its QoS parameters, and the rest of the EC is used for asynchronous traffic. This scheduling scheme can be very inefficient in its use of the available bandwidth, given that the difference between the maximum peak of synchronous traffic in a EC and the synchronous traffic of each EC in the MC can be significant, in this way introducing a large number of unused slots. At the moment of introducing multimedia traffic, some authors add another window for this, using a fixed scheduling scheme (Tovar et al., 2001) of the Equal Partition Allocation Scheme type. In our opinion the throughput of

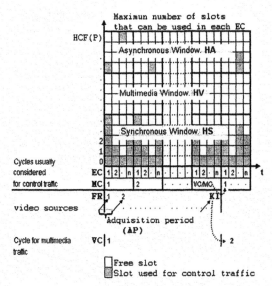

a) Fixed windows

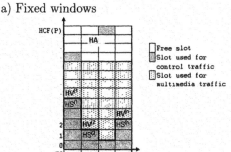

b) Variable and dynamic windows

Fig. 2. Cycles considered in the paper

multimedia transmission can be significantly improved through the use of a scheduling system which takes into account the dynamic characteristics of the video traffic.

In the proposed scheme we provide a mechanism which permits that the window sizes are not fixed but variable in the functions of the EC (Fig.2b). This will permit the optimization of bandwidth use and the capacity to adapt on-line to changes in the traffic sources requirements.

Video Cycles

To give support to the video traffic scheduling the Video Cycle (VC) is introduced. This presents a period depending on the characteristics of the frame acquisition system of the period of image acquisition (Know as AP, and usually in msec.) and the interval between keyframes (KI, usually 24-30 frames) being:

$$VC = AP x KI \qquad (1)$$

When this time has elapsed, the IPB frame pattern will keep repeating.

In fig.2a we can see the way in which the period between two frame requests (FR) can encompass several EC's. This will be taken advantage of by the scheduling algorithm when placing the load of each frame, given that a AP deadline will be considered for each frame.

2.2 Traffic Model

In the proposed system various types of traffic flows have to coexist. The existence of k stations, p synchronous real-time control flows in each station and m video flows is assumed.

Real-Time Control Synchronous Traffic

Each station $i (i \leq k)$ has $j (j \leq p)$ flows of synchronous messages $S_i^{(j}$ characterized by its period $P_i^{(j}$ and its deadline $D_i^{(j}$. It is assumed that the messages are of the maximum length provided by the protocol. (C_{max}=240 bytes in Profibus)

And that the deadline of each $S_i^{(j}$ is not $P_i^{(j}$ (which can introduce a high variability when considering the highest periods) but a determined number of EC's, in fact 1 EC.

We denominate as $HS_i^{(v}$ the space reserved in the station i for the synchronous real-time control traffic in the ECv. In order to satisfy the requirements of this traffic it is necessary to assign to each station with traffic control a synchronous capacity sufficient for the transmission of all the synchronous requests of the ECv, that is to say:

$$HS_i^{(v} \geq \sum_{j=1}^{p} eC_i^{(j}, where$$
$$e = 1, if vMOD(P_i^{(j}/HCF(P)) = 0,$$
$$e = 0, in other cases \quad (2)$$

We denominate as $HS^{(v}$ the space reserved in the ECv for the transmission of all the synchronous requests of all stations, that is to say:

$$HS^{(v} = \sum_{i=1}^{k} HS_i^{(v} \quad (3)$$

Real-Time Control Asynchronous Traffic

Each station or node i can generate any given asynchronous messages at any instant characterized by a deadline $D_i^{(asyn}$. It is assumed in the paper that all the asynchronous messages have the same deadline $D^{(asyn}$. Given that this type of traffic presents a low load, such as that typically used in alarms, etc. its duration CA_i in all stations is limited to CA bytes. In each EC a space will be reserved for the transmission of an asynchronous message in each station, denominated HA.

$$HA \geq \sum_{i=1}^{k} CA \quad (4)$$

Real-Time Video Traffic

Each one of the video flows m is characterized not only by the AP and KI previously mentioned but also by a quality index which determines frame load ($CI_{(m}$, $CP_{(m}$, $CB_{(m}$), as well as its distribution in a MPEG4 sequence (the order of the p frames type P and b frames type B which follow frame I). It will be necessary in order to transport these flows to associate with each one, the video windows, know as $HV_{(m}$, of different size in the function of the EC and of the frame type, for each of the $EC's$ that there are between two type I frames (that is to say, for VC)

$$HV_{(m} = \{HV_{(m}^1, HV_{(m}^2, ..., HV_{(m}^{VC}\} \quad (5)$$

in a way that each one of the frames can be delivered before its AP. Therefore, so that all the frames of a GOP can be transmitted before their deadline, the following must be fulfilled:

$$HV_{(m}^1 + HV_{(m}^2 + ... + HV_{(m}^{VC} \geq$$
$$CI_{(m} + CP_{(m} + CB_{(m} \quad (6)$$

whereas to meet the deadline of each frame, transmitted between AP EC's the following must be fulfilled:

$$HV_{(m}^n + HV_{(m}^{n+1} + ... + HV_{(m}^{n+AP} \geq C_{(m} \quad (7)$$

being n an EC multiple of the AP, and C the load of the frame I, P or B in function of the IPB distribution of this sequence and of the EC in which we find ourselves.

The window for the video traffic in each EC will then be the sum of the windows associated with each of the flows being:

$$HV^{(v} = \sum_{i=1}^{m} HV_{(i}^v \quad (8)$$

Scheduler Traffic

For the management of the scheduling policy, it is necessary to transfer request/response messages. These messages have a maximum length of CA bytes and the deadline is associated with the type of message it is dealing with. The same as in the case of asynchronous traffic, a HM window (Management Window) could be reserved for the transmission of this type of message. However, in this paper it is considered unlikely that the need exists to transmit at the same instant as much management traffic as asynchronous traffic through the same window used by both types of transmission. Nevertheless for the management of the protocol, it is vital that there is transmission of control packets for the management cycle, in order that the scheduler can guarantee the transmission of at least one type 5 message (see table 1) know as HMC (cycle management) for each EC.

2.3 Bandwidth limitation

In a token pass protocol, a parameter that is fundamental and common to all stations is the Token Token Rotation Time (TTRT), which shows the passage of time between two consecutive token arrivals. Its selection influences the time in which each station can keep the token (Agrawal et al., 1993), in the traffic that can be guaranteed on Profibus, and in the worst case, a high priority package (Tovar and Vasques, 1999). Nevertheless, given the necessity of guaranteeing large -sized flows, an outline is chosen which limits the synchronized capacity of each station, so that with a TTRT (Silvestre et al., 2002)

$$TTRT \geq \max_{i=1..VC}\{HS^i\} + HA + HV_{i,\forall i=1..VC}^{(m} $$
$$+ \max_{i=1..VC}\{HV_i^{(m}\} + HM_C + \Theta N \quad (9)$$

it guarantees the maintenance of the space which is assigned to each station, avoiding overrun situations. Furthermore, in order to meet deadlines, the following must be fulfilled:

$$HS^i + HA + HV_i + HM_C$$
$$\leq P_{min}, \forall i = 0..VC \quad (10)$$

which also optimizes the useful bandwidth of Profibus (Silvestre et al., 2002) by being used as a formula to calculate the admission control as well as periodical requests, such as video requests.

3. PROTOCOL

3.1 Set-up and run-time

The protocol gives greater priority to the control packets so that to begin with, in its initialization phase, only these can be dealt with, to determine thus the basic parameters of the network. In the set-up phase of the system, the SM emits a type 0 broadcast packet (see table 1) indicating the initialization state to the rest of the stations and releases the token. As this reaches the other stations, these will make their request to SM using the transmission of type 1 packets. A part of the SP (Setup Phase) having elapsed, the SM has sufficient information available to calculate the EC and MC, and so assigning the bandwidth required to satisfy all these requests (type 3 packets). Once the initialization phase has finished, the SM starts to generate EC cycles (SDN broadcast packet type 5) where each station transmits a number of

Table 1. Codification of management frames

Function	T	C				NB
Initialization	B	0				1
Sync. Req	U	1	A	N	period	4
Vid. Req.	U	2	A	N	QR	6
Answer 1	U	3	N	EC		3
Answer 2	U	4	N	QA		6
EC Change	B	5	EC no.		2	
EC Maint.	B	6	no. free			2
Control	-	7				1
Video	-	8				1

T: Type. C: Code. NB: no. of bytes. B: Broadcast. U: Unicast. A: Address. N: no. request. QR: Quality request. QA: Quality assigned.

packets not greater than the bandwidth assigned by the scheduler ($HS_{(i}$ and/or $HV_{(i}$, depending on the traffic sources of each master). If all the stations have finished transmitting their assigned traffic and the EC has not yet been completed, the scheduler will transmit type 6 packets indicating the free space which can be used by the other stations for the transmission of new synchronized requests or non-real-time traffic, of new requests or of pending video packets.

3.2 Video request

Once the initialization has finished, the monitoring station (SCADA) might wish to visualize video flows in the network with a determined quality. To do this, it transmits type 2 packets to SM, indicating the direction of the video source, the number of the associated requests and the quality parameters of the video. The scheduler will have to analyze if it is possible to deal with the request with the quality asked for. If this is not possible, it will be attended to, with an inferior quality. If this not possible, he will act as an admission control, disabling the transmission of this flow. In the cases where the request can be attended to with an inferior quality, the transmission windows of each type of frame will be set to the station which generated the video, using the transmission of the type 4 packet. The scheduling policy for video traffic is therefore FIFO type, since the requests are attended to according to the order of arrival.

It is assumed that the scheduler has information stored about the history of past sessions and/or by configuration on the index relation of requested quality, and bytes required for the I, P and B frame types. The transmission protocol of the images has been explained in (Sempere and Silvestre, 2001).

At any moment, using HA, the stations can transmit a type 1 message to ask for the transmission of new flows of periodic traffic or to cancel previous requests (indicating a 0 period). In its turn, SCADA could add, remove or modify the quality of the video stream, using type 2 packets.

Faced with these requests, SM has to reschedule the synchronous bandwidths associated with each master, using the transmission of 3 and 4 packets. Given the greater priority of control traffic over video traffic, the control requests have to be attended to with greater speed of reply than the video traffic. This does not create any drawback, as each modification in the synchronous traffic involves few changes in the assigned bandwidth, while in the case of video requests, important modifications may be involved, and these must be modified in a synchronized form. Furthermore, with control traffic we are on a msec. timescale, whereas the operator who monitors the video information is on a seconds scale (Lazar and Stadlelr, 1993).

3.4 Data structure to give support to the protocol

As a consequence of the requests granted by the scheduler, the masters with control traffic only need to know the bandwidth assigned to each EC (Silvestre *et al.*, 2002)in a MC. The masters with video traffic need to know the EC in which they have to transmit frame type I, the IPB distribution of a VC, and the bandwidth required for the transmission of frames type I, P or B.

4. SCHEDULING ALGORITHM

Algorithm 1. Synchronous traffic scheduling
```
01 if (state==SP) CalcBasicParam();
02 for (i=1;i ≤ NumPer;i++)
03    tper=period[i]/EC;
04 if CtrlAdPer(i)
05    for (j=tper; j <EC; j+ =tper)
06       HS[j]++
```

Algorithm 2. Video traffic scheduling
```
01 for (i=1;i¡=NumMul;i++)
02    if (CtrlAdMul(i))
03       for (i=1; 1¡KI; i+=AP/EC)
04          C=CalcType();
05          Free=SlotsFreeRF();
06          while (C > 0)
07             FreeEC=max-HS[i+t]-HV[i+t];
08             t=0;
09             while ((FreeEC>0) & (C>0))
10                FreeEC--; C--;
```

The algorithm of control traffic scheduling assumes a deadline of 1 EC in control traffic requests so as not to allow jitter in its transmission and therefore the necessary slots are arranged according to the period of each flow. This produces peaks in the bandwidth demand necessary for the transmission of control traffic at every moment (see example Fig. 3), which has to be taken into account in the scheduling of video traffic.

The scheduling algorithm 1 calculates the basic functioning parameters in each request (line 1,algorithm 1), in other words, the EC and MC, according to the periods of the received requests in the set-up phase.

Afterwards, for each request made, the admission control (line 4,algorithm 1) verifies the completion of (10) in the ECs of an MC, with the characteristic $HV_i = 0$. In other words, giving priority to control traffic so that the acceptance of new traffic of this type can reduce the quality of, or simply eliminate, a video flow. Up to now, it had not been foreseen that the introduction of new flows of periodic traffic during the run time could alter the basic parameters, such as the duration of EC and MC.

The scheduling algorithm of video traffic (algorithm 2) is based on the synchronization of type-B frames on a type-I frame of the first video request. This limits the number of simultaneous video sources with the distribution seen in point 2, at 5 sources, which is considered sufficient in an environment of video transmission for monitoring in a Profibus network.

For each video flow, first of all the admission is analyzed, which will be determined according to 10, but now considering HV_i if the video flow can be transmitted or not, and with what quality. Furthermore, it is considered that the frame C load (I, P or B) could be shared out between different ECs, specifically between AP/EC.

Subsequently, the type of packet which is to be sent is calculated according to the request number, which determines the load of the ECs which make up a AP (C in line 4, algorithm 2), and the free slots left by the periodic traffic in the ECs which occupy the frame transmission (line 5). Next, the algorithm shares out the C load of the frame, in as many ECs as are necessary, assigning the $HV_{(m}^{EC}$ to the video flow. This form of distributing the C load to each frame, tries to locate it in the fewest ECs possible, instead of using a uniform distribution between the ECs that satisfy a RF, so in cases where the real frame load is higher than the average used, these packets could be sent in the last EC's of the RF. This would be carried out in this way in the case where the SM sends type 6 packets with sufficient capacity to send the rest.

5. NUMERICAL EXAMPLE

Table 2. Synchronous traffic

Master 1		Master 2		Master 3	
no.	P	no.	P	no.	P
1-1	10	2-1	10		
1-2	30	2-2	30	3-1	30
1-3	60	2-3	60	3-2	60
1-4	90	2-4	90	3-3	90

Master 4		Master 5		Master 6	
no.	P	no.	P	no.	P
4-1	30	5-1	30	6-1	30
4-2	60				
4-3	90	5-2	90	6-2	90

Table 3. Video traffic

	high			low		
Type	I	P	B	I	P	B
Sequence 1	52	20	2	36	10	1
Sequences 2,3,4	36	10	1	23	5	1

Sizes given in no. of full packets. AP=30 ms. KI=30 frames. GOP's of the sequences obtained with DivX 3.0: IBPPPPBPPPPBPPPPBPPPPBPPPPBPPPPBPPP

Departing from an initial scenario with the periodic and video traffic shown in tables 2 and 3 (the data of the load of each one of the frames of a video sequence has been extracted from the analysis of 4 sequences of 5040 frames, 168 seconds), the fundamental parameters would be:

EC=10 msec. MC=180 msec. VC=900 msec.

Up to 8.1 msec. is reserved for periodic traffic (including the HMC of each start of a cycle), and the rest, 1.9 minus the token rotation latencies, for management and asynchronous traffic (up to 20 CA packets = 10 bytes, which supposes a duration of 90 $\mu sec.$), this is considered sufficient to reserve space for the transmission of an asynchronous packet and a request packet for each station as well as for the transmission of reply 3 and 4 packets for part of the SM. (The parameters of the Profibus network as well as the calculation of time can be found in (Silvestre et al., 2002).

This will generate distribution for synchronized traffic as can be seen in Fig. 3 With this distribution, the requirements of this type of traffic are guaranteed, a free space would also exist, that in the case of static assignation outlines, would serve for the transmission of 8 complete packets (slots 19 to 26 in Fig. 3).

Nevertheless, using a static outline (given the maximum peak of periodic traffic of 18 complete packet that are given in the 18 ECs), it is easy to see how the transmission of frame I of one type-1 sequence would require almost 9 ECs (only 8 slots per EC can be used), therefore losing frames 2 and 3. For the transmission of a type-P packet, 3 ECs are needed, and 1 EC for a type B packet. Nevertheless, adding whatever other video flow in these conditions will worsen the situation, causing the sequences to lose a large number of frames as well as being taken to a level of application with important delays, which could produce glitches in the resulting video.

By means of the scheduling outline proposed, four video sequences could be transmitted, using high quality in the first two sequences, and low quality in the last two. This would avoid loss of frames and delays in the deadline of each frame, as can be seen in Fig. 3

6. CONCLUSION AND FUTURE RESEARCH

The dynamic scheduling outline proposed allows the maximization of the number and quality of video sequences which can be transmitted, guaranteeing the behavior of real-time traffic whether periodic or not. Although it had a different aim from the monitoring, by the use of a static assignation outline, such as that used by the authors (Sempere and Silvestre, 2001; Silvestre et al., 2002), with the scenario of the numerical example explained, the valuation of transmission of images is 1.53Mbps, whereas in this way 2.15 Mbps is reached by inputting only one overhead of 1 byte in each packet. Nevertheless, the most important characteristic is to have the flexibility to change the configuration as well as the video and control traffic on-line.

REFERENCES

Agrawal, G., B. Chen and W. Zhao (1993). Guaranting synchronous message deadlines with the timed token medium access control protocol. *Proc. of ACM Multimedia.*

Almeida, L. (2001). Flexibility, timeliness and efficiency in fieldbus systems: the disco project. *Proc. 2001 8th IEEE Int. Conf Emerging Technologies and Factory Automation* 1, 159–167.

Almeida, L., E. Tovar, J.A. Fonseca and F. Vasques (2002a). Schedulability analysis of real-time traffic in worldfip networks: An integrated approach. *IEEE Transactions on Industrial Electronics.*

Almeida, L., J.A.Fonseca and P. Fonseca (1999). Flexible time triggered communication system based on the controler area network: Experimental results. *Proc. Of 3th IFAC Int. Conf. On Fieldbus Systems and their Applications FeT'99* 1, 24–31.

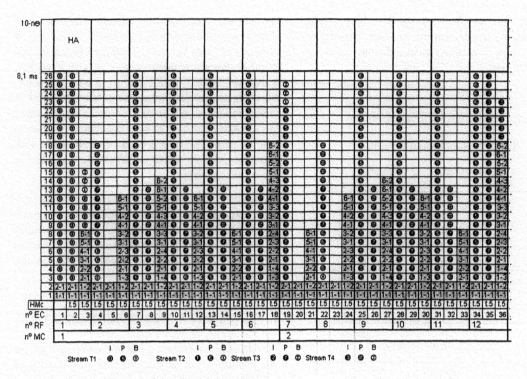

Fig. 3. Output of the scheduling algorithm with the input given in tables 2 and 3

Almeida, L., P. Pedreiras and J.A. Fonseca (2002b). The ftt-can protocol; why and how.. *IEEE Transactions on Industrial Electronics* 49(6), 1189–1201.

Chan, M., G. Pacifi and R. Stadler (1995). Managing real-time services in multimedia networks using dynamic visualization and high level controls. *Proc. of ACM Multimedia, November 5-9, San Francisco, CA.*

Feng, F., P. Zhao and A. Kumar (1996). Bounding application to application delays for multimedia traffic in fddi-based communications. *Proc. of Multimedia Computing and Networking 1996, San Jose, Jan.* pp. 174–185.

Han, C-C., K.G. Sing and C.J. Hou (2001). Synchronous bandwidth allocation for real-time communications with timed-token mac protocol. *IEEE Transactions on Computers* 50(5), 414–431.

Kee-Yin, J. (1999). Mpeg transmision schemes for a timed token medium access control network. *ACM SIGCOMM Computer Communication Review* 29(1), 66–80.

Lazar, A.A. and R. Stadlelr (1993). On reducing the complexity of management and control of broadband networks. *Proc. of the Workshop on Distributed Systems: Operations and Management.*

Pedreiras, P., L. Almeida and P. Gai (2002). The ftt-ethernet protocol: Merging flexibility, timeliness and efficiency.. *Proc. Of 14th Euromicro Conference on Real-Time Systems (ECRTS'02).*

Sempere, V. and J. Silvestre (2001). Image transport system in profibus networks. *Proc. Of 4th IFAC Int. Conf. On Fieldbus Systems and their Applications FeT'01* pp. 24–31.

Sempere, V. and J. Silvestre (2003). Multimedia applications in industrial networks: Integration of image processing in profibus. *IEEE Transactions on Industrial Electronics.*

Silvestre, J., V. Sempere and M. Montava (2002). Optimization of the capacity of profibus for the transmission of images and control traffic. *Proc. of 4th IEEE Workshop on Factory Communication Systems WFCS'02* pp. 133–140.

Stankovic, J et al. (1996). Strategic directions in real-time and embedded systems. *ACM Computing Surveys* 28(4), 751–763.

Thomesse, J.P. (1998). The fieldbuses. *Annual Reviews in Control* 22, 35–45.

Tovar, E. and F. Vasques (1999). Cycle time properties of the profibus timed token protocol. *Computer Communcation* 11(13), 1206–1216.

Tovar, E., F. Vasques, F. Pacheco and L. Ferreira (2001). Industrial multimedia over factory-floor networks. *Proceedings of the 10th IFAC Symposium on Information Control Problems in Manufacturing (INCOM '01), Vienna, Austria* pp. 20–22.

Wu, C. and J.D. Irwin (1998). Multimedia and multimedia communication: A tutorial. *IEEE Transactions on Industrial Electronics* 45(1), 4–14.

Zheng, Q. and K.G. Shin (1993). Synchronous bandwidth allocation in fddi networks. *Proc. of ACM Multimedia, 1993.*

ELSEVIER

IFAC
PUBLICATIONS
www.elsevier.com/locate/ifac

CORRELATING BUSINESS NEEDS AND NETWORK ARCHITECTURES IN AUTOMOTIVE APPLICATIONS – A COMPARATIVE CASE STUDY

Jakob Axelsson[1], Joakim Fröberg[2,3], Hans Hansson[3], Christer
Norström[3], Kristian Sandström[3], and Björn Villing[4]

[1]Volvo Car Corporation AB, [2]Volvo Construction Equipment
Components AB, [3]Mälardalen Real-time research Centre,
Mälardalen University, [4]Volvo Truck Corporation AB

Abstract: In recent years, networking issues have become more and more important in the design of vehicle control systems. In the beginning of the 1990s a vehicle control system was built up by 'simple' computer nodes exchanging 'simple' and relatively non-critical data. Today we have moved into distributed vehicle control systems with functions spanning several nodes from different vendors. These systems are running on communication architectures consisting of different types of communication buses providing different functionality, from advanced control to entertainment. The challenge is cost efficient development of these systems, with respect to business, functionality, architecture, standards and quality for the automotive industry. In this article we present three different architectures – used in passenger cars, trucks, and construction equipments. Based on these case studies with different business and functionality demands, we will provide an analysis identifying commonalities, differences, and how the different demands are reflected in the network architectures. Copyright © 2003 IFAC

Keywords: Automotive, Architectures, Networks.

1 INTRODUCTION

One of the initial driving forces for introduction of communication networks in automotive vehicles was to replace the numerous cables and harnesses and thereby reduce the number of connection points, cost and weight. Moreover, *multiplexing*, as the in-vehicle networking is commonly referred to in the automotive industry, is an important enabler of new and increasingly complex functions. Using software and networking it is today possible to create new functionality, such as an anti-skid system, that was considered unfeasible some ten years ago, both with respect to cost and functionality.

The vehicle industry strives to enable cost effective implementation of new functionality. A network enables reuse of sensor data and other calculated values. Moreover, distributed functions can be facilitated for truly distributed problems like coordination and synchronization of brakes.

Another driving force is the demand for increasingly efficient diagnostics, service, and production functionality. The network in a vehicle should provide functionality not only during normal operation of the vehicle, but also in the after market and during production. It should provide communication for diagnostic functions in control

units, and provide a single point interface to service tools to meet goals on more efficient service.

There is a wide span of requirements on the communication infrastructure in today's vehicles. The vehicle industry works with demands on product variation, branding, super structures, and extendability in aftermarket. This leads to high requirements on network flexibility in terms of adding or removing nodes or other components. Moreover, part of the functionality has stringent requirements on real-time performance and safety, e.g., safety critical control; whereas other parts of the functionality, such as the infotainment applications, have high demands on network throughput. Yet, other parts require only lightweight networks, as for example locally interconnected lights and switches. All of these varying requirements in vehicle networks are reflected in the architecture, implementation, and operation of a modern in-vehicle network.

In this paper, we try to describe the context in which today's vehicle industry work and develop distributed systems. This is uncovered in three case studies covering passenger cars, trucks, and construction equipment. Although the three different vehicle domains have very much in common there are also distinct differences, e.g., in production volume and the number of different models. In fact, these differences are so important that they result in three quite different network architectures.

The contributions of this article are the presentation of three case studies covering much of automotive industry and describing the business context and demands for automotive networking, together with an analysis of the relation between businesses needs and architectural design issues of networking in automotive vehicles.

The rest of this article is organized as follows: Section 2 provides a structuring and presentation of important issues and challenges in automotive networking. Section 3 presents the three case studies and in Section 4 we analyze the architectures with respect to the different business needs. Finally in Section 5 we draw some conclusion and links to site wherefurther information are provided in Section 6.

2 CHALLENGES

2.1 Functionality

Functionality in a vehicle is not limited to end-user functions, but includes also functions to support, for instance, production and service. In this section we will outline some important groups of functionality, both supportive and end-user functions that is often addressed in vehicle development.

Feedback control includes functions that control the mechanics of the vehicle, for example engine control and anti-lock-brake-systems. Several feedback control systems can be combined to achieve advanced control functions for vehicle dynamics. Examples are electronic stabilizer programs, ESP, and other chassis control systems like anti-roll systems. Furthermore, the vehicle manufacturers strive to achieve cheaper and more flexible functionality by going towards x-by-wire solutions. X-by-wire solutions, such as steer-by-wire, are achieved by replacing e.g., mechanical or hydraulic solutions by computer control systems.

Discrete control, in this context, includes simple functions to switch on or off devices, e.g., control of lamps. The challenges for this group of functions often relates to the sheer number of such simple devices and thereby the amount of traffic on the network.

Diagnostics and service. Functions in this group are used in vehicles to support maintenance of delivered products. Diagnostic functions provide means to diagnose physical components as well as software properties, such as version number and bus load. Service functions provide means for updating the electronic system by downloading new software and to provide feedback of vehicle operation. Because of the large amount of retrievable information, solutions are needed for automatic, or at least tool supported, diagnoses and service.

Infotainment. Information and entertainment systems are sometimes requested in today's vehicles. Examples are Internet connection and video consoles. This leads to requirements on high bandwidth for vehicle networks. Components like network controllers and software are often purchased off-the-shelf, and must be integrated in a harsh physical environment. Components must also be integrated without impacting safety critical functionality in the vehicle.

Telematics [1] refers to the set of functions that uses communication networks outside the vehicle to perform their task. This includes functions in the vehicle and outside the vehicle. There is a strong trend in the vehicle industry to increase the use of telematics. Examples include fleet management systems, maintenance systems, and anti-theft systems.

2.2 Cost

Providing cost efficient network architectures is a challenge in several respects. The architecture should exhibit properties that support a variety of business needs. Business needs in the context of vehicle network systems often include life-cycle aspects of development, production, maintenance, and service.

Fixed and variable cost. Building a vehicle is a process of finding the best compromise between conflicting aspects, and one of the most important trade-offs is to find the balance between cost, performance, and functionality that provides the best business case. The cost can be divided into variable cost (the cost of purchasing the physical components that go into the vehicle and the resources consumed during production) and fixed cost (the investments made in development, production facilities, tooling, after market support, etc.). There is always a trade-off to be made between the two, which depends heavily on the production volumes, and the relations between various cost factors [2].

Maintenance and service. To reduce the life cycle cost of the product it is often important to consider various aspects of maintenance. To facilitate maintenance it is desirable to develope architectures that allow future upgrades and extensions to the delivered product. Servicing delivered products require the ability to upgrade both software and physical components. Configuration management and distribution is then a crucial issue, e.g., to determine compatibility of new components in an existing configuration.

2.3 Standards

The use of standards is motivated by the need to meet goals related to:

- Cost reduction
- Integration of supplier components
- Increased reliability of components
- Commonality in tools used in e.g., development, diagnostics, and service.

One example is the Controller Area Network (CAN) [3] standard that has provided the vehicle industry with cheap CAN controllers. Due to the large volumes of these controllers, they are tested to a great extent (in the field) under diverse conditions, which increases reliability.

A challenge with respect to standards is to standardize properties of software to accommodate reuse of software components and to allow for common development tools.

2.4 Architecture

The IEEE has the following definition of architecture [4]: *"Architecture: the fundamental organization of a system embodied in its components, their relationships to each other and to the environment and the principles guiding its design and evolution."*

For automotive electronic architectures, the components are mainly the electronic control units (ECU), and the relationships between them are the communication networks. The environment is the vehicle itself, as well as the life-cycle processes that must support it.

Aspects of maintenance must be considered when designing the architecture in order to facilitate aftermarket service of ECUs and software. Also, environmental factors like temperature and EMC influence the location of the ECUs. As the product development is increasingly focused on platforms and commonality, the ability to create many variants from the same overall structure is important.

3 CASE STUDIES

3.1 Volvo Car Corporation

3.1.1 The car industry

The European premium car brands are driving the development of vehicle electronics, having both the demand for advanced functionality and the production volumes to support the costs associated with the introduction of new technology. Cars are consumer products, and the customers tend to be sensitive not only to the functionality of the car, but also to how it feels and its visual appearance.

Cars are typically manufactured in volumes in the order of millions per year. To achieve these volumes, and still offer the customer a wide range of choices, the products are built on platforms that contain common technology that has the flexibility to adapt to different kinds of cars.

The component technology is to a large extent provided by external suppliers, who work with many different car companies (or OEMs, original equipment manufacturers), providing similar parts. The role of the OEM is thus to provide specifications for the suppliers, so that the component will fit a particular car, and to integrate the components into a product. Traditionally, suppliers have developed physical parts, but in modern cars they also provide software. As the computational power of the electronic control units (ECUs) increase, it will be more common to include software from several suppliers in the same nodes, which increases the complexity of integration.

3.1.2 Functionality

The driving factor behind the development is increasing demands on functionality. There are several classes of functionality, and in the following paragraphs we provide examples in each of them, that represent some of the largest future challenges for the industry.

Feedback control systems were one of the earliest uses for electronics in cars, and the early applications were engine control, ABS brakes, and vehicle dynamics. These areas are still developing, and one of the main challenges is to cope with new environmental constraints, in particular related to the reduction of the level of CO_2 emissions. The systems are refined in the sense that more and more sensors are added, and new modes of interaction are included, thereby increasing the overall complexity of the functionality.

Discrete control systems are also common in current cars, in particular in body electronics. Applications include driver information, security, and lights. Due to the fact that the overall functionality increases, as well as the abilities of the owner to configure the car through various parameter settings, the complexity of the driver interface becomes a bottleneck. This is caused both by the physical space around the driver seat, and the ability of the driver to process information while driving the car. Novel ways of interaction is thus needed, but also more intelligent systems that in most cases can make correct decisions without driver intervention.

Safety critical control systems become more common as traditional mechanical solutions are replaced by electronics. For the functions currently implemented, there is always a natural fallback solution if the electronics fail, but future by-wire systems may not have that possibility, which increases the need for fault-tolerance in the electronics and communication. The first such application is likely to be brake-by-wire, and later steer-by-wire will follow. The driving factors behind this development are that the cost and weight could be reduced, but also that it enables new control systems to support the driver, e.g. to enhance safety. Again, this means a considerable increase in system complexity.

Diagnostic systems provide information about the status of the vehicle. Initially, this was driven by legal requirements that mandated monitoring of emission related components, but it is also an important factor in increasing the perceived quality of the system. The diagnostic system consists of an in-car part and a workshop tool. The former is usually distributed to all the nodes of the on-board network, and consists of fault detection routines and diagnostic kernels that interact with the workshop tool. As the number of sensors increase, so does the need for diagnostics, and there is also a wish to increase the intelligence of the on-board system, to e.g. detect the need for preventive maintenance so that the customer never experiences critical problems, and thereby gets a perception of high quality.

Infotainment systems implement entertainment functionality, extending from traditional radios to multimedia applications such as TV, video, and gaming, and also contains information functions such as navigation systems. As the number of devices for audio and video data increase, sharing of input and output is essential to bring down cost and conserve space, and this means that complexity moves from hardware to software and communication.

Telematic systems are used for wireless data communication with the world outside via a built-in mobile phone. The applications range from automatic emergency calls in case of an accident to Internet access, and many ideas exist for services that the car owner could be interested in. The area is still in its infancy, and the business cases are currently unclear, but the underlying technology is being developed rapidly.

As can be seen above, complexity is a keyword that must be handled in the development. (For an introduction to the nature of complexity in technical systems, see [5].)

3.1.3 Cost

In the car industry, the development cost is huge in absolute numbers, but still comparatively small in relation to the total variable cost of the production, or the investment in tooling. This means that it is usually profitable to invest in development cost to optimize the components, or to increase commonality between car models on the same platform. Since the cost of development is closely related to the complexity (i.e., the information that needs to be processed to describe the product), it is thus profitable to increase complexity to obtain more flexible components that can be used in many different cars.

For software, the cost relations are somewhat different. There is an indirect variable cost, in that the characteristics of the software influence the resource needs in terms of memory size and processing capability.

One way to decrease software development cost is to raise the level of abstraction when describing the functionality. At VCC, model based system development is being introduced [6], where the system is described using a tool chain based on UML, Statecharts, and data flow models. Code generation is then employed to reduce the cost of producing the final software. In a way, the complexity is moved from the specific applications to general tools that can be used over and over again. Another example is in network communication over the CAN buses. The Volcano system [7] provides tools for packing data into network frames, and for verifying the end-to-end communication timing to ensure the control performance.

3.1.4 Standards

As indicated above, the OEM's role is to integrate systems from suppliers into a product. This means that it is important to have well-defined interfaces so that the various systems fit together. One area where standardization has been particularly vivid is in communication protocols. The following protocols are now used or planned by VCC:

CAN [3] is used for backbone control-oriented peer-to-peer communication. It provides predictable timing and moderate bandwidth (up to about 500 kbit/s), and many microprocessors are available with built-in support circuits.

LIN [8] is a low cost alternative for control-oriented master-slave communication. Originally developed at VCC, it is now an international standard. The bandwidth is low (up to about 20 kbit/s), but controller circuits are considerably cheaper than for CAN.

MOST [9] is used for multimedia communication in the infotainment system. It is based on optical fibre technology, and provides bandwidth up to about 20 Mbit/s, with timing characteristics and services optimized for infotainment applications. However, the cost for circuits is substantially higher than for CAN or LIN.

Flexray [10] is expected to come in use instead of CAN for safety-critical applications where fault-tolerance is needed. The protocol is time-triggered, which makes it suitable for implementing control functions with high demands on jitter, and for redundancy.

Selecting a bus protocol is thus a trade-off mainly between cost, bandwidth, predictability, and fault-tolerance. Another area where standards are important is in diagnostics, where authorities mandate it so that they may check that a vehicle fulfils emission regulations, using a single tool.

The current development trends in automotive software also calls for increasing standardization of the software structure in the nodes. In particular, the use of code generation requires a clear interface between the support software and the application, and the need to integrate software from different suppliers in the same node also calls for a well-defined structure. The node architecture includes several important components:

Operating systems (RTOS) provide services for task scheduling and synchronization. Traditional real-time operating systems are usually too resource consuming to be suitable for automotive applications, and do not provide the predictable timing that is needed. Therefore the new standard OSEK has been developed. There are several suppliers of OSEK compliant operating systems exist.

Communication software provides a layer between the hardware and the application software, so that communication can be described at a high level of abstraction in the application, regardless of the low-level mechanisms employed to send data between the nodes. At VCC, the Volcano concept [7] is used for both CAN and LIN communication.

Diagnostic kernels provide an implementation of the diagnostic services that each node must implement to act as a client towards the off-board diagnostic tool. It relies on the communication software to access the networks and on the operating system to schedule diagnostic activities so that it does not interfere with the application functionality.

All these components interact with each other and with the application, and must therefore have standardized interfaces, and at the same time provide the required flexibility. To conserve hardware resources, the components are configurable to only include the parts that are really necessary in each particular instantiation.

3.1.5 Architecture

An example of a contemporary car electronic architecture is that of the Volvo XC90 (see Figure 1). Boxes and lines represent ECUs and communication busses and Figure 1 shows complexity rather than detail. The maximum configuration contains about 40 ECUs. They are connected mainly by two CAN networks, one for powertrain and one for body functionality. From some of the nodes, LIN sub networks are used to connect slave nodes into a subsystem. The other main structure is the MOST ring, connecting the infotainment nodes together, with a gateway to the CAN network for limited data exchange. Through this separation, the critical powertrain functions on the CAN network are protected from possible disturbances from the infotainment system. The diagnostics access to the entire car is via a single connection to one ECU.

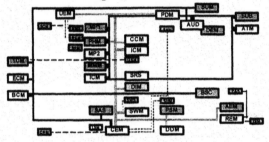

Figure 1. The electronic architecture of Volvo XC90.

In the future, when safety-critical functionality is introduced, the architecture will be extended with a Flexray based network, and this will again be isolated from the less critical parts using a gateway. Another important aspect is to create a more flexible partitioning. The main use for this is probably not to find the optimal partitioning for each car on a given platform, since that would create too much work on the verification side, but to allow parts of the software to be reused from one platform to the next. This puts even higher demands on the node architecture, since the application must be totally independent from the hardware, through a standardized interface that is stable over time. Therefore, further standardization work is needed, in particular for sensor and actuator interfaces.

3.2 Volvo Trucks

3.2.1 The trucking industry

The functionality in trucks has grown dramatically during the past 10 years. Earlier a separate stand-alone system for handling each function; today all these sub-systems are integrated into a single complete system. The development time has decreased and the vehicles are much more complex.

There are different demands depending on market in the truck industry. One example is the system voltage, which is 12V in the North American market and 24V in the rest of the world. Moreover, it is very common for the customers to choose their own driveline in the U.S., which puts special demands on OEMs to integrate engines and transmissions from various suppliers.

One way to obtain cost-effective solutions is to use a common platform covering both mechanical, electrical and software systems. The challenge is to have a shared platform and yet maintain the unique truck brands.

The market has changed from delivering vehicles, to providing a complete transport solution, which may include several different types of services, e.g. 'around the clock' support and online fleet management, etc. A complete transport solution could mean providing the full logistic routing system for a town.

3.2.2 Functionality

Trucks have many areas of use. The requirement on functionality can be split in three different segments.

- Goods transportation and logistics
- Building and construction
- City distribution and waste handling

'Goods transportation and logistics' means transportation of goods over long distances, e.g., food from southern Europe to Sweden. 'Building and construction' e.g. concrete trucks, crane trucks or gravel trucks operate under rougher conditions such as on a construction site, in mines or in roadworks. 'City distribution and waste' refers to local transport, for instance a garbage truck.

Feedback control systems were among the first electronic systems introduced in the beginning of the 80ies (e.g. electronic motor control and anti-lock braking system, ABS). These systems were complete stand-alone systems. Over time the systems became more complex and integrated. For example, the ABS system can command the engine not to apply the exhaust brake when ABS is activated. Furthermore, some sensors are shared between the systems and data can be exchanged through the vehicle network.

Discrete control systems include functions such as driver information, but also systems like climate control, exterior light, central locking, tachograph etc. With increasing amount of functionality and information on the network, the requirements on the Human Machine Interface (HMI) are getting increasingly complex. It becomes a challenge to support the driver in deploying the functionality the right way.

Superstructures. Trucks can be supplemented with superstructures, such as concrete aggregates, tipping devices, refrigerator units etc. One way to decrease the total cost of the vehicle is to have a well-defined interface between the electrical system and the superstructure, to allow, e.g., the crane equipment to control the engine speed to facilitate the right flow in the hydraulic pump.

Safety critical control systems. The increase in safety critical systems has been striking in the last few years. One driving factor for this is to prevent personal and property damage in case of accident Earlier mechanical systems are being replaced/supplemented. Many systems are common with the car industry, for instance ABS and airbags etc. One difference is that the gross combination weight (weight of vehicle and trailer) is greater. VTC also have many variants and it is common that trucks have more than 4 wheels.

Recently the EBS (Electronically-controlled Brake System) and ESP (Electronic Stability Program) were introduced. In the EBS, an electronically controlled valve located close to each wheel applies individual braking force to each wheel. There is also a centrally positioned ECU that controls the vehicle's braking effect, both in the disc brakes and the engine brake. There is a back-up system using pressurized air.

The ESP system is a supplement to EBS. By means of a YAW rate sensor, a lateral accelerator sensor and a steering wheel sensor the new functionally can be added. ESP is an active safety-enhancement system whose task is to stabilize the vehicle, eg, prevent jack-knifing when the driver makes a rapid avoidance manoeuvre.

Diagnostic system. An increasing part of the vehicle electronics has demands on efficient built-in diagnostics. Not only for the aftermarket, but also to check the mounting of components in production. The goal is to be able to check all components, not only ECU's but also very simple components like switches and bulbs[1].

There is an aftermarket tool that communicates with the control units. Through this tool it is possible to read out fault codes, sensor values etc. The tool is also able to run tests in the control units and download new software and parameters. Since the North American market requires support for 3rd part components, it must also be possible for the supplier's aftermarket tools to co-exist.

Infotainment has not been very common in trucks, but because many drivers that are transporting goods far away are living in their trucks, the need has increased. Because of the low product volumes compared to the car industry it is likely that the truck industry for cost reasons will inherit from the car industry.

Telematics in trucks is mainly used for traffic information and tracking of goods. Volvo has its own telematics system called Dynafleet. Tracking of goods gives the possibility enhance logistics. This type of system becomes more and more common due to the demand for just in time transports. Because of the growth of the Internet, the telematics systems

will become more integrated in all business systems. By using the information on the truck location and data about the goods (e.g. the temperature in a cold transport) the transports can be planned with a positive effect both on economy and the environment.

3.2.3 Cost

Cutting costs is becoming more important. The vehicles offer more functionality but the cost per function is decreasing. Because the truck industry has lower volumes compared to passenger cars it is important to get the right trade-off between development cost and product cost. Systems that are fitted in all truck models must have a low product cost. On the other hand, systems that are only available in some variants produced in a couple of 1000s/year are less sensitive for product cost. Here instead, is a gain to be made by using "general purpose" components that realizes more than one variant.

3.2.4 Standards

A truck is essentially built by integrating systems from many suppliers. Volvo Trucks develops core control units, but there are a lot of systems that are considered unprofitable to develop in-house. One such system is the ABS brakes. In this area there are a few big suppliers that have key-knowledge and their own development of systems.

It is important to agree on standards to make the integration in different brands as simple as possible. For heavy vehicle there are two standard protocols that Volvo Trucks is using in today's production:

SAE J1939 [11] use standard CAN 2.0B for communication and communicate with 250 kbit/s. J1939 also define data (signals e.g. vehicle speed) and the packaging in frames. J1939 also define how some control functions, e.g. the interaction between engine and transmission under gear shifting. J1939 also allows for some proprietary messages.

SAE J1708/J1587 [12] is an older protocol that has been available for a long time. It uses RS485 as base and communicates with 9600 bit/s, and is used in the vehicles for mainly diagnostics and for some fallback for J1939.

Other protocols that will be used or are under investigation at Volvo trucks include:

LIN [8] will be used for sensor networks but also for sub-busses.

MOST [9] is the optical ring bus that is used for infotainment by some manufacturers in the car industry

Flexray [10] is under evaluation for safety critical systems.

TTCAN [13] is a further development of CAN 2.0B. TTCAN offers services, such as global time and scheduling of communication. Because TTCAN is built on the well used CAN protocol, TTCAN might be very interesting for control intensive systems.

---

[1] Compare to the semiconductor industry where diagnostics is included on the chips for production tests only.

3.2.5 Architecture

The market pull in the US for the option to select truck components from different vendors introduces additional complexity, when deciding on a feasible architecture. To handle the electrical integration, a couple of OEMs and vendor suppliers have created a standard for communication between components in vehicles; the J1939 that is based on CAN. Since J1939 defines how the interaction between some components should be performed e.g. controlling the engine speed, it acts partly as architecture. Volvo trucks have also added some in-house strategies and guidelines to J1939 and for other electrical installations. Together with the standard this defines the architecture. The advantage is the proportionately simple way to integrate components from different vendor suppliers.

To be able to meet requirements on reduced development time it is important to find new solutions. One way is to use a reference architecture that covers all products and include both SW/HW.

The superstructure interface must also be included into the architecture. A well-defined interface decreases the cost and time for adding superstructures.

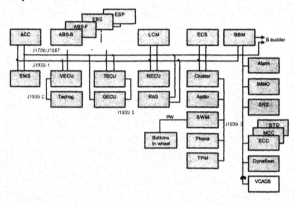

Figure 2. Volvo FH electronic architecture.

Progress in the last ten years has moved VTC vehicles from centralized computer systems to distributed systems, and much functionality has been added. The complexity of the systems is increasing at the same time as demands on shortened product development time and higher quality is increasing. One way to handle this situation is to use reference architectures supported by tools and better methods for product development.

3.3 Volvo Construction Equipment

Volvo Construction Equipment, VCE, develops and manufactures a wide variety of construction equipment vehicles, such as articulated haulers, excavators, graders, backhoe loaders, and wheel loaders. The products range from relatively small compact equipment (1.4 tons), to large construction equipment (52 tons). VCE is divided into product companies with focus on one type of equipment, and which typically manufactures product lines of similar products e.g. excavators, or wheel loaders.

Compared with passenger cars and trucks, construction equipment vehicles are equipped with less complex electronic systems and networks. Also, the focus in product development is somewhat different. The products are used in construction sites, and the most important aspect of the vehicle is to provide a reliable machine to increase production. A customer of passenger cars, on the other hand may be interested in a number of styling attributes related to the look and feel of the car, whereas the dominant requirements for customers of construction equipment are related to production goals. The requirements for reliability and robustness are equally high in both cases, but styling requirements are typically low for construction site machinery.

3.3.1 Functionality

Here we describe today's situation in functionality requirements for construction equipment at VCE. To each class of functionality we present some major drivers and examples.

Feedback Control. The main feedback control functionality in VCE products typically includes control of engine, gearbox, retarder brake, and differential lock. Besides engine control, the automatic gearbox is a complex feedback control system that can include control of clutches and brakes for the gear pinions as well as converter, lockup, and drop box. The functionality for an automatic gearbox includes logics for when to shift gears, minimization of slip, avoidance of hunting, various efficiency optimizations, and self-adapting solutions to accommodate a variance of mechanical properties. Feedback control systems also include the cooperation of brakes to minimize wear e.g. the cut in of a retarder brake, or exhaust brake.

Discrete Control. Like in the truck case, discrete control systems include control of driver information, wipers, lamps, and other on-off type devices. The challenges arise due to configuration issues rather than constructing a functional system.

Diagnostics and service. Diagnostic functions are used to determine status and operational history of electronic components. Some diagnostic functions reside in the on-board software and some in service tools. On-board diagnostic functions implement criterions for faults, and can send fault codes via the network. Diagnostic functions also include logging functions to store operational data e.g. fault history or general operation statistics on buffers, network load, or sensor input. The service tool can run tests to diagnoze faults either by invoking on-board functions, or by running test programs to verify functionality, but also download new software or parameters into the ECUs. The tool can be connected to a central configuration database that holds information on compatibility between versions and configurations. When downloading software or parameters, the central system is updated to reflect the current configuration.

Infotainment. As mentioned, customers of construction equipment purchase products with the intent of increasing production at construction sites.

There is currently little incentive to pay for entertainment systems and therefore VCE does not provide any such systems today. There are demands for information systems, but this does not usually include general systems such as Internet connection and video. Instead, tailored applications to increase production are requested.

Telematics. In the field of construction equipment, telematics can be used to achieve the increased production in several ways. Applications running on an office desktop computer can be developed to access information in a fleet of vehicles and to present and analyze data far away from the actual vehicles. Examples are fleet management systems, maintenance systems, and anti-theft systems. A fleet management system can provide information to increase efficiency of a fleet of vehicles. A maintenance system could, for instance, report on status of mechanically worn components like brakes and thereby reduce maintenance cost.

The challenge is to accommodate telematic functions in a cost efficient way. Construction sites could be located in remote regions and wireless technology like satellite or radio communication must be considered as opposed to the car industry where mobile phone communication seems to be sufficient. Fitted equipment and cost for communicating over commercial networks is expensive and may not be crucial to every customer.

The trend towards telematic systems is very strong and a variety of systems are expected to be available in a few years.

3.3.2 Cost

The electronic systems in VCE products are, in total, less complex and are sold in smaller volumes compared to cars or trucks. The final products are also generally more expensive than passenger cars, but the electronic content is a much lower percentage of the total product cost. Therefore, the cost for developing electronic systems in VCE is relatively large compared to the variable costs for production, at least compared to VCC and VTC. This implies that it is usually not profitable to optimize hardware to a large extent since it would generate increased complexity of the system and increased development cost e.g. need for configuration handling.

Accommodating commonality is considered very beneficial because VCE has a large number of products (although sold in smaller volumes). Reuse is beneficial for both hardware and software as it directly affects development cost. This results in VCE focusing heavily on commonality, even though it may mean that a lower-end product is produced with some spare resources in terms of electronics.

Compared to both cars and trucks, VCE builds on-board electronic systems that have a lower development cost, smaller volumes, and lower overall complexity.

Trends indicate that the electronic content (and complexity) will increase quite rapidly in construction equipment over the coming years. The

situation for construction equipment is likely to resemble the situation of trucks today and later maybe the situation for cars. However, the volumes will not equal those of trucks or cars. VCE have fewer products with higher price and will thereby not focus as much on optimizations, but rather at handling complexity and commonality. Commonality can also help reducing development time

3.3.3 Standards

VCE uses the same standards for communication protocols as VTC, i.e. SAE J1939 / CAN and SAE J1587 / J1708 (see Section 3.2.4 for details).

3.3.4 Architecture

The focus of VCE's electronic architecture effort is concerned with assuring system properties that are essential to provide a good business case. System properties include scalability to support product variation, reusability and partitioning of SW components to lower development cost, as well as safety and reliability issues. This means that methods for designing SW in control units and handling communication are also considered architectural issues. The goal is to have an architecture that helps designing on-board electronic systems with respect to system properties that are identified as the most important for a VCE business case.

As temporal behavior is considered very important, VCE uses a design process [14] that focuses on high-level design and temporal attributes of the system. VCE use the operating system Rubus, which provides a configuration tool allowing specification of temporal constraints, communication and synchronization. This method separates the design of timing characteristics from the design of functionality, and enables early verification of temporal behavior.

Partitioning, scalability, and commonality. VCE has a different setting than VCC and VTC in that there is almost no use of externally developed control units. This leads to the possibility to use the same software component model, operating system, and to reuse software components. By doing this, the partitioning of functionality is likely easier than in the case of a network with control units from many vendors. Easy partitioning gives the possibility to scale the system with respect to hardware and optimize hardware content in a specific product. For instance, a low-end product with fewer features requires less hardware resources, and can be realized by placing software components on other nodes and thereby reduce the number of control units. Thus, the architecture is reused, but the numbers of ECUs differ between products. (see Figure 3)

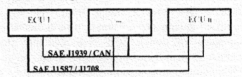

Figure 3. VCE Products network architecture

To have a situation with a large degree of in-house developed control units also give benefits in terms of commonality. The use of common design methods for control units gives in itself reduced cost, but also helps in decreasing the complexity of the system. (The overall information that must be processed during development is decreased). Reusing infrastructure like drivers, communication components and other software components is a goal in commonality that can be met as long as common design methods are used.

On the other hand, VCE platforms are used in a wider variety of different products. Facilitating commonality in very different products, such as an excavator and an articulated hauler presents a different situation compared to VTC or VCC.

For the future, VCE is likely to move towards the VTC situation with products including external vendor nodes. A new approach to accommodate scalability, partitioning and commonality will be required. Furthermore, the trend towards supplying services rather than vehicles is relevant to VCE. Currently the trend of rental products can be seen as a step in this direction, as VCE accepts responsibility for up time of products.

4 ANALYSIS

The different business demands on cars, trucks, and construction equipment lead to different focus in the design effort of the respective vehicles. In this section we present analysis of the correlation between different business and product characteristics and key properties of the resulting network architectures. In Table 1 below, the business and product characteristics are given for the three different organizations.

	VCC	VTC	VCE
Annual production volume, order of magnitude	10^6	10^5	10^4
Products	~8	~8	>50
Platforms	3	3	8
Number of physical configurations per product	Many	Very many	Few
Amount of Information	Huge	Very large	Moderate
Standards - Application level	Propriet. Volcano	J1587 J1939	J1587 J1939
Number of network technologies	~4	2	2
Hardware optimization	High	Medium	Low
Openess	None	High	Some
Safety critical	Yes	Yes	Yes
Advanced Control	Yes	Yes	Yes
Infotainment	Much	Some	None

Table 1. Business characteristics for each organization.

The case study has shown that the product volumes are different for the three organizations, and thereby also the focus on fixed cost and hardware optimization. The willingness to reduce fixed cost at the expense of variable cost increases with the product volume. One way of achieving a reduced fixed cost is to optimize vehicle hardware content to include a minimum of resources. Software components are not subject to the optimization profit, due to increase in variable cost but almost no gain in fixed cost. VCC that produces vehicles in the range of 10^6, can benefit to a larger extent by reducing fixed cost and therefore an increased cost for design of optimal hardware is more profitable than for VCE that has volumes in the range of 10^4.

The number of vehicle models sold for the respective organizations is indicated in the table by 'Number of products'. The number of products and the product volume directly affects the profitability of reusing components i.e. commonality, and this also includes software components. VCE has a high number of products, but smaller volumes, while VCC and VTC have high volumes. Thus, the effort to achieve commonality is emphasized in all the three organizations.

The 'Number of physical configurations' means the different options of network topologies that can be fitted in a certain product. VTC products, which may be configured in many ways, achieve a high extendability and can facilitate change of configuration in the aftermarket or adding superstructures by other vehicle developing organizations. The large amount of data and the many configurations put high requirements on management of different components, e.g., ECUs, connected to the network in different variants.

The large amount of information together with the requirements for optimization in the VCC case, imply that using several tailored networks for specific needs can be profitable. The use of LIN networks provide a cost effective network for handling locally interconnected lights and switches, and a high bandwidth MOST network serves the needs of infotainment applications. Especially for VCE, the increase in development cost for designing tailored networks for a certain purpose is deemed unprofitable and this is reflected in the small number of network technologies.

The use of in-vehicle networks open up the possibility for efficient diagnostics and service, by the ability to extract information via the bus. Although the amount of information varies in the three cases, the needs for diagnose and service are emphasized in all three organizations. The reason is that there is enough information in all systems to substantially ease analysis of the distributed system.

As mentioned, VTC needs to facilitate superstructures, and this is reflected in the large number of physical configurations. In order to support extensions to the network by other parties, standardized communication interfaces like SAE J1939 and J1708/J1587 are used. VCE uses the same standards, since both VTC and VCE belong to the Volvo group and there is commonality in service tools.

Safety aspects on networking include that messages are transferred with correct timing and without being corrupted. One step towards guaranteeing the real-time properties and integrity of messages related to safety critical functionality is to use communication protocols with support for deterministic and analyzable timing behavior. Examples are CAN and LIN and the coming protocols Flexray and TTCAN, which are all used or evaluated by the three considered organizations. Another step is to use several networks that are interconnected through gateways. Safety critical communication can thereby be separated from communication that is not trusted to the same degree.

5 CONCLUSION

In recent years, networking issues have become more and more important in the design of vehicle control systems. One of the initial driving forces for introduction of communication networks in automotive vehicles was to replace the numerous cables and harnesses and thereby reduce cost and weight. Today it is possible using software and networking to create new functionality that was considered unfeasible some ten years ago.

We have presented case studies of the context, use, and requirements of networking in passenger cars, trucks, and construction equipment. Furthermore we have identified challenges with respect to functionality, cost, standards, and architecture for development of vehicles. Based on these case studies with different business and functionality demands, we have provided analysis of the design principles used for the communication architectures in these domains. Despite a common base of similar vehicle functionality the resulting network architectures used by the three organizations are quite different. The reason for this becomes apparent when looking at different business and product characteristics and their affect on the network architecture. An important lesson from this is that one should be very careful to uncritically apply technical solutions from one industry in another, even when they are as closely related as the applications described in this paper.

6 REFERENCES

[1] Yilin Z., Telematics: Safe and fun driving, Intelligent systems, IEEE, Volume: 17 Issue: 1, Jan/Feb 2002 Page 10-14

[2] J. Axelsson. Cost Models for Electronic Architecture Trade Studies. In Proc. 6th International Conference on Engineering of Complex Computer Systems, pp. 229 -239, Tokyo, 2000

[3] Road Vehicles - Interchange of Digital Information - Controller Area Network (CAN) for High-Speed Communication. International Standards Organisation (ISO). ISO Standard-11898, Nov 1993.

[4] IEEE Recommended Practice for Architectural Description of Software-Intensive Systems. IEEE Std 1471, 2000.

[5] J. Axelsson. Towards an Improved Understanding of Humans as the Components that Implement Systems Engineering. In Proc. 12th Symposium of the International Council on Systems Engineering (INCOSE), Las Vegas, 2002.

[6] M. Rhodin, L. Ljungberg, and U. Eklund. A Method for Model Based Automotive Software Development. Proc. Work in Progress and Industrial Experience Sessions, pp. 15-18, 12th Euromicro Conference on Real-Time Systems, Stockholm 2002.

[7] L. Casparsson, A. Rajnak, K. Tindell, and P. Malmberg, *Volcano a revolution in on-board communications*, Volvo Technology Report. 98-12-10.

[8] LIN - Protocol, Development Tools, and Software Interfaces for Local Interconnect Networks in Vehicles, 9th International Conference on Electronic Systems for Vehicles, Baden-Baden, October 2000.

[9] MOST Specifcation Framework Rev 1.1, MOST Cooperation, www.mostnet.de, November 1999.

[10] FlexRay Requirements Specification, Version 2.0.2 / April, 2002, www.flexray-group.com

[11] SAE Standard, SAE J1939 Standards Collection, www.sae.org

[12] SAE Standard, SAE J1587, Joint SAE/TMC Electronic Data Interchange Between Microcomputer Systems In Heavy-Duty Vehicle Applications, www.sae.org

[13] Road Vehicles - Controller Area Network (CAN) – Part 4: Time-Triggered Communication. International Standards Organisation (ISO). ISO Standard-11898-4, December 2000.

[14] C Norström, K Sandström, M Gustafsson, J Mäki-Turja, and N-E Bånkestad. Experiences from Introducing State-of-the-art Real-Time Techniques in the Automotive Industry. In proceedings of 8th Annual IEEE International Conference and Workshop on the Engineering of Computer Based Systems (ECBS01), Washington, US, April 2001. IEEE Computer Society.

FORMAL VALIDATION OF THE CANOPEN COMMUNICATION PROTOCOL

M.Barbosa* M.Farsi** C.Allen** A.S.Carvalho***

Dep. Informática, School of Engineering
University of Minho, Portugal
**Dep. Electrical and Electronic Engineering*
University of Newcastle upon Tyne, United Kingdom
***Dep. Electrical Engineering and Computers*
Faculty of Engineering, University of Porto
ISR - Polo do Porto, Portugal

Abstract: A well written specification may be the element that makes a given fieldbus system stand out from among its competitors. Formal methods bring a comprehensive and unambiguous character to communication protocol specifications. This paper describes a formal specification and formal analysis of the CANopen communication protocol. The analysis uses the PROMELA Formal Description Language and Spin, a tool for the simulation and validation of PROMELA models. The results clarify several aspects of the operation of CANopen-based networks. They also demonstrate the advantages of performing a formal analysis of existing communication protocol specifications. *Copyright © 2003 IFAC*

Keywords: CAN, CANopen, validation, PROMELA, Spin, formal methods

1. INTRODUCTION

Behind a great fieldbus system there is a great communication protocol specification. In fact, a well written specification may be the element that makes a given fieldbus system stand out from among its competitors. The decision on which fieldbus system a given company implements in its products may come down to how easy it is for software developers to understand the specification. This will determine the time it will take to produce an acceptable implementation of the protocol.

A protocol specification that contains ambiguities, inconsistencies or is incomplete is susceptible to interoperability problems. Even if a fieldbus system is "protected" by a conformance and interoperability certification system, there is no guarantee that end users will encounter no problems with certified products. Conformance and interoperability testing systems must accept all implementations that do not violate the protocol specification. If the protocol specification allows incompatible or contradictory implementation options, products which are not actually interoperable may be certified.

Formal methods have since long been presented as one way of enabling developers to construct systems that operate reliably, despite their complexity. This is particularly true for communication protocols. In addition to the well-known advantages of using formal methods during system development, formal methods bring a comprehensive and unambiguous character to communication protocol specifications.

This paper describes a formal analysis of the CANopen communication protocol specification.

This analysis was performed as part of a broader study of this protocol, in the area of conformance testing. The results that are presented clarify several aspects of the operation of CANopen-based networks. They also demonstrate the advantages of applying formal specification and validation techniques to existing communication protocol specifications.

The following section includes a brief description of CANopen. Section 3 introduces PROMELA, the Formal Description Language (FDL) used in this work. The analysis and its main results are described in Section 4. Finally, Section 5 contains some concluding remarks.

2. OVERVIEW OF CANOPEN

CANopen is a CAN-based fieldbus protocol. It offers the characteristics of the Controller Areas Network (CAN) for real-time communication, enhanced with a set of higher-level communication services. These services constitute an object-oriented distributed environment for simplified system integration.

In Open Systems Interconnection (OSI) [ISO7498, 1984] terminology, CANopen defines a three-layer communication system: the Physical Layer, the Data Link Layer and the Application Layer are specified. The Physical Layer and the Data Link Layer functionalities are based on the Controller Area Network (CAN) [ISO11898, 1993]. They are implemented by CAN hardware implementations, which are usually called CAN controllers.

The CANopen Application Layer functionality is defined in the CANopen Communication Profile [DS301, 2000]. The CANopen Application Layer offers services for:

- Service data transfer – offline transfer of configuration data, of arbitrary size, for system configuration, between one Client device and one Server device (SDOs).
- Process data transfer – on-line transfer of limited-size data packets with real-time requirements, between one Producer device and one or more Consumer devices (PDOs).
- Action synchronization in two or more devices e.g. data acquisition or actuation.
- Error signaling.
- Global time synchronization.
- Network management – one Master device is able to control the operation of several Slave devices e.g. switching the whole network from off-line to on-line status.

The CANopen specification goes beyond the seven layers defined in the OSI Reference Model. In fact, part of the CANopen specification applies to

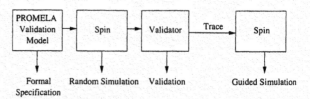

Fig. 1. Use of PROMELA and Spin in communication protocol development

functionality that is in the domain of the User Application. This part of the CANopen specification is defined in Device Profiles. Device Profiles ensure device interoperability by constraining the implementation of User Application functionality. This is usually referred as a User Layer [Hodson, 1997].

The CANopen Device Profiles define standard types of CANopen devices. Each Device Profile specifies how a given class of devices (e.g. I/O modules, motor drives, etc.) uses CANopen Application Layer services.

3. PROMELA: A LANGUAGE FOR PROTOCOL VALIDATION

PROMELA [Holzmann, 1991] is a notation for specifying the procedural rules that make up communication protocols and the correctness criteria associated with these rules.

PROMELA specifications are called validation models. They provide only a partial description of a communication protocol, focusing on its structure. This simplifies the verification of communication protocol completeness and logical consistency properties. A validation model defines only the interactions between processes or entities in a distributed system. It does not resolve implementation details such as how a message is to be transmitted, encoded, or stored.

[Holzmann, 1991] proposes an automatic tool that can be used, not only for the simulation of PROMELA validation models, but also for the generation of executable programs, called *validators*. These validators are capable of checking various types of correctness requirements embedded in a validation model, such as invalid end states, assertion violations and live-locks. This tool is called Spin, which is the short form for *Simple PROMELA Interpreter*. PROMELA and Spin can be applied at various stages in the development or analysis of communication protocols, as illustrated in Figure 1 [Holzmann, 1991].

Spin can be used to perform random simulations of PROMELA validation models. Alternatively, it may be used to generate the validators, which apply model checking algorithms to these models. When validators encounter a violation of the correctness requirements embedded in the

PROMELA code, they output a *trace*. This allows the user to perform a guided simulation of the validation model, again using Spin, so that the behavior that caused the violation can be analyzed.

4. VALIDATION OF THE CANOPEN APPLICATION LAYER

4.1 Analysis Method

As mentioned in Section 2, the CANopen Physical and Data Link Layers are provided by the CAN protocol. The properties of the CAN protocol are well known: its operation has been thoroughly studied and documented by other authors [Farsi and Barbosa, 2000]. Therefore, this formal analysis of the CANopen communication protocol focused on the services specified at the Application Layer level.

The study began with the formal specification of the CANopen Application Layer services, using PROMELA. According to the OSI Reference Model, the data transfer services provided by a given layer are provided by peer Protocol Entities, residing on both ends of the communication link. This formal specification of the CANopen Application Layer is, in fact, constituted by PROMELA models of the Protocol Entities that CANopen specifies in its Application Layer.

This initial specification process aimed to look for particular aspects of the CANopen specification that may bring difficulties to implementers and users of CANopen devices. These difficulties may arise from omissions or ambiguities, or simply because too many implementation options are allowed. The main conclusions that were drawn at this stage can be found in Section 4.5.1.

In order to analyze the Application Layer functionality it was necessary, before anything else, to establish the nature of the services offered by the layers under it. In this case, this means that the functionality of the CAN messaging services must be fully characterized before the CANopen Application Layer is analyzed.

Since CAN is not under analysis, a black box view of the services it provides is sufficient. This is the perspective of the Application Layer entities that use it. This black-box view must also include the data transfer errors that may be introduced at the Data Link and Physical Layer levels. The validation of the CANopen Application Layer functionality must assess the way in which such errors are dealt with. Section 4.2 contains a summary of the characteristics that were assumed for the services provided by CAN.

The PROMELA validation models that were used to analyze the CANopen Application Layer were

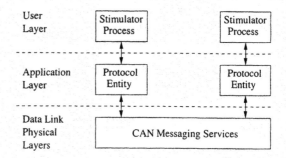

Fig. 2. Structure of the PROMELA validation models

obtained by combining the Protocol Entity models with a simplified model of the CAN subsystem. This model of CAN emulates the behavior described in Section 4.2. The structure of the resulting PROMELA validation models is illustrated in Figure 2.

Note that above the Application Layer Protocol Entities reside additional components. These components represent User Layer or Application entities, which use the Application Layer services. Their purpose and use is described in Section 4.3.

Finally, the correctness requirements for each of the services were added to each validation model using appropriate PROMELA tags. The models were then automatically processed using the Spin tool, using its random simulation, validation (model-checking) and guided simulation functionalities. The main results of this analysis are presented in Section 4.5.2.

4.2 Black-box view of the CAN messaging services

The CANopen Application Layer sees the underlying CAN network as a communication channel that provides services for transferring and requesting pieces of information across a shared medium. Each individual piece of information transferred over CAN is up to eight bytes in length, and it is identified with a unique number. This number also determines the priority of the message that carries it.

According to [Holzmann, 1991], the errors that can occur in data transfers can be divided in five categories: insertion, deletion, duplication, distortion, and reordering.

In this work, it was assumed that in CANopen systems, message duplication and message loss may occur at the Data Link Layer level. This is due to two well known problems of CAN-based implementations: the double receive "feature" of the CAN specification [Grifford *et al.*, 1998], and the message overrun possibility in CAN controllers [Farsi and Barbosa, 2000]. These problems must be considered in the analysis of the CANopen Application Layer functionality.

On the other hand, it was assumed that message reordering, message distortion and message insertion are not problems that the CANopen Application Layer must address, and were excluded from the analysis. The reasons supporting these assumptions are thoroughly discussed in [Barbosa, 2000].

4.3 Modeling the CANopen User Layer

In order to use the validation models described in Section 4.1 with the automated validation functionality that is provided by Spin, it was necessary to include processes that generate stimuli. These processes interact with the processes that model the Application Layer Protocol Entities and drive them into the points of operation that are being analyzed at a given simulation or validation run.

The stimulator processes emulate the behavior of User Layer entities using the services on both sides of communication, as shown in Figure 2.

Necessarily, for each service that was analyzed, suitable stimulator processes had to be developed. These processes were tailored to expose the Application Layer Protocol Entities to all the situations that are contemplated in their specification.

For example, consider a data transfer protocol that provides a confirmed service whose result may be *success* or *failure*. The stimulator processes must drive the Application Layer entities through both of these situations.

4.4 Modeling Example: Service Data Objects (SDO)

Figure 3 shows a block diagram of the PROMELA validation model for the SDO data transfer protocols. The core of the validation model is the formal specification of the SDO services, composed of the two PROMELA processes SDO Client and SDO Server. These processes model two peer Protocol Entities that provide the SDO functionality.

The Client process operates in a conceptual CANopen device that accesses a remote data store using the CANopen network. In CANopen terminology this data store is called an Object Dictionary [DS301, 2000].

The Server process operates on a conceptual device that provides access to its Object Dictionary through the CANopen network. An Object Dictionary is simply a list of data records. Each record provides access to the attributes of one object residing inside the CANopen device hosting the Object Dictionary.

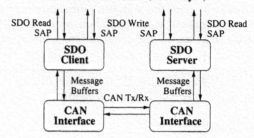

Fig. 3. SDO Validation Model

The interface between User Layer entities and the Application Layer Protocol entities is modeled as two independent Service Access Points (SAPs), each composed of two synchronous (rendez-vous type) PROMELA channels. One SAP provides access to the SDO Read service and the other to the SDO Write service. The operation of these SAPs is based on the OSI Abstract Service Primitives for confirmed services. The rendez-vous interactions in each of them represent function calls. This is how Application Layer service invocations work, in typical CANopen implementations.

The connection between the Application Layer Protocol Entities and the associated CAN interfaces is modeled as two finite-length message queues. These queues are emulated by two asynchronous PROMELA channels. Again, the choice of asynchronous channels to model this interface is based on the operation of typical CANopen implementations.

The PROMELA processes modeling the CAN Interfaces on both Client and Server side are identical. These processes incorporate the functionality discussed in Section 4.2, including the possibility of emulating message loss and message duplication.

4.5 Analysis Results

The results presented in this section are a summary of the conclusions that were drawn from the analysis that was described in Section 4.1. These results are documented in detail in [Barbosa, 2000]. A complete description of what was found is too lengthy for inclusion in this paper. These details are only relevant for implementers of CANopen devices and for the maintainers of the CANopen specifications. In this paper it is more interesting to focus on the general problems that were identified in the protocol specifications. Following this approach, it is hoped that these remarks may be of use to the developers and users of other communication protocols.

4.5.1. Specification Issues As mentioned in Section 4.1, a first assessment of the CANopen data transfer protocols was performed when producing a PROMELA formal specification of the associated Protocol Entities. Three specification issues were highlighted by this analysis, that are worth mentioning.

The first point is associated to what in the CANopen specification are called *Implementation Issues*: invalid messages and timeouts. The CANopen specification states that the way in which errors introduced by the Data Link and Physical Layers are handled is implementation-specific. However, it does provide two general guidelines on how to deal with these errors, in order to minimize their impact on Application Layer data exchange protocols. These guidelines are:

- A message that is received out of context during the execution of a communication protocol, i.e. an invalid message, should be ignored.
- The eventuality of an expected message not arriving should be accounted for using time-out mechanisms.

This type of approach to Communication Protocol specification is less than ideal. It is acceptable in the case of invalid messages, since this gives an indication to implementers on what to do when the specification does not state the action to be taken on reception of a given message. However, in the case of timeout mechanisms, it would be preferable to specifically identify, in the protocol definitions, the points where timeouts may occur, and the action to be taken when this happens.

One additional objective of this validation process was therefore to clarify in which cases and in what manner should these mechanisms be used to deal with errors introduced by the CANopen lower layers.

The second issue that deserves discussing concerns several inconsistencies that were found in the CANopen specification. One example is the use of a toggle bit mechanism in segmented service data transfers. On one hand, the CANopen specification states that if a toggle bit error is detected this would render the message invalid and, as such, it should be ignored. On the other hand, it states that if a toggle bit error is detected, it may optionally be signaled to the application, that may decide to abort the transfer. Furthermore, the service primitives defined for service data transfers do not specifically encompass the latter type of error indication. Therefore, it cannot be assumed that, on detecting a toggle bit error, a CANopen implementation will terminate the transfer. This type of inconsistency and ambiguity may lead to interoperability problems between CANopen implementations, due to different interpretations.

The final issue is related with the fact that the inclusion of parameters in some protocol messages is optional. One example of this is the indication of the size of the data that is transferred in service data exchanges. The implications of not including this information in the protocol messages is, however, not specified. This data size information is used in a consistency check of the type *"is the size of the data received equal to the size that was indicated during handshaking?"* It is clear that the robustness of the data transfer protocol is reduced if the data size is not included. Why leave it as an optional parameter? A conclusion that may be drawn from this discussion is that CANopen implementations will either require additional processing overheads to cope with these possibilities or totally disregard the data size parameters.

4.5.2. Validation Issues   The second stage in this analysis was the study of the CANopen data transfer mechanisms using the PROMELA validation models presented in the previous sections and the simulation and model-checking capabilities of Spin.

Firstly, the procedural rules associated with the CANopen transfer procedures were validated for operation under ideal conditions. The objective was to verify that, when no errors occur at the Data Link Layer and Physical Layer levels, these rules produce a satisfactory result. Part of this initial validation was based on correctness requirements embedded in the stimulator processes referred above, in the form of assertions. These assertions signal errors when there is a mismatch between the expected and run-time values of different operational parameters. Additionally, the valid end states in the different validation models were marked using the corresponding PROMELA tags.

The results of this first phase of the validation revealed that, as would be expected, no design errors exist in the procedural rules defined in the CANopen specification. In other words, it was found that in the absence of Data Link Layer and Physical Layer errors, the CANopen protocols produce the result for which they were designed.

At a second stage in the validation process, the possibility of sporadic message loss at the Data Link Layer level was introduced. Under these conditions it was confirmed that, when the implementation issues mentioned in the CANopen specification are not addressed, deadlocks may occur. The analysis then focused on the way in which timeout mechanisms could be used in the CANopen protocols to solve these deadlock problems. The critical points where these mechanisms should be employed were identified. Details on this will be omitted from this paper, but can be found in [Barbosa, 2000]

Two problems arising from message loss, however, could not be solved:

- Unconfirmed services – for services with simple, unconfirmed, protocols, it is not possible to detect a message loss: the receiver simply is not aware of the service invocation.
- Loss of the last confirmation message in confirmed services – in this case, the sender may detect a timeout and retry the transfer, even though the receiver may be unaware that a problem has occurred.

One way to solve these problems would be to associate unique numbers or time-stamps to service invocations. This is not done in the CANopen protocol.

These conclusions underline the importance of a fail-safe mechanism that can be implemented in all CANopen devices. This is the signaling message loss events at the Data Link Layer level using Emergency Messages. It is very important in detecting situations where the operation of the network has deteriorated, and minimizing the impact of message loss.

The final stage of the validation of the CANopen data transfer protocols consisted in evaluating the effect of message duplication errors introduced at the Data Link Layer level. It was found that in service data transfers, which use more complex protocols, duplicate messages will be detected and eliminated as invalid messages. However for services with simpler data transfer protocols, such as those used for process data transfer, duplicate messages are not detectable. This may lead to either a wrongful indication of an additional event occurrence or a wrongful acceptance of an additional command, depending on the data that is being transferred.

The consequences of this type of error will also depend on the nature of the process data. There might be no consequences whatsoever, for example, if the data is simply an updated reading of the value of a variable. However, this is not guaranteed e.g. for commands that operate on a toggle basis, such as toggling the values of the digital outputs of a device.

5. CONCLUSIONS

The authors present the results of a study conducted on the CANopen communication protocol. This study consisted on the formal specification and validation of this protocol.

The analysis focused on the correctness of the CANopen Application Layer communication mechanisms. Formal specifications of the CANopen communication protocols were developed using the PROMELA Formal Description Language. Validation models were developed, based on these specifications. The correctness requirements for the corresponding communication protocols were also embedded in these validation models. The Spin tool was then used to conduct the analysis on these models, using simulation, model-checking and guided simulation functionalities.

The main result of this work is the demonstration of the overall soundness of the design of the CANopen communication protocols. It was possible, however, to identify ambiguities, inconsistencies and omissions in the CANopen specifications. These results are presented as contributions for the work being done in communication protocol specifications, namely CAN-based protocols and, particularly, CANopen. These results may also serve as examples of the problems that the implementers and users of a given communication protocol may encounter when reading the corresponding specification.

6. ACKNOWLEDGMENTS

M. Barbosa was sponsored by the BD/13641/97 scholarship from the Fundação para a Ciência e Tecnologia, Programa PRAXIS XXI (Portugal).

REFERENCES

Barbosa, M.B. (2000). Conformance Testing issues with application to the CANopen protocol. PhD thesis. University of Newcastle Upon Tyne.

DS301, CiA (2000). CANopen Application Layer and Communication Profile.

Farsi, M. and M. Barbosa (2000). *CANopen Implementation: Applications to Industrial Networks*. Research Studies Press.

Grifford, D., B. Kirk and B. Leisch (1998). A Component Based Architecture for CAN Based Systems. In: *Proceedings of the 5th CAN Conference*. CiA. San Jose.

Hodson, W. (1997). Fieldbus to change DCS role, but death reports greatly exaggerated. *InTech*.

Holzmann, G.J. (1991). *Design and Validation of Computer Protocols*. Prentice-Hall.

ISO11898 (1993). Road Vehicles, Interchange of Digital Information - Controller Area Network (CAN) for high-speed Communication.

ISO7498 (1984). Information Processing Systems - Open Systems Interconnection - Basic Reference Model.

ELSEVIER

IFAC

PUBLICATIONS
www.elsevier.com/locate/ifac

A CANOPEN-BASED PROTOCOL FOR DYNAMIC ASSIGNMENT OF PRIORITIES

Salvatore Cavalieri

University of Catania, Faculty of Engineering
Department of Computer and Telecommunications Engineering
Viale A.Doria, 6 95125
Catania (Italy)
E-mail: Salvatore.Cavalieri@diit.unict.it

Abstract: The author deals with the problem of information flow management in a CANOpen network. CAN ISO IS-11898 physical medium access mechanism, adopted in CANOpen, features an arbitration protocol based on an identifier assigned to each message to be transmitted. The lack of a centralised scheduling for the identifier assignments in CANOpen does not allow achieving any guarantee to meet the deadlines of time-critical messages. The aim of the paper is to propose a protocol for dynamic and centralised assignment of identifiers to each message to be transmitted, in such a way that its transmission requirements are fulfilled. The main features of the protocol proposed are the operational flexibility it offers and full compatibility with the CANOpen standard. *Copyright © 2003 IFAC*

Keywords: CANOpen, Fieldbus, Identifiers, Scheduling, CSMA/CA

1. INTRODUCTION

The main feature of the CAN fieldbus is the Carrier Sensor Multiple Access/Collision Avoidance (CSMA/CA) mechanism based on the presence of a dominant bit (zero) (ISO IS-11898, 1993). This mechanism is applied to an identifier field univocally associated to each message to be transmitted.

CAN is actually used in wide range of industrial applications. Most of them are classified as Real-Time applications whose correctness depends not only on the logical results of computing, but also on the time at which the results are produced (Arvind, Ramamritham and Stankovic, 1991).

In Real-Time applications, messages must be properly scheduled to access to the network, in order to have their deadlines respected. Literature presents several scheduling algorithms in CAN. Many of them are based on the idea to link the message identifier with its transmission time features and requirements. One of the most known approach sets the identifier to a unique priority, according to the

Deadline Monotonic rule (Tindell, Hansson and Wellings, 1995) (Tindell, Hansson and Wellings, 1994). The reader is referred to (Tovar, 1999)(Ouni and Kamoun, 1999) for other examples of scheduling approaches in CAN.

When the identifiers of messages must be assigned according to a particular rule (e.g. deadline monotonic, as said before), the need for a protocol able to assign and to on-line manage the message identifiers (coping with reconfiguration, addition and deletion of nodes, addition and deletion of messages produced by a node already existing) arises. The definition of such protocol is much more difficult in a CAN network made up by a standard Application Layer above the ISO IS-11898 protocol, because the protocol for the dynamic assignment of identifiers must be compliant with the services available at that layer.

In the previous work (Cavalieri, 2001), the author proposed a protocol for the dynamic assignment of identifiers fully compliant with the CAL CiA standard (CiA/DS201, 1996). The aim of the paper is to extend this work to the CANOpen communication system (CiA/DS301, 1996)(Cenelec EN50325-4, 2001). Although it derives from CAL CiA standard, CANOpen offers different services and may not provide for some features present in CAL CiA. For example, in this last standard a particular entity, called DBT Master (CiA/DS204-1, 1996)(CiA/DS204-2, 1996), is responsible to distribute the identifiers to each message in the network. The protocol proposed in (Cavalieri, 2001) is fully based on this entity. The not mandatory presence of DBT Master in CANOpen requires the re-definition of an identifier assignment policy for the CANOpen system.

On the basis of what said, the aim of this paper is to propose a protocol able to cope with on-line management of identifiers in CANOpen communication system. The protocol is able to assign identifiers to new nodes, added to the network, and to reconfigure identifiers when necessary. Further, it is full compliant with the CANOpen standard (CiA/DS301, 1996)(Cenelec EN 50325-4, 2001). No modifications to the CANOpen are required, as the protocol proposed is just an add-on.

The paper will clearly describe the dynamic identifier assignment protocol and will demonstrate its compatibility with the CANOpen standard. Finally a performance evaluation will be presented in order to point out the capability of the protocol to assign the identifiers and to respect all the real-time constraints of the messages to be transmitted.

2. A BRIEF OVERVIEW OF THE PROTOCOL

The protocol here proposed is based on the definition of a centralised policy for the assignment of identifiers to messages in the CANOpen communication system. In this standard, the identifies are called Communication Object Identifiers (COB-ID). It is assumed that one of the communication nodes in a CANOpen acts as master and manages the assignment of the COB-IDs. The node responsible for this is called the Schedule Master (SM) and the other nodes are the Schedule Slaves (SS). They are the communication nodes connected to sensors/actuators/controllers, so producing information flow featuring both real-time (RT) and non real-time (NRT) constraints. As known, Real-time traffic may be further divided into hard real-time (HRT) and soft real-time (SRT).

The SM receives information from each SS about the traffic it generates (e.g. real-time features such as period and deadline). The SM executes a suitable scheduling algorithm able to assign COB-IDs to each SS in such a way time requirements of each SS are satisfied through the contention mechanism featured by the CAN Data Link Layer. Finally the SM will distribute the assigned COB-Id to each SS. If one or more time requirements cannot be met by COB-ID assignment, the SM will activate a procedure to reassign all or a part of COB-IDs, in order to meet the unfulfilled requests.

It's important to point out that no scheduling algorithm is here proposed, as it goes beyond the aim of the paper. Section 4.3 will give some details about the features that a scheduling algorithm must offer.

The proposed protocol features different behaviours according to the operational phase of the CANOpen communication system: at system start-up and run-time. On system start-up, when no nodes have as yet obtained COB-IDs, the SM collects the list of the time constraints to be fulfilled from each node connected to the network. Then the SM runs a scheduling algorithm in order to assign the COB-ID to each request and distributes the COB-IDs to the SSs. At run-time, when a new node is added to the network and/or an SS that is already active needs to obtain new COB-IDs for processes that have been added to it, the SM has to collect the COB-ID assignment requests from the new node or the existing one and has to assign and distribute the COB-IDs to it. At system run-time a re-assignment procedure must be also played by the SM, when one or more COB-ID don't fulfil time requirements of the relevant time critical processes.

3. COMMUNICATION IN CANOPEN STANDARD

According to (Cenelec EN 50325-4, 2001) and (CiA/DS301, 1996) communication in CANOpen is realised through two types of communication objects: Process Data Object (PDO) and Service Data Object (SDO).

The real-time data transfer is performed by means of PDO on a producer/consumer(s) basis. There are two kinds of use for PDOs. The first is data transmission (realised through Transmit-PDO) and the second data reception (through Receive-PDO). Devices supporting Transmit-PDO are PDO producers and devices which support the Receive-PDOs are called PDO consumers. Each PDO (Transmit and Receive) corresponds to entries in the device Object Dictionary hold by each device in a CANOpen network. The Object Dictionary is essentially a grouping of objects accessible via the network in an order pre-defined fashion. Each object within the dictionary is addressed using a 16-bit index. Each PDO is fully described by a Communication Parameter and a Mapping Parameter. Communication Parameter specifies, among others, the COB-ID used for the exchange of information. Mapping parameter specifies the objects mapped to the PDO-based information flow. Communication and Mapping Parameters corresponds to entries in the device Object Dictionary. In CANOpen for each device it's possible to configure at most 512 Transmit and 512 Receive PDOs. PDO transmission shall follow the producer/consumer relationship. One of the services allowing a producer to send data to consumer(s) is the WritePDO one, that will be used in the proposal here presented.

SDO is used to transfer larger data set, containing arbitrary large block of data. Another difference with PDO is that SDO communication is based on a client/server model. Basically an SDO is transferred as a sequence of segments. Like PDO, each SDO corresponds to entries in the device Object Dictionary. Two kinds of SDO entry are present in the Object Dictionary: the Server and the Client SDO Parameter. The last is used by the requesting node, while the first by the responding node. In CANOpen, for each device it is possible to configure at most 128 Server PDO Parameters and 128 Client SDO Parameters. Among other things, a Server/Client SDO Parameter specifies the COB-IDs used for the exchange of information from client to server and from server to client. One of the services used in the SDO-based communication is the SDODownload, that will be used in the proposal.

Like all CAN-based communication systems, all the messages sent in the network features a priority given by the COB-ID. Considering a 11-bit ID CAN network, the COB-IDs available for the communication are 1760, ranging from 1 to 1760.

The COB-ID assignment in CANOpen network may be pre-configured and/or dynamic. The adoption of the services defined in CAL CiA (CiA/DS201, 1999)(CiA/DS204-1, 1996)(CiA/DS204-2, 1996) to dynamically manage the COB-ID assignment is foreseen. In this case, a particular entity called DBT Master plays the role to assign and distribute identifiers to each communication node, maintaining these identifiers into a particular database. As can be found in (CiA/DS301, 1996), DBT Management is not mandatory in CANOpen communication system. For this reason, it wasn't considered in the proposal here presented.

4. A DETAILED DESCRIPTION OF THE PROTOCOL PROPOSED

The aim of this section is to present a more detailed description of the protocol proposed for COB-ID assignment. It will be divided into three parts, referring to the actions required for the assignment of new COB-IDs at system start-up, the assignment of COB-IDs to new nodes or to new processes added to already existing nodes, and to the COB-ID re-assignment procedure.

4.1 COB-ID Assignment at System Start-up

At the system start-up, according to the protocol here proposed, the SM is the only node in the network enabled to start the communication; all the other nodes (i.e. SSs) are in a stand-by condition, listening for any command sent by the SM and reacting to this command as explained in the following.

The SM polls all the nodes, sending to each one of them a particular message containing the information about the notification that the sending node is the SM, its address (i.e. node ID) and the command to

receive the COB-ID assignment requests. After the SM has sent this message to a particular node, it waits until a time-out expires. If the SM doesn't receive any request from the SS, it will pass to poll the next node in the network. The need for a time-out for the SM is clear as the CANOpen communication systems features the presence of 127 nodes, but a lower number of nodes may be present at system-start-up.

The service used by the SM to perform this protocol is the WritePDO service provided for by the Process Data Object (PDO) (CiA/DS301, 1996). With this Store-Events type service the SM will compel each SS to transmit COB-ID assignment request(s).

If the SS is connected to the network reacts to the previous service by preparing and sending one or more COB-ID assignment requests, using SDODownload service provided for the Service Data Object (SDO).

When the SM will collect all the requests coming from the SSs connected to the network, it will run a suitable scheduling algorithm, in order to evaluate the COB-IDs to be assigned to each SS, able to respect the relevant time requirements and constraints. Finally, the SM will distribute these COB-IDs to each SS using again the SDODownload service.

Figure 1 shows the sequence of messages generated on the network. As can be seen, the figure shows also the COB-IDs used to transmit the services WritePDO and SDODownload. As COB-ID assignment is not featured by real-time requirements, it was assumed to use COB-IDs whose values belong to an intermediate priority class, considering that COB-ID range is from 1 to 1760.

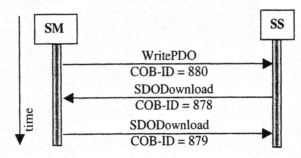

Figure 1 - Sequence of messages involved in the request and assignment of COB-IDs at system start-up.

4.2 Run-Time COB-ID Assignment

After the system start-up, if a new node is added to the network, it can't start any activities. It has to wait for an explicit command from the SM of the network. For this reason, the protocol here proposed foresees that at run-time the SM periodically polls all the nodes whose node ID is not in its list of active nodes. This list has been built at system start-up during the polling procedure described in the previous sub-section.

When the polling involves a new added node, this last can send all its time requirements and can receive the relevant COB-IDs from the SM. The exchange of information between SM and the new added node is quite similar to that depicted in Figure 1. In this case the WritePDO service is used by the SM to poll the nodes not included in the list of active nodes. The SDODownload services are used to collect COB-ID assignment requests from SSs and to distribute the COB-IDs to the SSs.

Let's now consider another scenario, featured by the presence of a SS already present in the network (i.e. whose node ID is included in the list of active nodes hold by the SM), requesting new a COB-ID assignment. This happens when one or more processes are started within the SS. The SS will notify this event to the SM by a WritePDO service. With this Store-Events type service, the SM will be aware of the COB-ID assignment need featured by the particular SS. After this notification, the SS prepares and sends an explicit COB-ID assignment request, using the SDODownload service.

When the SM will receive the requests coming from the SS, it will evaluate the COB-IDs and will distribute them to the requesting SS by the SDODownload service.

Figure 2 shows how the COB-ID request and assignment protocol at run-time is implemented in the case an already existing node requires other COB-IDs to be assigned, pointing out the COB-IDs used for the information flow.

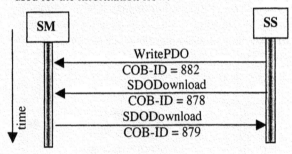

Figure 2 - Sequence of messages involved in the run-time request and assignment of COB-IDs.

4.3 Run-Time COB-ID Re-Assignment

COB-ID re-assignment procedure is needed when one or more "unsatisfactory" COB-ID assignments happen. Definition of unsatisfactory assignments can't be given if general criteria about the COB-ID assignment algorithm are not pointed out.

As said before, the definition of the scheduling algorithm played by SM for the COB-ID assignment is beyond the aim of the paper. Literature presents a lot of scheduling algorithms, and the reader can easily find the most suitable to be used in the proposal here presented. In order to do this, the main features the scheduling algorithm has to hold, must be clearly defined.

First of all, the algorithm must base the scheduling procedure on the time requirements of the requesting

process (i.e. the hard, soft and non-real time kind of information flow, its period and deadline, if any). Then, the COB-IDs must be assigned in such a way the highest priority COB-IDs (i.e. the COB-IDs with the lowest values) are assigned to the hardest real time processes with the shortest deadline.

It's clear that there is the possibility that a COB-ID value corresponding to a particular kind of traffic and a particular deadline is not available during COB-ID assignment procedure. For example let assume that COB-ID values belonging to the range [900-1000] have been already assigned to hard-real time process with deadline ranging from 7 to 10 msec. Further, let assume that no free COB-ID are available in this range. If a new hard real time process featured by a deadline of 8 msecs requests a new COB-ID assignment, a COB-ID outside the range of [900-1000] must be assigned.

In this case, the SM has two possibilities: a higher priority COB-ID or a lower priority COB-ID could be assigned to the requesting SS. There are no constraints on the choice performed by SM; choosing a higher priority COB-ID the SM will satisfy the time constraints of the requesting node, but it will introduce the possibility that more critical processes can't find free COB-IDs in the future. On the contrary, choosing a lower priority COB-ID means that the time constraints of the requesting SS may be not satisfied. In some cases, the SM is obliged to choose a lower priority COB-ID, for example when no higher priority COB-IDs are available at the moment. If the SM chooses to assign a COB-ID featured by a lower priority than that relevant to the assignment request then an "unsatisfactory" COB-ID assignment arises.

Coming back to the previous example, and assuming that the SM has assigned to the hard real time process featured by a deadline of 8 msecs a COB-ID higher than 1000, generally it's not possible to state that the deadline of the process will be certainly missed. Several techniques to ascertain this are available in literature. For example in (Tovar, 1999) a worst-case response time evaluation technique has been shown. It's clear that whatever procedure for investigating the respect of the deadline for each "unsatisfactory" assignment cannot be realised on-line during COB-ID assignment procedure.

So in the proposal, it was foreseen that periodically the SM evaluates all the previous "unsatisfactory" assignments. If a COB-ID not able to respect the deadline of the process is found, then a reassignment procedure is activated by the SM. The actions taken by the Schedule Master are as follows:

1 reassigning the "unsatisfactory" COB-ID to a higher priority value (if COB-ID values are now available), evaluating the respect of the deadlines;
2 sending to the SS relevant the reassigned COB-ID a new COB-ID, using the WritePDO service, as depicted in Figure 3. The figure shows the COB-ID used in the information flow.

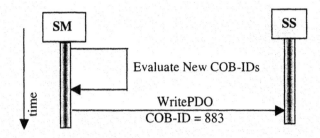

Figure 3 - Operations and messages on the network following a re-assignment event.

5. A DETAILED DESCRIPTION OF THE SERVICES REQUIRED TO IMPLEMENT THE PROPOSAL

The aim of this section is to give details of the services and parameters used to implement the dynamic identifier assignment protocol described in the previous section. As said, both PDO and SDO-based services available in (Cenelec EN 50325-4, 2001) have been used. According to the standard, their use implies the need to define particular entries in the Object Dictionary of the devices (SSs and SM) and to reserve COB-IDs for the information flow needed to implement the protocol here proposed.

The description of the PDO and SDO-based services used in the COB-ID assignment protocol will be made according to the various phases of the protocol.

5.1 System Start-up

As said in Section 4.1, the COB-ID assignment at system start-up has been realised using two services: WritePDO and SDODownload. Further, it was said that the exchange of the service WritePDO was realised using COB-ID 880, and exchanges of service SDODownload using COB-IDs 878 and 879.

WritePDO is used by the SM to notify to each SS its presence and to request COB-ID assignment requests from SS. WritePDO needs two arguments:

- the PDO Number of the TransmitPDO parameters in the Object Dictionary of the sending SM and of the ReceivePDO parameters in the Object Dictionary of each receiving SS;
- the data to be transmitted.

In the proposal the following arguments have been used:

WritePDO(1, event_collect)

The first argument, as said before, refers to the PDO Number of the Transmit/Receive-PDO. In the paper it was assumed to use the PDO Number 1 to realise the communication between SM and SS at system start-up. In particular, the SM has the Transmit PDO number 1 stored in its OD. It's featured by a PDO Communication parameter at the index 1800h and a PDO Mapping Parameter at index 1A00h. In the PDO Communication parameter the COB-ID number 880 is associated. Each SS has the Receive PDO number 1 stored in the OD. It's PDO Communication

parameter is at index 1400h and contains the COB-ID 880.

The second argument, event_collect, represents the data to be transferred. This data is a record made up by the fields: Node-ID of the sending SM and the Node-ID of the SS to which the service is sent.

SDODownload is used by each SS to send to the SM the COB-ID assignment requests. In this case it was assumed to reserve the COB-ID 878 for the exchange of this service. SDODownload is featured by several arguments, among which the most important to be defined are:

- the SDO Number of the ClientSDO parameter for the request from a client, and the SDO Number of the Server SDO parameter for the response from the Server; in the proposal it was assumed that the SM plays the role of Server and each SS that of Client;
- the data to be transmitted.

In the proposal the following arguments have been used:

SDODownload(1, schedule_data)

The first parameter refers to the ClientSDO Number stored in both SM and each SS. According to CANOpen (Cenelec EN 50325-4, 2001) the first ClientSDO Number corresponds to an entry at index 1280h in the Object Dictionary of each device.

The second parameter, schedule_data, represents the data to be transferred. It is a structure featuring the following fields:

- RTtype indicates the type of real-time service requested by the SS: HRT, SRT, or NRT;
- Deadline indicates the deadline of the periodic process (valid only for HRT and SRT services, ignored for NRT services);
- Period indicates the period of the periodic process. A value of 0 is used to indicate asynchronous processes;
- NodeID identifies the number of the module and can take a value between 1 and 127;
- ProcessID identifies the process on the node and can take a value between 0 and 65535;
- COBname is a string of 14 characters containing the name of the COB.

SDODownload service is also used in the proposal for the distribution of the COB-IDs to each SS by the SM, after it has collected all the COB-ID assignment requests from SS and has performed the relevant assignment. In this case it was assumed that the exchange of information from SM to SS is realised using the reserved COB-ID 879. Further the following arguments of the SDODownload service have been used:

SDODownload(2, schedule_result)

The first parameter refers to the ServerSDO Number stored in both SM and each SS. The second ServerSDO Number corresponds to an entry at index 1201h in the Object Dictionary of each device.

The second parameter, schedule_result, represents the data to be transferred from the SM to each SS. It is a structure featuring the following fields:

- NodeID identifies the number of the module to which the reply is addressed;
- ProcessID identifies the process on the node that has requested the COB-ID;
- COBname is the name of the COB;
- COBID is the identifier assigned by the SM.

5.2 Run-Time

As explained in Section 4.2, a new node added into the CANOpen network or an already existing SS, may advance COB-ID assignment requests. The protocol here proposed features the WritePDO service used by the SS to notify its need to the SM, and the SDODownload service to collect the COB-ID assignment requests from each SS and for the COB-ID distribution to each SS by the SM. Exchange of WritePDO service from SS to SM is realised using the reserved COB-ID 882, while the exchange of SDODownload is performed using COB-IDs 878 (from SS to SM) and 879 (from SM to each SS).

In the proposal the following arguments have been used for the WritePDO service exchanged from SS to SM:

WritePDO(1, event_novel_assignment)

The first argument, as said before, refers to the PDO Number of the Transmit/Receive-PDO. In the paper it was assumed to use the PDO Number 1 to realise the communication between SS and SM at run-time. In particular, each SS, which now is the initiating node, has the Transmit PDO number 1 stored in its OD. It's featured by a PDO Communication parameter at the index 1800h and a PDO Mapping Parameter at index 1A00h. In the PDO Communication parameter the COB-ID number 882 is associated. The SM has the Receive PDO number 1 stored in the OD. It's PDO Communication parameter is at index 1400h and contains the COB-ID 882.

The second argument, event_novel_assignment, represents the data to be transferred. This data is a structure made up by only one field represented by the Node-ID of the sending SS.

About the arguments of the services SDODownload used to collect COB-ID assignment requests from the each SS and to distribute the new assigned COB-IDs to each SS, they are the same used in the system start-up phase of the protocol.

5.3 COB-ID Re-assignment

During COB-ID assignment procedure, the SM may realise that one or more COB-IDs are not compliant with the real-time constraints of the requesting node. In this case it was assumed that periodically, the SM try to adjust the COB-ID assignments, changing one or more COB-IDs in order to better respect time constraints of a particular node. The exchange of information between SM and SS to send the new

assigned COB-ID is realised by the WritePDO service, as explained in Section 4.3. Exchange of WritePDO service from SS to SM is realised using the reserved COB-ID 883.

The following arguments have been used for the WritePDO service exchanged for the COB-ID reassignment purpose:

WritePDO(2, event_reassignt)

The first argument refers to the Transmit/ReceivePDO Number. The PDO Number 2 is used to realise the communication between SM and SS for the COB-ID reassignment. The SM, which is the initiating node, has the Transmit PDO number 2 stored in its OD. It's featured by a PDO Communication parameter at the index 1801h containing the COB-ID number 883, used for the information flow. Each SS has the Receive PDO number 2 stored in the OD. It's PDO Communication parameter is at index 1401h and contains the COB-ID 883.

The second argument, event_reassign, represents the data to be transferred. It is a structure featuring the following fields:

- NodeID identifies the SS to which the new COB-ID is sent;
- OLD_COB-ID is the COB-ID that must be changed;
- NEW_COB-ID is the reassigned COB-ID.

6. PERFORMANCE EVALUATION

The aim of this section is to present an assessment of the COB-ID assignment strategy here proposed. The assessment has been carried out taking into account that the main aim of the COB-ID assignment procedure has to be the respect of the time constraints (e.g. deadline) of the real-time messages. So the assessment was finalised to verify the capability of the proposal to respect all the real-time constraints of the requesting node in a CANOpen network.

As said in the Introduction, literature presents a lot of paper about scheduling solutions in CAN-based systems. However, these solutions are generally based on the CAN ISO IS-11898 protocol adopting a proprietary solution at the Application layer. On the other hand, the proposal here presented is based on a standard Application layer (i.e. the CANOpen). For this reason a comparison (in terms of capability to respect real-time constraints) between the scheduling solution proposed in this paper and those available in literature (based on CAN ISO IS-11898) hasn't considered. In fact, in the case of some kind of difference in the assessment, it would be very difficult to attribute this difference to the algorithm featuring the scheduling solution or to the number (and kind) of communication layers on which the scheduling solution is based.

Comparison with the optional COB-ID assignment strategy that could be used in CANOpen, have been assumed to be more meaningful in order to evaluate the scheduling solution here proposed. As can be found in (CiA/DS301, 1996), CANOpen may foresee the presence of a COB Database, containing registration of all the COB-IDs transiting through the network. A particular entity called DBT Master is responsible to assign COB-ID and make the relevant registration into the COB Database (CiA/DS204-1, 1996) (CiA/DS204-2, 1996). As already said, use of DBT Management is not mandatory in CANOpen communication system. So it could be very interesting to compare the capability in respecting the real-time constraints of the information flow in a CANOpen based on the use of DBT Master and in a CANOpen where the COB-ID procedure here presented runs. In this case, comparison is meaningful as both solutions are based on the same Application layer.

A CANOpen communication system, the proposed COB-ID assignment strategy and the optional COB-ID assignment strategy based on DBT Management have been modelled using Estelle (ISO 9074, 1997). The use of Estelle has also allowed to validate the COB-ID assignment protocol presented in this paper. Simulations have been carried out varying different parameters, like the number of variables to be transmitted and the relevant time constraints. For each simulation COB-ID assignment policies (that proposed in the paper and that offered by the CANOpen) have been evaluated verifying that all the time requirements taken into consideration have been fulfilled.

Performance evaluation and comparison between the proposal presented in the paper and the DBT management foreseen in the CANOpen has been performed according to the procedure described below.

Once a particular communication scenario made up by a bit rate value of the communication system, a certain number of variables to be transmitted, their time constraints (real-time kind, deadline, period) has been fixed, simulation is performed to obtain the COB-ID assignment realised according to the approach here presented and that offered by the standard CANOpen (using DBT services of CAL CiA).

On the basis of the COB-IDs assigned to each variable, both in the start-up phase and in run-time, the worst-case response time analysis presented in (Tovar, 1999) is applied. This analysis allows obtaining the worst-case response time value of each variable. This value has been compared to the relevant deadline, in order to check if this last has been respected.

In the following, some of the most significant results of the performance evaluation carried out will be presented.

Table 1 summarises the communication scenario taken into consideration. As can be seen, 600 variables have been considered, all featured by hard real-time constraints. Length, period and deadline of the variables are shown in the table. A CANOpen communication network running at 100Kbps has been considered. Then it was assumed that at system start-up the 200 variables featuring deadline of 220

msec and the 200 variables with deadline of 400 msecs were present. At run-time the other variables shown in the table were assumed to be added to the system.

Table 1 - Communication Scenario

Number of Variables	Period	Deadline	RT Type	Size bytes
200	220 msec	220 msec	hard	2
200	400 msec	400 msec	hard	2
100	125 msec	125 msec	hard	2
100	500 msec	500 msec	hard	2

Figure 4 shows the Worst Case Response Time (WCRT) versus the deadline of each variable (each value in the abscissa represents a variable, the first 400 of which are those present at system start-up, as said).

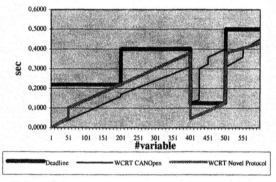

Figure 4 - Worst Case Response Time Evaluation

Worst case response time is depicted considering the proposal here presented (WCRT NovelProtocol) and the standard CANOpen (WCRT CANOpen). As can be seen, the protocol here proposed assures that all the time constraints are satisfied. Using DBT services of CAL CiA, some deadlines are not respected (in particular the 125 msecs deadline of the 100 variables added at runtime).

Another simulation was performed taking into account the communication scenario shown in Table 1. The only difference is that now it was assumed that at system start-up 150 variables featuring deadline of 220 msec and 150 variables with deadline of 400 msecs were present. At run-time, the other 50 variables featuring deadline of 220 msec, the other 50 variables with deadline of 400 msecs and the other variables shown in the table 1 were added to the system.

Figure 5 shows the comparison of the Worst Case Response Time considering the protocol here proposed (WCRT NovelProtocol) and that based on DBT services (WCRT CANOpen) versus the deadline of each variable. As can be seen, again the protocol here proposed assures that all the time constraints are satisfied. Using DBT services of CAL CIA, some deadlines are not respected.

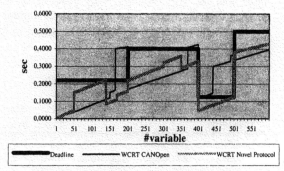

Figure 5 - Worst Case Response Time Evaluation

7. CONCLUSIONS

The paper has presented a protocol for dynamic assigning COB-IDs in a CANOpen communication system. The assignment strategy is an original one and is perfectly compatible with the Application Layer standard, as shown in the paper.

REFERENCES

Avind K., Ramamritham K. and Stankovic J.A. (1991). A Local Area Network Architecture for Communication in Distributed Real-Time Systems. Real-Time Systems. Vol.3, No.2.

Cavalieri S. (2001). A Protocol for Dynamic Assignment of Identifiers in Can Application Layer. Proceedings 4th IFAC FeT Conference Fieldbus Technology, Nancy, 15 - 16 November 2001, Palais des Congrès, pp.89-96.

Cenelec, EN50325-4 (2001). Industrial Communications Subsystem based on ISO 11898 (CAN) for Controller-device Interfaces. Part 4: CANOpen.

CiA/DS201. (1996). CAN Application Layer for Industrial Applications: CAN in the OSI Reference Model.

CiA/DS204-1. (1996). CAN Application Layer for Industrial Applications: DBT Service Specification.

CiA/DS204-2. (1996). CAN Application Layer for Industrial Applications: DBT Protocol Specification.

CiA/DS301. (1996). CAL-based Communication Profile for Industrial Systems-CANOpen. Version 3.0.

ISO IS-11898. (1993). Road Vehicle-Interchange of Digital Information-Controller Area Network (CAN) for High Speed Communication.

ISO 9074. (1997). Estelle: A Formal Description Technique based on an Extended State Transition Model.

Ouni, S., Kamoun, F. (1999). An Efficient Protocol for Hard Real-Time Communication on Controller Area Network. Proceedings 6th international CAN Conference '99. pp.04-16, 04-23.

Tindell K.W., Hansson H. and Wellings A.J. (1995). Calculating Controller Area Network (CAN) Messages Response Times. Control Engineering Practice. Vol.3, No.8.

Tindell K.W., Hansson H. and Wellings A.J. (1994). Analysing Real-Time Communications: Controller Area Network (CAN). Proceedings Real-Time Systems Symposium. pp.259-263.

Tovar E. (1999). Supporting Real-Time Communications with Standard Factory-Floor Networks. Ph.D. Thesis, University of Porto.

AN OBJECT-ORIENTED FRAMEWORK FOR CAN PROTOCOL MODELING AND SIMULATION

Arnaldo Oliveira[1] , Pedro Fonseca, Valery Sklyarov, António Ferrari

University of Aveiro/IEETA, Campus Universitário de Santiago, Aveiro - Portugal

Abstract: This paper presents an object-oriented software framework created with the aim of modeling and simulating distributed applications based on the CAN protocol. The created infrastructure consists of a modular simulator that allows describing all the system components, namely the interconnection bus and the distributed nodes. The object-oriented programming paradigm is used to simplify the modular implementation of the several protocol layers. The simulator was written in C++ and can be used as an executable specification to assist in the hardware implementation of a CAN controller. *Copyright ©2003 IFAC.*

Keywords: CAN, Modeling, Simulation

1. INTRODUCTION

Defined in the late 80's, CAN (Controller Area Network) (Bosch, 1991) found widespread acceptance in embedded distributed control systems, from automotive to industrial applications. In spite of its popularity, the application of CAN to safety-critical systems is, nevertheless, impaired by the event-triggered characteristics of the original definition. In CAN, a node can send a message at any time, provided there is silence on the bus (CSMA); the MAC mechanisms will handle the resulting collisions. As a consequence, a node sending a message has no guarantee in what concerns the delivery time of that message; depending on its priority, it may loose contention for several consecutive times, thus postponing the effective sending of the message.

For critical applications, time-triggered systems are preferred, due to their scalability, composability and dependability properties (Kopetz, 1997). The last few years saw the outcome of some proposals to improve the time characteristics of CAN (*e.g.*, TTCAN (Fuhrer *et al.*, 2000), FTT-CAN (Almeida *et al.*, 2002)). These take advantage of the fact that Bosch's and ISO specifications define only layers 2 and (partially) 1 of the ISO OSI model. With major or minor changes on the original definition, these new proposals impose some determinism in the message exchange behavior, namely by allowing a node to send its message at well defined instants in time. This is achieved by properly defining mechanisms in the layers above the original definition.

TTCAN (Time-Triggered Communication on CAN) started in ICC'98, the International CAN Conference, where an expert group, including CiA (CAN in Automation), chip providers, users and academia, joined the ISO TC22/SC3/WG1/TF6. The result was ISO 11898-4, part 4 of the ISO 11898 standard, that specifies time triggered communication on CAN ((ISO/TC 22/SC 3/WG 1, 2000)). FTT-CAN has been proposed at the

[1] Work supported by the FCT Ph.D. research grant SFRH/BD/3184/2000.

University of Aveiro as a mean to merge flexibility and timeliness in CAN systems. The aim is to achieve a communication paradigm that allows systems to be both timely, delivering the messages under the specified time constraints, and flexible, by not requiring the message set to be statically defined during system operation.

2. MOTIVATION

Both proposals for time triggered operation of CAN are built on top of the existing protocol, with little or no modifications (the aim of FTT-CAN is also to provide timely behavior with standard CAN controllers). The development of a new communication protocol requires its validation. Although simulation and formal validation play an important role here, they are not, on their own, sufficient. The last step in validation are always field tests, and these have to be performed with hardware devices. These tests should also involve the verification that the adopted solution is better than the alternatives. Ideally, we should have a communication controller that can be programmed to follow some specification and that can be modified.

Another issue to test is the robustness of the new protocols, specially in what concerns fault tolerance. To do this, faults have to be introduced in the system in a controlled and predictable way. Again, we meet the need for a controller that we can modify to our desire.

With these objectives in mind, a CAN system simulator started being constructed. More than a platform for validation and performing experiments, the developed simulator also acts as an executable specification to assist in the physical implementation of CAN and higher layers protocol controllers. By doing this, we aim at reducing the gap between the simulation model and the physical system: it is the specification that was used for validation that will be used for implementing the system. The initial requirements for the simulator were the following:

- Support for multiple nodes nodes attached to an interconnection bus;
- Modular implementation of the Application, Logical Link Control (LLC) and Medium Access Control (MAC) layers (other layers will be needed later for time triggered protocols);
- Simulation performed at time quantum or finer level.

The first objective means that it should be possible to integrate into the same application all the components of a typical application, namely an object representing the interconnection bus and as many objects as the number of attached nodes,

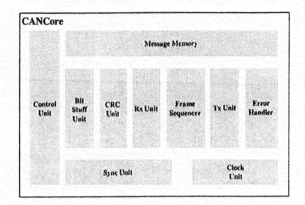

Fig. 1. *CANCore* architecture.

each with its own application layer. The second objective means a layered implementation of each node. However, it is important to refer that while the MAC layer requires an hardware implementation, being more easily obtained if described with an Hardware Description Language (HDL), the application layer is normally implemented in software. Finally, the third objective is very important in order to keep the simulator as close to real world as possible to effectively support the implementation.

The core component of a CAN node is a CAN controller, which implements the MAC and the LLC layers of the protocol. Thus, to meet the above requirements, a CAN controller started being modeled at RTL level using the EaSys language - a C++ based library and simulation kernel for hardware modeling at RTL level (Oliveira *et al.*, 2002). This approach was preferred because it allows avoiding the use of proprietary tools with co-simulation interfaces. However, due to the relative complexity of the CAN controller, some problems arisen while integrating and synchronizing the different modules of the core. This problem was solved moving the model to higher level of abstraction. As a result, a behavioral model of the CAN controller was written in C++.

3. CONTROLLER ARCHITECTURE

The developed model for the CAN controller implements the CAN 2.0 part A of the standard. Its architecture is depicted on fig. 1. It includes all the components required to implement the MAC layer specified on the standard. The developed model can be easily extended later to support the features of the CAN 2.0 part B of the standard. A short description of the modules is given next.

3.1 Bit Stuffing Unit

The *Bit Stuffing* Unit is used to: insert stuff bits on the transmitted bit stream and check/remove stuff bits from the received bit stream. It is shared by the transmission and reception parts of the core, because unless an error has occurred or the transmitter looses arbitration, within the stuffed fields the transmitted and the received bits must match.

3.2 CRC Unit

The *CRC Unit* calculates and checks the CRC sequence included in the frame. Similarly to the *Bit Stuffing Unit*, it is shared among the transmission and reception parts of the controller. In transmit mode, it calculates the CRC sequence during the *Start of Frame, Arbitration, Control* and *Data* fields. During the *CRC Sequence* field, the calculated sequence is shifted into the bus. In reception mode it compares the received sequence with the locally computed sequence in order to detect errors on the received bit stream.

3.3 Reception Unit

The *Reception Unit* latches the bus bit at the *Sample Point* and acknowledges a frame during the *Acknowledge Slot* field.

3.4 Transmission Unit

The *Transmission Unit* determines the bit to be transmitted by the node and latches it at the beginning of the bit time. The sources for the transmitted bit are the following:

- A message bit from the *ID, RTR, DLC or DATA* fields;
- A stuff bit;
- A CRC bit;
- A fixed polarity bit (recessive/dominant);
- An acknowledge bit generated by the *Reception Unit*;
- An error frame bit produced by the *Error Handler*.

3.5 Frame Sequencer

The *Frame Sequencer* plays a central role within the controller, performing the following tasks:

- Arbitration;
- Accepting requests to transmit messages;
- Detecting a start of frame in the bus;
- Sequencing fields in *Data* and *Remote Transmission Request* frames;
- Signaling the successful transmission of a message and the end of a message reception;
- Responding to overload frames.

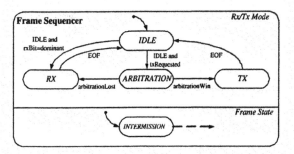

Fig. 2. *Frame Sequencer* state machine.

Fig. 2 shows a simplified behavioral specification of the *Frame Sequencer*. It consists of two parallel state machines: the *Frame State Machine* and the *Rx/Tx Mode State Machine*. The former defines the operating mode of the controller. The last establishes the sequence of fields for all frame types except error frames, which are generated directly by the *Error Handler*.

3.6 Error Handler

The *Error Handler* performs all activities related to fault confinement, error detection, counting and signaling. Internally it implements the mechanisms to detect the different error types and the error counters specified in the standard. When an error frame has to be sent, the transmission is performed by the *Transmission Unit*.

3.7 Message Memory

The *Message Memory* contains several message buffers. Four message buffers are available: two for transmission and two for reception. Two read/write ports are provided, one allowing access from the bus side (lower port) and the other from the application layer (upper port). Hardware implementations use a dual-port memory with different data port widths: 8/16/32 bits on the upper port and 1 bit on the lower port. This helps avoiding the large shift-register required in traditional implementations and takes advantage of the memory blocks currently available on commercial FPGAs.

3.8 Clock Unit

The *Clock Unit* generates all the clocks required to control and synchronize the activities of the other core components. Behavioral modeling of the CAN controller has shown that two clock signals are required for such purposes: a synchronization clock with frequency $f_{\text{SYNC}} = 1/T_Q$ (T_Q - Time Quantum period) and a control clock with frequency $f_{\text{CTRL}} = 2 \times f_{\text{SYNC}}$. This means that for a given time quantum frequency an input clock with only twice the frequency is needed.

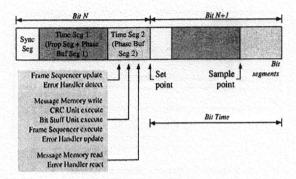

Fig. 3. Sequence of tasks within the *CANCore*.

3.9 Control Unit

The *Control Unit* generates all signals that control the other units, mainly enable and reset signals. The control sequence is not trivial and due to the lack of space, fig. 3 shows only a simplified view. Note that all tasks are performed during the *Phase Buffer Segment 2*, i.e. after the sample point.

3.10 Synchronization Unit

The *Synchronization Unit* generates the *sampleRxBit* and the *setTxBit* clocks used to latch the reception and transmission signals at the correct time instants, based on bus transitions and on the timing parameters of the node.

4. MODELING METHODOLOGY

After describing the controller architecture, let us take a look into the employed modeling methodology. As mentioned above, one of the targets is to support the hardware implementation of the CAN controller. To model hardware, several abstractions are normally employed, namely: modules, concurrent tasks, signals, ports and hardware specific data types. The adoption of an adequate modeling methodology is crucial to allow an easy migration from the high-level behavioral model in C++ (simulation) to an RTL model in VHDL (implementation). Thus, the following correspondences between hardware abstractions and software entities were established:

Modules are implemented in C++ using classes with internal attributes and interface methods.

Concurrent tasks are implemented in C++ using methods which must be explicitly called in order to model the desired behavior.

Clock and reset signals are modeled through the explicit invocation of methods.

Internal signals are internal variables of modules, implemented in C++ as class attributes.

Ports define the interface of a module. Once a module is described using one or more methods, its interface is implemented using the respective formal parameters.

At behavioral level, the operation of the system can be described as a set of sequential activities, which simplifies modeling. In fact, this approach was very important in order to define a correct sequence for the CAN controller events. Every object at behavioral level must have at least one operational method (e.g. Run) used to implement its functionality. It may also have an initialization method (e.g. Reset) used to put it in a well-defined state.

To illustrate these ideas lets take a look into the C++ model of the *Bit Stuffing Unit*. Its interface and implementation are defined in the class CBitStuffUnit shown on fig. 4. The *Bit Stuffing Unit* has two operating modes: *GENERATE* and *CHECK* used in transmission and reception modes respectively. Its attributes store the stuffing length, the polarity of the previous data bit and the number of consecutive equal bits. The input ports are the data bit, the operating mode and a enable signal. The output ports indicate if the next bit is a stuff bit, the respective polarity and if a stuff error has occurred. The CBitStuffUnit class constructor sets the attribute m_length with a concrete length value (5 for the CAN protocol). The Reset method initializes the class attributes and the module outputs. The Sync method models the synchronous part of the component implementing the counter of consecutive equal bits. Finally, in the Async method is implemented the calculation of the asynchronous outputs.

5. SIMULATOR STRUCTURE

For validation purposes, the executable specification of the CAN controller was integrated into a system simulator written in C++. The simulator consists of the following modules: *CANSim, CANBus* and *CANNode* (fig. 5).

5.1 CANSim

The *CANSim* module implements the main program where a bus and the required CAN nodes shall be created. The first action that must be performed is the instantiation of the bus object followed by the instantiation of the nodes. Fig. 6 shows an example of a main program which simulates a bus of length BUS_LENGTH with three nodes attached at different positions. The nodes share the same configuration parameters: *Baud Rate Prescaler, Time Segment 1, Time Segment 2* and *Synchronization Jump Width*. After creating

```
enum EBSUMode {BSUM_GENERATE = 0, BSUM_CHECK};

class CBitStuffUnit
{
public:
    CBitStuffUnit(uint length) : m_length(length){}
    ~CBitStuffUnit(){}
public:
    void Reset(bool& isStuffBit, bool& stuffError) {
        m_equalBitCount = 0;
        isStuffBit = stuffError = false;
    }

    void Sync(bool enable, bool dataIn) {
        if (enable)
        {
            if (dataIn == m_prevDataIn)
                m_equalBitCount++;
            else
                m_equalBitCount = 1;
            m_prevDataIn = dataIn;
        }
    }

    void Async(EBSUMode opMode, bool& isStuffBit,
            bool& stuffBit, bool& stuffError) {
        isStuffBit = (m_equalBitCount == m_length);
        stuffBit = !m_prevDataIn;
        stuffError = (m_equalBitCount > m_length) &&
                    (opMode == BSUM_CHECK);
    }

protected:
    const uint m_length;
    bool m_prevDataIn;
    uint m_equalBitCount;
};
```

Fig. 4. *Bit Stuffing Unit* interface.

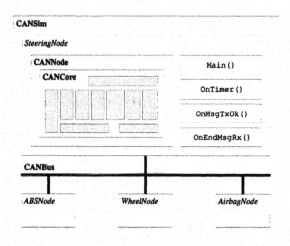

Fig. 5. Simulator structure.

the required objects and resetting the bus, the simulation is executed during 10000 cycles. The nodes are automatically initialized by the bus object.

```
#define MAX_NUM_NODES       16
#define BUS_LENGTH          8

#define BAUD_RATE_P_SCALER  16
#define TIME_SEG1           4
#define TIME_SEG2           2
#define SYNC_JUMP_WIDTH     1

int main(int argc, char* argv[])
{
    CCANBus bus(MAX_NUM_NODES, BUS_LENGTH);

    CWheelNode frontLeftNode(&bus, 1,
                    BAUD_RATE_P_SCALER,
                    TIME_SEG1, TIME_SEG2,
                    SYNC_JUMP_WIDTH);
    CWheelNode frontRightNode(&bus, 4,
                    BAUD_RATE_P_SCALER,
                    TIME_SEG1, TIME_SEG2,
                    SYNC_JUMP_WIDTH);
    CABSNode absNode(&bus, 7,
                BAUD_RATE_P_SCALER,
                TIME_SEG1, TIME_SEG2,
                SYNC_JUMP_WIDTH);
    bus.Reset();
    return bus.Run(10000);
}
```

Fig. 6. Example of a *CANSim* module.

Table 1. *CANBus* state when the node *absNode* starts sending a dominant bit.

Bus position	0	1	2	3	4	5	6	7
frontLeftNode	r	r	r	r	r	r	r	r
frontRightNode	r	r	r	r	r	r	r	r
absNode	D	r	r	r	r	r	r	r
Bus level	D	r	r	r	r	r	r	r

5.2 CANBus

The module *CANBus* represents a CAN bus. The bus is divided into elementary segments with null signal propagation time. The length of the bus corresponds to the number of elementary segments. A elementary bus segment represents the length a bit propagates during one simulation cycle. The minimum bus length is 1, representing a bus with null propagation time. Time is considered to elapse in the propagation from one segment to its neighbors. The bus is internally implemented with a bi-dimensional array of bits with $m + 1$ lines $\times$ n columns. Each column represents the bus state at each elementary segment. Each line represents the individual contribution of a single node into the bus state. An additional line contains the actual bus state, determined by the contribution of all nodes. If, after system reset, the node *absNode* starts transmitting a dominant bit, the state at the corresponding bus position is immediately changed while the remaining bus is kept at the same level (table 1).

243

5.3 CANNode

The *CANNode* module represents a generic CAN node attached to the bus. It is implemented by the class CCANNode, which cannot be instantiated because it is an abstract class. The main goal of this class is to implement all the functionality of the CAN protocol and provide a convenient programming interface to the application layer or higher layer protocols. The communication between the *CANCore* module and the application is based on the following primitives:

- void SetTimer(int nCycles); - used by the node application layer to program the timer used to generate interrupts at regular intervals;
- void ReqMsgTx(CCANMessage msg); - used by the node application layer to request the transmission of a message;
- void OnTimer(); - called by the simulator at regular intervals defined by SetTimer;
- void OnMsgTxOk(); - called by the simulator to notify the application layer of a successful transmission of a message (simulates the *Tx* interrupt of CAN controllers);
- void OnEndMsgRx(CCANMessage msg); - invoked by the simulator at the end of a successful reception of a message (simulates the *Rx* interrupt of CAN controllers).

A concrete CAN node must be derived from the CCANNode class, inheriting all the functionality implemented. The user has to implement the last three primitives, which are called by the simulator when the respective events occur. In addition to this methods, a Main function would be useful, but its implementation would require the use of threads in order to simplify its use.

5.4 CANCore

The *CANCore* module integrates all the blocks required to implement the MAC layer of the CAN protocol, whose function was already explained above. It provides a convenient simulation interface encapsulating in two methods (Execute and Reset) all the functionality of the CAN controller.

6. RESULTS AND FUTURE WORK

In the course of this work, a behavioral model for the CAN protocol was obtained. This model was tested against a commercial CAN controller and its correctness was verified in what concerns frame encapsulation/decapsulation, bit timing synchronization and error detection and signaling. Due to its complexity, the initial tests excluded the fault confinement mechanisms. The *Error Handler* was able to detect and signal errors, but the fault confinement mechanisms were not fully-implemented. The core's modularity simplifies the development of the simulator, allowing us to start with a crude definition and later refining it with further details.

For the purpose of validation and further implementation of both FTT-CAN and TTCAN, the upper layers (3 and above) will be modeled. The object-oriented approach simplifies this task: adding new layers to the existing model consists on deriving new classes, to which the attributes and methods corresponding to the new layers are added. The encapsulation principle allows all the details related to the simulation to remain hidden: the user needs only to concern with the application level.

7. CONCLUSION

In this paper, a time-quantum system simulator for the CAN protocol was presented. The cost of developing a new simulator is compensated by having a system description that is close to the hardware implementation. In this way, the validity of the results obtained during the simulation will be guaranteed in the hardware implementation which is, by all means, the final objective of this work. The simulator also provides a convenient platform for modeling and simulating a complete application based on the CAN protocol.

REFERENCES

Almeida, Luís, Paulo Pedreiras and José Alberto Fonseca (2002). The FTT-CAN protocol: Why and how. *IEEE Transactions on Industrial Electronics*.

Bosch (1991). CAN specification version 2.0.

Fuhrer, Thomas, Bernd Muller, Werner Dieterle, Florian Hartwich, Robert Hugel and Michael Walther (2000). Time triggered communication on CAN (Time Triggered CAN - TTCAN). In: *ICC 2000 - 7th International CAN Conference*. CiA - CAN in Automation.

ISO/TC 22/SC 3/WG 1 (2000). Road Vehicles — Controller Area Network (CAN) Part 4: Time Triggered Communicaction. Technical Report ISO/WD 11898-4. ISO.

Kopetz, Hermann (1997). *Real-Time Systems: Design Principles for Distributed Embedded Applications*. Kluwer International Series in Engineering and Computer Science. 1. ed.. Kluwer Academic Publishers.

Oliveira, Arnaldo, Valery Sklyarov and António Ferrari (2002). EaSys - a C++ based language for digital systems design. In: *Proceedings of the Fifth IEEE Design and Diagnostics of Electronic Circuits and Systems Workshop*. pp. 252–259.

ELSEVIER

IFAC

PUBLICATIONS
www.elsevier.com/locate/ifac

A CAN FIELDBUS BASED ARCHITECTURE FOR DISTRIBUTED CONTROL SYSTEMS IMPLEMENTATION

Alberto Bonastre, Juan V. Capella and Rafael Ors

Department of Computer Engineering
Polytechnical University of Valencia
Camino de Vera S/N - 46022 Valencia, Spain
{bonastre, jcapella, rors}@disca.upv.es

Abstract: The application of distributed systems theory to the implementation of industrial control systems is one of the best options to develop simple, scalable and physically distributed control systems.
Following this line, our group has proposed a new architecture based on Rule Nets (RN) as an HLP over CAN fieldbus for the implementation of Distributed Systems. This architecture has been applied to several control applications, obtaining excellent results.
The most outstanding feature of the architecture was its simplicity in the control systems design, without any efficiency loss.
New capabilities have been added to improve its features and develop new analysis tools, in order to increase its power and simplicity. In this work these new architecture features are presented. *Copyright © 2003 IFAC*

Keywords: Control architecture, distributed systems, CAN fieldbus, high level protocol, rule-based systems.

1. INTRODUCTION

Nowadays automatic control systems are indispensable in multitude of applications. Two of the most important challenges facing these systems are the physical distribution of the elements and the difficulty of programming them.

The fact that the control system elements are physically distributed might be solved by the application of distributed control systems. In these systems several distant nodes collaborate to each other to carry out the desired function. Since nodes need to communicate it is indispensable the existence of an interconnection network among them that enables the information exchange. The features obtained from the global system will be hardly dependent of the fieldbus selected. The properties of CAN (Bosch 1991) are unbeatable for the implementation of distributed systems.

Several problems appear in the programming of this kind of systems. The necessity to program different nodes, that could be different among themselves. The complexity of using a programming language to the general public, as well as the necessity to know the complex functions that manage the protocols to control the network access, without forgetting the coherence problems due to the parallelism, etc.

The use of rule based systems (RBS) for the implementation of distributed control systems offers some interesting advantages (Gupta and Sinha, 1998), not only offering certain intelligence level to the system,

but also being a good solution to solve the problems outlined before.

Accordingly, this work proposes an architecture that uses a RBS as application level, offering to the user a simple interface to specify the desired system behavior without the necessity of knowing any programming language (using a very similar language to the natural one). Besides, the expert system has the capacity to communicate to the user, in a completely automatic way, which nodes compose the system as well as the characteristics of each one of them. Once the user carries out the design it is possible to determine which dynamic properties will possess the system, as well as if the information coherence is guaranteed. Once the design has been validated, the user can order the automatic programming of each node through the network without having to program each one individually.

These characteristics make the proposed system easily profitable for users in general, who can start without important computer knowledge to program their own control system. This is very interesting in applications such as home automation, greenhouses, etc.

Rule nets are a tool for the design, analysis and implementation of rule based systems (RBS), that consist on a mathematic-logical structure which analytically reflects the set of rules that represents the knowledge of a human expert.

A centralized version of the proposed system (denominated *Rule Nets* or RN) has been successfully proved in the chemical analysis control (Peris, *et al.*,

1997), where it has demonstrated to have the characteristics exposed in the previous paragraphs. In (Bonastre, *et al.*, 1999) authors have already proposed a CAN based application protocol (HLP), called DESCAL, for the distribution of RN among a set of nodes interconnected by means of a CAN network. In this article a new version of DESCAL is presented, which includes new interesting features obtained in (Bonastre, 2001).

Firstly, the article introduces in section 2 the centralized version of the Rule Nets operation and later on its distributed one. The proposed protocol being presented with details in section 3. Developed tool for the protocol implementation and an application example are described in section 4. Finally, in section 5 the conclusions and future works are presented.

2. RULE NETS

Basically, Rule Nets (RN) are a symbiosis between Expert Systems (based on rules) and Petri Nets (PN), in such way that facts resemble places and rules are close to transitions. Similarly, a fact may be true or false; on the other hand, since a place may be marked or not, a rule (like a transition) may be sensitized or not. In the first case, it can be fired, changing the system state.

Like Petri Nets, RN admit a graphical representation as well as a matricial one; additionally, it also accepts a grammatical notation as production rules, which drastically simplifies the design, avoiding the typical problems associated with PN.

2.1 Introduction.

Every control system works with input and output variables, among other internal variables. In the case of RN, each variable has associated facts (corresponding to its possible state). For each associated fact, a set of complementary facts is defined as the one formed by the rest of facts associated with that variable.

Mathematically, a RN is a pair $RN = < F, R >$, where:

F is the set constituted by all the facts associated to all system variables (the ordinal of F is assumed to be f).

R is the set of rules. Each rule has one or several antecedents as well as one or several consequent facts (the ordinal of R is assumed to be r).

In order to represent a RN in matricial form, the following structures are defined:

Matrix A of antecedents. It is an $r x f$ matrix where element A_{ij} is 1 if fact f_j is antecedent in rule r_i, or 0 if it is not.

Matrix C of consequents. It is a $r x f$ matrix where element C_{ij} is 1 if fact f_j is consequent in rule r_i, or 0 if it is not.

Matrix D of unmarking. It is a $r x f$ matrix where element D_{ij} is 0 if element C_{ik} is 1 and fact f_j is complementary of fact f_k ; if it is not so, D_{ij} is 1.

Vector S of state. It is a f- components vector, so that each component corresponds to a fact of the system. In this vector, component S_i is 1 if fact f_i is true and 0 if it is false.

Rule r_j is sensitized in a determined states S_m if the two following conditions are met:

$$\overline{S_m} \text{ AND } A^j = 0 \qquad \overline{S_m} \text{ OR } C^j \neq 0$$

(where from now on X^j will denote the jth row of matrix X).

The firing of rule r_j sensitized in the S_m state yields a change in the state vector into a new one, S_{m+1}, which may be calculated from the equation:

$$\overline{S_{m+1}} = (\overline{S_m} \text{ AND } C^j) \text{OR } D^j$$

It can be easily demonstrated that all operations necessaries to run the RN are implemented by means of easy logical operations. It should also be remarked that there are other figures dealt with in the RN that are useful in the processes automation, as well as the possibility of obtaining the dynamic properties (liveness, ciclicity, etc.) owned by a system.

2.2 Rule Nets distribution.

Given a distributed system formed by n nodes, it is possible to implement a control system through a Distributed RN (DRN), so that each one posses a part of the global reasoning, *i.e.*, a part of the system variables and rules (Peris, *et al.*, 1998). For this purpose, each node N_i will handle a sub-set of variables and rules of the system (represented by a sub-vector S_i of the state vector S and three sub-matrices A_i, C_i and D_i of matrices A, C and D, respectively.

After defining the set of all system variables, two cases may arise. Some variables will be used in more than one node and others only in one. Those variables required in more than one node are *global variables* (will need a global variable identifier), whereas if only affect one node will be known as *local variables*.

Obviously, a variable will be local if all the rules in those that it appears are present in the corresponding node, whilst those appearing in rules of different nodes will necessarily be defined as global.

By minimizing the number of global variables, network traffic will be reduced. Besides, the variables referring to a physical input or output will necessarily be located in the node of the corresponding port. Taking this into account, there is an algorithm, explained in (Bonastre, 2001), that permits the optimum variables and rules assignment to each node (i.e. sub-matrices S_i, A_i, C_i and D_i). In spite of the algorithm complexity, non-optimal versions have been implemented successfully. It is possible to distribute the RN applying other criteria, such as fault-tolerance approaches that allow several nodes to update any global variable.

Write-Through propagation mechanism guarantees that if the rule net meets a series of properties, once it is distributed and working on a broadcast network, the information will be coherent and the whole system will run in a coordinated way.

3. DESCAL II PROTOCOL

The proposed system architecture consists on a distributed system is placed on a CAN network to which a group of nodes are connected. One of them is denominated Programming and Supervision Node (PSN). The rest corresponds to the so-called control nodes (CN), which are connected by means of the corresponding inputs and outputs to the process to be controlled.

3.1 Features.

Among others, the features offered by the protocol are:

- Self-identification of all the control nodes as well as the characteristics of each one, by means of the programming and supervision node.
- Simple graphical programming by the user of the required system behavior in the programming and supervision node. To carry out this programming, the user should not necessarily know any programming language, nor the fieldbus internal characteristics.
- Automatic distribution of the control algorithm among all the nodes.
- Allows implementing mechanisms for dependability, such as redundancy among nodes.
- Guided to be implemented in small-power microcontrollers.

3.3 Protocol stages and messages format.

By default, the CAN standard format (CAN 2.0 A) is to be used. The CAN extended version (CAN 2.0 B) can use the same philosophy, increasing the possibilities without any loss of compatibility, or keep reserved for future uses, as proposed in (Bonastre, *et al.*, 2002).

The protocol uses three types of messages:

Control messages: Their arbitration field always begins with 000, followed by a node identifier and, in the first 4 bits of the data field, the operation to be carried out, according to the table 1.

These messages can be directed to individual nodes (1 to 254) or use the diffusion address (0) to reach all the nodes. The address 255 is reserved for the PSN.

Table 1. Operation codes

Code	Operation	meaning
0001	RESET	Beginning of the protocol
0010	WHO	Identification phase start
0011	EOP	End of programming phase
0101	START	Execution phase start
1010	READ	Obtain an input variable value
1100	WRITE	Force an output variable value
1110	STOP	Halt
1101	ERROR	Error detected
Rest		Future features

Identification messages: Their arbitration field begins with the code 001, followed by the identifier of the involved CN. Two messages of this format are possible: *node identification* and *port identification*.

Load messages: Correspond to 010 and 011 codes, make possible the transmission of the variables and matrices from the PSN to each one of the CN's.

Updating Variable messages: Their arbitration field begins with a bit at 1, distinguishing it from the previous messages. Following this bit the code corresponding to the variable to be updated is placed.

The use of these types of messages is studied next with more detail according to the stage in which they are emitted.

Initialization. As shown in figure 1, when a CN starts, it is found in the so-called initial state, from which it will exit with the reception of a WHO message. At initial state, the CAN UART must be configured at 125 Kbps speed.

At the user's request (or after being started), the PSN will ask which nodes exist in the system with a global WHO message, as shown in figure 2. This message will

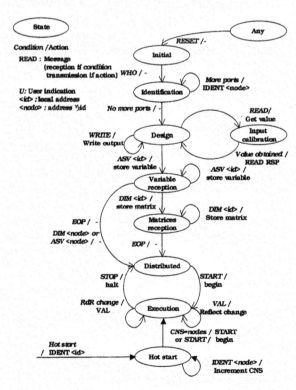

Fig. 1. CN state diagram.

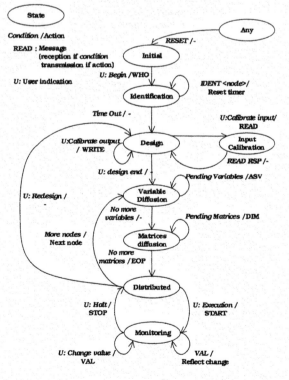

Fig. 2. PSN state diagram.

have the following format:

3	8	4
000	00000000	WHO
Identifier		Data

Against the reception of the WHO message every CN should compete for the network to transmit their node identification message (IDEN).

The node identification message will have the following format:

3	8	6	8	16	16
001	Node	000000	Portsnr	Signat	Speed
Identifier		Data			

Where *Node* corresponds to the identifier of the presented node. *Portsnr* indicates PSN how many ports the CN has, while *Signat* is a random 16 bits value to detect duplicate node addresses. Finally, *Speed* field is a bitmap where each bit reflexes a possible CAN speed for the node following the mask:

b_7	b_6	b_5	b_4	b_3	b_2	b_1	b_0
0	0	0	0	1 Mbps	500 Kbps	250 Kbps	125 Kbps

Every CN must be able to communicate at 125 Kbps, so minimum value of this field is 1.

After the successfully detection of every node, an individual identification of each port in these nodes is performed by means of port identification messages:

3	8	6	2	8	8
001	Node	Secnr	Type	Port	Resol
Identifier		Data			

Where *Secnr* is a sequence number (from 1 to 65), allowing message lost detection. Type indicates which operations are possible in this port.

Port is an address of the referred port inside the device, and finally *Resol* indicates number of bits of the ADC or DAC if existing.

To determine when the identification phase is over, the PSN keeps a timer from the last *ident* message seen. When this timer overpasses a fixed value (5 seconds) it is supposed that no other node wants to be presented.

Design. At this time, user should build its control system to fulfill the desired control features. The Rule Net creation, edition and debugging will be performed at the PSN, making the assumption that every port in the CN is locally controlled. At this phase, only remote inputs or outputs calibration is provided by the protocol.

The calibration consists on the remote control possibility over the inputs and outputs in a node. If the user wants to assign an analog magnitude level (for discrete conversion) with a variable value, these messages allow direct selection of the magnitude level by means of current values. For this purpose, the protocol defines two messages:

<u>READ</u>: It allows the PSN to force a determined CN to capture and transmit the value present in one of their input ports. The PSN generates a message with the following format:

3	8	4	8
000	Node	READ	Port
Identifier		Data	

When the node obtains the current input value and digitalizes it, the protocol answers previous message with a read response. Where *value* includes the obtained

measurement with as many bits as established at the identification phase.

3	8	4	8	*n* bits
000	Node	READ	Port	Value
Identifier		Data		

This is the only message where two nodes (PSN and CN) could send a CAN frame with the same identifier and different data field. In this case, a master/slave protocol is used, granting the absence of collision.

<u>WRITE</u>: It allows the PSN to force a specific CN analog output to adopt the analog value corresponding to the indicated one. The message format is the following one:

3	8	4	8	*n* bits
000	Node	WRITE	Port	Value
Identifier		Data		

Any problem associated with the calibration operation must be reported by means of an error message:

3	8	4	8	8
000	Node	ERROR	Port	Error code
Identifier		Data		

Where *Error code* identifies the event occurred.

Distribution. At the beginning of this phase, the PSN must execute the distribution algorithm, which assigns variables and rules to every node. A variable that only is known by one node is called *local variable*, in contraposition with the so-called *global variables*, that are relevant for more than one node, thus requiring that any change of state to be diffused through the communication network.

Relating to rules, best efficiency is achieved by means of single rule assignation (each rule is present in just one node), but fault tolerant approaches are possible performing a redundant distribution. The avoidance of coherence problems is granted by means of the CAN and Rule Nets properties.

After this protocol, each node has been assigned a subset of the whole Rule Net, being the transmission of these sub-networks required. This transmission is performed in four steps.

The first step consists on the transmission of a *dimensions distribution* message. The message format is the following one:

3	8	5	3	16	16
011	Node	00000	000	Facts	Rules
Identifier		Data			

By means of this message, the PSN indicates the CN (*node*) that it has been assigned a total of *Facts* and *Rules*, and it should be prepared to receive the next messages according of the transmitted dimensions

At the second step the PSN transmits as many *variable distribution* messages as variables have been assigned to the node.

3	8	8	10	6	8	$n \times (v-1)$
010	Node	Secnr	Var	States	Port	Values
Identifier		Data				

Secnr allows the CN to determine whether any message has been lost. The 10-bit *variable* identifier (*Var.*) only is meaningful in global variables. *States* allows until 32 states in a variable, *Port* relates the variable value with an input or output of the node and, in this case, the

248

Values fields indicates the bounds between states in the variable.

Matrix Distribution is the third step. By means of *distribute matrix* messages, the A, C and D matrices are transmitted to each node. The binary matrices must be raw-concatenated, and aligned to byte (adding zeroes). Several messages by matrix are possible, numbered by the *Secnr* field.

Antecedent matrix (A) is distributed by means of the following message:

3	8	5	3	Until 56
011	Node	Secnr	001	Values
Identifier			Data	

So does Consequent matrix (C):

3	8	5	3	Until 56
011	Node	Secnr	010	Values
Identifier			Data	

And Unmarking matrix (D):

3	8	5	3	Until 56
011	Node	Secnr	011	Values
Identifier			Data	

The forth step is the initial state distribution. By means of these messages, the initial value of every variable in the node is given.

3	8	5	3	Until 56
011	Node	Secnr	100	Values
Identifier			Data	

When every matrix has been received, the CN performs a boundary check, comparing the incoming matrices with the dimensions already received. If any error occurs, an *error* message is transmitted.

3	8	8	8
000	Node	ERROR	Error Code
Identifier		Data	

When implementing embedded control systems, matrices may be stored in ROM before distribution phase. In this case, subsequent executions of the system may use the so-called *EMS procedure*, explained later, which starts at this point with the START message.

Execution. The execution stage begins in each node when the PSN transmits a START message, which has the following format:

3	8	4
000	Node	START
Identifier		Data

Destination address of the message (*node*) can be the broadcast address, starting all node operation. When receiving this message, each node execute its RN.

When a CN should modify a global variable, because it has been read from an input or due to a deduction, it diffuses the new value of the variable through the net by means of the VAR message followed by the new value. Against this message, all the nodes modify their marking vector. The format of the message is the following:

1	10	8
1	Variable	State
Identifier		Data

Where *Variable* indicates the global identifier of the variable to update and *States* the new variable value.

In this stage the PSN can adopt monitoring functions or be removed off the system. However, in order to apply several of the enhanced mechanisms described below, it is indispensable that the PSN remains listening and actuate as a system controller.

3.4 Special features.

Several new features have been added to improve the protocol. These features are commented below.

EPROM mode start (EMS). This feature offers the possibility that a CN, previously programmed, begin the execution without a PSN. Any CN in the network must store current data on non-volatile memory when a STORE message is received. It also stores the total number of nodes at the system and a EMS bit set to '1'. After that, the node could be powered down.

When a CN is initiated, the EMS bit is checked. If set, an *IDENT <node>* message is transmitted without a *who* message, and the CN remains in listening mode. As every node wakes up and transmit its *IDENT*, a counter is increased in every node. If the counter in any node reaches the number of CN required means that the system is ready for execution and transmits an START message. An excessive amount of time without completing the count makes the CN to transmit a WHO command, which causes every node to transmit its *IDENT* message again.

Duplicate address test. To determine if more than one node is using the same identification a duplicate address test is provided.

After a *WHO* message, every node tries to transmit its IDENT <node>. This messages are seen by every node in the system because of the CAN network broadcast functioning. If any node receives an IDENT with its own address, must transmit a duplicate address message with the following format:

3	8	8	8
000	Node	ERROR	DUP (00000001)
Identifier		Data	

When receiving this message, the PSN will STOP all nodes and inform the user about the trouble.

It is possible that both nodes try to transmit its IDENT at the same time. In this case, and because the random sequence in the IDENT message, no one will transmit a valid CAN frame (CRC error). After several tries, nodes assume the duplicate address problem and transmit the previous message. No collision trouble arise because both nodes would transmit exactly the same message.

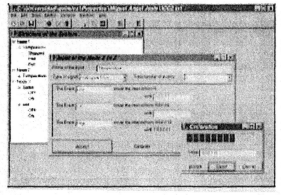

Fig. 3. Variable assignation and calibration.

Error messages. Several error messages are defined in order to communicate to the PSN, and thus the user, any problem that could arise. The error messages defined are shown in the table 2.

Table 2. Protocol error messages

Error	Code	Meaning
DUP	00100001	Duplicated node identifier.
ASV	00100100	Variable assignation error.
SECASV	00100101	Variable assig. sequence error.
ASVD	00100110	Var. has already been assigned.
ASVPORT	00100111	Port does not correspond with variable type.
DIM	00100010	Matrix dimension error .
SECDIM	00100011	Matrix assig. sequence error.
PORT	00110000	Port does not exists or is already used.
SIZEERR	00101110	CN out of memory .

Fault tolerance issues. This protocol allows the implementation of several fault-tolerance techniques. All of them need the PSN to remain as a watch dog of the system, and the adoption of several actions to deal with failure events.

In this case, PSN would send a WHO message periodically, which would be replied by means of an IDENT (ALIVE) message. If the IDENT message does not arrive, and after several tries, the PSN will assume a failure in corresponding node. This feature is not displayed in figure 1.

More sophisticated methods are contemplated. PSN knows which rules and variables belong to each node, so PSN can watch the network for variable updates. In this case, the PSN only would transmit a WHO message when no variable of one node has been updated for a fixed amount of time. It is also possible for the PSN to keep a copy of the distributed RN, simulate it and detect if an update that should have been transmitted is lost.

When the PSN notices that a node has fallen, several actions are available. PSN always will inform the user of the failure, but is also able to restart the system with the same RN, or distribute a new RN for degraded working mode.

Temporal issues. Mathematical form of Rule Nets allows the execution time evaluation of a determined RN. In this sense, and taking into account that this execution consists on simple binary operations (AND & OR) between bit matrices (usually implemented by bytes), a worst case (WCET) is defined, thus response time can be bounded.

Communication time on a CAN network can be limited, at the worst-case and in absence of transmission errors, because of the minimum latency between variable updates and deterministic collision resolution.

In this way, this protocol allows deterministic time responses, making it suitable for a great range of applications.

4. CONCLUSIONS

A new architecture for the distributed control systems implementation based on CAN network has been presented. This architecture offers several advantages, such as:

- Easy, simple but powerful programming of the desired behavior by means of logical conditional rules.
- Variable virtualization in logical values.
- Distributed nodes virtualization and auto-identification against the PSN.
- Control system centralized programming with possibility of physical calibration of magnitudes.
- Off-line evaluation of RN properties by state tree expansion and simulation-based debug tools.
- Rules and variables automatic distribution following any of the user's criteria (efficiency, fault tolerance, etc.)
- Centralized monitoring of the system.
- Fault tolerant mechanisms included on the protocol, such as redundancy, watch dog function, etc.
- Upper-limited response time of the control method.
- Suitable for its implementation on low-power microcontrollers.

Several programming environments have been implemented, one of them being shown in figure 3. These tools have been used for several control systems development, such as (Peris, *et al.*, 1997), obtaining very interesting results.

5. ACKNOWLEDGMENTS

This work is supported by the Spanish Comisión Interministerial de Ciencia y Tecnología under project number CICYT-TAP96-1090-C04-01.

REFERENCES

Bonastre, A., R. Ors, and M. Peris (1999). "A CAN Application Layer Protocol for the implementation of Distributed Expert Systems" *Proceedings of the 6th International CAN Conference.* Torino (Italy).

Bonastre, A. (2001). *Industrial Local Area Network: A new architecture for control distributed systems.* Ph. D. on Computer Science. Polytechnical University of Valencia.

Bonastre, A., R. Ors, J.V. Capella and J. Herrero, (2002). "Distribution of neural-based discrete control algorithms applied to home automation with CAN" *Proceedings of the 8th International CAN Conference.* Las Vegas (USA).

Bosch, R. (1991). *CAN Specification version 2.0.*

Gupta, M. and N. K. Sinha (1998). *"Intelligent control systems: theory and applications"*, IEEE Trans., Piscataway, NJ.

Peris, M., R. Ors, A. Bonastre and P. Gil (1997). "Application of rule nets to temporal reasoning in the monitoring of a chemical analysis process" *Laboratory Automation and Information Management,* Volume 33, Elsevier Science Publishers B.V.

Peris, M., A. Bonastre and R. Ors, (1998). "Distributed Expert System for the Monitoring and Control of Chemical Processes" *Laboratory Robotics and Automation,* Volume 10, 163 Elsevier Science Publishers.

DETERMINING END-TO-END DELAYS USING NETWORK CALCULUS

Kym Watson, * **Jürgen Jasperneite** **

* *Fraunhofer Institute IITB, Fraunhoferstr. 1,*
76131 Karlsruhe, Germany,
Fax:+49-721-6091-413,
e-mail: kym.watson@iitb.fraunhofer.de
** *Phoenix Electronics GmbH ,*
Automation Systems,
Dringenauer Strasse 30,
31812 Bad Pyrmont, Germany,
Fax:+49-5281-946-214 ,
e-mail:jjasperneite@phoenixcontact.com

Abstract: The a priori determination of temporal behavior is an essential requirement on the design of distributed real-time systems. It is shown how the analytical method Network Calculus can be applied to find hard upper bounds for transaction times in switched Ethernet networks. The analytical results are validated by simulation for an example scenario and the differences are discussed. *Copyright © 2003 IFAC*

Keywords: Realtime Communication, Ethernet, Network Calculus

1. INTRODUCTION

The increased deployment of Ethernet for field communication in automation has been actively debated for the last three years. One reason is the current series of addenda to the IEEE 802 standard, which make time critical data transmission appear to be possible. In particular, the extensions for switching technology and support for priorities must be mentioned, which make the previously used CSMA/CD procedure of Ethernet obsolete.

Before a communication technology originating in an office market can be introduced into industrial automation, several specific requirements must be met. The requirements range from industrial standard connectors and simple installation, up to a widely accepted application layer for interoperable communication. Furthermore, the question of real-time behavior must be answered. Real-time behavior means that given system functions are

executed in a time interval $[R, D]$ determined by the application environment. Here, R (Release) [1] denotes the earliest permissible completion time and D (Deadline) the latest permissible.

Due to its direct interaction with the technical process, the field level places time critical requirements on the communication. Exceeding an upper time limit may cause product parts or equipment to be damaged. This situation is aggravated by current developments in the area of transmission of safety critical data over fieldbuses (*INTERBUS Club*, web site, n.d.; *AS-Interface*, web site, n.d.; *Profibus Nutzerorganisation*, web site, n.d.; *Open DeviceNet Vendor Association*, n.d.).

Exceeding time limits in this case can endanger lives. The following simple example serves to demonstrate the immediate impact of the time

---

[1] The value of R is typically determined implicitly by the time instant of a function call.

behavior on the technical process. Fig. 1 shows the detection of a final position of a transport unit moving with velocity v and the resulting action (switching off a motor). The detecting sensor is at a distance s from the mechanical end of the transport path. The constraint $T_{PLC} + T_1 + T_2 \leq s/v$ must be adhered to to avoid damaging the equipment.

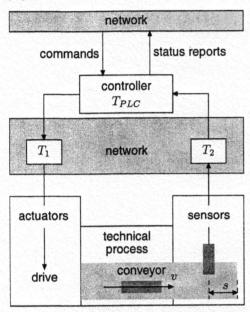

Fig. 1. Real-Time requirements on field communication

There is therefore a direct mutual dependence between the design of the mechanical construction, which has to maintain a given velocity v, and the automation design, due to the relationship between the minimal distance s and the processing time. Hence the automation design requires a priori knowledge of upper bounds for the various components of the processing time.

The Ethernet-based vendor concepts currently being defined IDA (IDA, 2001), Profinet (PNO, 2001), Ethernet/IP (*Ethernet Industrial Protocol (EtherNet/IP)*, n.d.) give no assistance with the determination of upper time bounds.

In the following we introduce a methodology which allows assured time bounds to be determined a priori for switched Ethernet networks. General guidelines and remarks on the practical application supplement the previous work of (Jasperneite *et al.*, 2002).

2. NETWORK CALCULUS

2.1 The principles of Network Calculus

Network Calculus is a widely applicable technique to assess the real-time performance of communi-

cation networks and is based on the fundamental work of (Cruz, 1991*a*), (Cruz, 1991*b*) and (Parekh and Gallager, 1993), (Parekh and Gallager, 1994). For a full introduction the reader is recommended to consult (Boudec and Thiran, 2001). Network Calculus basically considers networks of service nodes and packet flows between the nodes. Whereas traditional queuing theory deals with stochastic processes and probability distributions, Network Calculus involves bounding constraints on packet arrival and service. These constraints allow bounds on the packet delays and work backlogs to be derived, which can be immediately used to quantify the real-time network behavior. Traditional queuing theory, on the other hand, normally yields mean values and perhaps quantiles of distributions. The derivation is often difficult and upper bounds on end-to-end delays may not exist or be computable. The packet arrival process in Network Calculus is described with the aid of so-called *arrival curves* which quantify constraints on the number of packets or the number of bits of a packet flow in a time interval at a service node. Typical packet flows of interest are, for example, the packets of a given message type entering or departing from a service node. Let F be a packet flow and let $R(t)$ be the number of packets of F arriving in the time interval $[0, t]$. The flow is said to be constrained by or has arrival curve $\alpha(t)$ if, for all $0 \leq s \leq t$

$$R(t) - R(s) \leq \alpha(t - s) \tag{1}$$

whereby $\alpha(t)$ is a non-negative, non-decreasing function. Although one can consider packet arrival constraints, it is mostly more convenient to consider the corresponding work arrival curve. A packet P requires $W[sec]$ work in the service node N, if N needs this long to process P. Typically, W can be calculated from the packet length (including overhead), the transmission rate and interframe gap on the medium. Let $R_w(t)$ be the amount of work due to flow F which arrives $[0, t]$. Then F is constrained by a work arrival curve $\alpha_w(t)$ if, for all $0 \leq s \leq t$

$$R_w(t) - R_w(s) \leq \alpha_w(t - s)$$

A service curve describes the service of flow F in node N. Let $R'_w(t)$ be the amount of work completed in the time interval $[0, t]$. Then the non-negative, non-decreasing function $\beta(t)$ is called a (minimal) work service curve for F in N if, for all $t \geq 0$,

$$R'_w(t) \geq R_w \otimes \beta(t)$$
$$= \inf_s \{R_w(s) + \beta(t - s) : 0 \leq s \leq t\}$$

where $\otimes$ denotes the convolution operator. The work output flow is itself constrained by the arrival curve

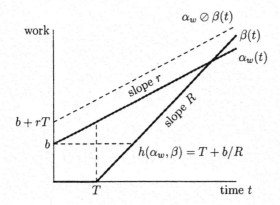

Fig. 2. The basic concepts of Network Calculus

$$\alpha_w \oslash \beta(t) = \sup_s \{\alpha_w(t+s) - \beta(s) : s \geq 0\}$$

where $\oslash$ denotes the deconvolution operator. In a packet network it is important to consider the packetization of the output. If packets flow from $\mathcal{N}_1$ to $\mathcal{N}_2$ then work arrives at $\mathcal{N}_2$ for further service only when the complete packet has been received. If $\beta(t)$ is a work service curve of a packet flow and δ_{max} is the maximum service time of packets of the flow, then $\beta_1(t) = \beta([t - \delta_{max}]^+)$ is a work service curve in $\mathcal{N}_1$ for the flow including packetization of the output.

The constraints given by an arrival and service curve for a flow suffice to calculate an upper bound on the delay of a packet in the service node and on the work backlog. The work backlog is bounded by the vertical distance between α_w and β, whereas, if the service order is FCFS (First Come First Served), then the packet delay is bounded by the horizontal distance:

$$work\ backlog \leq \sup_{t \geq 0} \{\alpha_w(t) - \beta(t)\} \quad (2)$$

$$delay \leq \sup_{t \geq 0} \{\inf\{s \geq 0\ with\ \alpha_w(t) \leq \beta(t+s)\}\} \quad (3)$$

Computations with arrival and service curves can be greatly facilitated if concave arrival and convex service curves can be applied. The most important representatives are the leaky bucket arrival curve $\alpha(t) = rt + b$ with rate r and burst or bucket size b and the rate latency service curve $\beta(t) = \max(0, (t - T)R)$ with latency T and rate R. Both are determined by just two parameters. The above delay bound in this case is simply $T + b/R$ (if $r \leq R$), which manifests the dependence on the service latency and the burst size. The output flow is constrained by the leaky bucket arrival curve $\alpha^*(t) = \alpha \oslash \beta(t) = \alpha(t+T)$ (cf. Fig. 2).

The packet generation in a flow F has burstiness $b > 1$ when more than one packet enters the service system in a very small time interval. This

occurs, for example, when a packet from an upper communication layer is segmented into several packets in a lower layer, e.g. at the interface from the IP layer to the MAC layer. As shown in figure 3, the output in flow F of a service system has a higher burst than the corresponding input flow when other flows have to be served as well.

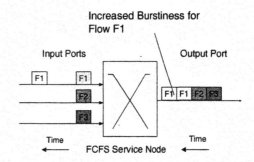

Fig. 3. Increased Burstiness at the output of a FCFS service node

2.2 Results for FCFS and PQ service nodes

In this paper two service strategies are considered: FCFS and non-preemptive priority queuing, denoted by PQ. The service order within a priority class in FCFS. We give two specific results on the Network Calculus for service nodes serving two or more flows according to a FCFS or PQ service strategy [refer to (Boudec and Thiran, 2001, corollary 6.2.1 and corollary 6.4.1) and (Watson, 2002) for these and more general results]. Denote $\max(x, 0)$ by x^+.

Proposition 2.1. [FCFS Split] Consider a service node $\mathcal{N}$ serving two packet flows F_1 and F_2 in FCFS order. Assume that $\mathcal{N}$ gives the aggregate flow a rate latency work service curve $\beta(t) = \max(0, (t - T)R)$ and suppose that F_2 is constrained by the leaky bucket work arrival curve $\alpha_2(t) = r_2 t + b_2$. Let $\theta = T + b_2/R$. Define β_1 as follows:

$$\beta_1(t) = [\beta(t) - \alpha_2(t - \theta)]^+\ for\ t \geq \theta$$
$$= 0\ for\ 0 \leq t < \theta$$

Then β_1 is a rate latency work service curve for F_1 with rate $R - r_2$ and latency θ.

Proposition 2.2. [PQ Split] A service node $\mathcal{N}$ gives prioritized packet flows F_i $(i = 1, \cdots, K)$ non-preemptive priority service. The priority of flow F_i is $p(i) \in \{1, \cdots, P\}$. F_i has higher priority than F_j whenever $p(i) > p(j)$. Let δ_i be the service time of packets of flow F_i for $i = 1, \cdots, K$. Suppose that each F_i is constrained by the leaky bucket work arrival curve $\alpha_i(t) = r_i t + b_i$. For $j = 1, \cdots, K$ define

$$L_j = \max\{\delta_i : p(i) < p(j)\}$$
$$r_j^H = \sum_{\{i: p(i) > p(j)\}} r_i$$
$$b_j^H = \sum_{\{i: p(i) > p(j)\}} b_i$$
$$\beta_j(t) = [t - r_j^H t - L_j - b_j^H]^+$$

If $r_j^H < 1$, then β_j is a rate latency work service curve for $G_j = \cup\{F_i : p(i) = p(j)\}$.

2.3 Network Calculus Techniques

One could in principle use the above results directly to do a network calculus of various station topologies and packet flows. However, the iterative application of the FCFS and PQ split calculations along a chain of service nodes yields unacceptably high delay bounds due to the rapid increase in burst and latency values. A chain of service nodes does not exhibit a linear behavior if new packets join the flow through the chain.

The first way around this problem is to consider the service curve offered to a flow in a complete subsystem. If flow F with work arrival curve α is given a work service curve β_1 in node subsystem S_1 before entering node subsystem S_2 where it is given work service curve β_2, then F is given the concatenated work service curve $\beta = \beta_1 \otimes \beta_2$ in the combined system $S = S_1 \cup S_2$. The Network Calculus can then be done using the input arrival curve α and the service curve β in S. Compare the "pay bursts only once principle" in (Boudec and Thiran, 2001, section 1.5.3).

Care has to be exercised in choosing suitable subsystems as one still requires "good" service curves for the flow under consideration using the split techniques described above as well a "tight" arrival curve describing the work flow into the subsystem. Note also that the horizontal distance between the arrival and service curves is a delay bound only if the overall service order for the flow in the subsystem is FCFS.

In systems where the packet flows take different paths through the system it is in general difficult to compute acceptable arrival and service curves for a given flow. Aggregation of flows wherever possible is strongly recommended.

An arrival curve for the output of a flow from a system S with known delay bound can be derived using the easily proven fact that

$$\alpha^*(t) = \alpha(t + D_{max} - D_{min}) \qquad (4)$$

is an arrival curve for the output flow, if α is an arrival curve for the input flow into S, and D_{max} and D_{min} are upper and lower bounds on the system delay of the flow in S respectively. This is very useful if a delay bound can be calculated (e.g. from an aggregate flow) but no service curve is known.

The selection of aggregate flows and subsystems for Network Calculus leads in general to several different approaches to computing delay bounds for flows. Typically the best approach varies, depending on the actual parameters (such as rates and burst values) of the flows in the system. For this reason it is advisable to combine delay bounds calculated for different subsystems to improve the delay bound for the total system.

For a detailed explanation of the application of Network Calculus techniques in the case of the star and line topologies with centralized communication, see (Watson, 2002). These two scenarios are considered below.

3. VALIDATION

3.1 Description

The Network Calculus results are now compared with those obtained by simulation for an example scenario in Fig. 4 with star and line topologies from (Jasperneite *et al.*, 2002). In the star topology all stations are connected to the same switch, whereas in the line topology each station is connected to a dedicated switch and the switches are interconnected in a linear chain. The number of

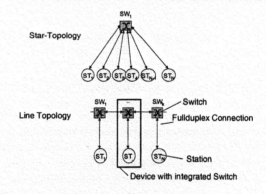

Fig. 4. Star and line topologies

stations is set at the moderate value of $N = 50$. In order to create a data traffic mix four message types (MT_i), which can be found in every industrial application, are defined from the application viewpoint (cf. Table 1). The information flow distribution defined by the application also has to be considered. It describes every communication relationship between the devices within a message type. The central information flow distribution is very important in the field of automation technology. This is the case if there is one special device in the system which is involved in all communication relationships, either as source or sink. In the simulation experiments, a uniform distribution of destination addresses is assumed. A central station generates all requests of types MT_1, MT_3 and MT_4, whereas all responses have this station as destination. The simulation model was realized with OPNET Modeler 7.0B from OPNET Technologies, Inc., by extending the standard models for switches and MAC to incorporate priority queuing (MIL, 1999). In the scenarios only full duplex transmission channels (FDX) with a transmission capacity of $C = 100Mbps$ are used. In the switches and devices the scheduling strategies FCFS (First Come, First Served) and PQ (Priority Queuing) are considered. Packet processing is asumed to be store and forward (i.e. a packet must be completely received before the transmission to subsequent switch or station may begin.) In addition, it is assumed that packet processing times in a switch or station are negligible. The burst size in this example depends on the system load ρ following the results of (Jasperneite and Neumann, 2000).

3.2 Analytical Delay Bounds

The analytical method presented above was applied to derive sure upper bounds on the transaction times. Fig. 5 shows the analytical results for the transaction time bounds as a function of the relative system load ρ (and implicitly the burst value b) for the star and line topologies as well as for the $PQ4$ and FCFS scheduling strategies.

Fig. 6 shows the upper time bounds determined by Network Calculus and by simulation. There are considerable differences between the curves, which decrease with higher loads. The deviation at lower loads is due to the fact that in the simulation model the bursts on the different connections are not simultaneous, whereas Network Calculus admits simultaneous bursts.

Table 1. Definition of message types

Type	Description
MT_1	Cyclic send and request of process data objects with response
MT_2	Transmission of events, e.g., alarms
MT_3	Acyclic transmission, e.g. for network monitoring, diagnostics or device configuration
MT_4	Acyclic transmission of datagrams for file transfer or IP-based applications

Table 2. Defining the workload

Feature	Message Type			
	MT_1	MT_2	MT_3	MT_4
% of packets	80%	10%	9.5%	0.5%
Length MAC payload [bit]				
request	368	368	1024	1024
response	368		1024	12000
user priority	Standard (5)	Medium (6)	High (7)	Low (4)

Table 3. Defining the values of the burstiness $b = f(\rho)$

ρ	0.05	0.2	0.5	0.8	0.9	1
b	2	3	6	8	9	10

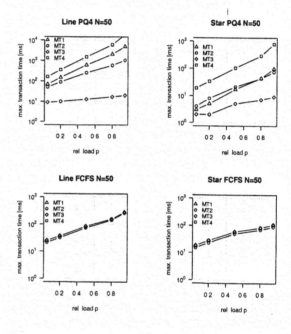

Fig. 5. Upper bounds on the transaction times determined by analysis with $b = f(\rho)$

4. SUMMARY

The methodology presented here can be used to calculate assured upper time bounds in switched Ethernet networks. The simulation results confirmed the applicability of these bounds for scenarios with equivalent source behavior.

It was shown that the real-time behavior at MAC level, i.e. that all transactions are always completed within a time span set by the technical process, can be verified with the technique *Network Calculus*. The following criteria have to be fulfilled:

- Switches with full duplex connections must be used throughout to avoid collisions (micro-segmentation)
- The arrival process of communication requests must be characterized for each connection. Here the so-called "leaky bucket" is

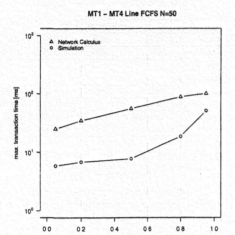

Fig. 6. Comparison between NC and Simulation

used with the parameters rate r and bucket size b as approximation of the arrival process.

- The maximum packet size must be known for each traffic class. A wide range of sizes in the same traffic class is unfavorable, since the maximum size determines the service latency.
- The service strategy must be known for each station and switch. In the case FCFS (First-Come-First-Served) no service system in the network may be overloaded. In the case of PQ (Priority Queueing) time bounds can be obtained at least for the higher priority classes if they are not overloaded.

For the practical application of this methodology it is necessary in a future step to derive the input parameters such as rate r and bucket size b from the configuration data of the automation design. Indeed, this data will have to be enhanced with performance and time-oriented information.

REFERENCES

AS-Interface, web site (n.d.). *http://www.as-interface.com.*

Boudec, J.-Y. Le and P. Thiran (2001). Network calculus: a theory of deterministic queuing systems for the Internet. In: *Lecture Notes in computer science*. Vol. 2050. Springer Verlag.

Cruz, Rene L. (1991a). A Calculus for Network Delay, Part I: Network Elements in Isolation. *IEEE Transactions on Information Theory* **37**, 114–131.

Cruz, Rene L. (1991b). A Calculus for Network Delay, Part II: Network Analysis. *IEEE Transactions on Information Theory* **37**, 132–141.

Ethernet Industrial Protocol (EtherNet/IP) (n.d.). *www.ethernet-ip.org.*

IDA (2001). *IDA Architecture Description and Specification V1.0.*

INTERBUS Club, web site (n.d.). *http://www.interbusclub.com.*

Jasperneite, J. and P. Neumann (2000). Measurement, Analysis and Modeling of Real-Time Source Data Traffic in Factory Communication Systems. In: *2000 IEEE International Workshop on Factory Communication Systems*. Porto, Portugal. pp. 327–334.

Jasperneite, J., P. Neumann, M. Theis and K. Watson (2002). Deterministic Real-Time Communication with Switched Ethernet. In: *4th IEEE International Workshop on Factory Communication Systems (WFCS'00)*. Vol. 2. Västeras, Sweden. pp. 11–18.

MIL (1999). *OPNET Modeler 7.0.*

Open DeviceNet Vendor Association (n.d.). *http://www.odva.org.*

Parekh, A. K. J. and R. G. Gallager (1993). A generalized processor sharing Approach to flow control in integrated services networks: the single-node case. Vol. 1. pp. 344–357.

Parekh, A. K. J. and R. G. Gallager (1994). A generalized processor sharing Approach to flow control in integrated services networks: the multiple-node case. Vol. 2. pp. 137–150.

PNO (2001). *PROFINET Architecture Description and Specification V 0.91*. ProfiBus Nutzerorganisation. *www.profibus.com.*

Profibus Nutzerorganisation, web site (n.d.). *http://www.profibus.com.*

Watson, K. (2002). Network Calculus in Star and Line Networks with centralized communication. Technical report. Fraunhofer-Institut für Informations- und Datenverarbeitung (IITB). Karlsruhe. Bericht-Nr. 10573; available at http://www.iitb.fhg.de/?4766.

ELSEVIER

IFAC
PUBLICATIONS
www.elsevier.com/locate/ifac

Design and Timing Analysis of a Two Way
Priority Transmission Scheme under the Total Frame Protocol

T. Erdner[†], W.A. Halang[†], K.C. Chan[‡], J.K. Ng[‡]

[†] *Faculty of Electrical Engineering, FernUniversität, 58084 Hagen, Germany*
[†] *thomas.erdner|wolfgang.halang@fernuni-hagen.de*

[‡] *Department of Computer Science, Hong Kong Baptist University, Kowloon Tong, Hong Kong*
jng|kcchan@comp.hkbu.edu.hk

Abstract: The data transfer under a priority controlled Total Frame Protocol on a ring bus system and its timing analysis are described. The priority scheduler is implemented by a programmable bypass channel in the slave units, which are controlled by function bits in the telegram status sequences. A combination of various signal encodings is employed in order to achieve higher redundancy of the data transmitted. Consequently, the different kinds of signal encoding guarantee safe detection of the data signals and more economic data transfer. Finally, the worst case transmission times and the maximum delay bounds for the two modes of data transmission are analysed. *Copyright © 2003 IFAC*

Keywords: Fieldbus, fault tolerance, Hamming encoding, signal coding, total frame protocol.

1. FIELDBUS STRUCTURE

Basically, a fieldbus is a master/slave system, which means that the master is connected to the slaves through individual connections. Similar to a shift register, within one bus cycle data are transmitted over the medium between the master and the slaves on a bit by bit basis. That is, output data are shifted from the master to the slaves and input data are shifted from the slaves to the master within the same bus cycle. Figure 1 shows the structure of the fieldbus considered, and the protocol adopted is called "Total Frame Protocol" (Baginski and Müller, 1994).

2. TOTAL FRAME PROTOCOL

The protocol frame is, in fact, a concatenation of the message frames issued by all devices attached, with a status sequence at the beginning, and an *End Of Transmission* (EOT) bit at each end (cp. Fig. 2). For security reasons the message frames are subjected to a Hamming encoding (Lint, 1992; Swoboda, 1973). The latter allows to detect, or even to correct, predeterminable numbers of errors in the data bits of telegrams, respectively. In the proposed design, three protection bits are provided for any four data bits, i.e., for any nibble. Thus, in any such data transfer element of seven bits, two single bit errors can be detected, or one single bit error can be corrected.

Table 1: Function bits of the status sequence

No.	Combinations f_0	f_1	f_2	Function
0	0	0	0	Initialisation
1	0	0	1	Data transmission to each slave (normal mode)
2	0	1	0	Data transmission to selected slaves (high priority mode)
3	0	1	1	Selection of slaves
4	1	0	0	-
5	1	0	1	-
6	1	1	0	Time synchronisation
7	1	1	1	Reset

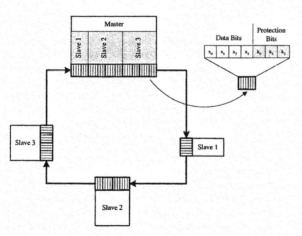

Figure 1: Example of the fieldbus structure

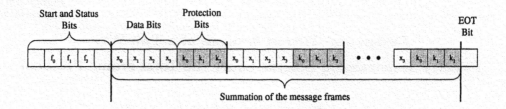

Figure 2: Frame of the total frame protocol

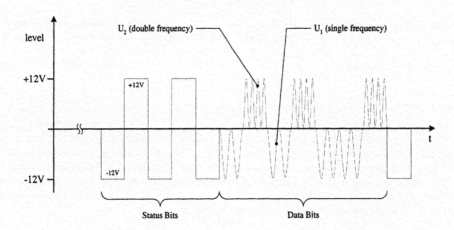

Figure 3: Communication telegram with status and data bits

The transfer mode of the total frame protocol is determined by the status sequence depicted in Fig. 4. This sequence determines the function of the subsequent data frames with the three function bits $f_0...f_2$ (cp. Tab. 1). The functions are described in Table 1. Each slave module can receive the status sequence with the corresponding function bits independently on its operating mode during every bus cycle.

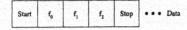

Figure 4: Function bits of the status sequence

3. SIGNAL ENCODING

Disturbances on transmission channels caused by electro-magnetical interference fields are normally impulses which superimpose the traffic signal and cause amplitude errors (Riegelmayer, 1995). Correspondingly, transmitted telegrams are falsified. For this reason, a modified Frequency-Shift-Keying (FSK) procedure is employed to transfer signals. Here, depending on the state, not only the frequency is changed, but the direct current (DC) part is changed as well. With this modification, the respective state can be obtained both via the frequency and via the DC parts (Erdner and co-workers, 2001). Therefore, the evaluation of the frequency is reliable in the presence of low frequency disturbances, since the DC part changes during these disturbances. On the other hand, the evaluation of the DC part is advantageous with respect to high frequency disturbances, since these can make themselves noticeable as a change in frequency.

In the course of transmission, data, status and synchronisation characters are exchanged. As depicted in Fig. 3, in this design the status characters and the data characters are clearly distinguished by their signal encoding. This distinction simplifies identification, and facilitates safe recognition of the two signals.

The synchronisation and status bits are transmitted as square waves, like the RS-232 or RS-485 standard signals. The advantage of this transmission scheme is the signal's very simple generation, and the easy evaluation of start or trigger marks, because square waves have very steep edges. Consequently, square waves are insensitive against changes of the DC potential, because the trigger point does not vary if the DC part is changing (cp. Fig. 6).

The data bits are transmitted as a combination of alternating current (AC) and DC part for secure communication. The pure AC part corresponds to the conventional FSK procedure, which uses two fixed frequencies for digital communication. Thus, this kind of signal encoding includes two information items, which can be evaluated simultaneously, and which considerably improve the recognisability in the presence of disturbances on the transmission channel.

Fig. 5 shows the signal encoding of the entire Total Frame Protocol. The synchronisation and status bits are transmitted as square waves, and the data bits as a combination of alternating current (AC) and DC part. This facilitates the recognition of the different telegram parts, as the receivers can very easily distinguish between them.

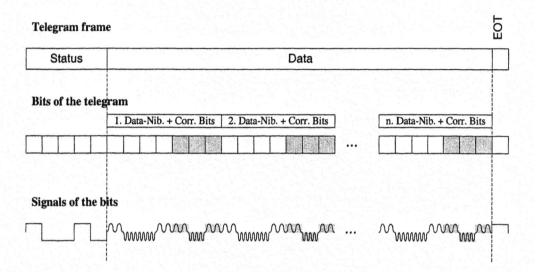

Figure 5: Signal encoding of the Total Frame Protocol

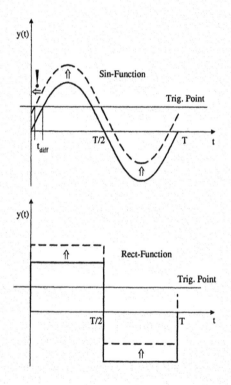

Figure 6: Sine and square waves during a change of DC potential

4. SLAVES MODULES

The slave modules have three-way-data-switches for the transmission way of the telegrams (cp. Fig. 7). The way through a slave module is determined by the previously adjusted operating mode and the status bits of the telegram. The status bits are transmitted directly into the status evaluation logic, and then to the subsequent slave modules. The evaluation logic adjusts the subsequent operating mode of the slave module. Depending on the operating mode, the data bits are transmitted either into the data registers or through the bypass. If a slave module is to receive data, the data line is switched into the slave registers. Otherwise the data bits will go through the bypass. In that case other slave modules are receiving the data in high priority mode. Therefore, the delay for data transmission in this slave is only the detection time of 1 bit for input signal encoding.

The synchronisation of the slaves modules' operation is dependent on the part of the telegrams. For reception, the slave modules are prepared by the status sequence (cp. Fig. 4). The start bit in this sequence introduces the transfer. The following function bits determine the operation mode of the subsequent data sequences, and the stop bit completes the status sequence. Hence, the duration of the status sequence is limited to 5 bits. Therefore, it is compatible to each asynchronous telegram of other serial communication systems. For this reason, synchronisation and reception of the status sequence are very easy.

However, the transmission time of the subsequent data sequences is dependent on the number of data nibbles to be transmitted. Since a longer transfer of digital information without any synchronisation can lead to problems in the receivers, it is required to arrange for a synchronisation between senders and receivers. For this purpose the pre-determined frequency components of the data bits are used. As the frequency components of the data bits can accept only two specific frequencies with a predefined phase position, two digital matched filters (Chen, 1982; Lüke, 1999) are required for the detection of frequency and phase position.

A matched filter is a signal fitted filter clearly identifying the frequency and also the phase position of a pre-defined signal. For this reason, the synchronisation times of the receivers can be determined from the specific phase positions of the frequency components.

The final EOT bit completes the transfer. As it is very short, no so-called Loop-Back-Word needs to be transmitted, since the master module recognises it as the final transmission item at signal decoding.

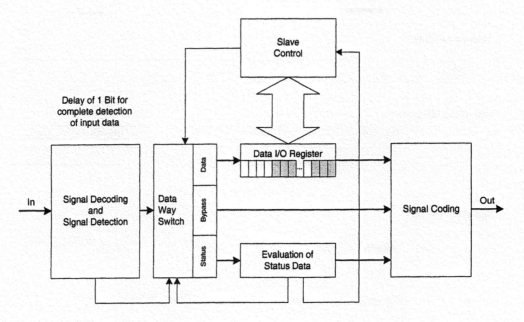

Figure 7: Three-way control of the data

5. PRIORITY CONTROL

Data transfer is possible in two ways. In standard mode, all slave modules are homogeneously provided with the data traveling on the bus (cp. Fig. 8). In high priority mode, the data are transmitted from the master to a pre-defined set of slave modules (cp. Fig. 9–10) or even to a single one, only. The other slave modules operate in the bypass mode. Therefore, the transmission time is reduced considerably by the high priority mode, and preferential data exchange with a subset of the slave modules is possible.

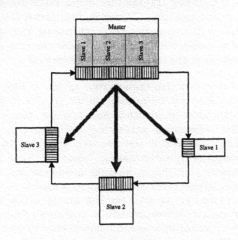

Figure 8: Standard mode data transfer

A further advantage of the programmable slave selection is the direct change between the preferred slave modules. Therefore it is possible to switch very easily between two different sets of preferred slave modules. For this switch-over to a new selection of preferred slave modules it is not necessary to switch all slave modules into the standard mode, first. Fig. 11 shows the initialisation phase from standard mode into high priority mode. For the selection of predefined slaves modules a initialisation telegram is required.

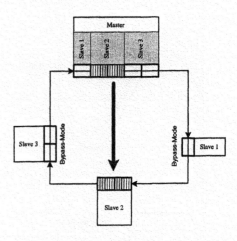

Figure 9: High priority mode data transfer

6. TRANSMISSION TIME OF THE TOTAL FRAME PROTOCOLS

The transmission time of a telegram completely transferred via the ring bus consists of different technical and physical components:

t_{trST}	Transmission time (standard mode)
t_{trHP}	Transmission time (high priority mode)
Nib_{ST}	Number of data nibbles[1] (standard mode)
Nib_{HP}	Number of data nibbles[1] (high priority mode) ($Nib_{ST} > Nib_{HP}$)
$Slaves$	Number of slaves[2]

---

[1] A nibble includes 4 data and 3 correction bits.
[2] Conversion time of 1 bit is required in each unit.

Total Frame Protocol

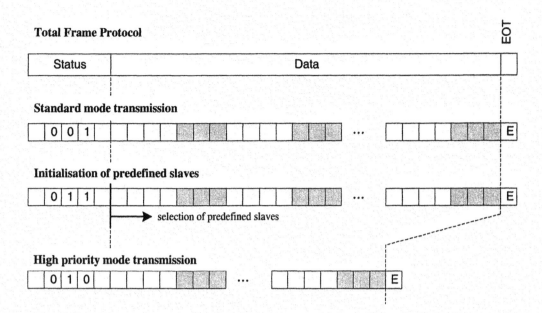

Figure 11: Initialisation of high priority mode

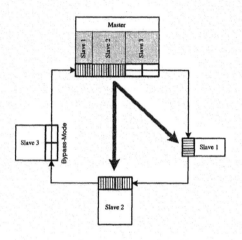

Figure 10: High priority mode data transfer with 2 slaves

Master	1 bit for the master detection unit[2]
Start	5 bit status data
End	EOT bit (End Of Transmission)
t_{Bit}	Duration of a bit
t_{SW}	Software delay
t_{Med}	Delay on the media (Cu: $t_{Med} = 0.016 \frac{ms}{km} \cdot l$)

The time required for a standard data exchange with all slave modules can be calculated as:

$$t_{trST} = (Nib_{ST} \cdot 7 + Slaves + Master +$$
$$Start + End) \cdot t_{Bit} + t_{SW} + t_{Med} \quad (1)$$

The speed difference between the standard and the high priority transmission methods is dependent on the number of data nibbles. Every data nibble excluded reduces the transmission time on the ring bus by the transmission time of 7 bits. Since the length of the communication line cannot be changed in either mode, the transmission time t_{Med} on the medium does not change. Furthermore, both modes use the same number of slaves, because the slaves' detection units must be used in each mode. Therefore, in each slave the delay for the detection is the one caused by 1 bit. Only the number of data nibbles is different, because the slaves do not receive any data nibbles in bypass mode. Thus, the time required for a high priority data exchange is:

$$t_{trHP} = (Nib_{HP} \cdot 7 + Slaves + Master +$$
$$Start + End) \cdot t_{Bit} + t_{SW} + t_{Med} \quad (2)$$

The time difference Δt_{tr} between the standard and the high priority data exchange is calculated as follows:

$$\Delta t_{tr} = (Nib_{ST} - Nib_{HP}) \cdot 7 \cdot t_{Bit} \quad (3)$$

7. WORST CASE TRANSMISSION TIMES

Since the fieldbus system starts operating in standard mode and switches to high priority mode if necessary, the worst case transmission time for a high priority message will be caused if

1. the requesting slave has to wait t_{trST}, because it just missed to set the status bit to inform the master to switch to high priority mode,

2. the slave has to wait another t_{trST} for the initialisation of the selected slaves, and

3. the slave can now send the high priority message with a transmission time of t_{trHP} if the message has up to 4 bits; otherwise it can take up to m rounds of t_{trHP} before the entire message is sent.

Hence, the worst case transmission delay for a high priority message is

$$t_{trHPmax} = 2 \times t_{trST} + m \times t_{trHP} \qquad (4)$$

where $m = \lceil \frac{max\{Msg_{HP}\}}{4} \rceil$ and $max\{Msg_{HP}\}$ is the maximum length of the high priority messages among all slaves.

On the other hand, the worst case transmission for standard mode occurs if

1. the master has just switched to operate in high priority mode and it will take t_{trST} for the initialisation of the slaves seleted,

2. there will be m rounds of high priority transmission before the high priority messages are exhausted, hence the waiting time is $m \times t_{trHP}$,

3. it will take another t_{trHP} to inform the master and the selected slaves to switch back to standard mode, and

4. the slave can now send the standard mode message with a transmission time of t_{trST} if the message has up to 4 bits; otherwise it can take up to n rounds of t_{trST} before the entire message is sent.

Hence, the worst case transmission delay for a standard message is

$$t_{trSTmax} = t_{trST} + m \times t_{trHP} + t_{trHP} + n \times t_{trST} \quad (5)$$

where $m = \lceil \frac{max\{Msg_{HP}\}}{4} \rceil$, $n = \lceil \frac{max\{Msg_{ST}\}}{4} \rceil$, $max\{Msg_{HP}\}$ and $max\{Msg_{ST}\}$ are the maximum lengths of the high priority and the standard messages among all slaves, respectively.

8. MAXIMUM DELAY BOUND

We assume here that all jobs are periodic and start in-phase to lead to the worst case scenario for message delays. A job of priority i can be represented by a tuple (π_i, T_i) where π_i is the period and T_i is the maximum time needed to send all priority i jobs within the period, with 1 being the highest priority. Furthermore, the jobs' periods have the following characteristics: $\pi_{i+1} = \pi_i \times 2$, for $i = 1, 2$. Let now Max_i be the maximum message delay to transmit all priority i messages in the system. Then, we have

$$Max_1 = T_1 \qquad (6)$$

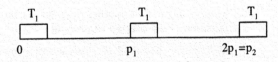

Figure 12: The schedule after inserting T_1 requests

Referring to Figure 12, if $T_2 \leq (\pi_1 - T_1)$, we can fit T_2 into the first *hole* of the schedule; otherwise the system will transmit the rest of T_2 after transmitting another set of priority 1 requests. Therefore, we have

$$Max_2 = \begin{cases} T_1 + T_2 & \text{if } (T_1 + T_2) \leq \pi_1 \\ 2 \times T_1 + T_2 & \text{otherwise} \end{cases} \qquad (7)$$

Thus, we can write down the general form of the maximum bound B_i for priority i requests as

$$B_i = \sum_{k=1}^{i} 2^{(i-k)} \times T_k \geq Max_i \qquad (8)$$

with 1 for high priority and 2 for standard messages.

9. CONCLUSION

The design of a two-way priority transmission scheme presented allows for data transfer under a priority controlled Total Frame Protocol on a ring bus system. The priority control of the slave modules is implemented by a programmable bypass channel, which is controlled by function bits in the telegram status sequence. Thus, the transmission time is reduced considerably in high priority mode, and a preferential data exchange with a selected group of slave modules is possible. The detection of various telegram parts is fostered by two different kinds of signal encodings in the telegrams. Therefore, these kinds of signal encoding allow for economic data transfer and safe detection of the data signals. The timing analysis revealed complete insight into the transmission time of the two Total Frame Protocols, the worst case transmission time and the maximum delay bound of the ring bus.

REFERENCES

Baginski, A., and Müller, M. (1994). *Interbus-S, Grundlagen und Praxis*. Hüthig.

Chen, C.H. (1982). *Digital Waveform Processing and Recognition*. CRC Press, Boca Raton, FL.

Erdner, T., Halang, W.A., Ng, J.K., and Chan, S.K. (2001). Secure Data Communication over Fieldbus Systems. In Proc. *FeT 2001 - IFAC International Conference on Fieldbus Systems and their Applications*. Nancy, 15 – 16 Nov. 2001, D. Dietrich, P. Neumann and J.P. Thomesse (Eds.), pp. 37 – 44.

Lint, J.H. (1992). *Introduction to Coding Theory*. 2nd ed., Springer-Verlag, New York.

Lüke, H.D. (1999). *Signalübertragung: Grundlagen der digitalen und analogen Nachrichtenübertragungssysteme*. 7th ed., Springer-Verlag, Berlin-Heidelberg.

Riegelmayer, W.P. (1995). *Verkabelungskonzepte, Grundlagen und Praxis*. Vogel Buchverlag.

Swoboda, J. (1973). *Codierung zur Fehlerkorrektur und Fehlererkennung*. R. Oldenburg Verlag, Munich.

ELSEVIER

IFAC
PUBLICATIONS
www.elsevier.com/locate/ifac

PREDICTION OF NETWORK LOAD IN BUILDING AUTOMATION

Mario Neugebauer, Jörn Plönnigs, Klaus Kabitzsch

Dresden University of Technology
Institute for Applied Computer Science
Dresden, Germany

Abstract: The process of designing building automation networks is a very challenging task due to the inherent complexity. During the design process performance evaluation is essential for high reliability and cost efficiency. Therefore, no tool is available up to now, although in other fields of research sophisticated methods for analyzing the system performance exist.

In this paper we introduce a method that comprehensively integrates the entire system to evaluate the performance fast and autonomous. By approximating the environment, automatically modelling the network components and analyzing the interrelations we achieve not only a specific load prediction, but also suggestions for optimization. *Copyright © 2003 IFAC*

Keywords: Performance Evaluation, Fieldbus, Load Forecasting

1. INTRODUCTION

Today, it becomes more and more common to apply fieldbus control systems for complex tasks in building automation. The need to provide different features cause service providers to mount their devices on one fieldbus system. Interoperability guidelines and standardization for building automation fieldbusses make it easy to install components independent of their functionality. This flexibility definitely is an advantage. Trades can act without the overhead of centralized coordination during the design phase, but each service provider can have different demands of bandwidth. Dimensioning the network can only be performed insufficiently if integration is done uncoordinatedly.

In current practice it turns out that particular system integrators need to do the overall integration and bear the responsibility for the entire network functionality. Proper working of the building automation network needs to be assured over the whole life-cycle of the fieldbus network. But, how about the tools for the integration? Manifold techniques for performance evaluation of networks already exist; even tools for analyzing IT-infrastructures are available. Also, the behavior of the building, has been examined in detail before. Further, design databases for implementing building automation networks are established. Unfortunately, there is no tool known which comprises all those features. We propose to cope with that insufficiency by merging all methods. The outcome will be a tool, able to perform the following tasks:

- assure sufficient use of medium
- identify dangerous load scenarios
- improve iteratively network performance
- assure long-term reliability
- parameterize for synergies

To reach those goals, the complexity of the entire network has to be reduced. Therefore, high demands according to formal analytics need to be met. This means to analyze the actual network as well as the environment. In addition, the results need to be abstracted for ease of use and there should not be any extraordinary hardware demands.

We propose automatic load calculation for large size building networks, with more than 100 nodes and several channels. On the one hand, it is inaccurate to use strong, abstract static methods. On the other hand, detailed analysis is not recommended to be done by hand. Thus, there is a need for an automatable algorithm.

To establish a basis for our work we will analyze the research done in the area of network load prediction. Subsequently, in the third part, a new concept for comprehensive network performance prediction will be proposed. Thereby, we will attempt to cope with the insufficiencies that occur in other approaches. In the fourth section we will discuss the mentioned concept based on an implementation for a certain fieldbus system.

2. RELATED WORK

2.1 Basic Load Calculation

In (Schmalek, 1995) the performance evaluation has been done for fieldbusses (i.e. LonTalk). The load behavior of networks of different sizes has been analyzed, but the network performance was not investigated starting with the load assumptions of real networks with different load behavior of the nodes.

Beside these works research for load prediction in LON has been done by Florstedt within a diploma thesis in 1999 (see (Florstedt, 1999)). This work comprises an approach to implement a tool for load prediction in field bus networks and resulted in a prototyped software package. It provided the ability to build and configure a simple field bus network (based on the LonWorks standard) and to perform a static load prediction. The network structure has to be managed in text files without support. Based on this proposal for solution we will build a more complex and efficient tool.

2.2 Simulation of Fieldbus Systems

In (Schwarz and Donath, 1997) and (Donath et al., 1997) a detailed example of a small automation task is discussed. As a basis for the whole simulation the hardware modeling language VHDL is used. With the results from (Miskowicz et al., 2002) we can get an idea of the channel behavior in different load scenarios (e.g. throughput, collision rate, etc.). If we consider all parameters influencing the channel load as the input space,

then all points in this space (and the corresponding channel reactions) cannot be determined in a reasonable time frame.

The simulation approach has certain disadvantages. First of all, achieving resonable results is always extremely time-consuming, especially for large-size networks. Furthermore, the networks of interest have to be edited without support. For fast performance evaluation of fieldbus networks in building automation, the estimation by simulation is inappropriate.

2.3 Commercial Tools for IT-infrastructures

HyPerformix (Inc., 2002) and OPNET Guru (OPNET Technologies, 2003) are representative software packages in this area. With these products it is possible to perform an evaluation of an enterprise-wide network structure. The IT-infrastructure can be modeled, managed, analyzed and optimized comprehensively. User models and a wide range of hardware models are available to emulate and finally predict the performance of a future IT-infrastructure. The existing tools cannot be adapted to the domain of fieldbus networks for building automation. But, we want to pick up the idea of *all-inclusive-prediction* and bear it in mind for our further work.

2.4 Queuing Networks

Queuing network analysis is a well-known technique to estimate the performance of systems. As long as the process to investigate is in some way to model in queues, this powerful analysis method can help to design, gain experience about and optimize it. The main idea of this theory is to describe the system with interconnected queues and servers. Of course, for large systems the representing queuing model becomes rather large as well.

To cope with these problems, tools for the automated queuing network analysis can be used. One example is the function library PDQ (pretty damn quick)(Gunther, 2000). It provides the ability to compute relatively large queuing networks in a short time. Though, generating and specifying a model is rather circumstantial. It has to be done in C-files. As a result of the computation, PDQ offers manifold information about the analyzed system (e.g. utilization of the servers, residence times and delays). But only with the knowledge of the environmental behavior and the corresponding impact on the queuing model, common-sense results can be achieved.

2.5 Traffic analysis and generation

Performance estimation for a wide range of networks is only as good as the traffic models, presumed for the application domain. As an example of traffic characterization in a different application domain the approach in (ETSI, 1998) remains to be mentioned. Another approach to generate traffic for experimental purposes is to simulate the process (building with physical properties). In (Metzger, 1999) for instance, a new method for the generation of flexible and complex simulations is described.

Research done in the area of building simulation delivers comprehensive results for physical building behavior. Though, simulating or measuring the physical network environment is too complex. Therefore, generalized models of the physical behavior are needed for further application in the *all-inclusive* performance evaluation.

3. CONCEPT

In our (all-inclusive) approach we want to meet the following boundary conditions:

- specialized in building automation
- advanced standardization of the devices and fieldbusses
- large-size networks that are only automatically processable
- no specialized or extraordinary hardware demands and fast computation
- easy to use with minimized pre-editing by the user

We propose a new method for performance evaluation of large-size networks. Our approach (see Figure 2) is based on a simple assumption: the mean load L_{mean} of a channel results from the number of messages $n_{i,j}$ sent by the ith connected device via the jth output connection and their individual message sizes $s_{i,j}$ over a representative period of time T. It results in

$$L_{mean} = \frac{1}{T} \sum_{i=1}^{N_D} \sum_{j=1}^{N_{out,i}} s_{i,j} \cdot n_{i,j}, \qquad (1)$$

with N_D for the number of devices connected to the channel and $N_{out,i}$ for the corresponding number of connections, going out from this device.

In queuing network calculation (see 2.4 above) the message number normalized to the time is called arrival rate $\lambda_{i,j}$, which is to determine by

$$\lambda_{i,j} = \frac{n_{i,j}}{T}. \qquad (2)$$

For a calculation based on this assumption comprehensive information about the network is needed.

First, the logical topology of the network has to be known. The physical dimension of the network is negligible as long as the maximum transmission time is much smaller than the sending time of a message and physical closed loops are excluded. Normally these demands are fulfilled in CSMA based networks.

To calculate the load automatically it is recommendable to **generate a database** ((1) in Figure 2) containing the logical topology of the network, for example extracted from an existing network design database.

The inherited logical **network topology** needs to be **analyzed** (3) to restore the way each message takes from the sender device to a destination device.

Furthermore, it is necessary to know the size of each sent message. In common building automation systems the message size is standardized. Based on these facts it is possible to group the message types in *size classes* with determined sizes.

Additionally, it is necessary to know the arrival rate of each class. Normally a generalized one is used (see (Gunther, 2000)), but in the focus of building automation a more accurate **arrival rate estimation** (2) is possible. It can be deduced from the behavior of the used devices since it depends on known physical processes.

Devices can be classified by their input-output arrival rate relation in three classes, *load sources* (output arrival rate is independent from the input), *load processors* (output arrival rate depends on the device input) and *load sinks* (only input). For example, a sensor typically is a load source, an event based controller a typical load processor and an actuator a load sink. This classification enables

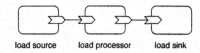

load source   load processor   load sink

Fig. 1. classification of devices

a bottom-up calculation for the whole network, as long as it is possible to estimate the load sources. However, the estimation of the arrival rate for load sources is difficult as the behavior of sensors among others depends on the environment. To estimate the arrival rate for all load sources of a building a complex environment model is needed. This model depends furthermore on the load sinks (e.g. HVAC units). Such a complex dynamic model extends automatic load calculation.

A collection of representative static cases is more functional.

The **load calculation** (4) of a network can be done from the resulting arrival rates, message size classes and the logical network topology easily.

Besides this load calculation the collected information can be used in a classical **queuing network** (5) (see 2.4). With the help of this model, the average transmission time of a message can be calculated.

Both models result in an abundance of data, no system integrator can cope with. Hence, the system needs to **analyze the results** (6) and extract the critical states, for example overloaded channels. Nevertheless, the resulting complexity enables more than just load analysis. Advanced predictions can be made, such as instable closed loop controls or wrong network parameterization. And, as the system possesses the complexity, it can suggest solutions as well. This data mining should be fast and reliable.

4. CASE STUDY

Currently, we are implementing the concept presented before for the LonTalk protocol in a project named *NetPlan*.

LonTalk (ANSI/EIA 709.1; ENV 13154-2) has been developed in the late 80's and has grown to one of the most popular building automation bus systems since then. It implements all 7 OSI layers as the only one beneath them. Beside the LonTalk protocol specification (see (*LonTalk Protocol Specification*, 1994)) some recommended standards exist. The LonMark Interoperability Association dedicated itself to elaborate standardized interfaces based on *Standardized Network Variable Types* and standardized behaviors with *Functional Profiles* (see (Dietrich *et al.*, 2001)).

The dataflow chart and the functional blocks of our tool is shown in Figure 2. It represents the concept introduced in section 3. The single steps will be explained in more detail in the following subsections.

4.1 Import to the NetPlan Database

A LON network can be developed using different software tools (see (Stöcklhuber, 1999)). A growing number of them use the standardized LNS database developed by Echelon. To offer an open interface for these tools we establish our approach on this database. The whole network topology is imported from the LNS database in an internal *NetPlan Database*. This separated database enables us to store results and allows the user to

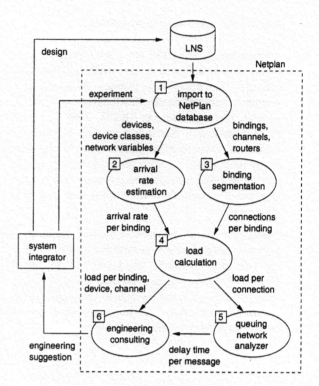

Fig. 2. dataflow chart of the implementation

experiment in our network, without changing the real one.

4.2 Arrival Rate Estimation

The network topology represented in the LNS database uses an application layer model including *Network Variables* and *bindings* to connect them. These Network Variables are normally Standardized Network Variable Types. Therefore, their size is known and enables categorization in size classes.

For each Network Variable of each device the arrival rate needs to be approximated. In the first step, this is done for the load sources, after classification of the network devices in load sources, load processor and load sinks.

Most of them use the parameter MinSendTime T_{minS}, MaxSendTime T_{maxS} and SendOnDelta d to influence their sending behavior. The SendOnDelta describes how much the input signal has to change for a new message. The MinSendTime is the minimum time that has to pass between two messages; the MaxSendTime is the maximum time without a message and enables a heartbeat cycle. The arrival rate results from these parameters in

$$\lambda = \min \left(\frac{1}{T_{minS}}; \max \left(\frac{1}{T_{maxS}}; \frac{|\Delta|}{d} \right) \right). \quad (3)$$

The adjusted values of T_{minS}, T_{maxS} and d are stored in the LNS database. Only the absolute rise $|\Delta|$ of the input signal needs to be determined. This has been done for some typical and special cases of characteristic input signals in building automation, e.g. outside temperature, humidity. These *characteristics* have to be associated to the Network Variables, by analyzing the relations between Functional Profiles and Standardized Network Variables Types. The arrival rate can be automatically estimated for all load processors and load sinks by input superpositioning.

4.3 Binding Segmentation

The application layer model needs to be transferred to a physical layer model to analyze the load on the physical channels. This is done in the binding segmentation.

A message between two devices can be transmitted in large networks over many channels. This transmission is controlled by routers. The information about the way a message takes, is not accessible in the LNS database and needs to be reengineered. We use the recommended routing algorithm in (*LonTalk Protocol Specification, 1994*) to rebuild the routing tables of the routers. After this, each binding is separated into communications. A *communication* is a directed connection between two network elements in the same channel involved in the transmission of a message over a network (e.g. two routers in one channel (see Figure 3)).

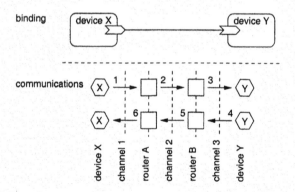

Fig. 3. binding segmentation

Additionally each sent application layer message can release protocol messages, as acknowledges, authentications or repetitions. These responses can be directed toward the information sender.

For each protocol message the way it takes is rebuilt with communications as well. In the end, a single binding can result in up to $(3 \cdot 127 \cdot 255) + 1$ (authentication; 127 group members; 255 channels) communications, with different message sizes and directions, representing the transmitted messages in all channels on physical layer.

4.4 Load Calculation

Load calculation can be done easily when the communications and arrival rates are known. Summation of the used connections (Equation 1) leads to specific load for all channels. Vertical summation over all channels in Figure 4 leads to the following result for the load at all channels:

$$L_1 = \lambda_{1,2}s_{1,2} + \lambda_{2,1}s_{2,1} \tag{4}$$
$$L_2 = \lambda_{1,2}s_{1,2} + \lambda_{2,1}s_{2,1} +$$
$$\lambda_{3,2}s_{3,2} + \lambda_{3,4}s_{3,4} + \lambda_{4,3}s_{4,3} \tag{5}$$
$$L_3 = \lambda_{1,2}s_{1,2} + \lambda_{2,1}s_{2,1} + \lambda_{3,2}s_{3,2}. \tag{6}$$

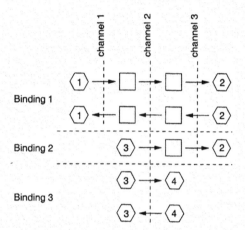

Fig. 4. load superpositioning

4.5 Queuing Network Analysis

With the results of the binding segmentation a complete queuing model will be built. Therefore, each component of the network (all devices, network elements, channels) is described in a service center, consisting of queues and servers. The several parts are connected properly to achieve a comprehensive representation as a queuing model for the calculation of the network performance (e.g. processing times, transfer delays times).

4.6 Engineering Consulting

All available results need to be reduced to the important facts. This is done with a set of simple rules applied on the results. The remaining simplified output is presented to the user. He will be supported by a cause analysis and appropriate solutions. Independently he can investigate in more details.

5. CURRENT STATE AND FURTHER WORK

At the moment we are still implementing the introduced concept. The first prototype of the load calculation has been proved in some test cases but a broad practical validation needs to be completed. Detailed research in the area of automatic association of the arrival rates to devices (see 4.2) is needed. Also, the queuing network model needs to be adapted to our requirements. The rules for Section 4.6 are only rudimentarily defined and need to be specified. We plan to detect the following things:

- top 10 of the load producing devices
- top 10 of the network bottle necks
- wrong parameters
- violated transfer time demands
- stability in cyclic closed loop controls
- detection of logical and load loops

Particularly, the stability of closed cycle controlling in event based network and controllers needs more investigation.

6. CONCLUSION

With our concept the mean load in each network element can be calculated automatically and fast. The included queuing network analyzing estimates even the mean delay time of each message and the queue length. An efficient data mining reduces the results to the important facts.

But, our concept of the *all-inclusive-prediction* offers more than simple performance evaluation. It can automatically detect problems and assist a system integrator to solve and avoid them. We think we will be able to:

- optimize positioning of routers
- tune device parameters to reduce load
- optimize parameters to fulfill time demands
- optimize controllers for better performance
- optimize network for use or emergency cases

This supports a system integrator in analyzing and optimizing his network and last but not least to save money.

7. ACKNOWLEDGEMENT

The project, the present report is based on, was promoted by the Federal Ministry of Education and Research under the registration number 13N8177. The authors bear all the responsibility for contents.

REFERENCES

Dietrich, Dietmar, Dietmar Loy and Hans-Jörg Schweinzer (2001). *Open Control Networks*. Kluwer Academic Publishers Boston. London.

Donath, Ulrich, Peter Schwarz, U. Hartenstein and Klaus Kabitzsch (1997). Simulationsunterstützung für den Entwurf von Feldbussystemen. In: *Proc. FeT 97 Conference Fieldbusysstems in Applications*. Wien. pp. 263–270.

ETSI (1998). Universal Mobile Telecommunications System; Selection procedures for the choice of radio transmission technologies of the UMTS. Technical report. ETSI.

Florstedt, Thomas (1999). Erstellung und Implementation von Algorithmen zur Vorhersage der Bandbreitenauslastung innerhalb von verteilten Rechnersystemen. Master's thesis. Dresden University of Technology, Institute for Applied Computer Science.

Gunther, Neil J. (2000). *The Practical Performance Analyst*. iUniverse.com Inc.. Lincoln, Nebraska.

Inc., HyPerformix (2002). http://www.ses.com.

LonTalk Protocol Specification (1994). 3.0 ed.. Palo Alto. www.echelon.com.

Metzger, Andreas (1999). An Interlink of Building Control Prototypes and the Lightning Simulation Lumina. Technical report. University of Kaiserslautern. Kaiserslautern.

Miskowicz, Marek, Maria Sapor, Marcin Zych and Wojciech Latawiec (2002). Performance Analysis of Predictive p-Persistent CSMA Protocol for Control Networks. In: *4th IEEE International Workshop on Factory Communication Systems*. Västeras, Sweden. pp. 249–256.

OPNET Technologies, Inc. (2003). http://www.opnet.com.

Schmalek, Richard (1995). Analyse des Zeitverhaltens von LONWorks. In: *Automatisierungskonzepte mit dezentraler Intelligenz (LONWORKS), Workshop und Ausstellung*. Dresden, Germany.

Schwarz, Peter and Ulrich Donath (1997). Simulation-based Performance Analysis of Distributed Systems. In: *International Workshop Parallel and Distributed Real-Time Systems*. pp. 244–249.

Stöcklhuber, Andreas (1999). Eine Übersicht über die wichtigsten LON-Programmiertools. In: *de-Special: Bussysteme für die Gebäudeinstallation*. pp. 143–146. Hültig & Pflaum Verlag GmbH & Co Fachliteratur KG. München / Heidelberg.

ELSEVIER

IFAC

PUBLICATIONS
www.elsevier.com/locate/ifac

VIRTUAL TOKEN-PASSING ETHERNET – VTPE

Francisco Borges Carreiro[1,2], José Alberto Gouveia Fonseca[2], Paulo Pedreiras[2]

{fborges@dee.cefet-ma.br, fborges@ieeta.pt}, jaf@det.ua.pt, pedreiras@det.ua.pt

[1]*DEE/ CEFET-MA – Brasil*

Av. Getúlio Vargas N⁰ 04 Monte Castelo

65025-001 São Luís - MA – Brasil

Phone: +55 98 218 9080   Fax: +55 98 218 9019

[2]*DET / IEETA – Universidade de Aveiro*

Campus Universitário de Santiago

3810-193 Aveiro Portugal

Abstract: The amount of information that must be exchanged in modern Digital Computer Control Systems (DCCS) and real-time industrial automation is now reaching the limits that are achievable with traditional fieldbuses. Currently, Ethernet is increasingly used in DCCS, however, the available solutions to guarantee real-time behaviour in Ethernet are not well suited to support the DCCS requirements and can't be implemented in low processor power devices. This paper proposes a real-time Ethernet-based protocol (VTPE) that supports real-time requirements, and implies just a small overhead in the system nodes. The proposed protocol is based on the virtual token-passing media arbitration and does not require specialized hardware. *Copyright © 2003 IFAC*

Keywords: Fieldbus, Ethernet, real-time, microprocessors, token-ring protocol.

1. INTRODUCTION

The quantity and functionality of microprocessor-based nodes in modern DCCS have been increasing steadily (Dietrich and Sauter 2000). This evolution has been motivated by new classes of more resource demanding applications, such as multimedia applications (e.g. machine vision), as well as by the trend to use large numbers of simple interconnected processors, instead of few powerful ones, encapsulating each functionality in one single processor (Pedreiras *et al* 2001). Consequently, the amount of information that must be exchanged among the network nodes has also increased dramatically over the last years and it is now reaching the limits that are achievable using traditional fieldbuses, e.g. CAN, WorldFIP, ProfiBus (Song 2001). Alternative protocols must be found to support this higher bandwidth demand while fulfilling the timeliness requirements of a real-time communication system. Well-known networks, such as FDDI and ATM, have been extensively analysed for both hard and soft real-time communication systems (Song 2001). However, due to high complexity, high cost, lack of flexibility and interconnection capacity (Song 2001), they have not gained general acceptance for the use at the field level.

Similar efforts have been done with Ethernet (Decotignie 2001), trying to take advantage of the availability of cheap silicon, easy integration with Internet, clear path for future expandability, and compatibility with networks used at higher layers in the factory structure. However, its standardized non-deterministic arbitration mechanism (CSMA / CD) prevents its direct use at field level, at least for hard real-time communications. Despite of this, there are many different approaches for achieving real-time behaviour on Ethernet. However most of the proposed solutions are costly in terms of processing power and memory requirements, and thus they are not well suited for small sensors, actuators and controllers.

There is a need to find Ethernet deterministic solutions, so that it becomes possible to take profit of its with higher data-communication capacity, and use it to replace fieldbuses to interconnect sensors, controllers and actuators at the field level. These solutions must be implemented with simple and efficient mechanisms to access the bus, in order to have reduced code (to fit in most microcontroller's memory) and to impose low communication overhead. The virtual token-passing scheme contains the main characteristics required to this Ethernet implementation.

This paper proposes the Virtual Token-passing Ethernet (VTPE), a new deterministic implementation to support real-time traffic. It uses a virtual-token passing scheme, similar to the one used in P-NET fieldbus (EN50170). However, it uses a producer-consumer cooperation model instead of P-NET's master-slave, in order to reduce the communication overhead. The sensors and actuators can be either directly connected to the microcontroler's ports or via other fieldbus sub network.

The remainder of this paper is organized as follows: Section 2 presents briefly the Ethernet standard, a set of arguments favouring the use of Ethernet at the fieldbus level, and relevant works to achieve real-time behaviour on Ethernet. Section 3 presents the VTPE protocol medium access, the frame format and the real-time parameters. Section 4 presents the VTPE real-time behaviour and a numerical example. Section 5 presents the conclusions and future works.

2. ACHIEVING R-T BEHAVIOUR ON ETHERNET

2.1 Ethernet standard

During the Ethernet evolution two fundamental properties have been kept unchanged: 1) a single collision domain, i.e., frames are broadcast on the physical medium and all the network interface cards (NIC) on it receive the message, and 2) the arbitration mechanism (CSMA/CD).

The use of a single broadcast domain and the CSMA/CD arbitration mechanism have created a bottleneck in highly loaded networks: above a certain threshold, as the load increases the bus throughput decreases (trashing). To overcome this and other disadvantages there are many approaches and techniques to achieve real-time behaviour on Ethernet as it will be discussed in section 2.3.

The standard Ethernet frame is shown in fig.1. It is composed of eight fields: Preamble, Start Frame Delimiter (SFD), Destination Address (DA), Source Address (SA), Type or Length Field, Data, Pad and Frame Check Sequence (FCS).

Alternating 1s/0s	SFD	DA	SA	Type or length	Data	Pad	FCS
Preamble				Frame length (min. 64 bytes e max. 1518 bytes)			

Fig .1 - Ethernet frame format

The destination and source addresses are each one six bytes long. The Type or Length field is two bytes long and specifies the type of protocol or number of data bytes inside of the Data field; the Data field can be between a minimum of 46 bytes and a maximum of 1500 bytes and the FCS is 4 bytes. When the Data field is smaller than 46 bytes it must be filled with zero (Pad) up to complete the minimum permitted Ethernet frame length.

2.2 Why using Ethernet at the fieldbus level

The main obstacle to using Ethernet in real-time communication is that, due to CSMA/CD access protocol, Ethernet cannot provide connected stations with deterministic channel access times and therefore guarantee that data delivery deadlines will be met (Lo Bello and O. Mirabella 2002). In fact, its designer has not envisaged this kind of applications and thus, some properties of this protocol, such as the non-deterministic arbitration mechanism, pose serious challenges concerning its use at this level.

Despite of this, many arguments favouring the use of Ethernet as a fieldbus are commonly defended such as:

- It is cheap, due to mass production;

- Integration with Internet is easy;

- TCP/IP stacks over Ethernet are widely available, allowing the use of application layer protocols such as FTP, HTTP and so on;

- Transmission speed has been steadily increasing in the past and is expected to continue in the future;

- Due to its inherent compatibility with the communication protocols used at higher levels, the information exchange at plant level becomes easier;

- Wide availability of technicians more familiar with this protocol;

- Wide availability of test equipment;

- Mature technology, well specified, equipment available from many sources, no incompatibility issues.

2.3. Achieving real-time on Ethernet

This section presents some efforts to make Ethernet real-time.

Modification of the Medium Access Control

This approach consists on modifying the Ethernet MAC layer to achieve a bounded access time to the bus, e.g. (LeLann, G, and N. Rivierre 1993), (Shimokawa, Y. Shiobara 1985) and (Court, R 1992). For instance, the solution presented in (LeLann, G, and N. Rivierre 1993), CSMA/DCR, consists in a binary tree search of colliding nodes, that is, there is a hierarchy of priorities. Whenever a collision happens the lower priority nodes voluntarily cease contending for the bus, and higher priority nodes try again. This process is repeated until a successful transmission occurs. Two main drawbacks can be identified:

- In some cases the firmware must be modified, therefore the economy of scale obtained when using standard Ethernet hardware is lost;

- The worst-case transmission time, which is the main factor considered when designing real-time systems, can be orders of magnitude greater than the average transmission time. This forces any kind of analysis to be very pessimistic and thus, leads to an under-utilization of the bandwidth.

Adding transmission control over Ethernet

Another way to achieve time-constrained communications over Ethernet consists in adding a layer above it, intended to control the instants of message transmissions, ending up with a bounded number of collisions or even a complete avoidance of them. The major advantage of this kind of approach, when compared to the modification of the MAC layer, is that standard Ethernet hardware can be used. Several different ways of doing transmission control over Ethernet are referred below.

Master/Slave. In this case, all ordinary stations in the system transmit messages only upon receiving an explicit command message issued by one particular station called master. This approach supports relatively precise timeliness, depending on the master, but introduces a considerable protocol overhead caused by the master messages (notice that the number of messages is duplicated). Moreover, with this approach, the handling of event-triggered traffic is normally inefficient because the master must first become aware of any request before inquiring the respective station.

FTT-Ethernet. This protocol supports both synchronous and asynchronous message exchanges over shared Ethernet (Pedreiras *et al* 2001). Each of these types of traffic is transmitted within specific time windows or phases (designated respectively synchronous and asynchronous). Nodes transmit synchronous messages after explicit request by a special node, designated by Master. This node centrally schedules the synchronous traffic and periodically broadcasts the schedule information using a trigger message (TM). The nodes transmit asynchronous messages autonomously, after application request. To achieve deterministic priority-based asynchronous message transmission, this protocol implements a mini-slotting scheme.

This protocol provides timeliness guarantees to both synchronous and asynchronous traffic. Moreover, the bandwidth used by the relaxed master-slave technique considerably reduces the overhead when compared with the regular master/slave transmission control, since a single TM may trigger the transmission of several data messages. Furthermore, this protocol supports arbitrary scheduling policies and on-line changes to the communication requirements without jeopardizing the timeliness guarantees. However, the complexity resulting from all these features requires resources (memory, CPU) that are not supported by low-end hardware frequently used in distributed embedded systems.

Token-Passing. This method consists on circulating a token among the stations. Only the station currently holding the token is allowed to transmit and the token holding time is upper bounded (IEEE 802.4 timed-token is one example). This scheme is still not very efficient due to the bandwidth used by the token and induces large jitter in the periodic traffic due to variations in the token holding time. Furthermore, token losses generally impose long periods of bus inaccessibility.

TDMA. In this case, stations transmit messages at pre-determined disjoint instants in time in a cyclic fashion. This approach requires precise clock synchronization and does not respond well to dynamic changes in the message set because the communication requirements are distributed and thus, changes must be done globally. On the other hand, it uses the bus bandwidth efficiently since there are no control messages beyond those to achieve clock synchronization.

Switched Ethernet

A solution for the problem of throughput decrease with load network increase, known as thrashing, has been proposed in the beginning of the 90's, consisting on the use of switches in the place of hubs. A switch creates a single collision domain for each node connected to it. This way, collisions never occur unless they are created on purpose for flow control. Switches also keep track of the addresses of the NICs connected at each port; therefore each NIC only receives the traffic addressed to it. This architecture allows the existence of multiple transmission paths simultaneously, between different network nodes. The global throughput increases significantly since, using switches, the devices on the network no longer share the bandwidth and collisions do not occur. The use of switches became very popular, recently, as a way to improve the performance of shared Ethernet.

When a node transmits a message, this one is received by the switch and then buffered in to the ports where the receivers of the message are connected. If several messages addressed to a given port arrive in a short interval, they are queued and then sequentially transmitted. Concerning the scheduling of messages waiting in an output port, 8 priority queues are available (IEEE 802.1p). The scheduling policy used at this level is a topic currently addressed in the scientific community, e.g, (Jasperneit, J. and P. Neumann 2001).

Unfortunately the use of a switch in an Ethernet network is not enough to make it real-time, in the general case. For instance, output buffers can be exhausted and messages lost if bursts of messages are sent to the same output port. This situation can occur more often than desired. For example, in distributed control systems, the producer/consumers model is typically used. According to this model, one producer of a given datum (e.g. a sensor reading) sends it to several consumers of that datum. This model is efficiently supported in Ethernet by means of special addresses, called multicast addresses. Each network interface card can define a local table with the multicast addresses related to the data that it should receive. However, the switch has no knowledge of these local tables, therefore treats all the multicast traffic as broadcasts, i.e., messages with multicast destination addresses are transmitted to all ports. Therefore, depending on the predominant type of traffic exchanged in a given application (unicast vs. multicast/broadcast), one of the main benefits of using Switched Ethernet, i.e. multiple simultaneous transmission paths, can be seriously compromised.

Other problems concerning the use of switched Ethernet (Decotignie, J. D 2001) are:

- In the absence of collisions the switch introduces an additional latency;

- The number of available priority levels is too small to support the implementation of efficient priority based scheduling;

- The switch only makes Ethernet deterministic under controlled loads.

Others existing proposals

There are many other solutions like the RETHER protocol (Venkatramani, C., T. Chiueh), the Virtual Time Protocol (Malcolm, N. and W. Zhao 1995), (Molle, M. and L. Kleinrock 1985) and the Windows protocols. None of them is well suited for microprocessor implementation because they require fast processors and high memory to implement them, so they are not discussed in this article.

3. VIRTUALTOKEN PASSING ETHERNET-VTPE

In the overview presented in the previous sections the characteristics of the most relevant approaches to achieve real-time communication over Ethernet have been discussed and their drawbacks have been highlighted. It is interesting to notice that such approaches either require specialized hardware, or provide statistical timeliness guarantees, or are bandwidth or response-time inefficient, or are inflexible concerning the properties of the network traffic as well as the traffic scheduling policy, or finally, they are costly in terms of processing power and memory requirements. Thus they are not well suited for use in small sensors, actuators and controllers with communications capability.

Based on these drawbacks and on the demand of distributed embedded systems, the following goals have been established to develop the VTPE.

- Support on the same bus of slow and cheap devices based in microcontrollers, as well as more demanding devices integrating powerful processors;

- Support for on-the-fly operational flexibility like: to add and to remove nodes, to add and to remove messages and change their parameters;

- Indication of temporal accuracy of real-time messages;

- Efficient use of network bandwidth;

- Efficient support of multicast messages;

- Low processing overhead in order to be implemented in microcontrollers with low processing power;

- Support of several data messages inside a single Ethernet frame.

3.1 VTPE medium access control

The VTPE is an Ethernet deterministic proposal based on implicit token rotation (virtual token-passing) like the one used in the P-NET fieldbus protocol. VTPE uses the producer-consumer cooperation model to exchange data over the bus. Producers, in terms of the bus, are active devices that can access the bus when they are allowed to do it. On the other hand, consumers are passive devices and can only consume the data on the bus[1].

The VTPE system architecture consists of a producer's logical ring like the one depicted in the figure 3. Each node consists of a processor or microcontroller attached to an Ethernet controller.

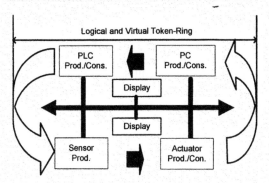

Fig. 3 The Virtual Token-passing in a VTPE system

In the example of figure 3, the distributed system is composed of six nodes, one producer (sensor), three producer/consumer (actuator, PC and PLC) and two consumers (Displays).

To implement a virtual token-passing schema using Ethernet, Ethernet's broadcast destination address must be used because all devices must read each frame dispatched on the bus. In a VTPE system each producer node has a node address (NA), between 1 and the number of producers expected within a system. All producers have an Access Counter (AC), which identifies the node that can access the bus in a specific time interval. Whenever a frame is sent to the bus, an interrupt must be generated in all producer nodes. After the interrupt all nodes increase their ACs and the producer node, whose AC value is equal to its own unique address, is allowed to access the bus. If the actual node doesn't have anything to transmit (or indeed is not present), the bus becomes idle and, after a certain time, all the access counters are increased by one. The next producer is then allowed to access the bus. If, again, it has nothing to transmit, the bus continues idle and the described procedure is repeated until a producer effectively uses the bus.

The procedure described in the previous paragraph accelerates token rotation time when producers have nothing to transmit. However, if it is used just like described, it can lead to a long idle time in the bus. The absence of bus activity can result in clock drift, which, in turn, could lead to AC inconsistencies among system nodes. To prevent this situation the VTPE forces the periodic transmission of a frame with K period to synchronize all access counters. To do this all producers must have an Idle Bus Counter

<hr />

[1] A device can be simultaneously producer and consumer

(IBC), which indicates how many times the bus became idle and no message was sent. All producers also have a *timer*, which can be programmed with time value t_1 or t_2. t_1 must be long enough to enable the slowest processor in the system to decode the VTPE frame (read the frame). t_2 is used to guarantee the token passing when one or more producers don't have something to transmit. t_1 and t_2 will be discussed further as well as the k value.

To explain the VTPE operation lets see the flow chart depicted in figure 4.

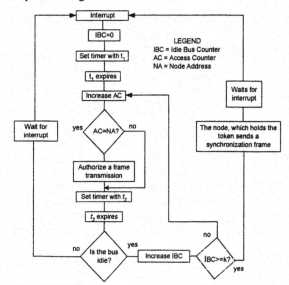

Fig.4 VTPE flowchart

After a frame transmission all producers must reset their IBCs to zero and initialise their *timers* with the t_1 value. After t_1 expires each producer node sets its *timer* with the t_2 value, increases its AC and checks if it is equal to its own node address. Two possibilities can occur:

- The node whose AC is equal to NA must immediately start a frame transmission if it has something to transmit and sets the timer with the t_2 value.

- The nodes with NA different from the current AC value must only set their timers with the t_2 value.

After t_2 expires each producer must check the Bus Status register of the Ethernet controller to verify if there is a frame being transmitted. If true, all producers will wait for the interrupt that will occur. If the bus is idle all producers increase the IBC and compare it with K. If IBC is smaller than K all producers increase the AC and repeat the last procedure until a producer does require access to the bus or the IBC becomes equal or greater than K. However, if IBC ≥ K, the node that holds the token must send immediately a special frame to synchronize the access counters. The use of the condition IBC ≥ K instead of only IBC = K solves the problem of an eventual absence of the node that would be holding the token when IBC = K. When the access counter exceeds the maximum number of producers, it is preset to 1 and the cycle is repeated again.

Although the VTPE uses the same bus arbitration principle as P-NET (EN 5070170 Volume 1), there are important differences, some due to new features of the protocol and others to the use of Ethernet as transmission medium:

- In VTPE the cooperation model used is the producer-consumer replacing the P-NET master-slave approach;

- In VTPE it is possible to send more than one message in the same packet;

- The VTPE data rate (10 or 100Mbps) is much greater than P-NET data transmission (fixed on 76.8Kbps).

- The VTPE may carry more data per packet (1500 bytes) max than P-NET (63 bytes) max;

3.2 The VTPE format frame

The VTPE protocol uses the MAC Ethernet frame encapsulating a special frame (VTPE frame) inside the Ethernet data field. This is shown in fig. 5.

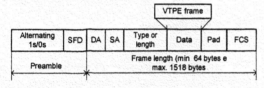

Fig. 5 Virtual Token-Passing Ethernet MAC frame

The VTPE uses the type field instead of length. It represents a reserved constant value, which must be used by all the VTPE messages on the network. The use of this field allows supporting the coexistence, in the same network, of other protocols. On frame reception, the nodes check the type field and only perform further processing if the frame is relevant. Nevertheless, the nodes producing non-VTPE frames must implement the VTPE access control, and transmit frames only if its AC is equal to its NA.

The VTPE frame carries one control field and one or more messages as it is depicted in the figure 6.

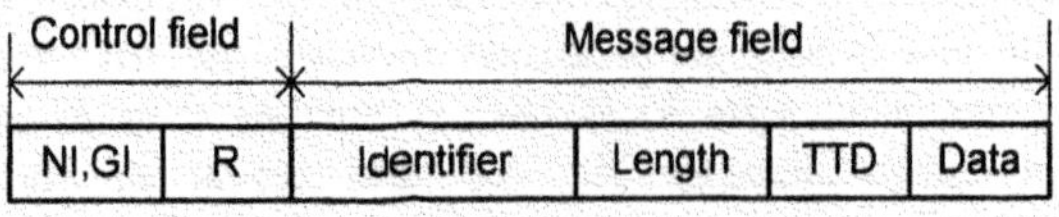

Control field		Message field			
NI,GI	R	Identifier	Length	TTD	Data

Fig. 6 - VTPE frame format

Since VTPE can send more than one message inside a single Ethernet frame more efficient bandwidth utilization is achieved due to the reduction on padding in case of small messages. Like it is shown

in figure 6, the VTPE frame is composed of two parts: the control field and the messages field.

Control field

The control field is two bytes long and the first byte is divided in two parts. The four less significant bits (NI) identify the number of messages inside the Ethernet frame (up to 16 messages). However this number can be reduced to bind the amount of information that each node must handle on frame reception, favouring the use of small processing power devices. The remaining four bits are the Group Identifier (GI), which will be used to create different producer groups, i.e, sub-networks. The GI idea permits to reduce processing overhead in the nodes by isolating devices that do not belong to the same group. In fact, on frame reception, the nodes check the GI field and only perform further processing if the frame is relevant. Nevertheless, the node must implement the VTPE medium access control.

The second byte of the control field (R) is reserved for future use.

Message field

The message field is composed of the identifier, the length, the TTD (Time To Deadline) and the Data. The identifier is unique and identifies the VTPE message in the system. It is 2 bytes long and thus can address 65536 different messages. The field is two bytes long and is reserved to contain an indication of the time remaining to the message's deadline. The length is two bytes long and indicates the number of bytes in a VTPE message. The VTPE data field is variable, so it can be so small as one byte or so long as 1492 bytes. Observe that the Length field is two-byte long and theoretically it can indicate 65536 bytes. However the maximum data possible per frame in a VTPE message is 1493 bytes (1500 bytes of the maximum data inside a single Ethernet frame minus 7 bytes of the control field and of the VTPE message's header).

To minimize overhead on small processing power devices the messages from and to these nodes must be compatible with their processing capacity. The maximum VTPE message length for these nodes will be fixed further.

3.3 The VTPE parameters t_1 and t_2

To establish these parameters it is necessary to determine the nodes processing workload to run the VTPE protocol, i.e, the workload of communication tasks on frame transmission and reception. This workload is presented next.

On frame reception

The host must execute three basic communication tasks: to attend immediately the Ethernet's controller request, to reset to zero the IBC, to program the *timer* to the value t_1, and, after t_1 expires, to increment the

AC and to check if it is equal to NA. The remaining activities depend of the protocol type, of the group identifier and of the number of VTPE messages. Table 1 resumes the remaining tasks after a frame reception.

Table 1 Tasks on received frame

Frame type	Tasks
No VTPE	Resets the Ethernet´s buffer controller an tramits if it is its chance (AC=NA)
VTPE	The host compares the GI in the frame incoming with the one programmed in its table. If equal, it continues and checks if any messages belong to its message's table. In this case, the messages are transferred to its buffers.

On frame transmission

To reduce the t_1 value the host transfers the VTPE frame to the Ethernet controller before holding the token. Then, when it holds the token, it must just authorize the Ethernet controller to send the frame.

The t_1 parameter is the time required by the host to decode the incoming frame, i.e, to execute the actions shown in table 1. Observe that the time t_1 is processor dependent as well as, indirectly, the number of messages inside the VTPE frame.

The second time, t_2 is the guard time needed to detect nodes absent from the network or that, despite being present, don't have anything to transmit. However, as the Ethernet controller response can differ from one controller to another one, some care must be taken. A higher value of t_2 leads to bandwidth spoiling, and a short value can be difficult to meet in low processing power microcontrollers. A value around 25μS should be adequate for most situations, but this parameter can be adapted according to the particular system characteristics.

4. VTPE REAL-TIME WORST-CASE COMPUTATION

The virtual token-passing idea facilitates to determine the MAC real-time behaviour. To explain the MAC real-time behaviour lets see the figure 7.

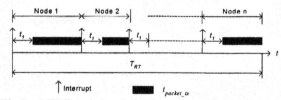

Fig.7 – VTPE real-time behaviour

It is shown in figure 7 that each node transmits a single frame per token holding time, starting at node 1. After node 1 transmission, node 2 gets the right of transmission, and so on up to the last node, the node *n*. After node's *n* transmission, node 1 gets the right of transmission again.

The T_{RT} is the time between two consecutive instants in which a specific node gets the right to transmit. Its value can be found based on the scenario depicted in the fig.7. Thus from the *Eq.1* it is possible to calculate the T_{RT}

$$T_{RT} = n * t_1 + \sum_{k=1}^{n} (t_{packet\_tx})_k \quad (Eq.1)$$

Where *n* is the number of nodes, t_1 is like mentioned before, and $(t_{packet\_tx})_k$ is the time to transmit a VTPE packet from node *k*.

The highest T_{RT}, token rotation time, occurs when all nodes transmit their maximum packet. So the maximum Token Rotation Time $maxT_{RT}$ can be calculated by equation (*Eq.2*).

$$\max T_{RT} = n * t_1 + \sum_{k=1}^{n} \max(t_{packet\_tx})_k (Eq.2)$$

The equations (Eq.1) and (Eq.2) shown that the VTPE MAC behaviour is deterministic, besides being very simple to determine the T_{TRT}.

4.1 VTPE example

This example shows how to calculate the VTPE parameter and the token rotation time. The system is composed of four processors, two of which are typical 8051 microcontrollers running at 12Mhz; the others are powerful processors. All nodes are supposed to be attached to a 10/100Mbps AX8876L Ethernet controller and the message set is shown in the table 2. In table 2 the node one has four messages and all can be filled and sent in the smallest Ethernet frame to improve throughput performance.

Table 2 VTPE message set

Node	Message Identifier	Consumer	Width (Bytes)	10Mbps (uS)	100Mbps (uS)
(1) 8051	1	(2)	2	57,6	5,76
	2	(2)	2		
	3	(3)	4		
	4	(3)	4		
(2) Processor	5	(4)	250	220,8	22.08
(3) 8051	6	(4)	16	57.6	5,76
(4) Processor	7	(2)	500	400	40

The node 3 has one message only and it can be sent in the smallest Ethernet frame too. In the case of node one and three the time to transmit their messages is 57.6μS and 5.76μS at 10Mbps and 100Mbps respectively. The time to transmit the messages from node 2 and 4 is 220.8μS and 400μS at 10Mbps, 22.08μS and 40μS at 100Mbps.

The messages 3 and 4 from node one are consumed by node three, so it must decode all VTPE frame, i.e. 36 bytes due to VTPE overhead (please see the VTPE format frame).

The messages from node 2 or 4 aren't important to node 1 or 3 (the nodes with microcontrollers) so they are only obligated to decode up to the message identifier.

To run this example in a 8051 it will be necessary 48 move instructions plus 5 compare with jump and plus 5 increment instructions. To execute each move instruction 24 clock cycles are necessary. To execute a compare with jump the processor needs also 24 clock cycles and 12 to the increment instruction. All totalize 1332 clock cycles, which is equivalent to 111µS when a 12Mhz crystal is used. The t_1 value is thus 111µS and the t_2 is 25µS such as suggested before.

The token rotation time can be calculated using the *(Eq.2)* and its value is:

$$T_{RT} = 1,832mS \quad at \quad 10Mbps$$

And

$$T_{RT} = 0,5176mS \quad at \quad 100Mbps$$

This example shows that the time impact caused by the 8051 over VTPE is not significant, which implies that it can be implemented in these processors.

5. CONCLUSIONS AND FUTURE WORKS

In this work a new Ethernet deterministic approach was presented. This protocol has been designed for use at field level, resource constrained devices like the ones typically founded in embedded distributed applications.

The most important conclusions over this protocol are:

- The VTPE system architecture can simultaneously rely on low processing power microprocessors or microcontrollers and on powerful processors.

- The introduction of low processing power devices has not a significant impact on the duration of the token rotation time, T_{RT}.

- There is no need for a specific chip-set when implementing the VTPE protocol. This provides the opportunity to use a common standard single chip microprocessor like the 8051 or like a device of the MICROCHIP PIC family.

- It has cheap implementation due to the use of COTS devices produced in large scale;

To continue the work on VTPE the main future developments are:

- Building of a VTPE demonstrator;

- Development of performance analysis;

- To make the VTPE protocol more flexible some additional functions will be included such as remote upload/download and configuration.

REFERENCES

Court, R.. Real-Time Ethernet. *Computer Communications*, **15** pp. 198-201. April 1992.

Decotignie, J-D. A perspective on Ethernet as a Fieldbus. *FeT'01, 4th Int. Conf. on Fieldbus Systems and their Applications*. Nancy, France. Nov. 2001.

Dietrich, D., Sauter, T.. Evolution Potentials for Fieldbus Systems. *WFCS 2000, IEEE Workshop on Factory Communication Systems*. Porto, Portugal, September 2000.

EN 50170, Volume 1- European Fieldbus Standard

Jasperneit, J. and P. Neumann. Switched Ethernet for Factory Communication. *ETFA'01, 8th IEEE Conf. on Emerging Tech. on Factory Automation*. Antibes, France. Oct. 2001.

LeLann, G, N and Rivierre. Real-Time Communications over Broadcast Networks: the CSMA-DCR and the DOD-CSMA-CD Protocols. *INRIA Report RR1863*. 1993.

Lo Bello, L., O. Mirabella., R.Caponetto. Fuzzy Traffic Smoothing: Another Step towards Statistical Real-Time Communication over Ethernet Networks. *RTLIA - Real-Time LANs in the Internet Age*. Vienna, Austria June 2002.

Malcolm, N. and W. Zhao. Hard Real-Time Communications in Multiple-Access Networks. *Real Time Systems* 9, 75-107. Kluwer Academic Publishers. 1995.

Molle, M. and L. Kleinrock. Virtual Time CSMA: Why two clocks are better than one. *IEEE*

Pedreiras, P., L. Almeida, and P. Gai, The FTT-Ethernet protocol: Merging flexibility, timeliness and efficiency, P*roceedings of the 14th Euromicro Conference on Real-Time Systems*, Viena, Austria, June 19-21, 2001.

Song, Y. Time Constrained Communication over Switched Ethernet. *FeT'01, 4th Int. Conf. on Fieldbus Systems and their Applications*. Nancy, France. Nov. 2001.

Shimokawa, Y. and Y. Shiobara. Real-Time Ethernet for Industrial Applications. *IECON'85*, pp829-834. 1985. *Transactions on Communications*. 33(9):919-933. 1985.

Venkatramani, C., T. Chiueh. Supporting Real-Time Traffic on Ethernet. *IEEE Real-Time Systems Symposium*. San Juan, Puerto Rico. Dec 1994.

PUBLICATIONS
www.elsevier.com/locate/ifac

PROFIBUS PROTOCOL EXTENSIONS FOR ENABLING INTER-CELL MOBILITY IN BRIDGE-BASED HYBRID WIRED/WIRELESS NETWORKS

Luís Ferreira, Eduardo Tovar, Mário Alves

Polytechnic Institute of Porto (ISEP-IPP)
Rua Dr. António Bernardino de Almeida, 431
4200-072 Porto, Portugal
E-mail: {llf@dei, emt@dei, malves@dee}.isep.ipp.pt

Abstract: Future industrial control/multimedia applications will increasingly impose or benefit from wireless and mobile communications. Therefore, there is an enormous eagerness for extending currently available industrial communications networks with wireless and mobility capabilities. The RFieldbus European project is just one example, where a PROFIBUS-based hybrid (wired/wireless) architecture was specified and implemented. In the RFieldbus architecture, interoperability between wired and wireless components is achieved by the use specific intermediate networking systems operating at the physical layer level, i.e. operating as repeaters. Instead, in this paper we will focus on a bridge-based approach, which presents several advantages. This concept was introduced in (Ferreira, *et al.*, 2002), where a bridge-based approach was briefly outlined. Then, a specific Inter-Domain Protocol (IDP) was proposed to handle the Inter-Domain transactions in such a bridge-based approach (Ferreira, *et al.*, 2003a). The major contribution of this paper is in extending these previous works by describing the protocol extensions to support inter-cell mobility in such a bridge-based hybrid wired/wireless PROFIBUS networks. *Copyright © 2003 IFAC*

Keywords: Fieldbus, Wireless, Real-time, Industrial Automation

1. INTRODUCTION

PROFIBUS is one of the most popular fieldbus protocols, with several hundreds of thousands of installations currently in operation worldwide. It was standardised in 1996, as EN 50170 (EN 50170, 1996), by CENELEC and, more recently in 2000, by IEC as IEC61158 – Fieldbus Standard for use in Industrial Systems.

In the last years, eagerness emerged concerning extending the capabilities of PROFIBUS to cover functionalities not previously considered: industrial wireless communications (Haehniche and Rauchaupt, 2000; Alves *et al.*, 2002; Rauchhaupt, 2002) and the ability to support industrial multimedia traffic (Pereira *et al.*, 2002).

The RFieldbus European project (Rauchhaupt, 2002) is just one example of that effort, where PROFIBUS was extended for encompassing hybrid wired/wireless communication systems.

In RFieldbus, interoperability between wired and wireless components is achieved by the use of intermediate networking systems operating at the physical layer level (i.e. as repeaters), resulting in a "broadcast" network with a single logical ring (just one token rotating between the masters). The main advantage of such a single logical ring (SLR) approach is that the effort for protocol extensions is not significant.

However, there are a number of advantages in using a multiple logical ring (MLR) approach to such type of hybrid systems. This concept was introduced and discussed in (Ferreira *et al.*, 2002) where a bridge-based approach (thus, layer 2 interoperability) was briefly outlined. In that work, references to how some complex functionalities (such as the handoff between adjacent wireless cells) could be supported with minimum protocol extensions and still maintaining the compatibility with legacy PROFIBUS technologies were briefly described.

The main advantage of a bridge-based solution is that it provides traffic segmentation, thus improved responsiveness for transactions between stations belonging to the same logical ring, and error containment within each domain. In (Ferreira *et al.*,

This work was partially supported by the European Commission under the project R-FIELDBUS (IST-1999-11316) and by FCT under the project CIDER (POSI/1999/CHS/33139).

2003a), all implementation details concerning an Inter-Domain Protocol (IDP) which is able to support Inter-Domain Transactions (IDT) are thoroughly described.

This paper extends previous works on the analysis and proposal of the required protocol extensions to support the inter-domain (inter cell) mobility of wireless nodes, therefore with a particular focus on the handoff functionalities.

The reminder of this paper is organised as follows. In Section 2 some fundamental aspects of the PROFIBUS protocol are presented. Then, in Section 3, we introduce the context and describe the main concepts related to bridge-based hybrid wired/wireless PROFIBUS networks. In Section 4, the mechanisms and protocols for supporting inter-cell mobility of wireless stations are described in detail. The concepts of Global Mobility Manager (GMM) and Domain Mobility Manager (DMM) are introduced. In Section 5, we discuss the approach proposed in this paper, namely concerning some timing characteristics. Finally, in Section 6, we draw some conclusions and outline the ongoing work.

2. RELEVANT ASPECTS OF PROFIBUS

This section addresses some features of the PROFIBUS protocol that are relevant for this paper.

2.1 Message Cycle

In PROFIBUS, master stations may initiate message transactions, whereas slave stations do not transmit on their own initiative but only upon (master) request. A transaction (or message cycle) consists of the request frame from the initiator (always a master station) and the associated acknowledgement or response frame from the responder (either a master or a slave station). The acknowledgement (or response) must arrive before the expiration of the *Slot Time*, otherwise the initiator repeats the request the number of times defined by the *max_retry_limit*, a PROFIBUS Data Link Layer (DLL) parameter.

A PROFIBUS master is capable of executing transactions during its token holding time (T_{TH}), which is given a value corresponding to the difference, if positive, between the target token rotation time (T_{TR}) parameter and the real token rotation time (T_{RR}). For further details, the reader is referred to (EN5017, 1996; Tovar and Vasques, 1999).

2.2 Ring Maintenance Mechanisms

In order to maintain the logical ring, PROFIBUS provides a decentralised (in every master station) ring maintenance mechanism. Each master maintains two tables – the *Gap List* (GAPL) and the *List of Active Stations* (LAS). Optionally it may also maintain a *Live List* (LL).

The *Gap List* consists of the address range from *TS* (*This Station* address) until *NS* (*Next Station* address, i.e., the next master in the logical ring). Each master station in the logical ring starts to check its *Gap* addresses every time its *Gap Update Timer* (T_{GUD}) expires. This mechanism allows masters to track changes in the logical ring: addition (joining) and removal (leaving) of stations. This is accomplished by examining (at most) one *Gap address* per token visit, using the *FDL_Request_Status* frame.

The LAS comprises all the masters in the logical ring, and is generated in each master station when it is in the *Listen Token* state after power on. This list is also dynamically updated during operation, upon receipt of token frames.

The *Live List* mechanism requires an explicit request from the PROFIBUS DDL user (via a management FMA1/2 request). This service returns the list of all active stations (masters and slaves).

2.3 Token Passing Procedure

The token is passed between masters in ascending address order, except for the master with the highest address, that must pass the token to the master with lowest address. Each master knows the address of the *Previous Station* (PS), the address of the *Next Station* (NS) and, obviously, its own address (*This Station* address - TS).

If a master receives a token addressed to itself from a station registered in the LAS as its predecessor, then that master is the token owner, and may start processing message cycles. On the other hand, if a master receives the token from a station, which is not its PS, then it shall assume an error and will not accept the token. However, if it receives a subsequent token from the same station, it shall accept the token and assume that the logical ring has changed. In this case, it updates the PS with the new address.

If after transmitting the token frame, and within the *Slot Time*, the master detects valid bus activity, it assumes that its successor owns the token and is executing message cycles. Therefore, it ceases monitoring the activity on the bus.

In case the master does not recognise any bus activity within the *Slot Time*, it repeats the token frame and waits another *Slot Time*. If it recognises bus activity within the second *Slot Time*, it assumes a correct token transmission. Otherwise, it repeats the token transmission to its next station for the last time. If still, there is no bus activity, the token transmitter tries to pass the token to the next successor of its LAS. It continues repeating this procedure until it founds a successor.

3. BASICS ON HYBRID WIRED/WIRELESS PROFIBUS NETWORKS

3.1 Network Components and Basics on Bridge Operation

A hybrid wired/wireless fieldbus network is composed by stations with a wireless interface (usually radio) that are able to communicate with wired (legacy) stations.

The wireless part of the fieldbus network is supposed to include at least one *radio cell*. Basically, a radio cell can be described as a 3D-space where all associated wireless stations are able to communicate with each other. Our architecture considers two types of domains.

A *Wired Domain* is a set of (wired) stations intercommunicating via a wired physical medium. A *Wireless Domain* is a set of (wireless) stations intercommunicating via a wireless physical medium. In the example of Fig. 1 the following set of wired PROFIBUS master (M) and slave (S) stations are considered: M1, M2, S1, S2, S3, S4 and S5. Additionally, the following set of wireless stations is considered: M3, S6 and S7. Within this set, only M3 and S6 are mobile. All wireless stations are assumed to be PROFIBUS stations with a wireless physical interface, capable of supporting radio communications and the mobility functionalities, like in RFieldbus (Rauchhaupt, 2002). Three bridge devices are considered: B1, B2 and B3.

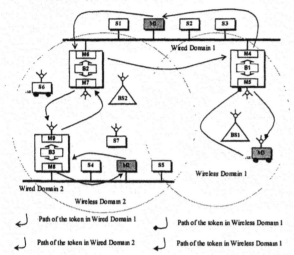

Fig. 1. Wireless PROFIBUS example network

In such a system, all communications are relayed through base stations: BS1 and BS2. Each base station uses two radio channels, one to transmit frames to wireless stations (the downlink channel), and another to receive frames from the wireless stations (the uplink channel). Since all frames in wireless domains are relayed through base stations, the downlink signal quality can be assessed by wireless stations to perform the inter-cell mobility (further detailed in Section 4). We will assume, in the remaining of the paper, that M5 and M7 include the base station functionalities in their wireless front-end, thus, structuring radio cells (wireless domains) 1 and 2, respectively.

Note also that the network operation is based on the Domain-Driven Multiple Logical Ring (MLR) schema, described in (Ferreira, *et al.*, 2002). Therefore, each wired/wireless domain has its own logical ring. In fact, each bridge includes two masters (Fig. 2): one belonging to the wired domain and the other belonging to the wireless domain.

In the example of Fig. 1, four different logical rings exist: {(M3 → M5), (M1 → M4 → M6), (M7 → M9), (M8 → M2)}. Obviously, our approach could be generalised to bridges interconnecting more than 2 domains.

We are also assuming that the network topology is tree-like, and that routing is based on MAC addresses. Traffic is relayed from one bridge master to the other if the destination address is included in the *Routing Table*

(RT) of the incoming side. Obviously, every bridge must include two tables (one for each bridge master). This approach imposes the use of a single address space, where every station in the network has a unique MAC address. This implies that bridge masters must read all frames, even if the destination address does not correspond to their own address.

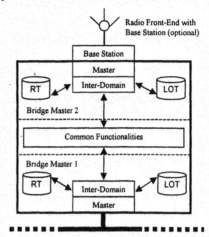

Fig. 2. Bridge components

3.2 Interoperability Between Domains

The communication between stations belonging to different domains; that is, Inter-Domain Transactions (IDT), is supported by the Inter-Domain Protocol (IDP) proposed in (Ferreira *et al.*, 2003a). The IDP not only defines the format of frames exchanged between bridges, but also specific bridge functionalities. In this section, we will just briefly describe the IDP.

When an initiator makes a request addressed to a station in another domain (an Inter-Domain Request), all stations belonging to the initiator's domain discard the frame, except the bridge masters (BMs) belonging to that domain. Only one of these BMs then handles the request frame. We denote this bridge master, i.e. the first bridge master in the path from the initiator to the responder, as BM_i, where i stands for initiator. The relayed frame, denoted as an Inter-Domain Frame (IDF), is coded using the IDP (Ferreira *et al.*, 2003a). Bridges perform routing based on the MAC addresses contained in the frames and on the routing table (RT) of the incoming side.

The IDF embeds the original request (or response) and additional information that allows both the decoding of the embedded frame and the matching between the request and the respective response. The BM_i is capable of matching a response to the related pending request, using the information contained in the IDF embedding the response, and by using the information contained in the *List of Open Transactions* (LOT). The LOT contains information about the request frame, such as destination and source addresses. It also contains a tag, the *Transaction Identifier* (TI), which must be included in the IDF related to the request and also in the respective IDF response.

The IDF embedding the request is relayed by the other bridges in the path until reaching the bridge master that connects to the domain the responder belongs to (the last bridge master in the path) - bridge master BM_r, where *r* stands for responder. Then, this bridge reconstructs the original request frame and transmits it to the responder, a standard PROFIBUS responder station (e.g., a wireless DP slave).

When BM_r receives the immediate response to that request, it encodes the frame using the IDP. This IDF will be relayed until reaching bridge master BM_i, where it will be decoded and stored.

In order to conclude the transaction, the initiator periodically repeats the (same) request until receiving the related response. When BM_i receives the (repeated) request, it responds to the initiator using the stored response frame meanwhile obtained. This mechanism is completely transparent from the point of view of the initiator, since BM_i emulates the responder in a way that the initiator station considers the responder station as belonging to its domain.

Considering the system scenario illustrated in Fig. 1, Fig. 3 represents a simplified timeline regarding a transaction between master M3 and slave S6.

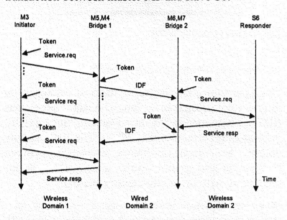

Fig. 3. Example timeline for an Inter-Domain Transaction (IDT) between M3 and S6

4. SUPPORTING INTER-DOMAIN MOBILITY

In RFieldbus, the inter-domain mobility depends on the assessment period (Alves *et al.*, 2002), which is periodically triggered by one of the masters (the beacon master). Each base station sends beacons in its radio channel, in order for the wireless stations to assess the quality of the different channels, after which, the wireless stations may switch to the channel offering the best signal quality. Note that as there is only one token rotating (single logical ring system) there is no message loss and no need for specific registration mechanism.

However, the approach described in Section 3 (bridge-based intermediate systems) requires a more sophisticated handoff procedure. The main reason is that the system has multiple logical rings. Mobile wireless stations must implement radio channel assessment and switching mechanisms, and also mechanisms to support stations joining/leaving the logical rings.

In (Ferreira *et al.*, 2002), the authors briefly described the possibility of using the native PROFIBUS ring management mechanisms to support inter-cell (inter-domain) mobility. However, additional mechanisms

must be added to guarantee no errors, no loss of frames or frame order inversion concerning inter-domain transactions (IDT).

Therefore, in this paper we propose a hierarchically managed handoff procedure that fulfils these requirements. One master in the overall system implements the global mobility management functionality – the *Global Mobility Manager* (GMM). In each domain, one master controls the mobility of stations belonging to that domain – the *Domain Mobility Manager* (DMM). Finally, the bridges must implement specific mobility services. The GMM must know the addresses of all the bridges and DMMs in the system. Each DMM must know the addresses of the bridges in its domain. For example, and concerning the scenario illustrated in Fig. 1, M1 assumes both the role of GMM and the DMM of wired domain 1. Bridge masters M5, M7 and M8 assume the role of DMMs for wireless domain 1, wireless domain 2 and wired domain 2, respectively.

The role of these management entities and the different phases involved in the proposed handoff will be described next.

4.1 Phases of the Handoff Procedure

The handoff procedure starts with a *Start Handoff Procedure* message sent by the GMM. This message is sent periodically, according to the mobility requirements (e.g. maximum foreseeable speed) of the mobile stations. All bridges in the network relay this message, which then triggers a sequence of actions that are briefly outlined in Fig. 4.

Phase 1

When the bridges receive this message, they stop accepting new IDTs from the masters belonging to their domains. Nonetheless, they keep handling pending IDTs (which are still present in their LOTs) and, importantly, they keep handling IDTs originated in the other domains. After completing all pending IDTs (those from their LOTs), the bridges transmit a *Ready to Start Handoff Procedure* message to the GMM. When the GMM receives such a message from all bridges, it broadcasts a *Prepare for Beacon Phase* message. Note that intra-domain transactions are allowed until this instant.

Phase 2

After the DMMs receive the *Prepare for Beacon Phase* message, and as soon as they receive the token, they do not pass it to other masters in their domains. Each DMM sends a *Ready for Beacon Phase* message to the GMM and starts an *Inquiry* service. During this phase, every DMM sends *Inquiry* frames addressed to bridge masters belonging to its domain. The bridges use the response message to transmit any mobility-related message from its output queue. This procedure minimises the communication latency between the GMM and the DMMs, and keeps small the inaccessibility period of the wired nodes, as it will be shown in Section 5. When a bridge master without domain management capabilities receives the *Prepare for Beacon Phase* message, it will only be able to

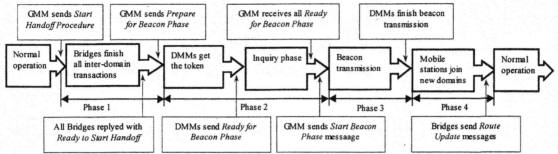

Fig. 4. Handoff Procedure phases (simplified)

communicate using the *Inquiry* service, and it clears all its routing table entries related to mobile nodes.

Phase 3

After collecting all *Ready for Beacon Phase* messages from the DMMs, the GMM starts the assessment phase by broadcasting the *Start Beacon Phase* message. Upon receiving this message, the DMMs start emitting beacons. Wired domains may resume intra-domain transactions, but they are not capable of performing inter-domain transactions (IDTs) while the bridges belonging to their domains do not receive the route update messages related to the mobile nodes. The mobile stations use the beacon frames to evaluate the quality of the different radio channels and to decide if they switch the radio channel (or not). So, every mobile station willing to handoff must switch to the new radio channel, before ending the beacon transmission.

Phase 4

After the end of the beacon phase, every wireless DMM (still holding the token) inquires all mobile stations in order to detect if they still belong to its domain. After this, mobile slaves are capable of answering requests, but mobile masters must still enter the new logical ring using the standard PROFIBUS ring management mechanisms. Since the routing table entries related to mobile stations have been cleared, only when the bridges receive updated routing information, at the end of the Handoff Procedure, they may restart routing IDTs related to mobile stations.

4.2 Details on the Handoff Procedure

State Machine for the GMM. The operation of the GMM is based on the state machine depicted in Fig. 5. We are considering that there is a mobility timer used to trigger the Handoff Procedure in a periodic fashion.

At power on, the GMM enters into the *INACTIVE* state, and the mobility timer is loaded with the Handoff Procedure period (which depends on the dynamics of the mobile stations). When the mobility timer expires (TIMER transition) the GMM state machine enters in the *WRSHP* state (Wait Ready to Start Handoff Procedure message) and the GMM sends the *Start Handoff Procedure* message.

In the *WRSHP* state, the GMM receives *Ready to Start Handoff Procedure* messages from all the network bridges (READYH transition). It will only enter into the *WRBP* (Wait Ready for Beacon Phase message) state when all bridges have replied (ALLRESP1 transition) and then it sends the *Prepare for Beacon Phase* message.

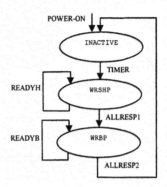

Fig. 5. State machine for the GMM

In the *WRBP* state, the GMM receives *Ready for Beacon Phase* messages from the network DMMs (READYB transition). When all DMMs have replied, the state machine enters into the *INACTIVE* state, and the GMM sends the *Start Beacon Phase* message.

State Machine for the DMM. The DMM is responsible for retaining the token, controlling the *Inquiry* service and transmitting beacons (only in a wireless domain). This DMM functionality can be embedded in any type of static master station, but for improved performance (in most cases) it should be located in a bridge master.

The DMM state machine (Fig. 6) goes into the *INACTIVE* state after power-on. Transition *SHP_MSG* is triggered when the DMM receives the *Start Handoff Procedure* message, and enters the *WPBP* (Wait Prepare for Beacon Phase) state, where the DMM waits for the reception of the *Prepare for Beacon Phase* message. This message triggers the transition (*PBP_MSG*) to the *WTOKEN* (Wait Token) state. In this state, the DMM waits until receiving the token from its predecessor, and then (*TOKEN_MSG* transition) it retains the token and sends the *Ready for Beacon Phase* message to the GMM. Following this, the DMM uses the *Inquiry* service in order to exchange mobility-related messages with the bridges in its domain. This service is needed in order to guarantee that all DMMs are able to communicate with the GMM. Nevertheless, if there are no other bridges belonging to the DMM domain, it transmits *void* frames in order to maintain network activity.

When the *Start Beacon Phase* message arrives to the DMM (*SBP_MSG* transition), the DMM starts transmitting beacon frames for a certain period. When this period ends, the DMM tries to detect if mobile stations are located in its domain, by inquiring them using *FDL_Request_Status* frames (*FDL_ST_MSG* transition).

When a DMM is responsible for a wired domain, it does not transmit any beacon frame and thus it passes from the *INQUIRY* state directly to the *INACTIVE* state (*WR_DOM* transition).

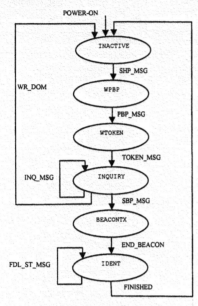

Fig. 6. State machine for a DMM

Other Bridges' Functionalities. The bridge's role during the Handoff Procedure is essentially to ensure that there are no pending IDTs during the Handoff Procedure and to relay mobility-related messages (when the DMMs are in the *INQUIRY* state). So, at power-on (Fig. 7), a bridge goes into the *INACTIVE* state, where it operates normally, relaying IDTs as described in Section 3.2. In this state the bridge can update its *List of Active Stations*, *Live List* or *Gap list*, and consequently its routing table according to the changes in the configuration of the system (*LAS_C*, *LL_C* and *GAP_C* transitions). These transitions also trigger the broadcast of a *Route Update* message. Also, when the bridge receives a *Route Update* message, it updates the routing tables and forwards that message (*RT_UDT* transition).

When a bridge receives the *Start Handoff Procedure* message (*SHP_MSG* transition) it goes into the *WIDT_END* (Wait Inter-Domain Transactions End) state, where the bridge waits until finalising all its open IDTs contained in the LOT. In this state, the bridge masters ignore new IDTs.

The completion of an IDT triggers the *IDT_FINISHED* transition. When all IDTs have been completed, the bridge enters into the *WINQUIRY* (Wait Inquiry message) state (*ALL_IDT_FINISHED* transition). In the *WINQUIRY* state, the bridge only communicates with its domain DMM, using the *Inquiry* service. In this state, when the bridge receives an *Inquiry* frame and it has mobility related messages, it responds (*RESP* transition), otherwise, no response is sent (*NO_RESP* transition).

When the beacon transmission starts, the bridge returns into the *INACTIVE* state and clears the entries related to mobile stations in its routing table (*START_BEACON* transition). Thus, all bridges must know the addresses of all mobile stations in the system.

From this point forward, the bridges are capable of relaying IDTs, if requested. Obviously, IDTs related to mobile stations will only be relayed when the bridge receives the related *Route Update* messages.

The description of the mobility related messages can be found in (Ferreira *et al.*, 2003b).

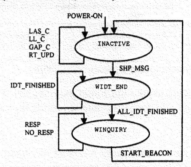

Fig. 7. State Machine for the mobility related functionalities in the bridges

5. EXAMPLE SCENARIO AND TIMING DISCUSSION

In order to outline an example of the Handoff Procedure, we are considering a network scenario as depicted in Fig. 1. For the sake of simplicity, we also consider that there is no additional traffic in the network, except for an Inter-Domain Transaction (IDT) between master M2 and slave S7, an Intra-Domain Transaction between M2 and S5, the token, and mobility-related messages. We are also considering that both M3 and S6 will execute a handoff. Fig. 8 supports the following description.

5.1 Example Scenario

The GMM starts the Handoff Procedure by broadcasting the *Start Handoff Procedure* message (M1.1). After receiving this message, bridges B2 and B1 having no open IDT immediately transmit the message *Ready to Start Handoff Procedure* (B1.1, B2.1).

Bridge B3 has an open IDT, related to request M2.1. Therefore it will only send the *Ready to Start Handoff Procedure* message (B3.1) when the transaction related to M2.1 is completed. After that, master M2 tries again to make the same transaction but bridge B3 ignores it. Note that the intra-domain transaction between M2 and S5 may still carry on.

After receiving the *Ready to Start Handoff Procedure* message from all network bridges (B1.1, B2.1 and B3.1), the GMM broadcasts the message *Prepare for Beacon Phase* (M1.2). When the DMMs M7, M8 and M5 receive that message, they start the *Inquiry* service. So, messages B2.2 and B1.2 are only transmitted when M1 (wired domain 1 DMM) sends the *Inquiry* frames M1.3 and M1.4, respectively addressed to B1 and B2. Message B3.2 is relayed in a similar way until reaching M1.

Also note that masters M8 and M5 do not have any other bridge belonging to its domain, thus they send void frames.

After receiving the *Ready for Beacon Phase* message from all the DMMs in the network (B1.2, B2.2 and B3.2), the GMM sends the *Start Beacon Phase* message (M1.5). When receiving this message, each DMM will start the transmission of beacon messages. The starting time of this phase is slightly different for the different domains due to communication latencies. Also, the duration of the beacon phase must be different for different domains so that all domains finish almost at the same time. The duration of the beacon phase must guarantee that all stations are capable of evaluating all possible radio channels and switch to a new one. Note that in wired domains it is not necessary to transmit beacon frames. Nevertheless, the bridges connecting to these domains must relay the *Start Beacon Phase* (M1.5) message to other wireless domains.

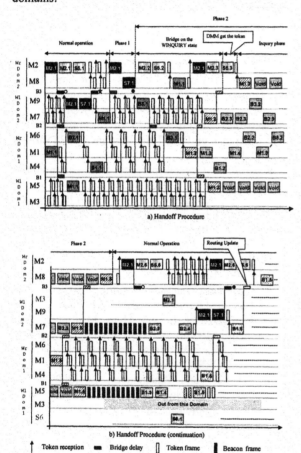

a) Handoff Procedure

b) Handoff Procedure (continuation)

↑ Token reception   ▬ Bridge delay   ▯ Token frame   ■ Beacon frame

o Open transaction   ● Open Trans. With response   ★ Open Trans. completion

Fig. 8. Timeline for handoff procedure

Before the end of the beacon phase, mobile master M3 and mobile slave S6 switch to the radio channels of wireless domain 2 and wireless domain 1, respectively.

After the end of the beacon phase, wireless DMMs M5 and M7 *Inquire* the mobile stations in the network (M3 and S6) in order to detect if they are located in its domain. M5 and M7 use *FDL_Request_Status* frames addressed to mobile stations S6 (B1.4 and B2.4) and M3 (B1.3 and B2.3).

From this point forward, slave S6 is capable of answering requests, but master M3 must still enter into the new logical ring using the standard ring management procedures. This is illustrated in Fig. 9.

Message B1.5 is the route update message related to station S6, but the message related to station M3 is only sent when M3 effectively enters the logical ring. When master M2 receives the token, it repeats request M2.1, but it will only be relayed by bridge B3 when M9 receives the token, after the end of the beacon phase. Nevertheless, intra-domain transactions in wired domain 2 are possible during this period.

When M3 enters into the new wireless domain, it detects that it was taken out of the ring and goes into the *Listen Token State*. M3 will only be able to enter the new logical ring when its predecessor station starts the *Gap Update* mechanism and subsequently passes the token to M3.

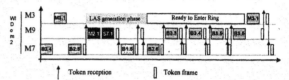

↑ Token reception   ▯ Token frame

Fig. 9. M3 entrance into the logical ring

Fig. 9 further highlights some details on how the entry of station M3 into the logical ring is performed. In fact, after the switching, station M3 is still on the *Active Idle* state. So, it can return an answer (M3.1) to the *FDL_Request_Status* request B2.3. After that, M3 detects that its predecessor station did not pass the token and enters into the *Listen Token* state, where it re-generates its LAS during two complete token rounds. During this phase M3 will not answer any requests addressed to it. After this phase M3 is ready to enter into the logical ring and is able to reply to any *FDL_Request_Status* frame (indicating its readiness).

M9 uses the *Gap Update* mechanism in order to include M3 in its logical ring, thus it sends *FDL_Request_Status* requests B3.3, B3.4 and B3.5, respectively addressed to stations with addresses 0, 1 and 2 (considering that station M9 HSA is equal to 9). Finally, M9 sends *FDL_Request_Status* request B3.5, which is addressed to M3, it replies with the *Ready to Enter Logical Ring* message, subsequently M9 passes the token to M3. To make the entry procedure fast, master stations must have a low *Gap update* factor.

5.2 A Discussion on Timeliness

Quantifying the duration of the different phases of the handoff procedure enables a notion of the latencies involved in this procedure.

In order to be able to obtain figures for the example presented in Section 5.1, we are making the following assumptions:

- data rate of 1.5Mbps;
- all data frames have equal duration (approximately 154μs);
- the beacon frames (the same type of frame as the token frame) have a duration of 30μs;
- bridge relaying latency is 200μs;
- *Slot Time* is 66μs;
- and wired/wireless bit rates and frame formats are identical.

Using these assumptions, Table 1 presents the results for the example scenario. Note also a second column (in shading) that contains the results for the case of 12Mbps.

The inaccessibility time for the wired domains is equal (for the depicted scenario) to the duration of the inquiry phase, 1640µs and 1450µs, respectively for wired domain 1 and wired domain 2. Only during this time it will not be possible to perform intra-domain transactions (IDTs) in the wired domains. IDTs will only be possible when the mobile nodes join their new domains. Thus, a station in wired domain 2 is not able to exchange messages with S6 for a time span of 5540µs, and with M3 for a time span of 7202µs.

Table 1 Handoff Procedure Timings (in µs)

Time Span	1.5Mb/s	12Mb/s
Time needed by bridge B3 to finish all of its IDT	355	53
Time during which B3 is in WINQUIRY state	1129	169
Time needed until the DMM of wired domain 2 obtains the token	183	28
Duration of the inquiry phase	1450	195
Duration of the beacon phase in wireless domain1	988	148
Duration of the beacon phase in wireless domain2	836	125
Inaccessibility for node S6	2953	443
Inaccessibility for node M3	5774	866

Note however that these are not worst-case values. They only reflect reasonable figures for the actual scenario mentioned in Section 5.1. In fact, these values result from a relaxed scenario in terms of number of concurrent transactions (both inter-domain and intra-domain). Nevertheless, the applications envisaged for wireless applications are not expected to require very tight control loops. Examples of such applications are handheld terminals, AGVs or multimedia devices (Pacheco et al., 2002). It is also obvious that if the bit rate increases to 12Mbps (PROFIBUS already supports it), the latencies involved in the handoff procedure are significantly reduced.

6. CONCLUSIONS AND ON-GOING WORK

In this paper, we have detailed and analysed mechanisms for supporting inter-domain mobility of mobile stations in hybrid wired/wireless bridge-based PROFIBUS networks. In such an architecture, the communication between the different domains is supported by an Inter-Domain Protocol. This protocol enables the use of standard PROFIBUS stations, since the additional functionalities are implemented by specific bridge devices responsible for emulating the behaviour of the responder stations.

In the proposed architecture, mobile/wireless stations may move between different wireless cells using a Handoff Procedure hierarchically managed by the Global Mobility Manager (GMM) and several Domain Mobility Managers (DMMs).

A crucial aspect of the proposed mobility protocol is its ability to cope with the timing requirements of distributed applications. Although in this paper we have discussed some aspects related to timeliness, it is an on-going work the development of a simulation tool, which implements the proposed protocols. This tool is now is now in the last stages of development and will enable further temporal characterisation of the proposed architecture. Another objective of this simulation tool is to assess possible enhancements in the protocol in order to increase its performance, e.g. in order to reduce the time needed by a master to enter into a new logical ring.

In this paper, and for the sake of simplicity, we did not make any references to any kind of error detection/recovery mechanisms, which obviously are necessary. This issue is also being addressed in the related on-going work.

REFERENCES

Alves, M., Tovar, E., Vasques, F., Roether, K. and Hammer, G. (2002). Real-Time Communications over Hybrid Wired/Wireless PROFIBUS-based Networks. In Proceedings of the 14th Euromicro Conference on Real-Time Systems (ECRTS'02), pp. 142-151.

EN 50170 – General purpose field communication system (1996). CENELEC.

Ferreira, L., Alves, M. and Tovar, E. (2002). Hybrid Wired/Wireless PROFIBUS Networks Supported by Bridges/Routers. In Proceedings of the 2002 IEEE International Workshop on Factory Communication Systems, pp. 193-202.

Ferreira, L., Tovar, E. and Alves, M. (2003a). Enabling Inter-Domain Transactions in PROFIBUS Networks. In Technical Report HURRAY-TR-0304.

Ferreira, L., Tovar, E. and Alves, M. (2003b). Inter-Domain Mobility in PROFIBUS Bridge-Based Hybrid Wired/Wireless Networks. In Technical Report HURRAY-TR-0305.

Haehniche, J. and Rauchhaupt, L. (2000). Radio Communication in Automation Systems: the R-Fieldbus Approach. In Proceedings of the 2000 IEEE International Workshop on Factory Communication Systems, pp. 319-326.

Pacheco, F., Pereira, N., Marques, B., Machado, S., Marques, L., Pinho, L. and Tovar, E. (2002). Industrial Multimedia put into Practice. In Proceedings of the 7th CaberNet Radicals Workshop, Bertinoro, Forlì, Italy.

Pereira, N., et al. (2002). Integration of TCP/IP and PROFIBUS Protocols. In WIP Proceedings of the 2002 IEEE International Workshop on Factory Communication Systems, Vasteras, Sweden.

Rauchhaupt, L. (2002). System and Device Architecture of a Radio Based Fieldbus – The RFieldbus System. In Proceedings of the 2002 IEEE International Workshop on Factory Communication Systems, Vasteras, Sweden.

Tovar, E. and Vasques, F. (1999). Real-Time Fieldbus Communications Using PROFIBUS Networks. IEEE Transactions on Industrial Electronics, vol. 46, no. 6, pp. 1241-1251.

ELSEVIER

IFAC

PUBLICATIONS
www.elsevier.com/locate/ifac

REALTIME COMMUNICATION IN PROFINET V2 AND V3 DESIGNED FOR INDUSTRIAL PURPOSES

Joachim Feld, Siemens AG, A&D AS RD 8, Germany

*Dipl.-Inf. Joachim Feld is system architect for SIMATIC at
Siemens A&D (Automation and Drives). He is working in the
areas communication and Ethernet.
He is member in the working group PROFInet of PI.
Authors Adress: Siemens AG, Automation & Drives, A&D AS RD 8,
Gleiwitzer Str. 555, D-90475 Nürnberg,
Tel. ++49 911 895-3847, Fax -15 38 4
E-Mail: Joachim.Feld@siemens.com*

Abstract: PROFInet is a concept devised by PROFIBUS International (PI) for either modular machine and plant engineering or distributed IO. Using a plant-wide multi-vendor engineering for modular machines, commissioning times as well as costs are reduced. PROFInet is based on a common object model which is implemented with the COM/DCOM technology. With version 2 PROFInet is enhanced with real-time communication on Ethernet. This enables the advantages of modular and multi-vendor engineering to be used even in applications with time-critical data transfer requirements. In the next step PROFInet V3 is extended with capabilities for high sophisticated motion control application. A common object model for PROFInet V2 and PROFInet V3 allows slightly migration from V2 to V3. *Copyright © 2003 IFAC*

Keywords: Automation, Realtime, Ethernet, Industrial control, Jitter, Packets,

1. INTRODUCTION

PROFInet is the initiative of PROFIBUS International (PI) to emerge Ethernet to the next generation of industrial automation. PROFInet consists of several topics such as distributed automation, decentralized field devices, network management, installation guidelines and web integration. All these different topics will help to make Ethernet easier to use in industrial automation.

Communication is a major part of PROFInet. This is true as well as for the distributed automation concept as well as for the decentralized field devices. In the distributed automation concept machines and systems are divided into technological modules, each of which is comprised of mechanical, electrical / electronics, and software. The functionality of the technological modules is encapsulated in the form of PROFInet components. From the outside the PROFInet components can be accessed via interfaces with standardized definitions. They can be combined with one another like building blocks as required and easily reused. The Distributed automation concept is based on COM. COM is a well known object

modelling technique and part of MS Windows. DCOM (Distributed COM) is the extension of COM used for transparent communication between COM objects, which are distributed on different devices. DCOM can be implemented in industrial automation systems (e.g. PLC or Motion control systems) and is also part of the PROFInet runtime SW. The PROFInet runtime SW, that can be loaded from the PI website for free by all PI members, is the base for the implementation of a PROFInet device.

2. REQUIREMENTS TO REAL TIME COMMUNICATION

But DCOM cannot fulfill all the real time requirements that are needed for the successful use in the shop floor. These requirements comprise of deterministic behavior, reaction times in the range of 5-10ms and a very small need of device resources e.g. processor load and memory use. Reaction time is the time needed to acquire a signal on a device A (aka provider), process it inside the user program, send it over the communication line to a device B and process it in the user program in device B (aka consumer).

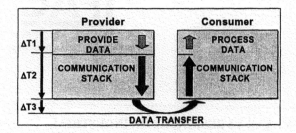

Fig. 1. Actualization rate consists of several factors.

Reaction time comprises of several parts:

- T1: Time to acquire data in the application on the provider resp. process it in the consumer
- T2: Time to process data inside the communication stack
- T3: Time to transfer data over the wire. This includes both the time on the wire and the delay of the network components.

Times to acquire or process data or the transfer over the wire are fixed and are mostly determined by the architecture of the devices or the network topology. So it is evident that the best savings can be achieved through optimizing the times needed in the communication stack.

Measurements of different TCP/IP stacks showed, that there can be latencies up to 200 ms (Klüger, 2001). Using TCP/IP or UDP/IP for real time communication would have some more characteristics that are typically not needed for real time communication:

- Through the use of a cyclic data transfer no connection oriented protocol is needed to control the communication partner
- Real time communication has not to be routed via subnet boundaries. The functionality of an ISO Layer 3 (IP) needs not to be supported, because delay of routing is a lot more than the reaction times needed in automation applications.
- The size of data packets that have to be transferred are typically in a range of 32-256 Bytes. Through the maximal size of 1532 bytes with Fast Ethernet no segmentation is needed. All protocol elements needed for segmentation can be omitted.

3. INFLUENCE OF NETWORK TOPOLOGY

Real time communication in PROFInet is based of Standard Fast Ethernet (100MB) with network components that support switching technology. Hub based switching has the following disadvantages:

- Hubs create a common collision domain. Network traffic between two devices is automatically sent to all other devices.
- The number of successive components and the length of the cables between the components is determined. Following IEEE a maximum of 10 hubs is allowed. Because hubs still create a common collision domain using CSMA/CD the maximum cable length is determined by the signal round trip time of 5120 ns (Held, 2002)
- Gigabit Ethernet allows no hubs anymore.

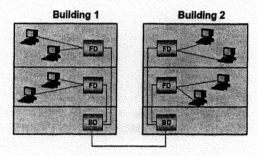

BD =Buildingdistributor, FD = Floordistributor

Fig. 2. Typical network topology used in office buildings

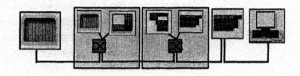

Fig. 3. Typical network topology used in automation applications

Automation applications with Ethernet are typically realized as line topologies in comparison to star or tree topologies used in office applications (see figure 2 and figure 3).

Typical automation applications consist of 64 up to 128 devices. Every device is connected to the Ethernet with a network component and every network component adds a delay to the actualization rate:

- *Delays caused by transferring data:* Network components with switching technology are responsible to sent the received data only to the port where the receiver is connected. This task can either be realized with "Store and Forward" or "CutThrough". "Store and Forward" receives the complete telegram, checks consistency and correctness and is sending out the telegram. "CutThrough" reduces the delay inside the network component to a minimum because a received datagram will be sent as soon as possible. After having recognized the target address the telegram will be sent to the selected port inside the switch.
- *Delays caused by queues:* Even with the CutThrough technology there will be queues where packets are stored if there is more than one packet at a time that should be transferred over a single port. Transferring a packet of maximum size of 1532 byte will last 125 µs in Fast Ethernet. This is why PROFInet uses priorization (IEEE 802.1Q 1998) to accelerate real time traffic inside the network components.

4. REALTIME CHANNEL IN PROFINET V2.0

PROFInet V2 is enhanced with a new communication channel that is especially designed to fulfil the above requirements for real time communication (SRT, Soft real time). The real time channel is based on a cyclic

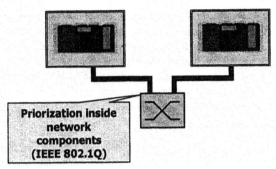

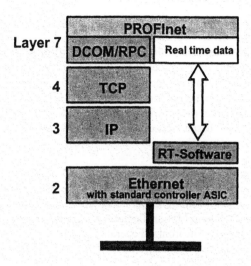

Fig. 4. Reducing delays in network components

Provider / Consumer architecture with Ethernet layer 2 frames for the fast and reliable data transfer between PROFInet devices. To ease the processing of real time frames inside the devices a specific ether type is used for PROFInet datagrams (see figure 5). The value of the ethertype is 0x8892.

To initiate a layer 2 communication a high level protocol is used to keep the layer 2 protocol as simple as possible. Inside PROFInet distributed automation DCOM is used to initiate the layer 2 communication. Inside decentralized field devices the initiation of the layer 2 communication is based on DCE RPC. Both concepts use the same Provider/Consumer concept and real time protocol to distribute real time data.

Real time data and standard data are scheduled by a middleware inside the device to keep the real time data at a high priority (see figure 6). PROFInet V2 can cooperate with IEEE 802.1 compatible network components. VLAN based prioritization is used to assure that real time telegrams are preferred during switching inside the network components compared to standard TCP/IP datagrams.

Fig. 6. An optimized real time channel enhances PROFInet V2

5. ISOCHRONOUS DATA TRANSFER WITH PROFINET V3.0

With the emerging use of Ethernet in industrial automation there are also emerging demands to use Ethernet in high end applications. Modular machine concepts, increasing axis separation and the demand for IT functionality up to the drive are current trends in drive engineering. Already today, decentralized drive systems replace conventional technology with mechanically coupled drive axis at many production machines. Modern servo technology makes it possible to replace the mechanical coupling of several drive axis over a line shaft by an electronic coupling. The electronic coupling takes place via the communication system.

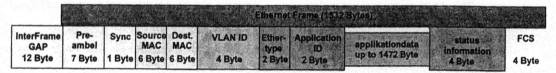

Fig. 5. Layout of the PROFInet real time PDU

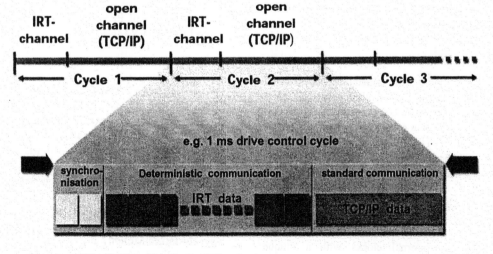

Fig. 7. Time based scheduling for isochronous data transfer (IRT)

287

Isochronous mode and Motion control are typically some of the sophisticated applications. PROFInet V3 will be capable of dealing with such problems. Based on the layer 2 protocol of PROFInet V2 there will be enhancements inside the network components and devices to synchronize up 100 axes in about 1 ms with a jitter of 1 μs. These enhancements are based on a time based scheduling using 100MB switched Ethernet.

That means improved performance demands to the communication system. In addition to the guarantee of determinism the communication system must fulfil the following Motion Control specific requirements:
- Clock synchronized operation of the communication system.
- Reduced cycle times regarding to the drive control cycle.
- Direct data interchange between synchronously coupled drives.

These requirements are not completely realizable by the software solution used with PROFInet V2. For high performance Motion Control applications a compatible alternative is offered with PROFInet V3. PROFInet V3 is based on a technology which is called "Isochronous Realtime Ethernet (IRT)". This technology operates with hardware support, but uses the same communication protocols as PROFInet V2. Thereby, a compatible scaling possibility is given to the users.

Using hardware support the host processor is released from communication tasks. That contains the composition of the data frames, the transfer scheduling and the error handling. The hardware support is realized by a new developed communication ASIC (Application Specific Integrated Circuit). This ASIC contains the complete Layer 2 functionality and integrates a full 4-port Ethernet switching functionality. Additionally, mechanisms for redundancy and configuration in run are supported by the ASIC functionality.

PROFInet V3 enables a hard real time communication on IRT at all circumstances:
- at any overload situations caused by additional ethernet communication.
- at any net topologies, also with many switches in a line (line topology).

This is reached by a time scheduling mechanism. Thereby, a time slot is exclusively reserved within a cycle for the deterministic communication (see figure 7). The rest of the cycle can be used for spontaneous TCP/IP traffic (FTP, HTTP, and others). The boundary between these two areas is exactly supervised by the ASIC.

The premise for such a mechanism is a clock synchronization of all network nodes supporting PROFInet V3. Thereby, the IRT network nodes synchronize there internal clocks with a unique master clock.

The mechanism of clock synchronization is based on the cyclic transfer of special synchronization telegrams. Clock synchronization in a switch based net is problematic because of the load dependent latency times of the switches. To achieve a load independent jitter less than one microsecond for a switched net the synchronization mechanism has to be supported by the ASIC.

6. USING REAL TIME COMMUNICATION

The real time communication in PROFInet is used for cyclic transfer as well as for acyclic transfer (events and commands).

During the cyclic transfer data is sent with a constant time schedule without acknowledgement. In SRT packet loss can be tolerated in up to parameterized number, because new data will arrive during the next cycle. For a reliable system some supervisions have to be established to recognize long lasting communication failures and to pass a state information to the application. Also substitute values are delivered in case of a communication failure. In IRT the packets are still unacknowledged. But for motion control applications it is necessary that every lost packet is recognized and a state information is sent to the application. Cyclic data transfer is established through a high level protocol based on DCOM (modular machines) or OSF DCE RPC (distributed IO).

Acyclic data transfer is the second possibility for data transfer. Though acyclic data transfer is used for events and commands data transport has to be acknowledged. Acylic data transport is based on state machines. The transport is based on the same datagrams as for the cyclic data transfer, so all benefits like VLAN priorities or preferred scheduling are also available for acyclic data transfer. The acyclic data transfer is also established using a high level protocol used in the same way as for the cyclic data transfer.

7. PERFORMANCE MEASUREMENTS

During the talk we will present the latest numbers and measurements of PROFInet V2 SRT and a look out into PROFInet V3 IRT.

8. CONCLUSION

With Version V2 PROFInet is supplemented with a real time channel based on a layer 2 communication beside the already existing standard TCP/IP and DCOM channel. This real time channel will allow update rates in the range of 5 – 10 ms by using COTS network components and allowing standard TCP/IP traffic beside the real time communication.

PROFInet V3 is qualified as a high performance communication system for sophisticated clock synchronized Motion Control applications. In contrast to specialized drive busses PROFInet V3 additionally supports standard IT functionality without any limitation. Real time data and standard TCP/IP communication can reside side by side in this concept, so all the benefits of standard communication like web based diagnosis or standard data transfer based on TCP/IP like ftp can be combined with the hard real time requirements for factory and process automation.

REFERENCES

Klüger, P. (2001), IDA im Praxistest, Interview, Computer&Automation, pp. 40

Held, G. (2002), Ethernet Networks. Design, Impelementation, Operation and Management, John Wiley & Sons

IEEE 802.1Q (1998), IEEE Standards for Local and Metropolitan Area Networks: Virtual Bridged Local Area Networks, , http://standards.ieee.org

IEEE 802.3 (2000) Carrier sense multiple access with collision detection (CSMA/CD) access method and physical layer specifications, Ausgabe 2000, http://standards.ieee.org/

EFFECTS OF NETWORK DELAY QUANTIZATION IN DISTRIBUTED CONTROL SYSTEMS

Gerald Koller*, Thilo Sauter**, Thomas Rauscher***

*Vienna Univ. of Technology, Automation and Control Institute
Gusshausstrasse 27/E376, A-1040 Wien, Austria
koller@acin.tuwien.ac.at

**Vienna Univ. of Technology, Institute of Computer Technology
Gusshausstrasse 27/E384, A-1040 Wien, Austria
sauter@ict.tuwien.ac.at

***LOYTEC electronics GmbH
Stolzenthalergasse 24/3, A-1080 Wien, Austria
trauscher@loytec.com

Abstract: When control loops incorporate fieldbus systems, typical quality of service parameters like delay and jitter play an important role with respect to the performance of the control. Traditionally, the desired quality of the control defined the constraints for the network. Today, it is desirable to use existing networks and make the control robust enough to cope with the deficiencies of the fieldbus. This paper investigates the influence of a network with bounded delay distribution on the performance of a discrete time control loop. We find that delays of about the sampling period give rise to a particularly sensitive behavior of the controller. Furthermore, the overshoot amplitude (as a quality indicator) seems to depend linearly on the delay distribution of the network, which offers a convenient design criterion for the controller. *Copyright © 2003 IFAC*

Keywords: Distributed control, Networks, Real-time control, Fieldbus

1. INTRODUCTION

Over the last two or three decades, networks have changed the world of automation. Fieldbus systems and more recently also IP-based networks (typically using Ethernet as transport medium) are today an integral and indispensable part of nearly all fields of automation. Their main purpose clearly is to cope with the enormous demand for data exchange between the nodes of a contemporary distributed control system. However, there is one aspect to automation where networks are only reluctantly used: closed-loop controls. With the exception of a few dedicated systems, fieldbusses are not normally located inside a control loop. Classically, such control loops are still placed directly within the field

devices, and only set points and configuration data are set over the fieldbus.

Still, as automation networks get more and more powerful, it becomes interesting to deviate from this old model and make systems really distributed, i.e., close the control loop over the fieldbus. The communication systems that are used, e.g., in the x-by-wire approaches in the automotive area are a step in this direction.

For the engineering of control systems, this paradigm change poses new challenges. Traditionally, if some sort of network had to be used for whatever reason, it was up to the control engineers to define "how good" the network had to be in terms of data rate, transmission delays, error behavior or the like.

Today, the task rarely is to design a new network for a control application, but to use an existing one. The best (e.g. cheapest) solution need not necessarily be a real-time fieldbus. For some application a LAN could be sufficient. Consequently, the approach needed today is to examine the effects a network has on a closed-loop control and to see which measures and strategies can be taken to improve the robustness of the control accordingly.

The significance of the topic is underlined by various new control theory and network engineering approaches. Husmann (1997) describes an integrated approach for discrete time controllers taking the deviation of sampling and actuation times of control architectures into account. An application of this approach to Profibus FMS/DP, Interbus-S and CAN demonstrates the influence of each fieldbus. Nielson (1998) presents an analysis of control loops suffering from network delays. Here, not only static delay distributions assumptions are made, but an underlying Markov model provides a more flexible way to model time-variant delay distributions. Much work has also been done on the real-time aspects of computer-controlled systems and the associated question of scheduling (Marti et al., 2001a, Marti et al., 2001b). Work on network scheduling has been done by Yong Ho Kim et al. (1996).

Another possible way to tackle the problem or at least to formulate it is to investigate the dynamic properties of the fieldbus. In wide area networks and lately also in LANs, these properties have become famous in the recent past as "Quality of Service" (QoS) (Shenker et al., 1994, Blake et al., 1998, Ferguson et al., 1998). Originally intended for voice data transmission, this loosely defined set of parameters describes how well a packet-oriented network, like TCP/IP or ATM, can mimic a connection-oriented one, like the analog telephone system. This is exactly what is needed in control loops. Evidently, QoS is particularly important when we try to send control data over IP-based networks that connect, e.g., remote fieldbus segments (Soucek et al., 2001). Still, QoS parameters are also applicable to fieldbus systems that normally have a more constraint dynamic behavior. In this paper, we will use such parameters to study their influence on a closed-loop control system.

A serious problem in designing closed control loops are dead time elements, especially varying ones. Moving from a non-distributed control loop to a distributed one requires treating network-induced delays and jitter as additional dynamic components in the closed control loop. These effects and optimized controller models are examined in detail in (Husmann, 1997, Nielson, 1998, Marti et al., 2001a, Marti et al., 2001c). A general overview on various mathematical approaches handling time-varying delays is presented in Ray (1989).

A less apparent, but important effect is observed whenever the delay distribution is not limited to a single sampling interval and the controller input is taken at discrete times. A simple non-predictive controller design faces virtual packet loss in this case, since a packet can be delayed into the next sampling period. In this case the older packet is discarded because it is overwritten by the newer packet.

The paper is organized as follows. In section 2, the QoS parameters used in this article are defined. Section 3 explains the chosen controller model and the basic effects induced by stochastic network delay. The simulation of the model in presented in section 4, while section 5 discusses the simulation results in detail. Section 6 presents a conclusion of the simulation results and mentions open topics requiring further research.

2. QOS PARAMETERS

There is no unique and comprehensive definition of quality of service. Likewise, there is no universal definition of the parameters that are relevant for QoS. What has to be considered largely depends on the application, i.e., on the data that are sent over the network. Voice data distribution requires different performance than file transmission, and control systems are different as well. For closed-loop systems, mostly low-level, network-related parameters are relevant. Therefore, we will consider in this article the following QoS parameters (Soucek et al., 2001):

End-to-end delay D_i: This is the total delay packets suffer from the emission by the sender until the reception over the network. A typical QoS restriction defines the delay for any packet i is lower than a maximum delay with a certain probability.

$$P(D_i \leq D_{max}) \geq Z_{min} \tag{1}$$

For hard real-time networks, Z_{min} equals 1, so that every packet must arrive within a finite maximum delay.

Delay Jitter J_i: The packet delay will in fact vary for each packet due to different processing times in intermediate nodes, different queue lengths, or network congestion. As a result, a packet stream transmitted in equidistant instants in time will be distorted while it travels over the network. Applications may want to bound the delay jitter J_i to a maximum value J_{max} with a probability equal or greater than U_{min}.

$$P(J_i \leq J_{max}) \geq U_{min} \tag{2}$$

Other QoS parameters like throughput or reliability are of little importance in this work. Throughput corresponds to the available bandwidth and is especially important for applications with constant data streams. Reliability measures the probability of

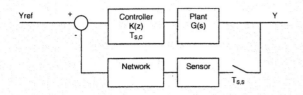

Fig. 1. The example controller setup.

packet loss. In real-time systems, packet loss may either not occur at all, or must be resolved by the transmission protocol within the real-time limits in each case. The application used in this work assumes a real-time network so that reliability issues can be neglected.

QoS parameters can change over time for example because of network congestion. Wang et al. (2001) suggest an adaptive method for internet delay estimation and compensation. The key idea is to use fuzzy logic to describe the network QoS parameters in unloaded, normal and seriously loaded situations.

3. MODEL

The goal of this work is to discuss the influence of stochastic QoS parameters on quality parameters of a control loop. Especially the interrelation between the delay distribution and the distribution of the step response will be examined in detail.

We assume a continuous time plant being controlled by a discrete time controller (Fig. 1). A sensor samples the plant output in equidistant intervals TS,S. Each measurement sample is immediately sent via a packet communication network which constitutes a dead time. In a general packet-switched network, packets can be arbitrarily delayed, dropped and reordered. In this article, however, packet loss and disordering will be neglected. Disordered packets can be easily detected, for example by sequence numbers. Furthermore, if we are considering single-segment networks (which is a typical case for many installations), there is no way that packets in the same data stream can overtake each other. Packet loss can be ignored if the network has real-time capabilities. In non-real time data transmission systems, packet loss forces the controller to either keep the latest received value or to estimate the values contained in the dropped packets.

Packets received by the controller are stored in a single-entry input buffer in our model. The controller peeks its input buffer in equidistant intervals $T_{S,C}$ and uses its actual contents at the sampling time to calculate the controller output for the next sampling interval. The limitation to a single-entry input buffer avoids additional delays introduced by the queuing of packets. Walsh and Hong Ye (2001) demonstrate that moderate packet loss has less performance impact than large queues.

The restriction to a single entry buffer matches common implementations in low-memory/low

performance fieldbus nodes, see also Chan and Özgüner (1995). This also allows to concentrate on the examined quantization effect. However, systems using communication systems with larger queues can be handled with methods described by Chan et al. (1995). In case of more than one sensor, this scheme can be extended to the multiple queue case, as shown by Chan and Özgüner, 1994.

The effect of missing a sample is called vacant sampling. The case in which a packet is dropped because two or more packets arrive within the sampling period is often called data rejection. See also Seoung Ho Hong (1995) for network scheduling which can reduce these effects.

The computation of the controller output for the n-th packet takes some time $\tau_{C,n}$. The selection of a timed controller requires some motivation. An event-based controller would provide less average dead time and allow data to be processed as soon as possible. Event-based systems however require more sophisticated schedulers and well-tuned scheduling if real-time requirements must be met.

The k-th packet P_k is sent by the sensor at $\tau_{Sout,k} = k \cdot T_{S,S}$. It arrives at the controller at the time $\tau_{Cin,k}$, delayed by $\tau_{SC,k}$ modeling the sensor-controller delay of the k-th packet. The controller samples its input buffer at $n \cdot T_{S,C}$ and thus uses the contents of the last packet received before $n \cdot T_{S,C}$. The controller output for the n-th received packet is set at $\tau_{Cout,n}$. In general, the controller computation latency for the n-th received packet $\tau_{C,n}$ must also be taken into account. In this paper we assume that $\tau_{C,n}$ is smaller than $T_{S,C}$, so that the computed output is valid already at $\tau_{Ccalc,n}$ which is always before $\tau_{Cout,n}$.

$$\tau_{Sout,k} = kT_{S,S}$$
$$\tau_{Cin,k} = kT_{S,S} + \tau_{SC,k}$$
$$\tau_{Ccalc,n} = nT_{S,C} + \tau_{C,n}$$
$$\tau_{Cout,n} = (n+1)T_{S,C}$$

(3)

The input buffer in the controller reduces the network timing effects to two dynamic effects:

1. Delay: Each measurement will be delayed by at least one time step. If the network delay increases, the overall delay will also increase in multiples of the controller sample time $T_{S,C}$.
2. Sampling period: Whenever a packet does not arrive at all or is overwritten by a newer packet arriving within the same sampling period (which can happen due to jitter effects), the controller uses the same input value more than once. This is equivalent to a temporarily increased sampling period.

Fig. 2 shows the impacts of possible dynamic effects on the packet sequence:

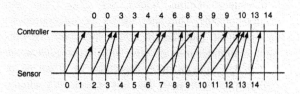

Fig. 2. Possible effects on the processing sequence of the data packets.

- Packet 1 is lost, which results in an increased sampling period because packet 0 is used twice.
- Packet 3 arrives faster than expected, so it overwrites packet 2 in the input buffer.
- Packet 4 arrives in time, so that the contents of packet 3 are used twice.
- Packet 5 is delayed, therefore packet 4 is also used twice. Packet 6 overwrites the contents of packet 5, so packet 5 appears as a lost packet.
- Packets 7 and 8 display the effects of disordering, the more recent value is used and the older one is discarded.
- Packets 10 and 11 show the effects of increased packet delay ($3T_{S,C}$), the contents of packet 9 are used twice.
- Packets 13 and 14 show the effects of decreased packet delay ($1T_{S,C}$). Packets 11 and 12 are overwritten in the controller.

In each case, the resulting packet sequence increases monotonically (no reordering takes place). Even if packet loss cannot occur on the network, packets can be dropped in the input buffer of the controller, which yields the same effect.

In a typical setup, $T_{S,C}$ and $T_{S,S}$ are identical, but generally, synchronized sensors and controllers cannot be assumed. When $T_{S,C}$ and $T_{S,S}$ differ only slightly, a modulation of the average delay with the beat frequency $(T_{S,C}-T_{S,S})^{-1}$ can be observed. Due to the sampling of the data stream, there is a quantization in the actual delay in the control loop. In the best case, only the real network delay is experienced by the controller. In the worst case, the packet delay probability density function is shifted towards higher delays by a controller sampling period $T_{S,C}$. However, the network can also be used to synchronize sensor and controller using time protocols to avoid the modulation effects. The advantages of synchronizing control loop elements are discussed in (Nielson, 1998).

In fact, the network delay is not fixed, but stochastically distributed. Given such a packet delay probability density function (delay pdf), there is a probability for each packet to arrive on time (within the foreseen sampling period) or to arrive in later periods. Fig. 3 illustrates this situation. When a packet fails to arrive on time, the controller has to

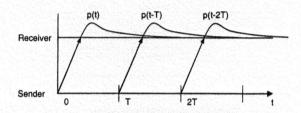

Fig. 3. Delay distribution. The arrows indicate the minimum delay.

retain the old input value in the simplest case, which is tantamount to a sample&hold behavior.

Three static cases can be used to benchmark the step response of the stochastic system:

- Minimum delay: Each packet requires the smallest possible time interval to arrive at the controller. In this case, the dead time is at its smallest boundary.
- Maximum delay (for real-time systems): Each packet requires the largest possible time interval to arrive at the controller. In this case, the dead time is at its largest boundary.
- Alternating min/max delay: Each odd-numbered packet experiences the smallest possible delay, while each even-numbered packet experiences the largest possible delay. Thus, every other packet is overwritten in the controller input buffer. The sampling time is virtually doubled. Such a behavior can occur if the mean delay is close to the sampling period. A relatively small amount of jitter will then cause the actual delay to alternate between the two extremes.

4. SIMULATION

The setup for the simulation consists of a discrete time PI controller and a continuous time PT₃ plant. The output of the plant is sampled by a sensor and transmitted to the controller via a loss-free network. The transfer functions of the components shown in Fig. 1 are defined as

$$K(z) = \frac{0,1z}{z-1}, G(s) = \frac{1}{1+0.1s} \cdot \frac{1}{1+0.1s+0.01s^2}, T_S=0.1 \quad (4)$$

Plant and controller are intentionally kept simple to examine a system in which dead time effects clearly show up.

Real-life applications of course will take serious attempts to take an estimated delay into account. Interesting concepts are state predictors which estimate the plant output in case of missing controller input caused by network congestion. A detailed analysis of such an approach is discussed in Beldiman et al. (2000).

However, a dead-time optimized controller would also face the quantization effects described in section 3, so we concentrate on the simpler case. The simple design of the controller (cancellation of poles and

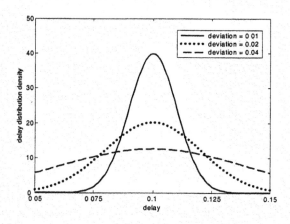

Fig. 4. Truncated Gaussian distribution density function.

zeros) is done in continuous time to get plainest results. For further simulation the corresponding discrete controller was used. Lower sampling time would cause less results on the system (getting closer to continuous transfer) while higher sampling time would cause instability ($T(s)/2+D_{i,max} > 0.67$).

The delay distribution is modeled by a modified Gaussian distribution. Its probability density function is shown in Fig. 4. The mean μ represents the average packet delay; the variance σ models the delay jitter. To model a real-time network and to avoid packet disordering, the delay pdf is truncated at μ-T_S/2 and μ+T_S/2 and renormalized afterwards. Thus, the arrival times cannot overlap and packet disordering is impossible.

$$f(x) = \begin{cases} c\dfrac{1}{\sqrt{2\pi\sigma^2}}e^{\frac{(x-\mu)^2}{2\sigma^2}} = c\varphi(x,\mu,\sigma^2) & |x-\mu| < \dfrac{T_S}{2} \\ 0 & |x-\mu| \geq \dfrac{T_S}{2} \end{cases} \quad (5)$$

$$c = \dfrac{1}{\dfrac{1}{\sqrt{2\pi\sigma^2}}\displaystyle\int_{\mu-\frac{T_S}{2}}^{\mu+\frac{T_S}{2}}\varphi(x,\mu,\sigma^2)dx}$$

The mean of the modified distribution is identical to the mean of the corresponding Gaussian distribution while the standard deviation changes due to the truncation.

A distribution of this type provides an attractive analysis method, since average and deviation can be freely chosen. A low standard deviation σ represents predicable networks while for $\sigma \to \infty$, the distribution degenerates to a rectangular distribution ranging from μ-T_S/2 to μ+T_S/2. The worst case jitter sensitivity appears when μ approaches a multiple of T_S since a slight delay jitter determines whether the packet is received before or after T_S. If the average delay is well below T_S, the probability for a packet to be received after T_S is very low for a small delay jitter. The same applies for average delays well above T_S.

Here, the probability for receiving the packet before T_S is low.

The simulation is done on a standard workstation with Simulink under Matlab 6.1. For each mean delay with a given standard deviation 10,000 time vectors are built. The delay is simulated with a single buffer, which is written and read asynchronously. From the resulting step responses, the mean overshoot and its deviation are calculated.

5. RESULTS

The simulation calculates the unity step response overshoot for a realization of the modified Gaussian delay distribution. The overshoot amplitude $\Delta h(\mu,\sigma)$ is chosen as the control performance criterion. From the large number of individual simulation runs, we determine the mean and standard deviation of the overshoot probability density function. The effects of network parameter changes are investigated in two different ways:

- The standard deviation of the delay pdf is kept constant while the mean delay is modified. In practice, this might be related to changes in the length of the network segment or to the introduction of repeaters.
- The mean delay is kept constant while the standard deviation of the delay is modified. This could stem from a higher work load of the communication interface, which in turn will result in larger deviations of the processing time.

The non-distributed case is not contained in the diagrams directly but is approached for $T_S \to 0$ in which the dead time influences the system dynamics only marginally.

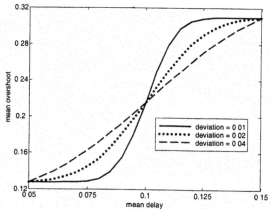

Fig. 5. Mean overshoot vs. mean delay, parameter = delay deviation.

5.1 Mean Overshoot

When μ equals T_S/2, the packet arrives before the sampling time in any case, so that the mean overshoot $\overline{\Delta h}(\mu,\sigma)$ equals the overshoot of a non-stochastic system with a delay of T_S. The same argument can be used for $\mu = 3T_S/2$ where the packet never arrives before T_S, so that the mean overshoot equals the overshoot of a non-stochastic system with a delay of $2T_S$. Any other mean delay shifts the non-zero parts of the delay pdf over the sampling time limit T_S, therefore the probabilities for a packet to arrive in the same or the subsequent sampling period are non-zero.

$$p(t < T_S) = \int_{\mu - T_S/2}^{T_S} f(x)dx \qquad (6)$$

$$p(t \geq T_S) = \int_{T_S}^{\mu + T_S/2} f(x)dx$$

In these cases, some sensor values experience a control loop with a higher dead time and an increased sampling time when a packet is overwritten in the controller input buffer. Therefore increasing the mean delay from $T_S/2$ to $3T_S/2$ means that more and more sensor values are delayed by $2T_S$ instead of T_S. The mean overshoot increases monotonically from the minimum delay case overshoot to the maximum delay case overshoot, as shown in Fig. 5. For low delay jitter (narrow delay pdf), the average overshoot rises from the minimum delay case to the maximum delay case in a small interval around the sampling time, reflecting the high sensitivity of the system if the delay equals the sampling period. When σ approaches zero, the graph degenerates to a step function at the sampling time. For high delay jitter, on the other hand, the graph is smoother. When σ approaches ∞ (which amounts to an constant delay pdf of the delay), $\overline{\Delta h}(\mu, \sigma)$ becomes linearly dependent on μ.

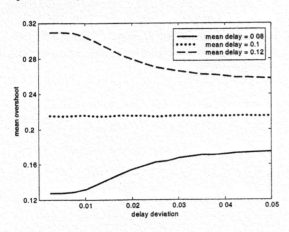

Fig. 6. Mean overshoot vs. delay variation, parameter = mean delay.

Fig. 6 shows the dependency of $\overline{\Delta h}(\mu, \sigma)$ on σ for constant μ. For a μ of T_S, $\overline{\Delta h}(\mu, \sigma)$ is nearly constant, irrespective of the jitter. If the average delay matches the sampling period, any deviation will obviously result in the same probability for a packet being on time or being late. Thus the average overshoot does not depend on the delay deviation in this case. For a $\mu < T_S$, $\overline{\Delta h}(\mu, \sigma)$ depends on σ significantly because a distribution with a higher deviation increases the probability for late packets contributing to the overshoot and vice versa. Therefore high standard deviations of the delay tend to shift the results towards the symmetrical case $\mu = T_S$.

5.2 Overshoot Deviation

The dependency of the standard deviation of $\Delta h(\mu, \sigma)$ on the mean delay is illustrated in Fig. 7, with the width of the distribution as parameter. For each parameter value, the peak of the overshoot deviation appears for $\mu = T_S$. To either side of the maximum, the overshoot deviation drops monotonically to 0. This result is not unexpected. When the average delay matches the sampling time, the packet is correctly received for any jitter ≤ 0 and dropped for any jitter > 0. Thus, this situation is most sensitive to delay jitter resulting in a large variation of the overshoot. When the average delay is greater or less than the sampling time, more packets will be treated alike, i.e., the probability for a packet to arrive before or after the sampling time is asymmetrical, which in turn reduces the variability of the overshoot.

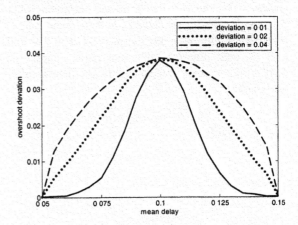

Fig. 7. Overshoot deviation vs. mean delay, parameter = delay deviation.

In Fig. 8, the overshoot deviation in respect to the delay deviation is shown. For $\mu = T_S$, the overshoot deviation is nearly constant. Like before, this behavior results from the delay time quantization and the symmetric delay pdf. Irrespective of the width of the delay pdf, the probability for a packet to arrive before T_S equals probability to arrive after T_S. For $\mu \neq T_S$, the overshoot deviation increases monotonically with the delay deviation because the larger width of the delay pdf makes the distribution more symmetrical around T_S. Therefore the delay quantization plays a more important role.

5.3 Further Results

Apparently, the influence of the delay distribution on the control loop is determined by the Bernoulli distribution describing whether the packet arrives before or after the sampling time. If the packet does not arrive on time, the sampling period is increased and thus the control loop shows higher overshoot. The probability distribution of on-time and late packets determines how often the more accurate normal sampling time case and the higher-overshoot increased sampling time case appear in a step response.

With the same sensor-controller setup and a more complicated delay pdf we obtain a strong hint on how the delay distribution influences the average overshoot. A two-peak delay pdf (Fig. 9) can occur in the network delay if the interface card of the transmitter or receiver incorporates a FIFO and the processing of the data varies grossly, so that the filling level FIFO fluctuates between completely full an empty. If this distribution is applied, we obtain similar results as before. Depending on the relative position of the mean delay with respect to the sampling period, more or fewer data packets arrive late, and late arrivals increase the mean overshoot. The smoother the delay pdf gets, the more the relation turns into a strictly linear one.

This last example clearly demonstrates that there seems to be a direct linear relationship between the distribution function $\int f(x)dx$ and the mean overshoot. While this is intuitively reasonable in view of the reasoning given before, a rigorous analytical proof seems very difficult. Even for a controller facing an arbitrary fixed dead time, the closed loop overshoot amplitude can only be expressed by approximation formulas. The considered system experiences stochastically changing delay times and the examined output parameter is the mean overshoot, thus approximation formulas also have to take the delay distribution into account.

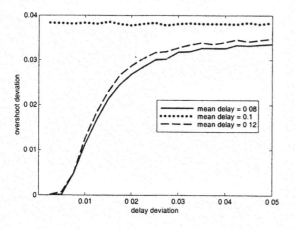

Fig. 8. Overshoot deviation vs. delay deviation, parameter = mean delay.

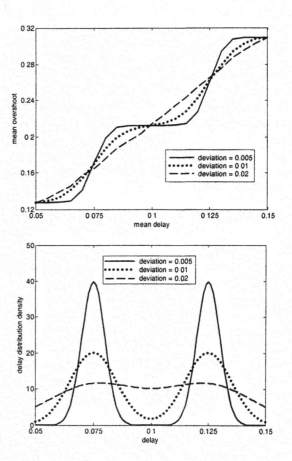

Fig. 9. Delay pdf with two peaks and resulting mean overshoot.

Branitzky et al. (2000) show that stability analysis in case of delays below the sampling period is analytically possible. Longer delays in their model yield a time-variable system matrix depending on the exact input schedule.

Another interesting observation made in the course of the investigation is that in the simulations, the controller output of the random system never left the corridor defined by the deterministic minimum and maximum delay cases. This behavior seems reasonable but yet a formal proof faces some non-trivial difficulties: The deterministic minimum and maximum delay cases are given by the deterministic unity step response. Although the unity step response can be calculated for each delay realization, there is no obvious way to prove that all realizations lead to a unity step response within the deterministic border cases. The simulation results and the presented reasoning suggest that the minimum and maximum delay cases can be used to estimate the best/worst case performance without the need of conducting a time-consuming Monte Carlo analysis, which would be a significant alleviation for the design.

6. CONCLUSION

Delay and jitter play an important role in the control quality of discrete time control systems. While this statement is quite universally true, it is particularly relevant for control systems incorporating automation networks. Their quality of service parameters have a significant influence on the performance of the control system as a whole. However, dead time quantization effects introduced by the inevitable data buffers needed in the controller can have an even more detrimental impact. Our investigation showed that as soon as the mean overall delay comes close to the sampling period of the controller, typical parameters such as the overshoot change suddenly and significantly.

On the other hand, the results also suggest that if the delay pdf is bounded, deterministic worst and best case evaluations of the controller output define limits for the behavior of the actual, stochastic system, which may help in the design of the control parameters. Finally, there is an apparent relationship between the delay pdf and the overshoot of the controller output, which is often used as a design criterion. This relationship exists at least for the comparatively well-behaved, stable system investigated here. It would be interesting to know whether this is a universal "quality of service" for any discrete time control application, as this would allow for a simple estimation of the control performance for a given network or, alternatively, for a deduction of the admissible network QoS parameters for a given control quality. To answer this question, however, will be the subject of further research.

7. BIBLIOGRAPHY

Beldiman, O., G.C. Walsh and L. Bushnell (2000). Predictors for networked control systems. American Control Conference 2000, Vol. 4, pp. 2347-2351.

Blake, S. et al. (1998) An Architecture for Differentiated Services, Internet RFC 2475.

Branicky, M.S., S.R. Phillips and Wei Zhang (2000). Stability of networked control systems: explicit analysis of delay. American Control Conference 2000, Vol. 4, pp 2352-2357.

Chan, H. and U. Özgüner (1994). Control of interconnected systems over a communication network with queues. 33rd IEEE Conference on Decision and Control, Vol 4, pp. 4104-4109.

Chan, H. and U. Özgüner (1995). Optimal control of systems over a communication network with queues via a jump system approach. 4th IEEE Conference on Control Applications, pp. 1148-1153.

Ferguson, P. and G. Huston (1998). Quality of Service: Delivering QoS on the Internet and in Corporate Networks, John Wiley & Sons, New York.

Seung Ho Hong (1995). Scheduling algorithm of data sampling times in the integrated communication and control systems. IEEE Transactions on Control Systems Technology. Vol. 3, Issue 2, pp. 225-230.

Husmann, H. (1997). Ein Beitrag zur Unterstützung des dynamischen Verhaltens feldbusgestützter Regelkreise, Fortschritt-Berichte VDI, VDI Verlag GmbH. Düsseldorf.

Nielson, J. (1998). Real-Time Control Systems with Delays, PhD Thesis, Department of Automatic Control, Lund Institute of Technology.

Marti, P., Villa, J.M. Fuertes and G. Fohler, (2001a). Stability of On-line Compensated Real-Time Scheduled Control Tasks. IFAC Conference on New Technologies for Computer Control, Hong Kong.

Marti, P., R. Villa, J.M. Fuertes and G. Fohler, (2001b) "On Real-Time Control Tasks Schedulability", IFAC European Control Conference, Porto, Portugal.

Marti, P., J.M. Fuertes and G. Fohler (2001c). An integrated approach to real-time distributed control systems over fieldbuses. 8th IEEE International Conference on Emerging Technologies and Factory Automation, Vol. 1, pp. 177-182.

Ray, A. (1989). Introduction to networking for integrated control systems. IEEE Control Systems Magazine, Vol. 9, Issue 1, pp. 76-79.

Shenker, S., R.Braden, and D. Clark. (1994) Integrated services in the Internet architecture: an overview, Internet RFC 1633.

Soucek, S., T. Sauter, and T. Rauscher (2001). "A Scheme to Determine QoS Requirements for Control Network Data over IP", 27th Annual Conference of the IEEE Industrial Electronics Society (IECON), Denver, Colorado, pp. 153-158.

Walsh, G.C. and Hong Ye. (2001). Scheduling of networked control systems. In: IEEE Control Systems Magazine, Vol. 21, Issue 1, pp. 57-65.

Wang, Q.P., D.L. Tan, Ning Xi and Y.C. Wand. (2001). The control oriented QoS: analysis and prediction. IEEE International Conference on Robotics & Automation, Vol. 2, pp. 1897-1902.

Young Ho Kim, Wook Hyun Kwon and Hong Seong Park (1996). Stability and a scheduling method for network-based control systems. 22nd International Conference on Industrial Electronics, Control, and Instrumentation, Vol. 2, pp. 934-939.

CONVERGENCE BETWEEN IEEE 1394 AND IP
FOR REAL-TIME A/V TRANSMISSION

Manfred Weihs * Michael Ziehensack *

* Institute of Computer Technology
Vienna University of Technology
Gusshausstraße 27-29/E384
A-1040 Wien, Austria
{weihs,zie}@ict.tuwien.ac.at

Abstract: Within the home environment there are different networks to which
consumer electronic devices, computers and peripherals are connected. This paper
gives a brief overview of multimedia networking in the home and explains the need
of convergence between these networks with respect to real-time transmission of
audio/video data focusing on IP and IEEE 1394. As a special example it outlines
a possible implementation of a gateway between IP and IEEE 1394/HAVi which
provides audio streams available on IP based networks to HAVi devices on the
IEEE 1394 bus. Copyright © 2003 IFAC

Keywords: Communication systems, Conversion, Multimedia, Real-time
communication, Synchronization, Transcoders

1. INTRODUCTION

There are many different kinds of network in a
typical home environment, which fulfil purposes
like home automation or connection of computer
equipment and consumer electronics. Examples
are field busses like EIB (Dietrich *et al.* 2000)
or LonWorks (Dietrich *et al.* 1997), IP based
networks, IEEE 1394[1] (IEEE Computer Society
1995), USB and many more. They are usually
optimised for some application and differ significantly with respect to some features (bandwidth,
security, maximum distance, possible number of
devices, configuration effort, media etc.).

One application, which is of increasing interest, is
the real-time distribution of audio and video data.
This means that some source transmits a data
stream at the same speed that will be used for
presentation on one or more sink devices. Therefore the sink need not save the complete stream
to a file but only has to buffer small portions. Of
course devices like digital video recorders might
also save the complete stream. That is completely
different from "download and play", which does
not impose real-time requirements (see section 2).

This paper focuses on distribution of real-time
multimedia data on IEEE 1394 and IP based networks, because they are the most common types
of network in the home that can be used for that
purpose. Some other kinds of network available in
a home (like field busses or some wireless technologies like Bluetooth) are not well suited for
real-time multimedia transmission, because they
suffer deficiencies like too little bandwidth or very
variable availability. There are other kinds of net
work (for example USB 2, ATM), which could in
principle be used for multimedia transmission, but
are not covered in this paper.

---

[1] also known as FireWire or i.Link

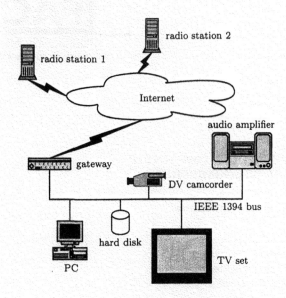

Fig. 1. Setup of the gateway between Internet and IEEE 1394

In both kinds of network there can be sources and sinks of audio and video transmission. For example there might be a digital satellite receiver streaming TV and radio programs on an IEEE 1394 bus[2], an IEEE 1394 enabled audio amplifier that can play audio streams received over IEEE 1394[3], and there exist radio stations that provide live streams of their program on the Internet.

The problem addressed by this paper is that source and sink of a connection must be part of the same network. It is not possible to play an audio stream offered in the Internet on an IEEE 1394 enabled audio amplifier (unless it is also part of the IP network). Therefore it makes sense to try to merge those networks. Fig. 1 gives an impression of the situation: On one hand there is the Internet with (among many other services) a few radio stations offering live streams, and on the other hand there is a typical IEEE 1394 network providing (among other consumer electronic devices) an audio amplifier. And between those networks, there a gateway. The aim of the present work is to make multimedia content that is available in one network within the home also available in the other one.

2. TECHNICAL BACKGROUND

Real-time transmission of A/V data puts some constraints on the network, the most important are:

- There must be sufficient *bandwidth* available. Audio/video data streams usually have a

fixed data rate, that is known in advance (it is also possible that a stream adapts the data rate, if the available bandwidth changes).
- In case of interactive two way communication (e. g. video conferences) *latency*, i. e. the average transit delay, should be below a certain limit. However for one way transmission latency is usually no problem.
- The *transit delay variation* (jitter) should be limited. The limit depends on the buffer used by the sink device to compensate the jitter.
- *Error rate* and *packet loss* should be low. Depending on the format of the data stream it might be more or less sensitive concerning data corruption.

If these requirements are guaranteed, the system provides *Quality of Service* (QoS). If that is not fulfilled, disruptions in the transmission might happen. An overview over several definitions of the term *QoS* is given by van Halteren *et al.* (1999), while Vogel *et al.* (1995) discuss QoS parameters.

Furthermore it is desirable, that the system supports multicast, if there is more than one sink in the network.

2.1 Real-time Multimedia Data on IEEE 1394

IEEE 1394 is very well suited for transmission of real-time audio and video data. The isochronous transport mode was exactly designed for that purpose. Timing is guaranteed and bandwidth can be reserved and is then guaranteed[4] (QoS is provided). Furthermore the transmission of multimedia data over IEEE 1394 is well standardised by the series of IEC 61883. This covers transmission of video and audio data typically received from a VCR or camera (IEC 1998a, IEC 1998b, IEC 1998d), video and audio typically received via digital television systems (IEC 1998c, IEC 2001a) and plain audio (IEC 2002).

Therefore suitable data formats for real-time distribution of A/V data on IEEE 1394 can be found easily. Audio and video data can be distributed as (possibly partial) MPEG-2 transport stream (IEC 1998c). This is very convenient to distribute TV programs received via a DVB tuner or to distribute the contents of DVDs, because both use MPEG-2 streams and therefore conversion is rather easy. A digital TV set with access to the IEEE 1394 bus should be able to present this kind of stream. Audio can be distributed according to

---

[2] A prototype of such device was developed within the project InHoMNet (InHoMNet Consortium 2002)
[3] One example of such device is Sony STR-LSA1 (a component of Sony LISSA HiFi set).

---

[4] The isochronous bandwidth is only guaranteed, if all nodes obey the rules of IEEE 1394 and reserve bandwidth and channel before they are used. Since this is not enforced by hardware or software, nodes violating these rules can compromise QoS.

IEC (2002). That makes it very easy to transmit uncompressed digital audio, which is available on CDs or from terrestrial radio stations (after an A/D conversion). Consumer electronic audio equipment with IEEE 1394 connector usually is able to handle that data format. DVD Forum (2002) specifies how to use these two transmission formats in order to distribute the content of a DVD in an IEEE 1394 network and make it possible to watch or listen to it on a digital TV and/or HiFi device (both IEEE 1394 enabled).

The third popular data format on IEEE 1394 is DV, which is typically used by digital IEEE 1394 enabled VCRs (IEC 2001b). An advantage of this format is a lower complexity of encoding and decoding. On the other hand it imposes rather high bandwidth requirements (about 25 Mbit/s for an SD format DV stream), because it does not use the sophisticated inter frame compression techniques known from MPEG-2.

2.2 Real-time Multimedia Data on IP

The other very important kind of network within the home are IP based networks. These networks were not designed to allow multimedia transmission and therefore have some weaknesses (multicast is just an add-on, that is not widely supported, Quality of Service is very limited) in regard of real-time data transmission. Nevertheless these networks are very common. Since they are available almost everywhere and fulfil the basic requirements needed for audio and video transmission (although they are usually not guaranteed), it is very desirable to use these networks for that purpose. One problem concerning IP networks is, that transmission of audio and video is not very well standardised. There is a wide variety of data and transmission formats, open and proprietary ones, which are not compatible with each other.

In IP based networks many different protocols are commonly used for real-time transmission of multimedia. The most primitive one is HTTP. There are implementations that use HTTP to stream real-time audio data (e. g. icecast, shoutcast; both are server programs, that are freely available in the Internet). But besides some advantages (very simple, works well with proxies and firewalls) it also has many disadvantages. It does not support multicast, but only unicast. Therefore if more than one node listen to the stream, then there are several connections, that use the corresponding multiple of the necessary bandwidth. Furthermore HTTP is based on TCP, which is a reliable protocol, i. e. it does perform retransmission of packets, that were not transmitted correctly. This is completely inadequate for real-time transmission,

because the retransmitted data will be late and therefore useless.

Hence this paper prefers the use RTP (Real-time Transport Protocol) proposed by Schulzrinne et al. (1996), which is usually based on UDP and in principle supports multicast as well as unicast, for transmission of multimedia data. RTSP (Real Time Streaming Protocol) can be used for setting up the transmission (Schulzrinne et al. 1998). By doing so the use of proprietary protocols like MMS (Microsoft Media Server) is avoided.

Unfortunately there is no common format of audio and video data used in this context. Many different (open and proprietary) formats for audio and video exist and can be transmitted in an RTP stream. Concerning audio the most important open formats are uncompressed audio 16 bit, 20 bit or 24 bit (Schulzrinne 1996, Kobayashi et al. 2002) and MPEG compressed audio streams (Hoffman et al. 1998). For video Hoffman et al. (1998) specified how to transmit MPEG video elementary streams. If audio and video are to be transmitted in one stream, MPEG-2 transport streams can be used as specified by Hoffman et al. (1998). The most important proprietary formats are RealAudio and RealVideo.

3. RELATED WORK

Since the digital distribution of multimedia data replaces analog techniques more and more, there is much work in progress bringing together sources and sinks of that data on different networks.

One issue is the transmission of DV based video over IP. Ogawa et al. (2000) showed how to transmit DV based video over RTP enabling the connection of two DV camcorders or VCRs (with IEEE 1394 access) via the Internet. This basically couples two IEEE 1394 networks containing these devices by the use of two gateways.

Saito et al. (2001) proposed two gateways, a "home gateway" and a "wireless gateway". Both receive DV streams over IEEE 1394 and transcode them before transmitting them on the IP network. The home gateway sends an MPEG-4 stream over RTP/IP, whereas the wireless gateway transmits MPEG-2 over RTP/IP.

At the 1394 Trade Association there is a wireless working group, that tries to connect IEEE 1394 networks by establishing an IEEE 1394.1 (Johansson 2002) compliant (distributed) bridge. In essence that approach tunnels the IEEE 1394 protocol over an IEEE 802.11 compliant link (Bard 2001). That of course only allows the connection of IEEE 1394 devices and excludes devices connected to other networks.

Johansson (1999) and Fujisawa and Onoe (2001) specified how to transmit IPv4 and IPv6 over IEEE 1394. Hence it would be possible to make all services that exist in the Internet available to the IEEE 1394 network by simply using an IP router, which connects the IEEE 1394 bus to the Internet. However that approach does not take advantage of the special features of IEEE 1394 and furthermore excludes typical consumer electronic devices which support IEEE 1394, but not IP.

4. CONVERGENCE BETWEEN IEEE 1394 AND IP

To achieve the aim of the present work, multimedia content available in one network should be made available to devices, that are part of another one. Here this is limited to IEEE 1394 and IP. For that purpose a kind of gateway is needed. Of course it is also possible to have more than one gateway, each offering some services, that are available in the other network, and a service might also be offered by several gateways.

The design of the gateway should take into account that on both networks data shall be transmitted in a form which is suitable and common on that particular network. It should not just retransmit data in a format that is used on the source network, but is unusual on the target network, because then usual consumer electronic devices would not be able to deal with that data.

There are many different kinds of gateway necessary to achieve convergence between IP networks and IEEE 1394 (for plain audio, for video, for combined video and audio, for each direction). This document focuses on one very special example, which is illustrated in Fig. 1. Many radio stations provide live streams of their programs on the Internet. The gateway described here should make these streams available in the IEEE 1394 network. That would enable an amplifier connected to the IEEE 1394 bus to play that stream. Unfortunately most of those stations use some proprietary format (e.g. RealAudio). Here the assumption is made, that radio stations provide their streams within RTP and use some data format, that can be decoded.

The gateway must perform several tasks:

- It must handle control information from both sides.
- The (audio) stream must be converted from RTP (IP based) to IEC 61883 (on IEEE 1394). This involves two parts:
  · conversion of the data format
  · reconstruction of timing information

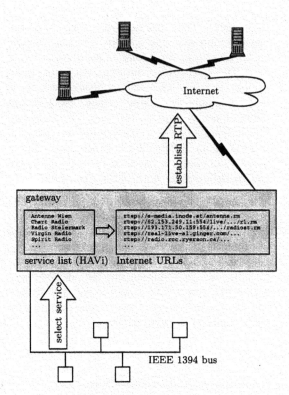

Fig. 2. Mapping between services (in the HAVi network) and URLs (in the Internet) within the gateway

4.1 Control

On the IEEE 1394 bus some middleware protocol is used to control the devices on the bus. Within this work HAVi (HAVi Inc 2001) is used, but of course the implementation for other protocols like AV/C (1394 Trade Association 1998) or UPnP (Microsoft 2000) could be performed in an analogous way. Therefore the gateway will appear within the IEEE 1394 network as a HAVi device and will supply a tuner FCM (functional component unit). That tuner FCM provides some radio programs (the URLs of the offered services might be taken from some configuration file or found by some discovery mechanism, that is without the scope of this document) and makes them public via service lists. Fig. 2 illustrates that mapping. When a service is selected (by issuing the appropriate HAVi command), the gateway will establish the RTP connection by use of RTSP (Schulzrinne et al. 1998) and start buffering (see section 4.3). When the corresponding IEC 61883 connection is established (that is done by manipulation of plug control registers), it will start streaming on IEEE 1394.

4.2 Conversion of Data Format

An audio stream on IEEE 1394 compliant to IEC (2002) contains uncompressed audio. That is viable, because on that bus within the home

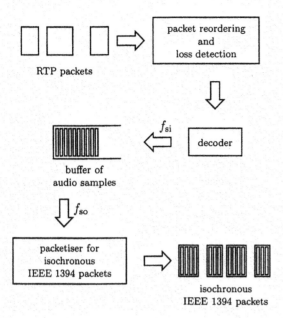

RTP packets

packet reordering
and
loss detection

f_{si}

decoder

buffer of
audio samples

f_{so}

packetiser for
isochronous
IEEE 1394 packets

isochronous
IEEE 1394 packets

Fig. 3. Processing and buffering of audio data in the gateway

there is usually enough bandwidth available and it makes it easier for the amplifier to play the stream (no decoding necessary). On the other hand the RTP stream arriving from the Internet might as well contain uncompressed audio as defined by Schulzrinne (1996) or Kobayashi *et al.* (2002), but due to limited bandwidth in public networks, it will usually utilise some compressed format, e. g. MPEG as outlined by Hoffman *et al.* (1998) or some proprietary format. Therefore some decoder (as shown in Fig. 3) is needed to obtain the plain audio samples, but its implementation is without the scope of this document.

4.3 Reconstruction of Timing Information

The RTP data stream contains a timestamp in the RTP header of each packet. This timestamp is intended for two types of synchronisation purposes. In conjunction with the mapping between NTP timestamps [5] and RTP timestamps contained in the sender's report (SR) of RTCP (RTP Control Protocol, see Schulzrinne *et al.* 1996) it can be used for *intermedia synchronisation*, which usually means ensuring an audio stream to be lip sync with a video stream. Since here only single streams [6] are used, that is not needed.

---

[5] The NTP timestamps used in RTP refer to wallclock time (absolute time).
[6] Even if a multiplexed stream (e. g. an MPEG-2 transport stream containing audio and video) is transmitted via RTP, that synchronisation mechanism is not used, because it is only one single RTP data stream. The synchronisation between video and audio is done by the decoder (exactly: the demultiplexer) in that case.

Intramedia synchronisation is the other issue, and this has always to be performed. In essence it consists of three parts:

- A drift between the sampling clocks at source and sink must be compensated.
- The arrival jitter of RTP packets must be handled.
- Lost packets and silence periods must be treated correctly.

In an RTP data stream the RTP timestamps could in principle be used to eliminate a drift between the sampling clocks. That timestamp reflects the sampling instant of the first octet in the RTP data packet (Schulzrinne *et al.* 1996). However, doing so would introduce the new problem, that the reference clocks must be synchronised. Of course that could be achieved, but it is also possible to use a much simpler approach and send the samples using the average arrival rate f_{si}. This can be done, because no intermedia synchronisation is necessary.

The same applies for the arrival jitter. Instead of working with timestamps, it is easier to maintain a buffer, that outputs the samples at the average arrival rate f_{si}. This of course requires, that the samples have to be played at a constant rate, i. e. the time between two samples must be constant. For audio this is fulfilled, but there are other kinds of stream, that do not meet this requirement.

Discontinuities in the RTP timestamps might be caused by lost packets (in this case there would also be a discontinuity in the sequence numbers) or silence periods (which are usually marked by a special bit in the RTP header). In case of radio silence periods are not usual, they are rather used in conferences. In both cases, an appropriate number of empty (or interpolated) samples must be inserted into the buffer. That number can be determined by calculating the difference between the timestamp in a new packet and the expected timestamp (which is the timestamp of the last packet increased by a number corresponding to the duration of the audio signal in the last packet). If no packets arrive for some time, then empty samples must be created as well to avoid a buffer underrun.

The gateway will start filling the buffer (see Fig. 3) with audio samples, after the connection is started and RTP packets arrive. It must queue samples for at least the expected maximum arrival jitter, but the delay should be small enough to make it acceptable to the user. Therefore a buffer size in the range of one or two seconds might be a good compromise. Then the gateway will calculate the time $T_{so} = 1/f_{so}$, which should elapse between two samples (in units of the IEEE 1394 cycle time, i. e. $1/24{,}576\,\text{MHz}$ which is approximately $41\,\text{ns}$), either by using the nominal sample rate or the measured arrival rate f_{si}. Then after starting with some cycle time for the first sample, it is possible to calculate the cycle time for each sample, when it should be presented at the receiver[7]. That calculated time gives a hint how to packetise the stream on IEEE 1394: Roughly all samples whose cycle times yield the same cycle count are put into one isochronous packet. For some samples (depending on the $SYT\_INTERVAL$) IEC (2002) requires timestamps in the SYT field of the CIP header. For that purpose the lower 16 bits of the calculated cycle time are used.

If the buffer of audio samples in the gateway tends to fill more and more in the long-time average, the gateway transmits too slowly on the IEEE 1394 bus. Therefore it should slightly decrease T_{so}. Vice versa it should slightly increase T_{so}, if the buffer seems to run out of data. The algorithm to adapt T_{so} to make it as smooth as possible is beyond the scope of this document.

5. CONCLUSION

The wide variety of separate networks in home environments makes convergence between them desirable. This paper outlined how to build gateways between different network types in order to overcome the borders with respect to real-time audio/video transmission. However, to achieve real convergence many more gateways will be necessary. Besides transformation of different kinds of multimedia data, other kinds of network (in addition to IP and IEEE 1394) should be taken into account. Aside from real-time A/V transmission, the borders between different middleware protocols (HAVi, AV/C, UPnP etc.) should be relieved.

REFERENCES

1394 Trade Association (1998). *AV/C Digital Interface Command Set, General Specification, Version 3.0*. 1394 Trade Association. Austin, Texas, USA.

Bard, Steve (2001). Wireless convergence of PC and consumer electronics in the e-home. *Intel Technology Journal.*

Dietrich, Dietmar, Dietmar Loy and Hans-Jörg Schweinzer (1997). *LON-Technologie. Verteilte Systeme in der Anwendung*. Hüthig.

Dietrich, Kastner and Sauter (2000). *EIB Gebäudebussystem*. Hüthig.

DVD Forum (2002). *Guideline of Transmission and Control for DVD-Video/Audio through IEEE 1394 Bus, Version 1.0*. DVD Forum.

Fujisawa, K. and A. Onoe (2001). Transmission of ipv6 packets over IEEE 1394 networks. RFC 3146. Internet Engineering Task Force.

HAVi Inc (2001). *Specification of the Home Audio/Video Interoperability Architecture*. HAVi, Inc.

Hoffman, D., G. Fernando, V. Goyal and M. Reha Civanlar (1998). RTP payload format for MPEG1/MPEG2 video. RFC 2250. Internet Engineering Task Force.

IEC (1998a). Consumer audio/video equipment – digital interface – part 2: SD-DVCR data transmission. Standard IEC 61883-2. IEC. Geneva, Switzerland.

IEC (1998b). Consumer audio/video equipment – digital interface – part 3: HD-DVCR data transmission. Standard IEC 61883-3. IEC. Geneva, Switzerland.

IEC (1998c). Consumer audio/video equipment – digital interface – part 4: MPEG2-TS data transmission. Standard IEC 61883-4. IEC. Geneva, Switzerland.

IEC (1998d). Consumer audio/video equipment – digital interface – part 5: SDL-DVCR data transmission. Standard IEC 61883-5. IEC. Geneva, Switzerland.

IEC (2001a). Consumer audio/video equipment – digital interface – part 7: Transmission of rec. itu-r bo.1294 system b transport 1.0. Standard IEC 61883-7. IEC. Geneva, Switzerland.

IEC (2001b). Recording – helical-scan digital video cassette recording system using 6,35 mm magnetic tape for consumer use (525-60, 625-50, 1125-60 and 1250-50 systems) – part 1: General specifications. Standard IEC 61834-1. IEC. Geneva, Switzerland.

IEC (2002). Consumer audio/video equipment – digital interface – part 6: Audio and music data transmission protocol. Standard IEC 61883-6. IEC. Geneva, Switzerland.

IEEE Computer Society (1995). IEEE standard for a high performance serial bus. Standard IEEE 1394-1995. IEEE Computer Society. New York, USA.

InHoMNet Consortium (2002). In-Home high speed Multimedia Network (InHoMNet) based on IEEE 1394 – final public report. Report IST-1999-10622. IST.

Johansson, P. (1999). Ipv4 over IEEE 1394. RFC 2734. Internet Engineering Task Force.

Johansson, Peter (2002). P1394.1 Draft standard for high performance serial bus bridges. P1394.1 Draft 1.04. IEEE Computer Society.

Kobayashi, K., A. Ogawa, S. Casner and C. Bormann (2002). RTP payload format for 12-bit DAT audio and 20- and 24-bit linear sampled audio. RFC 3190. Internet Engineering Task Force.

---

[7] It is very important to note, that the cycle time is synchronised between all nodes on the bus by IEEE 1394.

Microsoft (2000). *Universal Plug and Play Device Architecture, Version 1.0.* Microsoft Corporation. http://www.upnp.org/.

Ogawa, Akimichi, Katsushi Kobayashi, Kazunori Sugiura, Osamu Nakamura and Jun Murai (2000). Design and implementation of DV based video over RTP. In: *Packet Video 2000.* number 31 In: *Proceedings.* University of Cagliari. Forte Village Resort (Ca), Italy.

Saito, Takeshi, Ichiro Tomoda, Yoshiaki Takabatake, Keiichi Teramoto and Kensaku Fujimoto (2001). Gateway technologies for home network and their implementations. In: *2001 International Conference on Distributed Computing Systems Workshop.* IEEE. Mesa AZ, USA. pp. 175–180.

Schulzrinne, Henning (1996). RTP profile for audio and video conferences with minimal control. RFC 1890. Internet Engineering Task Force.

Schulzrinne, Henning, A. Rao and R. Lanphier (1998). Real time streaming protocol (RTSP). RFC 2326. Internet Engineering Task Force.

Schulzrinne, Henning, S. Casner, R. Frederick and V. Jacobson (1996). RTP: a transport protocol for real-time applications. RFC 1889. Internet Engineering Task Force.

van Halteren, Aart, Leonard Franken, Dave de Vries, Ing Widya, Gloria Túquerres, Johan Pouwelse and Phil Copeland (1999). QoS architectures and mechanisms. Deliverable 3.1.1. AMIDST project. Reference: AMIDST/WP3/N005/V09.

Vogel, Andreas, Brigitte Kerhervé, Gregor von Bochmann and Jan Gecsei (1995). Distributed multimedia and QOS: A survey. *IEEE MultiMedia* **2**(2), 10–19.

ELSEVIER

IFAC
PUBLICATIONS
www.elsevier.com/locate/ifac

DISTRIBUTED CONTROL SYSTEM BASED ON INTRANET

Huang Wei

Department of Automation, Southwest China Normal University, Chongqing, Beibei 400715,China

Abstract: This paper describes a dispatching and control system for City Gas Transmission and Distribution (CGTD), which is consisted of dual Dispatching and Control Centers (DCC), 82 Remote Terminal Units (RTU), Intranet communication base on Wide Area Network (WAN) and Local Area Network (LAN). Via the network, the DCC collects the data from every RTU, and sends the commands to each one. Real time control is ensured by each RTU, and real time communication is ensured by DNP3.0. This principle can be applied to more kind of trades. Such as oil, power transmission, water supply, forest fireproofing, and so on. *Copyright © 2003 IFAC*

Keyword: Distributed control, Supervisory Control and Data Acquisition (SCADA), Redundant network, Wide area network (WAN), Local area network (LAN), Database.

1. Introduction

In china, government is encouraging people to use cleanly energy source, and has built a long pipeline to transmit gas from west area to east. Along with this project implementing, the development of CGTD management is going into a new tidemark. Aiming at this growing up trade, an in-depth study is not only can fulfilling the requirement of market but also can benefiting to use the technology in other trades and domain. CGTD has some characters for its own. Such as bigger scale, extensive distribution, higher security requirement, higher calculation veracity requirement, complex structure, wider specialty scope, linked closely with people living. Thus the requirement of automation is very hard. As the basic condition building, every kind of new technology and new method must be adopted to ensure the advantage, practicability, security, and real time character of system.

In CGTD management automation system, the faster data transmission speed and the stronger process capability are needed. The development of information technology (IT) provides more ways to solve it. Intranet provides us the quick, steady, and opening communication network. Database technology provides us the seamless link between

with real time database and relation database. Those data not only can be used in distributed control system, but also can be used in other management system. This system is not an isolated and closed real time control system any longer, it is a part of full-integrated automation system in corporation. From the technology point of view, the correlative technologies are covered with automation, network communication, network security, computer, information handling, geography information handling, etc. Automation technology is included of automatic instrument, control system, control software, prevent burst, simulation online, and redundant control, etc. Network communication technology is included of WAN, LAN, communication protocols, and communication redundancy, etc. Network security technology is included of anti virus, anti hack, network channel redundancy, and Internet Protocol (IP) tunnel, etc. Computer technology is included of computer, server, storage device, and software programming technology, etc. Information handling technology is included of real time database, relation database, data exchanging, data storage, and data access, etc. Geography information handling technology is included of professional geography information system (GIS), basic city GIS, and waterpower calculation, etc. It means that all kinds of correlative

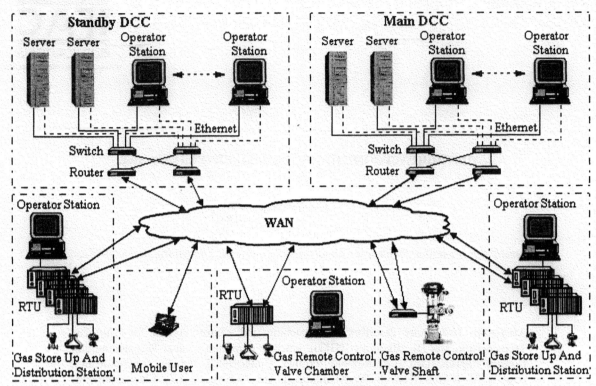

Fig. 1. Structure of System

technology are integrated together to form a high reliability and high practicability platform. One foundation is established to realize full-integrated automation and satisfy the requirement of corporation development. This paper presents the following solution to accomplish it.

2. System Collectivity Structure

2.1 Principle of the system design

In this system, a series of design principles are stood by at following aspects.

- Practicability

The practice need of user must be fulfilled, and the high compare value of performance and price must be provided.

- Advantage

The system must accord with development trend of IT, and ensure it not draggled in short period.

- Reliability

Redundant technology is adopted to improve the reliability.

- Real time capability

The system must satisfy the condition of real time control to ensure the veracity and quickly performance.

- Open capability

Seamless link must be accomplished whether among every kind of network, every kind of system, or between the real time database and relation database.

- Security

Some technologies such as firewall,

authentication, virtual subnet, communicating tunnel, and anti virus are adopted to improve the security of this system.

- Expansibility

This system must give attention to the expand capability, and suit the need of user in the future.

2.2 Structure of system

Considering sufficiently above principals, CGTD automation system integrates a lot of new technologies to establish a full-distributed control system. Four remote engineering is realized which is consisted of remote measure, remote signal transmission, remote control, and remote dispatching. It plays important action for improving security, reliability, work efficiency, and manage level in corporation management.

For monitoring and control these remote stations entirely, this distributed control system is consists of three parts. They accomplish different function severally, such as centralized monitor and management function, distributed data acquisition and control function, and network communication function. Structure of system is showed in Fig.1.

- The first part is DCC. It collects the data from every RTU, monitors the operation of those remote devices, sends the dispatching commands to them, and ensured the whole system working in normal state. It is consisted of main DCC and standby DCC. Dual redundant DCC are working at the same time, but only the main one is activated to output command,

and the standby one does not do it.

- The second part is included of each RTU and every kind of measuring, transmitting, and executing instrument in field. These devices acquire the real time running parameters and execute the control tasks to realize the real time control function in each site. RTU has processor for itself. Thus it works solely to ensure the real-time performance.

- The third part is network communication. It provides reliable channel for data transmission and exchange. Thus it is the core to support this system. It includes of WAN that is established on data communication main channel and standby channel among the DCCs and each RTU, and LAN that is based on full-exchanged Ethernet. For improving the real-time performance of network communication, switch technology, DNP3.0 protocol, star type connection are applied in it.

The whole distributed control system is established on backbone LAN and WAN. LAN is a full-exchanged Ethernet. It is not only the base to realize the data exchanging and sharing in this system, but also is the base of the connection to other management system. Through the LAN, control system connects with a GIS. GIS can provide the display of gas pipeline map, and support DCC to accomplish waterpower calculating and make dispatching decision. Other Management Information Systems (MIS) are connecting on the LAN too, such as Office Automation (OA), Consumer Service Center (CSC), and e-business, etc. Among these non-real time systems, communication protocol is adopted Transfer Control Protocol/Internet Protocol (TCP/IP). Thus this distributed control system not only must accomplish the real-time control, but also must accomplish the conversion of communication protocol. WAN is used to connect the DCC with each RTU. Digital Data Network (DDN) and Public Switched Telephone Network (PSDN) are used to accomplish the physical connection. DDN is main channel. PSTN is backup channel. The reason is that each remote station's communication case is differently but only PSTN can get it everywhere. They adopt router or modem to connect into this network. Communication protocol uses DNP3.0 that supports Report-By-Exception (RBE) and poll to ensure the real time performance. From the network partition point of view, this network is belonged to inside-network. It cannot be entered or accessed for not authorized user. Thus it is Intranet although it has used more Internet technology (Gal, *et al*. 1998).

RTU is distributed at each station to acquire the parameters of running device and execute the real time control task automatically in field. The local flow accumulation is realized but the final result can be corrected in DCC. Once failure occurring, RTU is able to accomplish the most fault identification and record them. Then these data that are consisted of every kind of parameters and processed result are sent to DCC. According to the dispatching decision of DCC, remote control commands of important valve can be sent to RTU directly and executed it in field. Each maintaining command is only sent from DCC to pipeline-maintaining company and branch office. Then the pipeline-maintaining company and branch office arrange person to remove these problems.

3. DCC and RTU

3.1 DCC

There are dual redundant DCC in this system. The main DCC is setting in head office of corporation. It is consisted of two redundant system servers, six redundant operator computers, one projector, one network printer, and two redundant data storage systems. These devices are connected together by the redundant secondly layer switches and redundant 100M full-exchanged Ethernet. Then through the 1000M-backbone network and redundant firstly layer switches, other management system are connected. At the same time, the firstly layer redundant switches connect with two routers to go into WAN. Through WAN, the standby DCC and RTUs are connected. GIS connects into 1000M Ethernet from its own secondly layer switch. Then through the firstly layer switch transferring it connects with other system, such as dispatching control system and other information management system. The firstly layer switches are the cores of LAN. Routers are the cores of WAN. This system is built on Intranet, and the core is network.

In the whole system, the remote control priority of DCC is the highest. DCC can confirm the gas transmission scheme, optimize the flow management of pipeline, control or adjust the flow and pressure of each station, turn on or off the important valve. There are a lot of functions on it such as remote control, flow accumulation, load forecasting, offline decision, and Human Machine Interface (HMI). In ordinary day, the running and dispatching commands are sent directly from DCC to each station. In the same time, each field control station sends the new data to DCC. As the fault presented in each station, DCC sends the information to the failed station in field and the branch office that manages the failed station. Maintain order is sent out from the branch office to failed station. The composition of main DCC is shown in Fig.2.

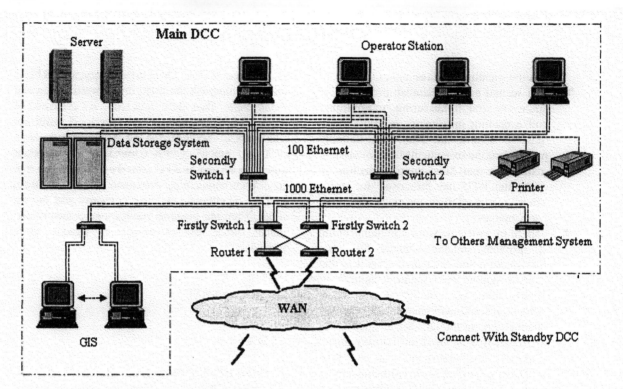

Fig. 2. Composition of Main DCC

For ensured the reliability of main DCC, it is necessary to build a standby DCC base on WAN. The best place of standby DCC has following characters. The geography place is locating at different region, power supply line is provided differently with main line, WAN communication line is separated with

accomplished by main DCC. Main DCC and standby one are working in synchronized state, but standby one do not sending any operating commands and it's

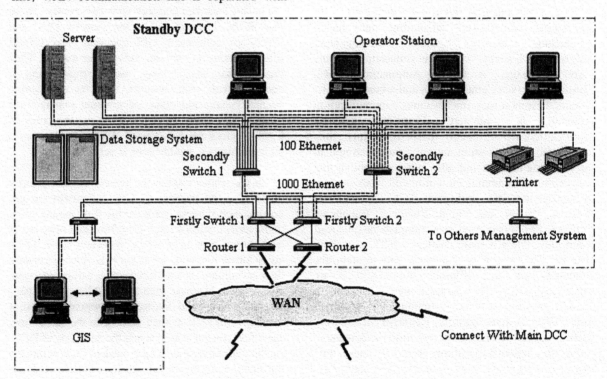

Fig. 3. Composition of Standby DCC

main line. The composition of standby DCC is shown in Fig.3. The consistency and the non-disturb switch between redundant DCCs are ensured by synchronization technology, database copying technology, data resuming technology, etc. In natural working case, all of dispatching and control tasks are

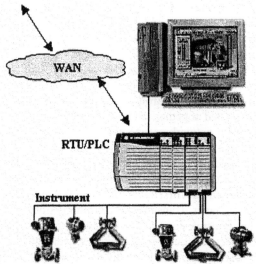

Fig. 4. Composition of Gas Store Up and Distribution Station

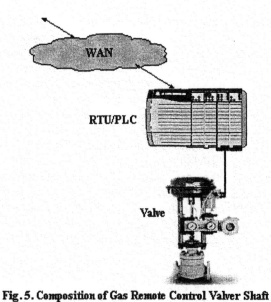

Fig. 5. Composition of Gas Remote Control Valver Shaft

priority is lower than main DCC. As anyone of the following condition appearing, such as the LAN or WAN of main DCC presenting fault, the equipment of main DCC presenting fault or needed to be maintained, and irresistible forces like the earthquake or fire occurring, main DCC will switch automatically to standby DCC, and the role of main DCC will be replaced of standby DCC. After the fault has removed and the system has got right, main DCC will resume its highest priority in dispatching and control role. In each DCC, there is a database copy for other. The integrality and synchronization of data are ensured.

In this system, the standby DCC is setting at branch office. Its system composition is the same as main DCC. Equipments are the same too. But the scale of other management system that connects on standby DCC is smaller than in main DCC. From the redundant mode point of view, the main DCC is master and the standby DCC is slaver.

3.2 RTU

From the technics point of view, RTUs are separated into five types. They are gas gate stations, gas store up and distribution stations, gas distribution station, gas remote control valve chambers, and gas remote control valve shafts. In these kinds of station, the gas store up and distribution station's control scale is the biggest, and the gas remote control valve shaft's control scale is the smallest. In field, there is an absolute controller to realize the data acquiring, local controlling, data storing, and alarm or fault handling. As the WAN communication is failed or interrupted, it can accomplish the real time control for devices in field by itself. In exigent case, operator in field can interrupt the automation program and choose the man-operating mode to run the device.

The control system of gas store up and distribution station realizes the decentralized measuring and centralized monitoring for those technics parameters, running status, leakiness, and flow. Such as temperatures and pressures of gas storing globe holder, the gas leakiness of compressor room, oil pressure and water pressure of compressor, and temperatures and pressures of gas pipeline, etc. Some important parameters are controlled automatically, such as Gas filling and outing, the starting or stopping of gas compressor, the fan of gas compressor room. Gas flow of each pipeline is cumulated and recorded. It is built as somebody kept watching, but satisfies the function of nobody kept watching. One computer, one RTU or Programmable Logic Controller (PLC), and a series of process automation instrument consist this system. Among the process automation instrument and RTU or PLC, 4 to 20mA DC signal or fieldbus communication are used. In the same time, RTU or PLC connect with WAN directly to transmit and receive the data. The important valve can be operated directly in DCC to open or close it. In field, computer connects with RTU or PLC through high speed Ethernet, accomplished HMI displaying, history recording, and control task. The composition is shown as Fig.4.

The composition of gas gate station, distribution station, and gas remote control valve chamber control system is similar as in store up and distribution station, but Input/Output (I/O) amounts and controlled devices are lesser than in storing up and distribution station. There is not gas storing globe holder at gas gate station and distribution station. There are not gas storing globe holder and compressor at gas remote control valve chamber. Thus their system scale is reduced. Their composition of system can refer Fig.4 too.

The gas remote control valve shafts are distributed under the ground of the city. Control equipment is built on aground. Its composition is shown in Fig.5. Both the open degree of valve, leakiness of gas, pressure of pipeline, seepage case of valve shaft, and status of valve shaft cover opening or closing are measured, alarmed, and monitored. Pressure of gas is control automatically. It is different with above four remote control systems that it is belong nobody kept watching control system. There are one RTU/PLC, one remote control valve, leakiness-measuring instrument, seepage-measuring instrument, and opening or closing status of valve shaft cover measuring instrument to consist this system. RTU/PLC connects with WAN directly to realize the data transmitting, and receive the commands given by DCC. The data of technics parameters, running status, leakiness, and flow are only display in DCC. There is not monitor device in field. The valve can be operated directly in DCC.

4. Network Communication

Network communication is the backbone for this system. It decides the availability and the reliability of system operating. Through the network, data exchanging and transmission are realized. The whole data communication is consisted of two parts. One is WAN data communication, and other is LAN data communication.

4.1 Data communication on WAN

DCC connects with each RTU by WAN to accomplish the data remote acquiring, remote transmitting, remote dispatching, and remote controlling function. Main line is used DDN that borrowed from local professional digital communication company. Backup line is used PSTN that borrowed from China Telecom company. If the main line fails, system will switch automatically to backup line for transmitting information. After the failed main line has resumed, communication will switch automatically back to main line. In this system, there are 6 storing up and distribution stations, 35 distribution stations, 10 remote control valve chambers, 31 remote control valve shafts, one DCC, and one standby DCC. Some remote gas stations are distributed very wide. The farthest distance from station to DCC exceeds fifty kilometers. Now in china, only the PSTN of China Telecom company can get everywhere. But If PSTN is used as main line for DCC to poll every RTU once, the time is very long, and the real time performance cannot be ensured. Thus PSTN only can be used as backup network communication line. Considering the development of IT in the future, Other kind of network can be used as the main line for different station, such as Frame Relay (FR), Wide Band Network (WBN), Wireless Network (WN), and so on.

Since the network communication lines of WAN are borrowed from the professional digital communication company, it means that Internet has been connected in. But in DCC and RTU, the network is belonged LAN. Thus routers and modems are needed to realize the connection for these two kinds of different network. Routers are installed in DCC and modems are installed in RTU. Router is working in third layer of ISO/OSI model that is network layer. Through it, different networks are connected together, and independence performance of each network can be ensured. Each network can use different topology, medium, and protocol. Router provides the route of data exchanged among with every source node and end node. DDN connection is belonged the first layer of ISO/OSI model that is physic layer. Transmission medium is fiber. In DCC, the transmission speed of DDN main line that connects in router is 20Mbit/s. Wavelength division technology is used to divide it as one hundred 200Kbit/s. Thus they can be seemed as one hundred logic channels. Each channel provides absolute data transmission and network connection. It reduces the crowd of channel and the collision of data. These dummy channels can provide 'Always On' connection among with DCC and each RTU. There is no dialing, authenticating, and logging on, etc. The transmission speed of each modem in RTU and network connection line is 64Kbit/s. The time delaying of once transmission between DCC with RTU is lesser than 450us. It can ensure the real time performance of network communication. This method solves the disadvantage that traditional DDN do not support switch function, and reserves the advantages of transmission speed quick, network time delaying small, and full-transparency.

In this distributed real time control system, communication protocol is used DNP3.0 Protocol (Triangle MicroWorks, DNP3 Overview). Communication modes support polling and RBE. DNP3.0 is an open, intelligent, robust, and efficient modern SCADA protocol. There are some characters such as request and respond with multiple data types in single messages, segment messages into multiple frames to ensure excellent error detection and recovery, include only changed data in response messages, assign priorities to data items and request data items periodically base on their priority, respond without request, support time synchronization and a standard time format, allow multiple master and peer-to peer operations, and allow user definable objects including file transfer. These characters suit this distributed real time control system very much. DNP3.0 is a layered protocol. DNP3.0 adheres to a simplified 3-layer standard proposed by International Electrotechnical Commission (IEC) that is called Enhanced Performance Architecture (EPA). However, DNP3.0 enhances EPA by adding fourth layer, a pseudo-transport layer that allows for message segmentation. In physical layer, it handles state of the media (clear or busy), and synchronization across the media (starting and stopping). It supports using fiber as media and over an Ethernet connection to implement it. Its control protocol is used CSMA/CD that is the same as Ethernet. In data link layer, it began each data link frame with a data link header, and insert a 16-bit CRC every 16 byte of the frame. A frame is a portion of a complete message communicated over the physical layer. The maximum size of a data link frame is 256 bytes. A data link control byte is contained in the data link header. It indicates the purpose of the data link frame, and status of logic link. According to it, if a data link confirmation is requested, the receiver must respond it. This method enhances determinism and robustness of network communication. The pseudo-transport layer segments application layer messages into multiple data link frame. For each frame, it inserts a single byte function code that indicates if the data link frame is the first frame of the message, the last frame of a message, or both (for single frame message). This method permits single-function messages larger than a data frame, and enhances the efficiency of data transmission. In application layer, it responds to complete messages received and builds message based on the need for or the availability of user data. Once messages are built, they are pass down to the pseudo-transport layer where they are segmented and passed to the data link layer and eventually communicated over the physical layer. When the data to be transmitted is too large for a single application layer message, multiple application layer messages may be built and transmitted sequentially. There is application control code that contains an indication if fragment is one of a multi-fragment message, contains an indication if an application confirmation is requested for the fragment, contains an indication if the fragment was unsolicited, and contains a rolling application layer sequence number.

In the normal time, DCC used the polling mode to communicate with RTU. Remote dispatching information exchanging alignment is controlled by DCC. It make the necessary command to search data, accomplish error checking, ensure the validity of received data, provide error removing, and accomplish the data exchanging. For preconcerted status or events, corresponding station uses RBE mode to communicate. It is an unsolicited reporting mode, where RTU can report field event without being polled by the DCC. This is useful when a high-priority condition occurs at a site that is normally polled at a very low rate. If an alarm condition such as pump failure or leak is detected, a response message can be immediately sent to DCC without waiting for next cyclic poll. To further improve efficiency it permits a method of operation where only changes are reported, reducing communication usage. Thus data can be monitored with a faster response time. It ensures the real time performance of data transmission (Triangle MicroWorks, Using DNP3 & IEC 60870-5 Communication Protocol In The Oil & Gas Industry).

Data transmission exist time delay. Especially as the communication present fault, the delay is longer. It requires that time is indicated for data at data source. For ensuring the data time indication veracity, clock of each device has synchronized. At the set range, the time error between with DCC and each RTU can not exceeding one second in thirty days. Through a centrality clock that has favorable performance, DCC proofread the clock of each RTU and standby DCC automatically and periodically, and ensured everyone to be synchronization. Events to be time-tagged are permit so that the sequence of events occurring in the field can be accurately identified. The information that is transmitted from RTU to DCC is included of event presenting and disappearing. Analog signal needs to add the time indication too. As the data changing exceeds threshold value, it is sent out at once. This can avoid the invalid using and crowding of communication channel. A "sequence of events" history for alarms (binary data), measured quantities (analog data), and counters (volume per unit time, custody transfer, etc.) is supported too. It means that even if a RTU is polled infrequently, all significant changes in the data since the previous poll can be reported at that time, possibly including time stamps that indicate the precise order of the field events. It is necessary for gas flow accumulation (Baruch and Leonard, 1997).

Communication parameters are display in DCC. They are included of the parameters that each RTU connects with DCC, and the system periodicity tested parameters that working in standby mode. They are included of communication times, checking error times, overtime error times, invalid response times, retrying times, and error orientation. The parameters are displayed as horary values and daily gathered value. The daily gathered values are chosen to display as today file or yesterday file.

4.2 Data communication on LAN

In this system, LAN communication is used in DCC and some site. In site, it is very simply. Only two devices are connected into 10M/100M Ethernet, such as computer and RTU (Jonas, Ethernet in Process Control). Thus the emphasis is in DCC.

In DCC, router uses its Ethernet port to connect with the first layer switch and enter the LAN. The LAN has two layers. The first layer is full-exchanged 1000M Ethernet and the second layer is full-exchanged 100M Ethernet. Big-capacity and high-powered 1000M Ethernet switch and 1000M fibers form the high-speed and steady network backbone. According with the tested result that given by West Virginia University of USA, if the Central Processing Units (CPU) of computer and server are used Xeon Pentium □ 500MHz and the Operate System software is used Windows 2000, the highest rate of Gigabyte Ethernet (GE) that can be realized in practice is 158Mbps, and the usage rate of bandwidth is 16%. It shows that the speed of Ethernet has improved to satisfy the faster communication. Through the 100M Ethernet switch and over five species cable, fast 100M Ethernet connects with each computer and server directly. The reliable usage rate of bandwidth in 100M Ethernet is 20%. It means that only 20Mbps speed can be used in practice. If the standard Ethernet mode is used in this system, a lot of problems will present such as determinism and robustness. The Carrier Sense Multiple Access with Collision Detection (CSMA/CD) is the essential reason to arouse these problems. For improving this problem, the switch is used to replace the hub, and the star connection mode is used to replace the bus connection mode. Using the star connection modes. Absolute channel is provided to each node that connects on switch port. Using the switch, the same bandwidth is provided to each channel too. In this time, the bandwidth of each channel can get 20Mbps. Among these nodes that connect on one and the same switch, channel resource and bandwidth resource competition is not exist. Thus the Full-exchanged Ethernet that is consisted of star mode connection mode and switches reduces the accessing collision, and improve determinism in Ethernet communication. It provides the better performance and robustness of network communication, and ensures the absolute channel and bandwidth for each user.

Switch is the core of LAN. All of exchanged information is transferred by it. Thus the reliability request is very high. Each one has configured redundant power, redundant processor, redundant backboard, redundant bus, redundant exchanging module, and redundant clock. Thus it has high processing capability, high exchanging capability, high backboard bandwidth, and high controlling capability. Across cable connection between with the 1000M Ethernet switch and 100M Ethernet switch is used to provide the cable redundancy. The 1000M Ethernet switch has 100M Ethernet port to connect with local network management device.

Two kinds of communication protocol are used in this system. For the distributed real time control system, the DNP3.0 protocol is used. For the data exchanged with other non-real time system, the TCP/IP protocol is used. The protocol conversion is accomplished in the server of control system. Absolute channel is provided to connect from server to second layer switch. Then through the first layer switch, the logic connection with other second layer switch of non-real time system is realized.

For improving security of network, virtual subnet technology, firewall technology, and route security technology are adopted (Ray, 1998; Soh and Yong, 1998). Virtual subnet provides effectual method to limit the broadcasting of the whole network and more nodes. It realizes the isolation among with the real time control system and non-real time control systems. Through its access control table, the access control for different virtual subnet is accomplished. Via switch, each virtual subnet can be connected together. The security is controlled by switch configuration. Firewall is used to isolate inside network with outside network. Address and data packages are filtrated. It can prevent the non-authorized user entering and authorized user doing some non-authorized operation. Base on the route security control character, access control table and access types are used to filtrate the IP address and control the access.

5. Conclusion

This solution has successful applied at China Chongqing gas Ltd. It establishes an opening and high reliability's distributed control system that not only realizes the data collection and real time control for gas supply, but also provides an opening and perfect platform for city gas management company to manage and maintain information. It has already become an organic part of cosmically information management system in corporation. The whole system has high reliability and flexible expansibility. It satisfies the development request of modern company and city.

Reference

Baruch Awerbuch, Leonard J Schulman. (1997). The maintenance of common data in a distributed system [J]. *Journal of the ACM,* 44(1), 86-103.

Gal I. P., Varga J.B. and Hangos K.M.. (1998). Intergrated structure design of a process and its control system[J]. *Journal of Process Control,* 8(4), 251-263.

Jonas Berge Smar. Ethernet in Process Control. *www.industrialethernet.com*

Ray Hunt. (1998). Internet/Intranet firewall security-policy, architecture and transaction services [J]. *Computer Communication,* 21, 1107-1123.

Soh, B.C. and Yong, S. (1998). Network system and world wide web security [J]. *Computer Communication,* 20, 1431-1436.

Triangle MicroWorks, Inc. Using DNP3 & IEC 60870-5 Communication Protocol In The Oil & Gas Industry. *www.TriangleMicroWorks.com.*

Triangle MicroWorks, Inc. DNP3 Overview. *www.TriangleMicroWorks.com.*

ELSEVIER

IFAC
PUBLICATIONS
www.elsevier.com/locate/ifac

SUPERVISION AND CONTROL SYSTEM OF METROPOLITAN SCOPE BASED ON PUBLIC COMMUNICATION NETWORKS

Sempere, V.* Albero, T.* Silvestre, J.**

Dpto. Communications. UPV.
**Dpto. Computer Engineering. UPV.*

Abstract: The present systems of control based on polling by radio frequency must evolve to be able to carry advanced telecommunication services, such as capture and multimedia information management as well as control information. In this paper a supervision and control system based in public networks that complements a classic polling system allowing to incorporate new information services and eliminating the typical long delays of these kinds of systems is described. Two alternatives of communication are evaluated, point to point connections and Virtual Private Networks, VPNs/IP on ISDN, for the interconnection of the stations and the control center. For this, images are transmitted as well as the control information through both types of connection, the throughput of the transmission is evaluated and its behavior is observed. Compression techniques are used and the different parameters that can influence the compressed image quality are analyzed. The viability of this type of solution in the proposed scenarios is demonstrated. *Copyright ©2003 IFAC*

Keywords: Images compression, Communication networks, Supervision, Control system.

1. INTRODUCTION

In this paper a newly designed system is introduced to modernize the control and supervision of a large city purification network. The form in which it operates is over public communication networks and it is able to manage control and multimedia information centrally. It is designed to be installed independently, or jointly with existing control systems since it is able to interchange control information directly from the installation sensors and actuators or point to point from any command equipment (PLC, terminal, etc), to capture images of critical areas of the installation selected by the operator and to transmit them to a central station where they are processed in real time.

The purification networks are normally governed by installations distributed of metropolitan scope based on PLCs that use polling (Ibe and Trivedi, 1990; Haverkort, 1999) by means of radio frequency for the communications between the central station that houses a SCADA (Luque *et al.*, 1996) and each and every one of the remote nodes. This type of solution supposes a bottleneck during the period on which new services and applications are being implemented. The reasons that have suggested these actions are:

The installation is constantly growing and the procedure of polling cannot offer the time guarantees that are required, since to greater number

---

[1] supported by UPV (Polytechnic University of Valencia. Spain)

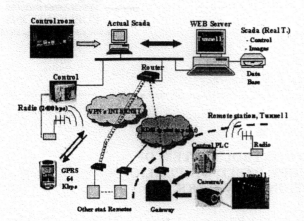

Fig. 1. System general scheme

of stations to be consulted by the central station, greater is the time lapse between successive consultations. The new system has to handle the later incorporation of new remote stations without any time increase.

It is necessary to equip the system with redundant communications to increase their reliability, implementing new advanced monitoring and control services that facilitate both, the inspection of any critical zone within the installation and the remote teleoperation The communications have to be protected against any possible unwanted third party interference.

In this paper a global description of the architecture system is presented, followed by a detailed evaluation of the processing, transmission and storage of the images. The performance of the system is analyzed, obtaining the transmission rate of images by means of two public networks communication alternatives:

(1) VPN/IP (Virtual Private Network) (Kosiur, 1998)
(2) point to point connection.

The paper is structured in the following way: In section 2 the different parts of the new system are described and the chosen solution is justified. In section 3 the treatment process for the transmission images is described. Later in section 4, the method of storing and accessing the information is described. Finally, the developed application and some basic measurements which demonstrate the viability of the proposed solutions are shown.

2. SYSTEM COMPONENTS

It is made up of three fundamental parts (see Fig. 1):

• Central station.
• Remote Nodes/gateways.
• Communication network.

2.1 Central Station

This is the part charged with managing the information coming from all the installation and controlling the remote communication between the central and remote stations. It communicates simultaneously (using several connections TCP/IP) with all the remote stations, eliminating the slow cycles of polling and contributing in joint installation cases with remote control systems, a redundant communication channel which equips the installation as a whole with a high degree of reliability and availability.

The central station uses equipment which does the SCADA in the new network and that houses the states and images data base (where it is used like a SQL Server 7.0. motor). A background process is charged with preserving coherence between the old data base (updated by polling) and the new one. The central station also houses a Web server which permits the data base which registers the information of each remote to be consulted locally and through the Internet.

By means of the local processing in the central station (developed with Borland C++ Builder 5), it allows the visualization of images coming from one or more installations simultaneously, as well as obtaining control data, greatly facilitating the tasks of tracking incidences.

The Web application has been developed with Javascript, VBScript, ASP and HTML. It is possible to make the Web connection from fixed or movable nodes using GPRS (Hoymann and Stuckmann, 2002), as a Web page has been designed that at the moment of connection detects and adapts to the type of user fixed (PC) or movable (PDA, Compaq iPAQ 3970 has been used). It is important to emphasize that for security reasons the Web usuary can only consult information and does not have control of the system. If the client connects itself from a PC, it is allowed to see data control and images in real time, furthermore the stored images. However if it accedes through a PDA it can see realtime images as well as the information of the states.

In Fig.2 the block architecture of the central station can be seen, where the arrows indicate the incoming/outgoing flow of information.

There are three types of subprocesses existing in the local processing in the central station as well as in the remote processing node; those executed at the request of the user (manual), those that have previously defined parameters and offer information in periodic form, and those that are not predictable and notify of some change of aperiodic form.

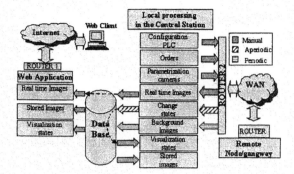

Fig. 2. Block architecture of the central station

2.2 Remote Node/Gateway

This is the equipment located in each of the remote stations to be governed. It is the gateway between the communication network and the local control system of each remote station. It is constituted of an industrial micro PC with operating system W-2000 Professional and a powerful applications developed with Borland C++ Builder 5 and Visual Microsoft C++ which makes easy adaptation to any installation.

It can also capture information coming from cameras (up to 4) located in the installation, process it and jointly transmit it to the central station by the communication language available in each case. It is able to communicate with any PLC on the market (the protocol of Simatic S5/S7 has been used) bidirectionally and to map in a simple way the main variables of the process. The remote processing node determines when susceptible variables being sent to the central station have taken place and executes an automatic update process.

The parametrization of the remote Node/Gateway is done centrally through the central station. The information provided is as follows:

- Variables to read/write from the PLC.
- Resolution, quality and rate of compression of the images.
- Position of the cameras.
- Intervals of capture of the images.

In Fig.3 the block architecture of the remote gateway can be seen, where as in the central station the arrows indicate the information flow.

2.3 Communication network

This has to give communication support in concurrent form to all the remote stations, allowing the incorporation of new services in real time such as the transport of images, states and orders. The present system of polling by means of radio frequency is a redundant system acting as backup to operate in case of failure of minimum services.

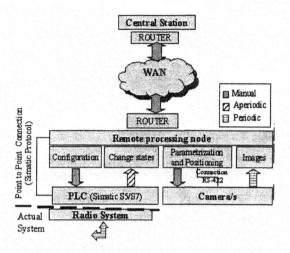

Fig. 3. Block architecture of the remote station

After a detailed study of the diverse possibilities two possible alternatives for the interconnection of the stations have been studied. These are:

- Use of point to point connections between the central station and each of the remote stations through ISDN (Integrated Services Digital Network)
- VPN (Virtual Private Network) through Internet with ISDN access.

In the result section a comparison between the two is made.

The criteria for determining this choice are as follows, in the first place, the adjustment of the service offered by the network (guaranteed bandwidth, asymmetry absence), secondly the area of coverage in urban zones. Other alternatives such as HFC (Hybrid Fiber Coax) or ADSL have not been considered in this paper, but they will be reason for study in later papers.

The central station has a ISDN router with sufficient basic accesses (2B+D) or a primary one (30B+D) according to the needs. The traffic between the central equipment and the remote stations is guided through this. A local area network infrastructure carries the communication between the new central station, the router, the old central station and the central PLC responsible for the polling tasks. At each remote station a basic access ISDN arrives. In the central station the control information and the images are simultaneously collected from all the remote stations.

When the communication channel is available, all exchange of information between the central station and the remote ones is made by means of TCP/IP, facilitating the incorporation of new services and the movement toward other futures possibilities with QoS (Quality of Service) on Internet (Ferguson and Huston, 1998).

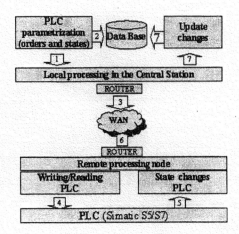

Fig. 4. Protocol for the control data processing

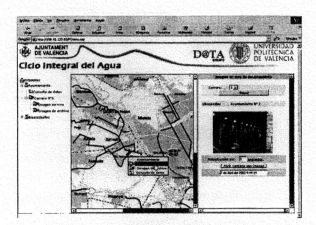

Fig. 5. Visualization of realtime images from the application Web

3. CONTROL INFORMATION TREATMENT

In Fig.4 the processes that take part in the control data processing can be seen, furthermore the arrows are also numbered to indicate the order of execution.

The central station does the variables parametrization(1) to read (states configuration) and to write (orders configuration) from the PLC. In this configuration, the position in the memory zone (for the Simatic PLC, DB (Data Block), DW (Data Word) and bit of that word) or the mark zone, (MB (Mark Block), MW (Mark Word), and bit) that contains the state/order information, is indicated. When the parametrization in the central station is finalized, the configuration is stored in the data(2) base and this parametrization is immediately sent(3) to the remote station, which by means of the corresponding process establishes the memory or mark zone and does the state reading(4) in the PLC continuously. Also from the central station orders can be generated(3) to the remote station, by means of this action the writing(4) is done in the PLC.

The remote station communication with the PLC is made by means of the PlcS5 ActiveX Control, that uses the Siemens Simatic S5 AS511 protocol by a serial connection. This control, is able to detect any changes(5) that can happen in the PLC (the change can be caused by an incidence in the system or by an order executed from the central station). When this happens, from the remote station the alarm is sent(6) to the central station, then the appropriate measures can be taken. Furthermore, the state change is stored in the central station data base(7).

The control information travels in IP packages independently of the images. As the state variability is not elevated and only the changes are transmitted, this traffic does not affect the system throughput.

4. IMAGE TREATMENT AND TRANSMISSION

Images are transmitted compressed from the remote stations to the central station as this helps to reduce the bandwidth consumed by the transport of images although at the expense of image quality.

Due to the low speed of the transmission channel (64 Kb/s), streaming of video is very difficult (Woods, 2002). Since it is an application of monitoring, where a great number of images per second is not necessary, JPEG2000 has been chosen for the compression which has a very good quality/compression relationship compared with JPEG.

Furthermore, codec has been chosen based on values obtained in the following parameters:

- Quality.
- Necessary/available bandwidth.
- Compression/decompression time.

After studying different codecs which are not going to be detailed in this paper, due to the good results given, the Morgan Multimedia M-JPEG2000 is used.

The transmission/reception of images can be required by a local user in the central station or by a usuary Web (see Fig.5), although the local processing in the central station will be the one that demands and receives the images by means of the protocol that is outlined next. In Fig.6 the transmission/reception protocol can be seen in schematized form. The arrows indicate if it is dealing with traffic of communication between the processes (handshaking) or data. These are also numbered to indicate the order of execution. The remote station has two processes "CapturaImag" and "ClienteImag", the first is the one charged with capturing the camera images and to compress them (Morgan M-JPEG2000), also acting as a server. "ClienteImag" is the client and is

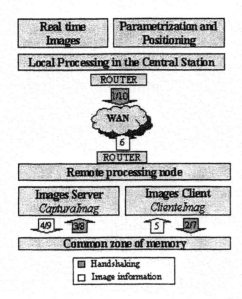

Fig. 6. Protocol for transmission of images

charged with transmission of images to the central station and also to receive the parametrization and positioning information. Both processes share a common zone of memory that allows them to write or to read information in synchronous form.

The central station is that which makes the parametrization and indicates when the transmission/reception can begin by means of message(1) to the "ClienteImag" process. When "ClienteImag" receives this message, it informs(2,3) "CapturaImag" that a compressed image can be collocated(4) in the common memory zone when it is available. From the common memory zone "ClienteImag" takes(5) the image and transmit(6) it to the process of the central station. When the complete image has been transmitted, it informs(7,8) "CapturaImag" that it can again put(9) a new one in the memory zone. It is also indicated from the central station when the transmission/reception of the images must conclude(10) by means of another message to the "ClienteImag" process.

5. IMAGE STORAGE AND RECOVERY

The parameters of images storage are defined from the central station, differentiating into two types: a) storage in background, b) manual storage.

The first type is independent of the current situation, the parametrization has been done previously and the image capture is done with constant periodicity throughout the day. Manual storage is done at any moment when the situation could be critical (strong rain). If the operator is visualizing these images on the screen and considers it necessary, he can store these images in the data base for later study.

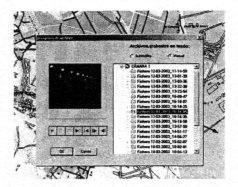

Fig. 7. Access to images stored in the central station

Table 1. Test Sequences

	Size(KB)	Duration(min)
C2	341.192	2.34
C3	364.450	2.36
C4	472.302	3.16

Table 2. Result of quality parameters after compression of sequences

	PSNR (dB)	Size (bytes)	CT (s)	DT (s)
C2-FQ87	40.12	8051.41	0.158	1.13E-5
C3-FQ87	39.70	8317.10	0.108	1.09E-5
C4-FQ87	40.34	8292.52	0.110	1.13E-5
C2-FQ56	37.67	5334.26	0.151	1.13E-5
C3-FQ56	37.19	5323.36	0.099	1.11E-5
C4-FQ56	38.00	5444.46	0.102	1.16E-5
C2-FQ30	34.26	2849.93	0.146	1.12E-5
C3-FQ30	34.00	2851.76	0.151	1.09E-5
C4-FQ30	34.90	2861.08	0.096	1.12E-5
C2-FQ18	31.77	1708.01	0.144	1.09E-5
C3-FQ18	31.57	1703.14	0.148	1.09E-5
C4-FQ18	32.48	1710.79	0.093	1.16E-5

For the capture of these images it is necessary to select: camera, quality, resolution and length of recording (maximum one hour). The storage is completed once the established time has passed or when the operator deems it correct to do so.

Later, the stored images can be recovered (background or manual) to be viewed and analyzed as can be seen in Fig.7.

6. EXPERIMENTS AND RESULTS

The tests to ascertain the efficiency of the system have been made with three video sequences stored on disc (see table 1) and not with real time images, to thus analyze a same test set. For ISDN communications, a single channel of 64 Kb/s has been used.

In order to limit the number of tests two important parameters have been used: Forced Quality (FQ) and Constant Bit Rate (CBR) which influence the compression quality (PSNR), the size, the compression time (CT) and the decompression time (DT) of each image decisively. As can be seen in table 2, for each sequence three quality levels

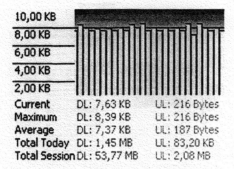

Current	DL: 7,63 KB	UL: 216 Bytes
Maximum	DL: 8,39 KB	UL: 216 Bytes
Average	DL: 7,37 KB	UL: 187 Bytes
Total Today	DL: 1,45 MB	UL: 83,20 KB
Total Session	DL: 53,77 MB	UL: 2,08 MB

Fig. 8. Monitoring of the connection through ISDN

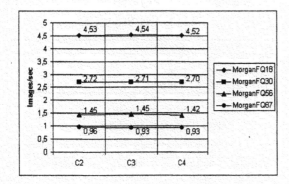

Fig. 9. Images/sec in the point to point connection, for different levels of compression

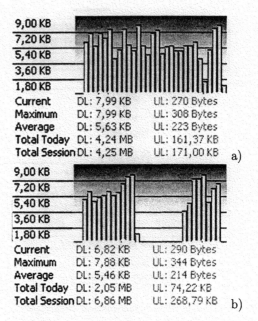

Current	DL: 7,99 KB	UL: 270 Bytes
Maximum	DL: 7,99 KB	UL: 308 Bytes
Average	DL: 5,63 KB	UL: 223 Bytes
Total Today	DL: 4,24 MB	UL: 161,37 KB
Total Session	DL: 4,25 MB	UL: 171,00 KB	a)

Current	DL: 6,82 KB	UL: 290 Bytes
Maximum	DL: 7,88 KB	UL: 344 Bytes
Average	DL: 5,46 KB	UL: 214 Bytes
Total Today	DL: 2,05 MB	UL: 74,22 KB
Total Session	DL: 6,86 MB	UL: 268,79 KB	b)

Fig. 10. Monitoring VPN/IP

6.2 Connection by VPN/IP on ISDN

Videos have been sent with an interval of one hour throughout different days and samples have been taken every two seconds (see Fig.10). A grater speed variation has been observed than in the point to point connection (see Fig.10a), since the speed of the connection depends on: the speed that the ISP (Internet Service Provider) offers and the state of the Internet. In Fig.10b, it is possible to observe that there are moments at which the communication is interrupted and then reinitiated after a few seconds. This creates a decrease in the average value of the speed of the channel and its reliability

Owing to the three sequences having similar image sizes, an average has been used to calculate the image per second rates in different hour strips for the different grades of compression.

As can be seen in Fig.11, for each of the compression grades the number of images/sec which can be transmitted is always smaller than with the point to point connection. There are also reliability problems that can cause the instability in the channel speed.

are going to be used. These have been obtained with different values from FQ (87, 56, 30 and 18). The compression and decompression time is insignificant against the time it takes to transmit an image considering that it has a theoretical speed of 64Kb/s, and so is not considered in the calculations.

6.1 Point to point connection through ISDN

Diverse measurements have been taken in different hour strips, taking samples every two seconds and it has been observed that the obtained speed values of the channel are practically constant, with very little variation and without cuts (see Fig.8). For this reason an average value has been taken to calculate the number of images.

It is evident (see Fig.9) that with a greater compression, a greater amount of images can be sent with a cost of quality loss, but is necessary to consider that it is a monitoring application, where a high quality is not necessary. Therefore, it is necessary to reach a compromise between quality and compression level. After impartial evaluation by various different observers, one conclusion has been reached, the MorganFQ30 compression (PSNR of 34,38 dB) offers an acceptable quality for a monitoring application and control, in which almost three images/sec. can be transmitted.

7. CONCLUSIONS

With the proposed new system, the state and alarms of each of the remote stations is obtained simultaneously since it is now unnecessary to wait for any polling cycle. The system has a greater immunity against radioelectric interference which increases its safety and reliability. The new network provides an improvement in many aspects, keeping the polling system as a backup system in

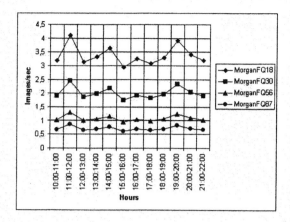

Fig. 11. Images/sec in the connection by VPN/IP, for different levels of compression

the case of failure in the public communication network.

After a carrying out the experiments the chosen solution for communication between the central station and the remote stations is point to point connection through RDSI. A greater number of images can be transmitted at a constant speed while with the VPN/IP a great variation in speed and even cuts in transmission have been observed, which is not advisable in remote control due to cuts in transmission being unacceptable in cases of emergency. It would be necessary to negotiate a determined quality of connection with the service provider to implement the use of this solution.

For the case of remote consultation through the Internet, the great flexibility from a communication point of view make this a powerful tool for the supervisors of large installations since the sending and receiving of critical information in certain incidences is now possible through various solutions such as mobile telephones and PDA's with adequate time responses. In this case, the tested solution of VPN/IP is perfectly applicable given that the consulting of information is limited.

REFERENCES

Ferguson, P. and G. Huston (1998). *Quality of Service: Delivering QoS on the Internet and in Corporate Networks*. Wiley.

Haverkort, B.R. (1999). Performance evaluation of polling based communication systems using spns. In: *Application of Petri Nets to Communication Networks: Advances in Petri Nets* (J. Billington, M. Diaz and G. Rozenberg, Eds.). 1rd ed.. pp. 176–209. Springer. Berlin.

Hoymann, C. and P. Stuckmann (2002). On the feasibility of video streaming applications over gprs/egprs. *IEEE Global Telecommunications Conference*.

Ibe, O.C. and K.S. Trivedi (1990). Stochastic petri net models of polling systems. *IEEE Journal on Selected areas in Communications* 8(9), 1649–1657.

Kosiur, D. (1998). *Building and Managing Virtual Private Network*. Wiley.

Luque, J., J.I. Escudero and I. Gomez (1996). Determinig the channel capacity in scada systems using polling protocols. *IEEE Transactions on Power Systems* 11(2), 917–922.

Woods, D. (2002). Shriking the video: How codecs works. *Network Computing* pp. 77–79.

ELSEVIER
IFAC
PUBLICATIONS
www.elsevier.com/locate/ifac

WEB-BASED CONTROL EXPERIMENTS ON A FOUNDATION FIELDBUS PILOT PLANT

R. P. Zeilmann , J. M. Gomes da Silva Jr., A. S. Bazanella, C.E. Pereira.

Electrical Engineering Department - DELET
Universidade Federal do Rio Grande do Sul - UFRGS
Av. Osvaldo Aranha 103 - Porto Alegre - Brazil - 90035-190
E-mails: {zeilmann, jmgomes, bazanela, cpereira}@eletro.ufrgs.br

Abstract: This paper describes the development of a Web-based remote access system to a Foundation Fieldbus pilot plant. The main goal of the developed system consists in providing a platform for control and automation education which can be remotely accessed by students. In this sense, the plant operation is fully configurable and programmable via a supervisory application executed on a Web server. Among other features, the supervisory interface allows the remote on-line configuration of fieldbus parameters, elucidation of main fieldbus concepts – such as function blocks structuring - and a real-time graphical visualization of all relevant plant variables. These features allows students to execute open and closed-loop control experiments that illustrates control concepts and techniques as well as concepts concerning the Foundation Fieldbus protocol. *Copyright © 2003 IFAC*

Keywords: Remote control, Process Control, Fieldbus, Internet, Laboratory education.

1. INTRODUCTION

There is a clear trend towards distributed architectures for automation and control of industrial plants, where process control and automation activities run in distributed, intelligent field devices, which communicate with each other over industrial communication networks, the so-called fieldbus technology.

The combination of fieldbus and its application layers with Web technologies makes possible the remote access to industrial automation systems through the Internet. The remote access to industrial plants provides flexibility for process configuration, parameter tuning, and remote supervision, as well as the physical security of operators.

On the other hand, the realization of experiments through the Internet has brought a whole new perspective to this distributed control paradigm. This technology has a strong appeal for research and teaching. It allows the use of experimental facilities to illustrate concepts taught in classroom and serves as a powerful tool for distance teaching. It also makes learning material available to a much larger audience of students, giving them a greater flexibility in terms of defining the speed and sequence of subjects when learning.

Following this trend, many institutions around the world have been developing experimental settings accessible over Internet. Systems aiming at teaching and research in several different areas have been proposed, such as digital process control (Poindexter and Heck, 1998), (Ramakrishnan, *et al.*, 2000), aerospace applications (Poindexter and Heck, 1998), PID control (Batur, *et al.*, 2000), predictive control, embedded communication systems (Schmid, 1998), and real-time video and voice applications (Johansson, *et al.*, 1999). Mostly, these experiments utilize customized devices and software to perform textbook style small scale experiments.

This paper presents a Web-based control experiment, which allows the control and supervision via Internet of an Industrial Pilot Plant containing smart devices based on the Foundation Fieldbus protocol. This experiment serves as a platform for research and teaching - both distance and in situ – of control systems and industrial automation. Contrary to most of the above mentioned systems, our pilot plant uses industrial scale devices, which makes it ideal for research and is also proving to be quite effective for teaching.

The Pilot Plant consists of three interconnected tanks where it is possible to configure level, temperature, and flow control loops. All instruments are smart sensors and actuators connected over a Fieldbus Foundation communication network. By means of supervisory software, which runs concurrently with a Web server, it is possible to remotely program and perform an on line modification of control parameters. To remotely access the system and run the experiments, remote users only need a standard Internet browser with Java runtime enabled.

The paper is organized as follows: In Section 2 the pilot plant is physically and conceptually described, with a brief discussion of the features of the communication protocol which is used. The software structure implemented in the system is described in Section 3. Issues concerning access control are discussed in Section 4. In Section 5 the experiments already available on the Internet are described and their didactic and research possibilities are discussed. Some concluding remarks are given in Section 6.

2. SYSTEM DESCRIPTION

2.1 The Pilot Plant

The pilot plant (see Fig. 1) consists of three interconnected tanks, four AC pumps, two of which driven by vector control drivers, and five control valves to control the liquid flow in the system. The liquid in the tanks can also be heated by resistive actuators. It is therefore possible to configure many different single-variable and multivariable control loops, involving tank level, liquid flow and temperature as process variables, with various choices of control inputs. All sensors and actuators in the plant are smart devices, which communicate through the Foundation Fieldbus protocol.

Fig 1. Pilot Plant.

2.2 The Foundation Fieldbus Network

The Foundation Fieldbus is an industrial protocol which allows the distributed control of processes in a network. It presents many benefits regarding conventional automation systems, concerning mainly the larger amount and higher quality of the control and extra control information contents in the network.

In a Foundation Fieldbus network the control is based on function blocks. The system configuration is made by connecting these blocks and setting their parameters. A major feature of these systems is that these function blocks are device independent, that is, a given function can be assigned to run in different physical devices, the physical configuration and logical configuration being performed at different times. In the physical configuration the operator assigns to each function block a device where its function will be ran, whereas in the logical configuration the logical connections among the function blocks are specified. This feature makes the system configuration easier to perform and provides high flexibility and enhanced reliability.

In conventional automation systems the information content can not go much beyond the volume assigned to control information. In Fieldbus systems, the digital communication allows a much larger volume of extra control information. Economic benefits also result from fieldbus technology, mainly due to the smaller amount of hardware in fieldbus based plants and its easy software and hardware (re)configuration.

From a research and teaching perspective, the use of a highly innovative yet industrial-oriented technology brings a new and very strong perspective. Practical interest is raised in the students, as the experiments become visually and conceptually closer to the professional practice.

Also a wider variety of concepts can be illustrated and several levels of difficulty and complexity/challenge can be explored.

3. SOFTWARE STRUCTURE

The software structure was developed on a Windows NT platform. This is a natural choice since the tools for configuring the industrial Foundation Fieldbus network protocol are available only for this platform.

The main applications running in the system and the interfacing among them are shown in Fig. 3. We shall describe the functionalities of the applications and their interfacing in the sequel.

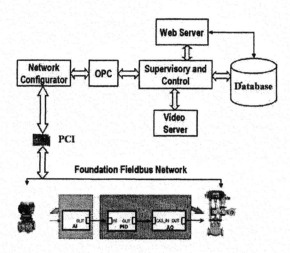

Fig. 2. System Architecture

3.1 The OPC interface

The fieldbus network provides a data server OLE (Object Linked and Embedded) for Process Control (OPC). This interface contains the database of all configuration parameters and variables associated to the control strategy implemented by the network devices. Hence, the OPC data server provides all the plant data to the supervisory application, which runs as its client.

3.2 The supervisory system

The supervisory system has been developed with the Elipse SCADA software Watcher (Elipse Software, 1998), which is an OPC client. Hence, the supervisory applications developed for the pilot plant can import the OPC data base provided by the fieldbus network. All the parameters and process variables are assigned with tags by the OPC interface. These tags are then conveniently manipulated by the supervisory applications in order to generate real-time graphics and reports. On the

other hand, the supervisory system receives parameters from the remote user and writes on the correponding tags of the OPC interface (e.g. the gains of a PID controller running on the fieldbus network).

The supervisory system is also responsible for providing the dynamic Web-pages that will be sent to the remote user and for managing the plant images. These tasks will be discussed in detail in the following sections.

3.3 Connection between the Supervisory System and the Web-server.

The communication between the supervisory and the Web server is based on a client/server structure, which allows any browser (client) to access the screens of the supervisory (server).

The communication is implemented by means of Java Applets. The task sharing between client and server is shown in Figure 3. The browser loads the Web page created by the supervisory using the HTTP protocol. This Web page contains the supervisory screen, with realtime graphics of the process variables, images of the running process and status information, as well as fields for entering control parameters to be sent to the pilot plant (see Figure 4). An Applet (which follows with the page) is executed in the client browser, establishing a connection with the supervisory. The users interacts with the browser, sending parameters to the plant. The information is processed and an answer is sent back to the user, closing the cycle.

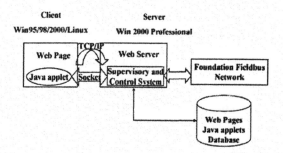

Fig. 3. Communication architecture

The Web-Server used is the Microsoft Information Server 5 (IIS 5). This server is a native application of the Windows NT platform. In this case, all the security features of that platform are extended to the Web-Server. Furthermore, this Server supports ADO and ASP technologies that are used for data exchange between the remote client and the Web-Server.

3.4 Image transfer system

Images of the plant are captured through two Webcam's, allowing users to follow visually the system's variables evolution.

The Elipse Software has a module for monitoring systems with capturing, recording and transmission resources, the Elipse Watcher (Elipse Software, 1998). It supports several different image formats, including MPEG, allowing the visualization in windows whose quality and size can be programmed. Hence, the images provided by the cameras are captured by the supervisory developed system by means of a video object. This object is placed on the supervisory experiment screen that is sent to the Web client through a Java applet. This applet is located in a html page. When the client loads this page, the applet starts to capture the supervisory screen periodically.

3.5 Interface between Supervisory System and the Data Bank

Since the supervisory does not have the ability to make its database available to the internet, an auxiliary data bank is implemented. This data bank stores the informations obtained by the supervisory system from the fieldbus network and also the parameters sent by the user through the Web-server.

The connection between the supervisory system and the data bank is made through an ODBC (Open Data Base Connectivity) interface. The database contains tables concerning specific sets of data, such as process variables, security and access control variables and auxiliary variables. The supervisory system can read and write on these tables by means of an ODBC connection string.

The Data Bank was implemented with the Microsoft Access. This choice was motivated because the amount of data is relatively small. Other Data Bank supporting the ODBC interface could be used without additional changes in the system.

3.6 Sending Information over the Internet

All the information sent by the user browser to the Web-server should be stored in the data bank to be read and processed by the supervisory system and,

eventually, send to the fieldbus network (such as the control parameters).

The access of the Web-server to the Data Bank is done by means of ASP pages containing scripts and methods for accessing the ODBC interface. ADO (ActiveX Data Objects) are then used to get, visualize, change, include and exclude data stored in the data bank. Hence, when the user sends information through the access of an ASP page in the Web-server, the ASP engine running on the server processes the information and updates the database.

4. ACCESS CONTROL

The access control system has the following objectives: user validation, experiments scheduling, monitoring session's duration, parameterization of experiments and experiments initialization.

The first step to be followed in order to run an experiment is to validate and authenticate the user. In this first version, this is done by providing a login and password for each user. This information will be asked and verified by the system at each time the user wants to access the system.

Once the login and the user's password are validated, he/she can have access to a scheduling system or, if an experiment has been previously scheduled, the experimental setup can be directly accessed. The scheduling system consists of a Web page where a calendar is available. The user can identify the available time slots to access the experiments of the system and select one of them. Each time slot has 30 minutes duration. If the user tries to access the experiments Web pages at a non scheduled time, access is denied and the scheduling page is automatically loaded.

The time of each session is monitored by the supervisory system. When the scheduled time is finished, the system automatically logs out the user. A system initialization is then carried out in order to make the experiment available for the next scheduled user. The user can visualize the current time of the experiment by means of a clock placed on the right foot of the supervisory Web-page (see Figure 4).

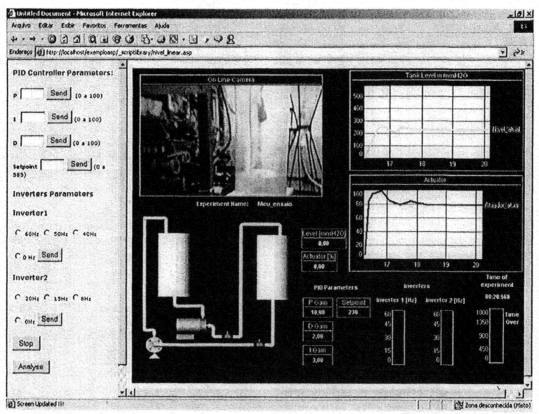

Figure 4. Snapshot of system's Web page

In case of communication failure, the supervisory system will proceed with the experiment initialization, leading the plant to a safe state.

5. RESEARCH AND TEACHING POSSIBILITIES

Many different control experiments can be performed in the Pilot Plant. At this point single-loop level and temperature control are available for access through the internet.

When the user connects to the Web server, his or her browser shows a logon screen. After logging on, a Web page is shown where he or she can pick from a variety of experiments. These experiments involve either the control of the tank level or liquid temperature in the tank. Four pumps, two of which driven by variable speed vector controlled drives, and five valves control the liquid flow in the tubulations and hence determine the liquid level in each tank. In the experiments currently available the control inputs are one of the control valves and a 5kW thiristor controlled heater. In each case three basic experiments are available:

- Open-loop step in the process input
- Closed-loop PID control
- Identification experiment with pseudo-random noise input

The open-loop step experiment can be used to identify system parameters, generating data which can be dowloaded by the user. From these historical data it is possible to tune a PID controller for the process using, for instance, the Ziegler-Nichols tuning rule. Then a closed-loop experiment can be performed to check the quality of the tuning. This sequence of experiments has been repeatedly applied for teaching in our lab.

Figure 4 shows the Web page corresponding to the result of a closed-loop PID experiment for the level control using valve opening as the control input. The screen is divided into two frames. The lateral frame is used for changing the control paremeters on line. In this case, an ASP page refreshes the database and the supervisory reads the new value and sends it to the fieldbus network via the OPC interface. In the main frame, the user can see the real-time graphic of the process variable (level) and the control variable (valve opening); the on-line image of the tanks; a graphical animation of the process; the value of the variables of the process; the current time of the experiment. In the open-loop experiments, the user sets the value of the process input (valve opening or heater power) in a lateral frame of the supervisory Web page, similar to the one of Figure 4. The main frame is also similar to the one of Figure 4.

In the closed-loop experiment, the PID controller is implemented by a functional block in the fieldbus network and can be configured to run physically in any of the smart devices in the network. This configuration is transparent to the system user.

These experiments represent typical practical industrial applications, yet they are simple enough to illustrate basic concepts in control systems. This is very positive from a didactic point-of-view, as theoretical concepts are directly connected to practical applications. On the other hand, from the point of view of research, the system provides an industrial platform for experimentation of new control and automation algorithms remotely. This plays an important role in research cooperation among different institutions.

6. CONCLUDING REMARKS

The system is available at the site: http://automation.eletro.ufrgs.br/Plantaweb.htm. Application for obtaining an access code for the execution of experiments can be made following the instructions in this homepage.

The main contribution of this work regarding previously described internet-based experimental facilities is the use of industrial scale equipment and a normatized industrial network protocol - the Foundation Fieldbus. Also, the system requires no customized software running on the users machine, which makes it easier to use. The advantages of such an experimental setting for research and teaching are manifold (experiment flexibility, enhanced student motivation, etc). On the other hand, it can become a powerful tool for remote education, which is an important future perspective of this work.

Some issues are still to be solved. The speed of the image transmission is slow and variable, sometimes making difficult to follow the experiment visually. A solution to this problem has been proposed in (Ramakrishnan, et al., 2000), in which good image transfer performance is achieved by separating the image transfer from data transfer. However, it requires extra hardware and software running in the user's machine. There are also safety constraints, due to which the presence of a person at the laboratory is still required in order to watch over the experiment to turn it off in case of a severe failure or bad operation. On the other hand, the access control has already been handled in a satisfactory way.

ACKNOWLEDGEMENTS

This work has been partially supported by the Brazilian research agencies FINEP, CNPq, and FAPERGS

REFERENCES

Batur, C., Q. Ma, K. Larson and N. Kettenbauer, (2000). Remote tuning of a PID position controller via internet. *Proceedings of the American Control Conference 2000*, 4403 – 4406, Chicago, USA.

Johansson, K., A. Horch and O. Wijk, (1999). Teaching multivariable control using the quadruple tank process, *38th IEEE Conference on Decision and Control*, 807 – 812, Phoenix, USA.

Poindexter, S. and B. Heck, (1998). Using the web in your courses: the how-to's and the why's, *Proceedings of the American Control Conference 1998*, 1304 – 1308, Philadelphia, USA.

Ramakrishnan, V., Y. Zhuang, S.Y. Hu, J.P Chen, K.C Tan, (2000). Development of a web-based control experiment for a coupled tank apparatus, *Proceedings of the American Control Conference 2000*, 4409 – 4413, Chicago, USA.

Schmid, C. (1998). The virtual control lab VCLAB for education on the web, *Proceedings of the the American Control Conference 1998*, 1319 – 1325, Philadelphia, USA.

Elipse Software, (1998). *Supervisory and Control System, User Manual*, Elipse Software, Brazil

ELSEVIER
IFAC
PUBLICATIONS
www.elsevier.com/locate/ifac

IT-BASED MANAGEMENT OF MAINTENANCE PROCESSES IN FIELDBUS-BASED AUTOMATION SYSTEMS

Martin Wollschlaeger, Thomas Bangemann

ifak Institut für Automation und Kommunikation e.V.
Steinfeldstrasse 3, D-39179 Barleben, GERMANY
Tel.: +49 (39203) 810-51
Fax: +49 (39203) 81100
Email: {mw | thb}@ifak.fhg.de

Abstract: Maintenance operations ceased to be a "necessary evil" and are currently considered as a part of a global EAO (Enterprise Asset Optimisation) policy implemented by a growing number of industrial organisations. One of the possible maintenance strategies is predictive maintenance, which requires online data acquisition as a basis for prediction arising problems. The paper presents a possible solution for integrating online status information into maintenance management systems by applying web enabling technologies. Since the multitude of web technologies offers a broad variety of suitable solutions, heterogeneous and incompatible implementations have to be expected. Using a concept based on context-depending XML (Extensible Markup Language) descriptions, linked together in a framework, this problem can be solved. The paper shows concepts, description methods, software tools, and migration paths in XML-based frameworks. Special focus is put on Web Services. In addition to the approach an practical example is presented. *Copyright © 2003 IFAC*

Keywords : Fieldbus, Maintenance, Management, Data Acquisition, XML, Web

1. INTRODUCTION

State of the art automation solutions, especially in process industry, can be characterised as complex technical systems. Depending on the production requirements, field devices, PLC, DCS, and HMI and operator stations are usually fully optimised in order to meet economic criteria. An enormous potential for future improvements can be identified in maintenance, trying to reduce maintenance costs and systems' downtimes. Therefore, the assets of automation systems have to be identified and managed.

Fieldbuses are an integral part of automation systems. While simple field devices like binary switches, proximity sensors etc. can be found in many manufacturing processes, the field devices in process industry are rather complex. Implementing complex functionality, they provide – and require - extended functions for parameterisation and configuration. With the ongoing use of fieldbus systems like PROFIBUS PA or Foundation Fieldbus a new quality in handling and maintaining field devices is required. Simple tools are not more sufficient for managing the complex devices. On the

other hand, the functionality of the devices allow new methods to be introduced, like parameterisation, download of complex functions, functional decentralisation, or test on demand. Thus, field devices evolve from simple sensors and actuators to important assets in automation and control systems.

The often used term "Asset Optimization" means using optimally the potentials of plant assets. There is a close relation between maintenance and Asset Management. In the focus of Online Asset Management are such assets, which current statuses can be accessed using the plant communication network, e.g. field devices. The tasks are in general the synthesizing of plant data for an economical optimization of technical plant support, for example optimizations of planned maintenance and thus minimization of occurring disruptions, improvement of process diagnosis and control, and the increase of production throughput by using existent reserves [1], [2].

To increase availability and productivity of their production processes, enterprises need to spend high costs for engineering their maintenance and online asset management applications. The reasons are, on one side, extended configuration requirements due to the increased functionality of the field devices. On the other side, a generic, manufacturer-independent and complete model of field devices is lacking. This results in a divergent variety of control concepts and maintenance or optimization tools for field devices.

Furthermore, different maintenance strategies can be found in industrial solutions. The most promising among them is predictive maintenance. However, this is also a complex solution. Several conditions have to be fulfilled for efficient predictive maintenance:

- Devices and systems have to be continuously investigated by extracting an application-specific set of measurement data and events, and by watching their changes and trends.
- A maintenance management has to be defined, providing the required logistic support. It enables remote access to technical documentations or to database systems, and to on-line modeling and expert knowledge.
- A use-case depending data presentation, considering personalization for different user groups, and for different capabilities of user interfaces and access methods.

These requirements are supported by general management concepts. Management services are used for transportation of relevant information, an appropriate definition of so-called managed objects specifies the properties used for the management process and guarantees an adequate presentation.

With respect to online asset management, this means to define the relevant management information, to model it according to asset optimization requirements, and to identify the data sources, their relations to supporting information, and necessary transactions to generate the required management information.

So maintenance system and online asset management system has to be able to automatically extract actual values, measurement data, and set-points, but also structural information, from a given automation and control system. These raw data have to be filtered and recombined and then presented in a suitable way for asset management functions. The extraction of the data has to be performed without a significant engineering effort, and – much more important – without consuming additional system resources. An extremely important task is the generation of maintenance information based on the extracted data. Using this kind of information, predictive inspection and maintenance can be scheduled. As a consequence, this will lead to a minimization of the system's down times.

2.   MANAGEMENT METHODS

According to ISO 7498-4, the term management is defined as the facility to control, co-ordinate and monitor the resources which allow communications to take place [3]. This definition is widely used in office networks and has to be adapted to be used in the context of industrial automation. For example, communication functions and application functions are in many cases still mixed and need to be separated.

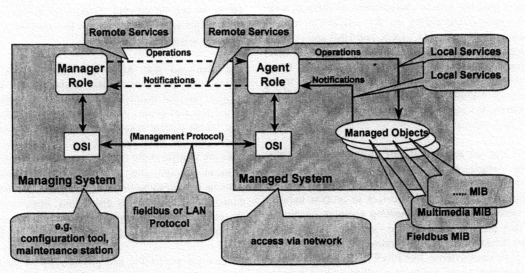

Fig. 1: Manager-Agent relations according to ISO

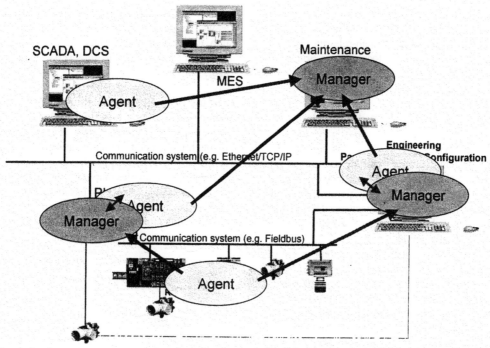

Fig. 2: Hierarchical management model

Not all of the management methods defined are currently used in automation systems. While Failure Management, Configuration Management, Security Management or Performance Management play an important role, until now Accounting Management has less meaning. Considering actual trends of globalisation, outsourcing and specialisation in combination with Internet based technologies, Accounting Management should be paid much more attention. Especially in the maintenance area, it enables smaller service providers to offer accountable online services. With respect to the growing decentralisation of automation systems and their components – both in vertical and in horizontal directions – the application of appropriate management methods helps to identify and to use existing functional preserves.

Starting point for online asset management is the internal structure of a complex automation system, as well as the operations and tools required for planning, configuration, parameterisation, operation, diagnosis and maintenance. Each of the components generate a certain amount of data, depending on its function.
- configuration data are generated by configuration tools and are partly stored in the device
- resources of a device are described by a set of data (e.g. device description)
- a field device cerates and/or processes data relevant for the technical process

- functions of devices and processes can be described by data

In addition, a large number of data exists, that are not explicitly of technical nature:
- documentations extending the facilities, like general descriptions, repairing instructions, lists of spare parts etc.
- general information like order numbers, inventory data, customer relations etc.

The data described above are heterogeneous by nature and have to be characterized into management relevant data and process data. Both together describe the complete data set of the system. The resources applicable to management processes are defined as Managed Objects.

The managed objects can be accessed as shown in Fig. 1. A managing object (a manager) asks the agent for a specific object by invoking an appropriate operation. This triggers the agent to read the object data from the Management Information Base (MIB) and passes them to the manager in a response operation. In case of events or failures the managed object can invoke a notification to the manager. For the management of complex systems, multiple layers of manager-agent relations can be set up.

As shown in Fig. 2, this structure can be applied to automation systems. Usually a System Manager exists, which controls all the resources of the system. For external applications, the System Manager fulfills the agent role by providing all system-relevant information. Therefore it accesses either local information, or it requests information from underlying agents. This process appears transparent for the user.

Information can be stored at different levels of the system:

- in field devices

  The system manager requests a managed object from a fieldbus server. This server reads the object from the field device via the fieldbus protocol and returns it to the system manager. This guarantees consistency, but has less performance.

- in the fieldbus server

  The fieldbus server refreshes a local image of the managed objects. Thus, the system manager accesses a copy of the real managed objects. With respect to the first method, performance is increased. However, consistency between fieldbus server and field devices has to be guaranteed.

- in a local data base of the system manager

  The system manager acquires managed objects from the underlying system and updates a local image of the managed objects. Every access from an external agent is a local access with high performance. In addition, local management information can be stored in the same data base. Consistency aspects become more important, since they have to be guaranteed system-wide.

- at referenced locations

  The system manager stores references to static management information. The managed object is accessed via the reference upon request of an external agent.

The functionality defined in ISO 7498-4 is only partly implemented in fieldbus based networks. Specific, rather complex management applications (managers) would be necessary. The following shows a possible use of web-based technologies for management methods as an alternative to conventional solutions.

3. WEB BASED MANAGEMENT FRAMEWORK

A web based management framework [4] relies on the vertical integration of an automation system into LANs. The framework implements the Hypertext Transport Protocol (HTTP) as a network-wide unique transport layer. Web servers (HTTP servers) act as data sources, while web browsers (HTTP clients) are used as user interfaces. An integration instance in the web server coordinates the data exchange between the automation system and the documents provided by the web server, as well as the integration of interactive tool components into these documents. Thus, the web server provides managed objects as a fieldbus server or a system manager.

The use of XML descriptions as a basis for the generation of the information provided allows to set up a heterogeneous description system (Fig. 3), which appears to be homogeneous to applications outside. This description can be used to create visualization and interactive dialogs. It enables mapping of management relevant functionality to both server based scripts, as well as to client side interactive software components. The application context defines the preferable method.

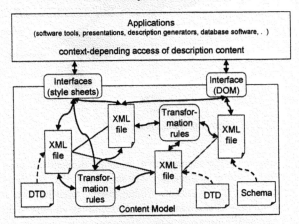

Fig. 3 A content model with structure and access methods

Using the flexible linking concept in XML, general description information can be extended by context-specific information (Fig. 4). This is especially important, if management functions contain web components like scripts or applets.

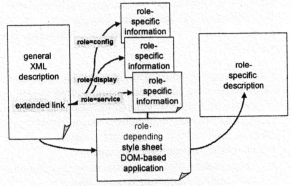

Fig. 4 Context-depending generation of web pages

4.   XML BASED WEB SERVICES

The solutions described above are very useful to access data objects and present them to the user. In principle, the user can perform actions by adding interactive elements like forms, JAVA applets or ActiveX controls. These elements open specific communication channels between the server and the user interfaces. The main drawback of this interaction is, that nearly every connection implements its own protocol and command sequence. In addition, security considerations may prevent such a solution (access to TCP/IP port numbers may be restricted).

Since interactive functionality, based on specific protocols, still remains a main user demand, a generic solution has been developed. These so-called Web-Services are implementations of specific XML-based protocols. The schema for defining the protocol is provided by the World Wide Web consortium (W3C).

The XML protocol working group defines a generic architectural model for implementing services [5]. The services map specific functions to a well-defined interface. Both client and server exchange messages to perform a function call or to return the results of that call. The messages are coded in XML. Since XML is not an application protocol, the messages have to be embedded into a transport protocol. Typical transport protocols are HTTP and the Simple Object Access Protocol (SOAP) [6]. While HTTP is mostly used when a human user requests the service, SOAP is preferred for applications requesting the service. The principle architecture of a Web-Service implementation at a server is shown in Fig. 5.

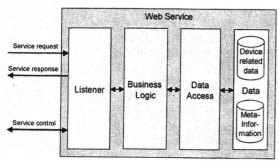

Fig. 5 Web-Service implementation schema

Returning back to the management model, the definition of manager agent relations is done by defining web service operations, while the managed objects are described by other (usually already existing) XML files and associated schemas of the content model as shown in Fig. 3

Some of the benefits using Web-Services are platform-independence, a generic definition framework for the messages by the Web Service Description Language (WSDL) [7], the use of HTTP as a transport protocol (a firewall-friendly protocol), and the possibility of performing validation and security checks before invoking the service. Of course, the drawback is the coding overhead of an XML-coded message, and the required computational resources. Since there is an increasing performance of modern hardware components, this should have a minor influence in the future. The definition and description of services allows standard XML tools to be used. Especially an automatic WSDL generation from already defined schemas is quite important.

The Web-Service listener is invoked, when a request has been passed to the Web Server. The listener itself runs in a servlet engine such as Tomcat or JSERV. Depending on the service, a local function of the object model interface is invoked. It interacts with the objects and returns the result of the function as an XML file. Cocoon then transforms this temporary file using an XSLT style sheet. This way, the result is created and is sent back to the client.

Taking a first look at this scenario, one might not notice the differences with respect to an explicit function call. However, the main difference is the existence of a generic service description. It allows a platform-independent access to the service, since there are no dependencies between the interface described in WSDL and the implementation of the listener.

Different strategies can be used for implementing the listener. Besides generic, but complex solutions like Microsoft's :NET [8] or SunONE from Sun [9], a consequent orientation towards open source software is also possible (Fig. 6). A server based implementation scenario is used in this case, where Apache Cocoon is chosen for server-side transformation. An interface to an object model (the managed objects) was implemented in JAVA. It provides access functions that can be invoked by Cocoon, initiated by a web service request. The resulting data the JAVA interface creates are XML-coded. They are passed back to the Cocoon transformation engine. An XSLT transformation is performed to create the WSDL response messages. While the required format is specified in the WSDL file, the style sheet defines the transformation rules. Since the management solution is used in an automation system context with fieldbus based communication, the managed objects are implemented in the field devices and in other system components. A schemas for coding these data in the web service responses is derived from the content model, e.g. from information in fieldbus profiles, data types, parameter names etc.

It is important to notice, that the same server structure is used for implementing the classical access by means of HTTP or WAP protocols. The creation of the required HTML or WML pages containing the data from the managed objects are generated by simply applying other style sheets [10]. So Cocoon is used as a multi-purpose, context depending transformation engine.

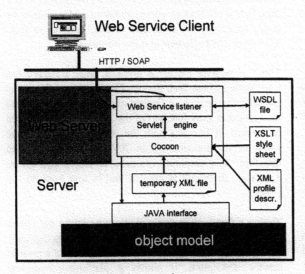

Fig. 6 Exemplary Web-Service implementation

On the other hand, the client can retrieve the WSDL file from the server and can use it to create an appropriate user interface. Since the WSDL file contains information on number, names, data types, and ranges of the parameters in the service request, a user interface can be generated on the fly. In addition, a validation of the parameters supplied can be performed before sending the request. This is an outstanding feature, since it can guarantee, that only valid requests will be sent.

5. PRACTICAL REALISATION EXAMPLE

A scalable multi-vendor configuration has been implemented at ifak to proof the concepts described above. The following aspects have been in the focus of the installation:
- integration of a heterogeneous system based on different network structures and with network gateways
- realization of typical application environments
- implementation of as products on the market available devices, which are typical in a given application context

- existence of classical solutions and approaches for the management (including configuration, parameterization, operation, administration), that can be used as a reference
- access methods to internal details of the used devices, tools and technologies.

The installation contains two separate parts, a continuous process with the adequate equipment used in process control, and a demonstration of a manufacturing process. The network topology of the process control part is shown in Fig. 7.

The PROFIBUS technology used in the installation enables new chances for an online parameterization and for administration and maintenance of field devices. For these tasks supporting solutions are developed at ifak, partly in cooperation with other partners.

An example is the development of an Asset Management Box (AMBox), together with RWTH Aachen [11]. This system is used for an extraction of information relevant for service, maintenance and asset management from the underlying process. The complete parameter set is read from the devices, using the block structure definitions in the PROFIBUS PA profile [12]. The data are assigned to specific objects of a runtime object system. The semantic meaning of the data is also defined in the profile. The objects represent agents for the managed objects. The manager roles are performed by application modules, which are specific for different management scenarios, e.g. parameterization, administration, service etc. A web access to the objects is implemented as shown above. So the AMBox acts as a management system as well as a maintenance data server.

In the installation described above, Web Services are used for tool-based access from outside. Only HTTP is used as a transport protocol. This overcomes the limitations that still exist in DCOM, CORBA, or even the built-in ACPLT/KS communication services, since these solutions require extra communication ports. The integrative effects of the solution are the providing of Web Services based on the ACPLT object model, which contains meta information on object classes derived from appropriate XML description files.

Several other research activities at ifak focusing on technologies mentioned above rely on this structure.

6. CONCLUSION

An online monitoring of the equipment used in automation systems is an important prerequisite for a prediction of possible failures or downtimes. It is necessary to integrate the processing of actual status information into a maintenance and online asset management system. In this context, the paper presented concepts and solutions for an integration of IT-based technologies (methods, protocols, meta languages, interface definitions) and discussed possible implementations in a practical environment.

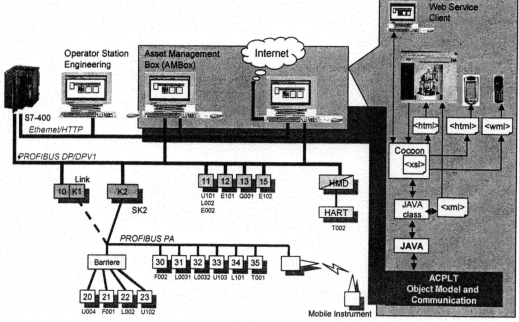

Fig. 7 network topology of the process control system

BIBLIOGRAPHY

[1] E. Nicklaus, H.-P. Fuß: Online Asset Management, atp - Automatisierungstechnische Praxis 42, Heft 5, pp.30-39, 2000.

[2] NAMUR, NAMUR-Recommendation NE91: Requirements for Online Plant Asset Management Systems, First Issue 11/2001.

[3] Seitz, J.: Netzwerkmanagement. Internat. Thomson Publ., Bonn, 1994.

[4] Wollschlaeger, M.: Planning, Configuration and Management of Industrial Communication Networks using Internet Technology. IEEE GLOBECOM '98, Sydney, Proceedings 2235-2240.

[5] n.n.: "XML Protocol Usage Scenarios" W3C Working Draft, 2001, http://www.w3.org/TR/xmlp-scenarios

[6] n.n.: SOAP Version 1.2 Part 1: Messaging Framework. W3C WD 2002, http://www.w3.org/TR/2002/D-soap12-part1-20020626/

[7] n.n.: Web Services Description Language (WSDL) V1.2. W3C WD 2002, http://www.w3.org/TR/2002/WD-wsdl12-20020709/

[8] Short, S: Building XML Web Services for the Microsoft .NET Platform. Microsoft Press, 2002

[9] Basha, S.: Professional Java Web Services. Wrox Press, 2002.

[10] Wollschlaeger, M.; Diedrich, C.; Simon, R.: Web Integration of Factory Communication Systems using an XML Framework. IEEE International Symposium on Industrial Electronics (ISIE'2002), L'Aquila, Proceedings.

[11] Wollschlaeger, M.; Müller, J.; Diedrich, C.; Epple, U.: Integration of Fieldbus Systems into On-line Asset Management Solutions based on Fieldbus Profile Descriptions. WFCS'2002 4th IEEE Workshop on Factory Communication Systems. Västeras, 27.-30.08.2002, Tagungsband S. 89 96.

[12] PNO, PROFIBUS PA: Profile for Process Control Devices, Version 3.0, October 1999.

AUTHOR INDEX